AF566464

PERGAMON MATERIALS SERIES
VOLUME 8

Thermally Activated Mechanisms in Crystal Plasticity

PERGAMON MATERIALS SERIES

Series Editor: Robert W. Cahn FRS
Department of Materials Science and Metallurgy, University of Cambridge, Cambridge, UK

Vol. 1 **CALPHAD** by N. Sauders and A. P. Miodownik
Vol. 2 **Non-Equilibrium Processing of Materials** edited by C. Suryanarayana
Vol. 3 **Wettability at High Temperatures** by N. Eustathopoulos, M. G. Nicholas and B. Drevet
Vol. 4 **Structural Biological Materials** edited by M. Elices
Vol. 5 **The Coming of Materials Science** by R .W. Cahn
Vol. 6 **Multinuclear Solid-State NMR of Inorganic Materials** by K. J. D. MacKenzie and M. E. Smith
Vol. 7 **Underneath the Bragg Peaks: Structural Analysis of Complex Materials** by T. Egami and S. J. L. Billinge
Vol. 8 **Thermally Activated Mechanisms in Crystal Plasticity** by D. Caillard and J. L. Martin

A Selection of forthcoming titles in this series:

Phase Transformations in Titanium-and Zirconium-Based Alloys
by S. Banerjee and P. Mukhopadhyay
Nucleation by A. L. Greer and K. F. Kelton
Non-Equilibrium Solidification of Metastable Materials from Undercooled Melts
by D. M. Herlach and B. Wei
The Local Chemical Analysis of Materials
by J. W. Martin
Synthesis of Metal Extractants
by C. K. Gupta
Structure of Materials
by T. B. Massalski and D. E. Laughlin
Intermetallic Chemistry
by R. Ferro and A. Saccone

PERGAMON MATERIALS SERIES

Thermally Activated Mechanisms in Crystal Plasticity

by

D. Caillard

CEMES/CNRS-BP4347, F 31055 Toulouse Cedex

J. L. Martin

IPMC/EPFL-CH 1015 Lausanne

2003

PERGAMON
An Imprint of Elsevier
Amsterdam – Boston – London – New York – Oxford – Paris
San Diego – San Francisco – Singapore – Sydney – Tokyo

ELSEVIER Ltd
The Boulevard, Langford Lane
Kidlington, Oxford OX5 1GB, UK

First edition 2003

Library of Congress Cataloging-in-Publication Data
Caillard, Daniel.
Thermally activated mechanisms in crystal plasticity / by Daniel Caillard, Jean-Luc Martin.
p. cm. – (Pergamon materials series; 8)
Includes bibliographical references and index.
ISBN 0-08-042703-0
1. Materials at high temperatures. 2. Crystals–Plastics properties. I. Martin, Jean-Luc, 1938-II. Title. III. Series.

TA417.7H55C35 2003
620.1'1296–dc21 2003053563

British Library Cataloging in Publication Data
Caillard, Daniel
Thermally activated mechanisms in crystal plasticity. -
(Pergamon materials series ; 8)
1. Dislocations in crystals 2. Crystals - Thermal properties
3. Crystals - Plastic properties
I. Title II. Martin, Jean-Luc
548.8'42

ISBN: 0 08 042703 0

⊗ The paper used in this publication meets the requirements of ANSI/NISO Z39.48-1992 (Permanence of Paper).
Printed in The Netherlands.

Series Preface

My editorial objective in this Series is to present to the scientific public a collection of texts that satisfies one of two criteria: the systematic presentation of a specialised but important topic within materials science or engineering that has not previously (or recently) been the subject of full-length treatment and is in rapid development: or the systematic account of a broad theme in materials science or engineering. The books are not, in general, designed as undergraduate texts, but rather are intended for use at graduate level and by established research workers. However, teaching methods are in such rapid evolution that some of the books may well find use at an earlier stage in university education.

I have long editorial experience both in covering the whole of a huge field – physical metallurgy or materials science and technology – and in arranging for specialised subsidiary topics to be presented in monographs. My intention is to apply the lessons learned in 40 years of editing to the objectives stated above. Authors (and in some instances, as here, editors) have been invited for their up-to-date expertise and also for their ability to see their subject in a wider perspective.

I am grateful to Elsevier Science Ltd., who own the Pergamon imprint, and equally to my authors and editors, for their confidence, and to Mr David Sleeman, Publishing Editor, Elsevier Science Ltd for his efforts on behalf of the Series.

Herewith, I am pleased to present to the public the eighth title in this Series, on a topic of great current concern.

ROBERT W. CAHN, FRS
(Cambridge University, UK)

Preface

The authors decided to start this joint venture, during the International Conference on the Strength of Materials (ICSMA11), in Prague, in August 1997. The idea was to gather experimental results and their physical interpretation about various dislocation mobility mechanisms. These were part of their respective lines of research, performed more or less independently through the years.

Later on, they were lucky enough to meet Professor R. W. Cahn, FRS, who became enthusiastic about the idea and very patiently encouraged them to realize their project.

The correct description of dislocation mobility mechanisms and related activation parameters, requires:

(i) The selection of the relevant experimental facts. This is sometimes a difficult task, given the abundance of available informations at different scales and levels of resolution.
(ii) A proper derivation of the related theory.

The reader will find a blend of data and related interpretations; we hope that the physics of the processes is not hidden by the equations.

It is a pleasure to acknowledge the helpful comments we have received from several colleagues and friends. At an early stage of the writing, Prof. F.R.N. Nabarro suggested several important points to include. All through the years repeated contacts and discussions with a variety of individuals have tremendously helped to refine our views on different subjects. In EPF-Lausanne, JLM would like to thank particularly F. Nabarro, G. Saada, and M. Kleman, who came several times as visiting scientists, F. Lévy, for his good advices, J. Bonneville for his long and active collaboration on several of the topics presented here, T. Kruml for numerous discussions and a long list of former postdocs, particularly M. Mills, K. Hemker, M. Cieslar and former PhD students among whom are M. Morris, P. Anongba, N. Baluc, P. Spätig, B. Viguier, B. Lo Piccolo (Matterstock), and C. Dupas (Charbonnier), whose articles are duly referenced. Many thanks are also due to Ms S. Lovato who has typed most of the manuscript in addition to her heavy daily tasks, and to M. O. Bettler and L. Heinen for their very careful handling of all the figures.

In CNRS Toulouse, DC is indebted to his colleagues and friends from the laboratory and from abroad, A. Couret, G. Molenat, M. Legros, F. Mompiou, G. Vanderschaeve, N. Clément, V. Paidar and E. Conforto for their contributions to several of the experiments and theoretical developments presented.

Grateful acknowledgement is made to Professor G. Margaritondo, Dean of the Faculty of Basic Sciences at EPFL, for providing favorable conditions for this venture, to the Swiss

National Science Foundation, for the financial support of most of the research performed in Lausanne.

Last but not least, the authors are greatly indebted to Prof. J. Friedel who, several years ago, showed them the way.

DANIEL CAILLARD, JEAN-LUC MARTIN

December 2002

Reader's Guide

Organization of the material: the arrows indicate a recommended reading order.

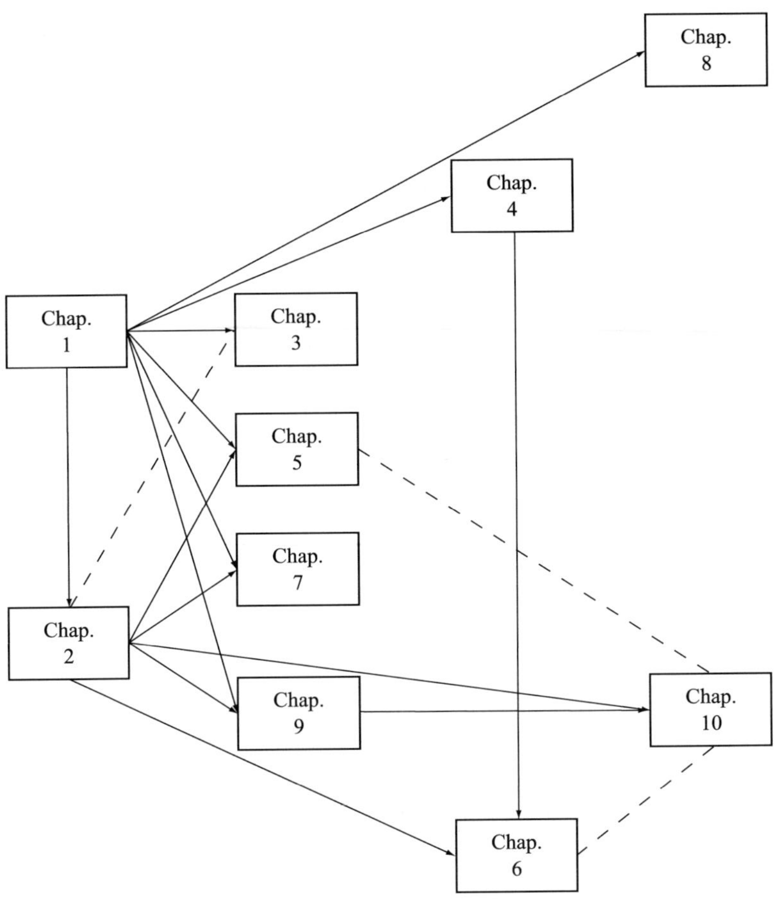

Contents

CHAPTER 3
INTERACTIONS BETWEEN DISLOCATIONS AND SMALL-SIZE OBSTACLES

CHAPTER 4
FRICTIONAL FORCES IN METALS

CHAPTER 6
EXPERIMENTAL STUDIES OF PEIERLS–NABARRO-TYPE FRICTION FORCES IN METALS AND ALLOYS

Chapter 1
Introduction

Chapter 1
Introduction

The understanding and the prediction of mechanical properties of materials implies a detailed knowledge of the elementary mechanisms that govern plasticity. In particular, those which control dislocation mobility and how this mobility changes under the influence of stress and temperature are of key importance. This active field of physics constitutes the core of the present review.

This introductory section is divided into two parts. The first one defines the authors' intentions, while the second one recalls some useful aspects of the theory of thermally activated dislocation dynamics, which will be used throughout the book.

1.1. SCOPE AND OUTLINE

The flow stress of a crystal can be decomposed into two main components. The first one reflects the long-range elastic interactions of mobile dislocations with the microstructure. It results from dislocation patterning to various extents and accordingly depends on the "sample history". The second component is the stress necessary to push dislocations over local energy barriers, which oppose their motion. These barriers can be of different nature: small obstacles, an intrinsic lattice resistance or an unpropitious dislocation core configuration.

The first component will not be studied here in detail. Only average values are estimated. At a given strain, it is a slowly decreasing function of increasing temperature, following the change of the crystal elastic constants. The dependence of flow stress on temperature predominantly reflects the properties of the second component. Short-range interaction of dislocations with energy barriers takes place in such a small volume that it is strongly influenced by thermal vibrations. Thermal activation helps dislocations to overcome these barriers, thus resulting in a reduction of stress as the temperature rises. These short-range thermally activated processes govern almost all the temperature dependent mechanical properties of materials.

These processes are obviously of fundamental importance for the understanding and modelling of strength of structural materials. New materials have rather complex structures, in which dislocation mechanisms are more difficult to identify than in single-phase metals and alloys. Fortunately, even in complex crystals, plasticity is usually controlled by a small number of elementary dislocation mechanisms.

We also hope that the proposed improved descriptions will constitute useful guidelines for the numerous attempts of multiscale modelling of crystal behaviour— a very active and promising field nowadays (see, e.g., the MRS Symposium Proceedings, Kubin et al., 2001 on this topic).

It is thus extremely important to possess a comprehensive description of the basic mechanisms, critically assessed by the best experimental results, in fairly simple situations.

A book by Kocks et al. (1975) was, to our knowledge, the first attempt in this direction. It synthesized the understanding of mechanical properties in terms of thermally-activated dislocation glide in a crystal containing a given distribution of microscopic obstacles to flow around. However, it did not include any review of experimental results. In the concluding remarks, the authors listed a few problems "that appeared to stand out as worth solving". More than 25 years later, we try to evaluate the answers brought to these questions, taking advantage of new techniques for investigations.

The state-of-the-art 20 years ago can be found in a book entitled "Dislocations et Déformation Plastique" (Groh et al., 1980). Several aspects of dislocation mobility mechanisms in connection with crystal plasticity are also covered in a book by Suzuki et al. (1991).

Here, we attempt to describe, in a consistent way, a variety of microscopic mechanisms of dislocation mobility. The principles of these mechanisms are recalled and the equations of dislocation motion are revisited to try and avoid reference to complex results scattered in the literature. The corresponding theoretical developments are borrowed from the inevitable reference books by Friedel (1964) and Hirth and Lothe (1992), whilst making certain important additions. The most complete and comprehensive experimental results are reviewed and analyzed in terms of these theoretical models.

The identification of the true mechanisms which operate relies on experimental techniques which have appeared or have been refined during the last 30 years. These include mainly in situ deformation experiments in the transmission electron microscope (TEM) on the one hand and accurate macroscopic transient mechanical tests on the other hand. The former experiments provide quantitative information about dislocation mobility mechanisms (viscous, jerky, with or without point obstacles), local stresses, mobile dislocation densities and velocities. The latter consist of stress relaxation experiments and transient creep tests. These tests yield average values of the activation energy of the dislocation velocity, its stress dependence. They also provide valuable information about mobile dislocation densities, a poorly documented parameter. Other types of experimental results will also be considered such as "post mortem" TEM observations of dislocation structures. Recent slip trace characterization taking advantage of the high resolution of the atomic force microscope will also be examined. In situ observations of deformation mechanisms in thick samples are now accessible using the intense photon beam of a synchrotron. These will be described in a few examples.

In the light of these new experimental results, several former models of dislocation mobility are revisited and some new ones are proposed. These mostly describe friction forces on dislocations, such as the lattice resistance to glide or those due to sessile cores, as well as dislocation cross-slip and climb.

The various aspects of this review are presented as follows. While the present chapter recalls some fundamentals of thermal activation of dislocation motion together with the definition of meaningful parameters, Chapter 2 introduces the experimental techniques that the authors consider to characterize dislocation mobility mechanisms. Among these mechanisms, Chapter 3 describes the interactions of dislocations with fixed extrinsic obstacles, namely forest dislocations and solute atoms in solid solutions. The modelling of Peierls–Nabarro type forces in metals is the subject of Chapter 4, while Chapter 5 provides information about dislocation cross-slip in face-centred cubic metals (the theoretical description and the available experimental data). Chapter 6 contains experimental results illustrating the predictions of Chapter 4. The Peierls–Nabarro forces in covalent crystals together with corresponding data are the subject of Chapter 7. Chapter 8 presents an exhaustive description of climb mechanisms under various circumstances, together with the few available data. Informations about dislocation sources and multiplication processes and the rate at which they exhaust are exposed in Chapter 9 together with estimations of the subsequent mobile dislocation densities, in connection with work-hardening. Finally, Chapter 10 is devoted to the understanding of the mechanical properties of some ordered intermetallic compounds. These include the anomalies in strength and work-hardening measured at an imposed deformation rate, as well as their creep behaviour. A reader's guide at the beginning indicates how the chapters are interrelated while a glossary at the end lists the symbols used throughout the book.

In this type of review, it is not possible to cover all the pertinent references in the literature and apologies are expressed if some important ones have been omitted.

1.2. THERMAL ACTIVATION THEORY: A SUMMARY

Exhaustive treatments of the subject can be found, for example, in Evans and Rawlings (1969), Poirier (1976), and Nabarro (1980), while the historical aspect of its development is exposed in Nabarro (1967).

The use of thermodynamics to describe thermally activated dislocation mobility poses some questions. Indeed, plastic deformation is an irreversible process, the dislocated states of the crystal being out of equilibrium up to the melting temperature. However, the microscopic processes of dislocation motion can be analyzed in terms of Eyring thermodynamics of viscous flow (Eyring, 1936). It is possible to define and measure some activation parameters of dislocation dynamics. These, at least, allow one to classify the dislocation mechanisms as a function of the material and the deformation conditions.

In many cases, the energy barriers that control dislocation movements are of the order of one electron-volt or less and involve some hundreds of atoms only. Under such conditions, the thermal energy favours the overcoming of these energy barriers. Consequently, the stress that deforms the crystal at a given strain-rate decreases as the temperature rises.

The following analysis of thermal activation of dislocation motion parallels those of Schoeck (1965) and Hirth and Nix (1969) which provide more details. The thermodynamic system is the whole crystal, so that the analysis is made in terms of the applied stress.

The derivation of the strain rate starts from the Orowan equation (Orowan, 1940). This kind of transport equation simply says that the deformation rate $\dot{\gamma}$ is proportional to the mobile dislocation density ρ_m, the average dislocation velocity v and the Burger's vector b:

$$\dot{\gamma} = A\rho_m b \mathrm{v} \tag{1.1}$$

where A is a geometrical coefficient. This law was originally established in the case of dislocation glide but it is also valid for deformation by climb (Nicolas and Poirier, 1976).

This expression of the strain rate can be refined as follows. We call y the coordinate along dislocation motion. The mobile dislocation segments of average length l are held up by the shear stress τ against the energy barrier at position y_0 and y_c is the threshold position of the dislocation while by-passing this barrier. If ν is the vibration frequency of the dislocation segment, the probability for this segment to jump over the energy barrier can be expressed, using a Boltzmann factor, as:

$$P = \nu_{id} \exp(-\Delta G/kT)$$

where k is the Boltzmann constant and T the absolute temperature. Here ΔG is the change of Gibbs free energy of the sample as the dislocation moves from y_0 to y_c. ΔG is used to characterize the energy barrier opposing dislocation motion.

ν_{id} can be expressed considering that the wavelength of the vibration scales with l. It is worth noting that, depending on the conditions, l can be constant, or stress-dependent (Section 3.3.1). Alternatively, the segment length λ involved in the thermally activated process can be smaller than l. This is the case, for example, with the kink pair mechanism in metals (Chapter 4). Using l as the characteristic length, for sake of simplicity:

$$\nu = b\nu_D/l \tag{1.2}$$

where ν_D is the Debye frequency $\approx 10^{13}\ \mathrm{s}^{-1}$. The average segment velocity is therefore:

$$\mathrm{v} = PA'/l$$

where A' is the area swept by segment l between two successive obstacles. Consequently, the strain rate is:

$$\dot{\gamma} = A\rho_m b(A'/l)\nu_{id} \exp(-\Delta G/kT) \tag{1.3a}$$

This expression is usually condensed into:

$$\dot{\gamma} = \dot{\gamma}_o \exp(-\Delta G/kT) \tag{1.3b}$$

with $\dot{\gamma}_o = A\rho_m b(A'/l)\nu_{id}$.

The identification of dislocation mobility mechanisms consists of determining experimentally the value of ΔG, which is the energy barrier for the studied conditions.

A difficulty arises in comparing expressions (1.3) of the strain rate with experimental data. Indeed, as emphasized by Schoeck (1965), the experiment yields the enthalpy change ΔH during the activation event.

Some useful thermodynamics quantities and relations are recalled which will be used later.

The first and second principles of thermodynamics allow one to write:

$$\mathrm{d}\Delta G = (\partial \Delta G/\partial \tau)_T \mathrm{d}\tau + (\partial \Delta G/\partial T)_\tau \mathrm{d}T$$

The partial derivatives in the above expression are successively:

- The activation entropy: $\Delta S = -(\partial \Delta G/\partial T)_\tau$
- The activation volume: $V = -(\partial \Delta G/\partial \tau)_T$

Alternatively, the activation area is sometimes used and equal to V/b. In the case of a mechanism with constant activation volume, the expression of ΔG is:

$$\Delta G = \Delta G_0 - \tau V \tag{1.4}$$

where ΔG_0 is the barrier activation energy at zero stress.

The experimental measurement of V is of prime importance, because V is related to the area swept by the dislocation during the thermally activated event. V exhibits small values for the kink pair mechanism (Chapter 4), larger ones for dislocation solute interactions (Section 3.3.2, Figure 3.14) and very large ones for the forest mechanism (Section 3.3.1).

Another useful relation is:

$$\Delta H = \Delta G + T\Delta S$$

The estimation of ΔS is therefore a key step towards the determination of ΔG from enthalpy values. Different sources of entropy change can be imagined. However, the evolution of the shear modulus with temperature which results from a change in the atomic vibrational frequencies is usually considered as the main component of ΔS. Under this assumption, Schoeck has derived the following relation between ΔG and ΔH :

$$\Delta G = [\Delta H + (T/\mu)(\mathrm{d}\mu/\mathrm{d}T)\tau V]/[1 - (T/\mu)(\mathrm{d}\mu/\mathrm{d}T)] \tag{1.5}$$

In addition, the Maxwell relations allow one to write:

$$TV(\mathrm{d}\tau/\mathrm{d}T) + \Delta H = 0 \tag{1.6}$$

In this relation, $(\mathrm{d}\tau/\mathrm{d}T)$ is meant at constant strain-rate and constant structure.

Experimentally, all data refer to a constant structure, i.e. are measured at constant strain in monotonic tests, ideally at yield. The yield stress can be determined as a function of temperature; V can be measured by strain rate jump experiments (Section 2.1.1) or by stress relaxation or creep transient tests (Sections 2.1.4 and 2.1.5, respectively). The activation enthalpy can be found using relation (1.5), while ΔG is obtained with Eq. (1.4). Cagnon (1971) gives an example of the determination of activation parameters, between 4 and 400 K in irradiated lithium fluoride using such a procedure. The latter is assessed by the following results:

- ΔH data measured by temperature jump experiments in creep or using Eq. (1.6) in constant strain-rate tests, are along the same curve as a function of temperature.
- ΔG values obtained from Eq. (1.5) yield a linear variation with temperature, which goes through the origin. This is predicted by relation (1.3b), which can be written:

$$\Delta G = \alpha kT \tag{1.7}$$

where α is the logarithm of the strain-rate ratio of relation (1.3b). Evaluating this ratio with reasonable orders of magnitude for ρ_m, b, A', l and ν, Cagnon (1971) finds $\alpha \approx 21$. This can be understood, considering the time scale used for laboratory experiments: common strain rates are from 10^{-6} to $10^{-2}\ s^{-1}$ and temperatures are high enough to prevent too high stresses. Therefore, values found for ΔG at one temperature are quite comparable whatever the crystal. In fact, the meaningful quantity to characterize the obstacle to dislocation motion is ΔG_0, i.e. the free energy at zero stress. Since it is obtained via relation (1.4), this shows once again the importance of the correct measurement of V, which is the real signature of the dislocation mechanism. Fortunately, methods exist that allow one to measure this parameter safely (Section 2.1.4.2).

Let us note that such a linear variation of ΔG as a function of T has also been found in Fe, either of high purity or containing various amounts of C additions, by Cottu et al. (1978) and Kubin et al. (1979), between 18 and 350 K. This is another experimental confirmation of the validity of relation (1.7).

Finally, combining relations (1.4) and (1.7) yields:

$$\tau = -\alpha kT/V + \Delta G_0/V$$

This indicates that, at a given strain, mechanisms with a small activation volume correspond to a rapid decrease of stress with temperature while those with a large V have a slower decreasing stress with temperature.

REFERENCES

Cagnon, M. (1971) *Phil. Mag.*, **24**, 1465.
Cottu, J.P., Peyrade, J.P., Chomel, P. & Groh, P. (1978) *Acta Metall.*, **26**, 1179.
Evans, A.G. & Rawlings, R.D. (1969) *Phys. Stat. Sol.*, **34**, 9.
Eyring, J. (1936) *J. Chem. Phys.*, **4**, 283.
Friedel, J. (1964) *Dislocations*, Pergamon, Oxford.

Groh, P., Kubin, L. & Martin J.L., Editors (1980) *Dislocations et Déformation Plastique*, Les Editions de Physique, Orsay.

Hirth, J.P. & Lothe, J. (1992) *Theory of Dislocations*, 2nd Edition, Krieger Publ. Company, Malabar.

Hirth, J.P. & Nix, W.D. (1969) *Phys. Stat. Sol.*, **35**, 177.

Kocks, U.F., Argon, A.S. & Ashby, M.F. (1975) in *Thermodynamics and Kinetics of Slip*, Progress in Materials Science, vol. **19**, Eds. Chalmers, B., Christian, J.W. and Massalski, T.B., Pergamon, New York.

Kubin, L., Louchet, F., Peyrade, J.P., Groh, P. & Cottu, J.P. (1979) *Acta Metall.*, **27**, 343.

Kubin, L.P., Selinger, A.L., Bassani, J.L. & Cho K., Editors (2001) *Multiscale Modeling of Materials*, Materials Research Society, Warrendale, p. 653.

Nabarro, F.R.N. (1967) *Theory of Crystal Dislocations*, Dover Publ. Inc, New York, p. 704.

Nabarro, F.R.N., Editor (1980) *Dislocations in Solids*, vol. **3**, North-Holland, Amsterdam, p. 61.

Nicolas, A. & Poirier, J.P. (1976) *Crystalline Plasticity and Solid State Flow in Metamorphic Rocks*, Wiley, London, p. 108.

Orowan, E. (1940) *Proc. Phys. Soc. London*, **52**, 8.

Poirier, J.P. (1976) *Plasticité des Solides Cristallins*, Eyrolles, Paris, p. 96.

Schoeck, G. (1965) *Phys. Stat. Sol.*, **8**, 499.

Suzuki, T., Takeuchi, S. & Yoshinaga, H. (1991) *Dislocation Dynamics and Plasticity*, Springer Series in Mater. Sci. Ed. Gonser, vol. **12**, Springer-Verlag, Berlin.

Chapter 2

Experimental Characterization of Dislocation Mechanisms

Chapter 2

Experimental Characterization of Dislocation Mechanisms

Three classes of techniques are described here, which can be used to characterize dislocation mechanisms in a variety of materials over a range of deformation conditions. These techniques consist successively of transient mechanical tests, of in situ observations of dislocations moving in a sample strained in a transmission electron microscope (TEM) or in a synchrotron beam, and slip trace analysis at the specimen surface.

The transient tests include strain-rate jumps, stress relaxation and creep experiments performed along a stress–strain curve during a monotonic deformation test. These are first described in Section 2.1. Section 2.2 recalls the difficulties, advantages and limitations of in situ TEM experiments, while Section 2.3 reports some preliminary results of deformation tests performed in situ in the intense photon beam of a synchrotron source.

At first glance, the relationship between transient tests and in situ experiments is not straightforward. However, it can be understood if the Orowan equation (see relation (1.1)) is considered. As illustrated in Sections 2.1.2–2.1.5, successive stress relaxations or creep experiments aim at separating the respective contributions of ρ_m and v to the plastic strain-rate with an acceptable accuracy. In such tests, macroscopic parameters, such as stress and strain-rate, activation energies and volumes, are known. In situ experiments directly show how many dislocations are moving and at what velocity, at least in local areas of the sample. However, macroscopic stresses and strain-rates are not measured directly. They also provide some hints about the corresponding dislocation mobility mechanisms. This can help in the interpretation of activation energies and volumes of mechanical tests.

It is worth noting that dislocation velocity measurements through etch-pitting have also provided valuable information about the corresponding mobility mechanisms (see e.g. Johnston and Gilman, 1959).

The combination of observations at these macroscopic, microscopic and sometimes mesoscopic scales provides a safer way of interpreting mechanical behaviour in terms of dislocation properties.

2.1. TRANSIENT MECHANICAL TESTS

A variety of such tests has been developed over the years. The idea is to slightly alter the conditions imposed on the sample at a given point of the monotonic test and observe the type of response. Provided the transient is not too long (around 30 s in modern tests), the corresponding deformation mechanisms are not too different from those operating in the monotonic test and useful information can be gained about them. Deformation parameters can be deduced that are not accessible during constant strain-rate loading. In what follows the principles of such transients are exposed, together with the interpretation of the material's behaviour and the related assumptions.

Resolved shear stresses and strains, τ and γ, respectively, will be considered for single crystals deforming along the primary system. However, normal stresses and strains (σ and ε, respectively) will be used as well for experiments on polycrystals. Subscripts or superscripts r and c refer to relaxation and creep conditions, respectively.

2.1.1 Strain-rate jump experiments

During a monotonic deformation test at strain-rate $\dot{\gamma}_1$, the latter parameter is suddenly increased (or decreased) to $\dot{\gamma}_2$, and the resulting change in stress, $\Delta\tau$, is recorded (see e.g. Makin, 1958). The idea is to characterize the material's response by an apparent activation volume defined as:

$$V_a = kT \frac{\partial \ln \dot{\gamma}}{\partial \tau} \tag{2.1}$$

where k is the Boltzmann constant and T the absolute temperature (see e.g. Gibbs, 1966). Parameter V_a rationalizes the change in stress associated with an imposed change in strain-rate. V_a is determined experimentally by the formula:

$$V_a = kT \frac{\ln(\dot{\gamma}_1/\dot{\gamma}_2)}{\Delta\tau} \tag{2.2}$$

Alternatively, another parameter obtained through the same experiment is also used, namely the strain-rate sensitivity of the stress (see e.g. Thornton et al., 1970) given by:

$$S = \frac{1}{T} \frac{\partial \ln \tau}{\partial \ln \dot{\gamma}} \tag{2.3}$$

It is worth noting that slightly different definitions of S are used, according to the authors. Comparing relations (2.1) and (2.3) yields

$$V_a = \frac{k}{\tau S}$$

The measurement of V_a (or S) allows us to classify deformation mechanisms according to the material's response to a strain-rate change. However, the microscopic interpretation of V_a is not straightforward. Indeed, combining the Orowan equation with Eq. (2.1) shows that

$$V_a = kT\left(\frac{\partial \ln \rho_m}{\partial \tau} + \frac{\partial \ln \mathrm{v}}{\partial \tau}\right) \tag{2.4}$$

Under such conditions, it is not possible to decide whether the observed change in stress is a consequence of the variation of ρ_m or of v or of both parameters as the strain-rate is altered.

Figure 2.1 illustrates various observed responses of the material during the jump. The ideal situation corresponds to Figure 2.1(a) in which $\Delta\sigma$ is easily measured. Complications arise if the work-hardening coefficient is different before and after the jump, or if the transient corresponds to the schematics of Figure 2.1(b). In this case, a small yield point is observed which likely corresponds to dislocation multiplication subsequent to the strain-rate increase. In this case, two apparent activation volumes can be computed according to relation (2.2), which correspond to $\Delta\sigma_{tr}$ or $\Delta\sigma_{ss}$, respectively.

To conclude, strain-rate jump experiments are easy to perform with almost any type of deformation set-up. They provide the apparent activation volume V_a.

2.1.2 *Stress relaxation tests*

In this other type of transient, the machine cross-head is stopped. Subsequently, the sample is maintained at a constant total strain, while the stress is observed to decrease (by an amount $\Delta\tau$—a negative quantity) as time proceeds. Figure 2.2 illustrates the procedure.

This type of test is, in principle, easy to perform on any type of straining machine. However, in practice, a high thermo-mechanical stability is required to record a relevant relaxation curve (stress–time curve).

Abundant literature can be found on this subject (see e.g. Feltham, 1961; Guiu and Pratt, 1964; Gibbs, 1966; Hamersky et al., 1992 and a review by Dotsenko, 1979). Single relaxation tests will be described first.

As illustrated by the arrows in Figure 2.2, the stress decrease during relaxation is accounted for by an increase in plastic strain γ_p which relieves the elastic strain τ/M, where

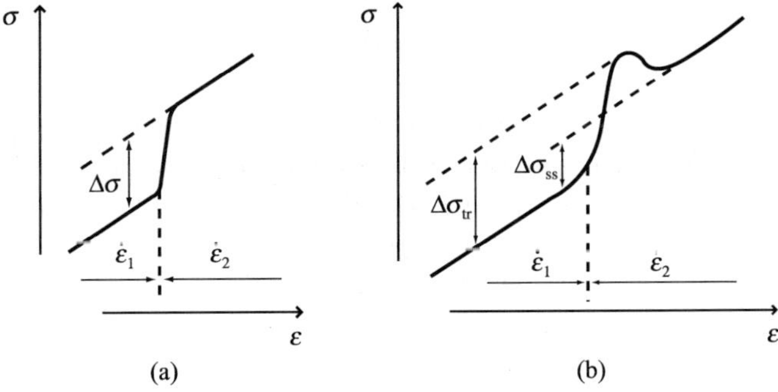

Figure 2.1. Schematics illustrating two types of transients subsequent to a strain-rate increase from $\dot{\varepsilon}_1$ to $\dot{\varepsilon}_2$. The observed response depends on the deformation mechanisms and the material as well as on the imposed conditions (see text).

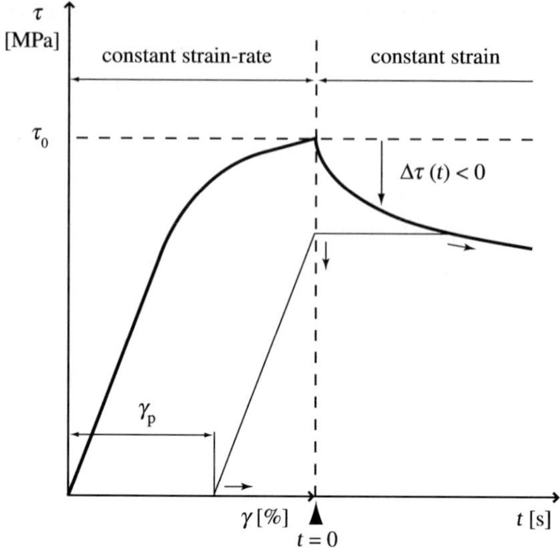

Figure 2.2. Schematic representation of a stress relaxation test. Arrows indicate how some parameters evolve with time.

M is the elastic modulus of the specimen and machine. The equation of the specimen–machine assembly is:

$$\gamma = \tau/\mathrm{M} + \gamma_\mathrm{p} \tag{2.5}$$

Its derivative with respect to time is:

$$\dot{\gamma} = \dot{\tau}/\mathrm{M} + \dot{\gamma}_\mathrm{p} \tag{2.6}$$

Under relaxation conditions, relation (2.6) becomes:

$$\dot{\gamma}_\mathrm{p} = -\dot{\tau}/\mathrm{M} \tag{2.7}$$

Relation (2.7) allows us to determine the plastic strain-rate, at time t, by measuring the slope of the relaxation curve.

In practice, the relaxation curves can exhibit a logarithmic variation of stress with time or a non-logarithmic one. The former case has been reported for a range of materials and deformation conditions and corresponds to a relaxation curve of equation

$$\Delta\tau = -\frac{kT}{V_\mathrm{r}}\ln(1 + t/c_\mathrm{r}) \tag{2.8}$$

where c_r is a time constant and V_r has the dimension of a volume. Therefore the relaxation curve is fully determined knowing c_r and V_r. Taking the time derivative of Eq. (2.8) and using Eq. (2.7) yields the plastic strain-rate during relaxation:

$$\dot{\gamma}_\mathrm{p} = (kT/\mathrm{M}V_\mathrm{r})(1/(c_\mathrm{r} + t)) \tag{2.9}$$

It is worth noting that V_r is the apparent activation volume during stress relaxation. Indeed, relation (2.9) allows one to calculate $\ln \dot{\gamma}_p$. Taking the time derivatives of $\ln \dot{\gamma}_p$ and $\Delta\tau$, respectively, (relations (2.8) and (2.9)) and then dividing the two expressions yields

$$V_r = kT \frac{\partial \ln \dot{\gamma}_p}{\partial \tau} \tag{2.10}$$

which is very similar to relation (2.2).

Considering relations (2.7) and (2.10) yields:

$$V_r = kT \frac{\partial \ln(-\dot{\tau}/M)}{\partial \tau} \tag{2.11}$$

A check of the logarithmic nature of the relaxation consists of plotting $\ln(-\dot{\tau})$ as a function of $\Delta\tau$. In the logarithmic case, a straight line is obtained, the slope of which is proportional to V_r.

An example of such a relaxation curve is shown in Figure 2.3. The relaxation curve $\Delta\tau(t)$ is superimposed on Figure 2.3(a) with a fitted curve according to relation (2.8). Figure 2.3(a) shows a very good agreement between the experimental and the logarithmic curves. The alternative checking procedure is illustrated in Figure 2.3(b) where a linear dependence is evident. The values of V_r are very similar using the two methods.

On the contrary, non-logarithmic relaxations are reported for other materials and deformation conditions. In this case relation (2.8) is not valid anymore. Figure 2.4 provides examples of such behaviour in a $[\bar{1}23]$ Ge single crystal at different stages of the stress–strain curve. In Figure 2.4(a), the stress decreases according to a linear dependence on time. This is due to important dislocation multiplication phenomena that take place before the upper yield point. Figure 2.4(b) shows a non-linear dependence of $\ln(-\dot{\tau})$ as a function of $\Delta\tau$ for the same type of crystal deformed after the lower yield point. According to relation (2.11), V_r is not constant along the relaxation curve, which is not logarithmic.

To conclude, for a material relaxing logarithmically, a single relaxation curve provides two parameters, namely, an apparent activation volume V_r and a time constant c_r. The usefulness of these parameters will be illustrated in Section 2.1.4.

To improve the understanding of how the mobile dislocation density and velocity contribute to the strain-rate, the *repeated relaxation test* has been invented. The procedure most commonly used nowadays (Spätig et al., 1993) is illustrated in Figure 2.5 (alternative procedures can be found, for example, in Sargent et al., 1969 and Kubin, 1974). The first relaxation starts at stress τ_0, over a time interval Δt, with a corresponding stress decrease $\Delta\tau_1$. The specimen is then reloaded to τ_0 fast enough to obtain quasi elastic conditions. It is then allowed to relax by an amount $\Delta\tau_2$ during Δt, then reloaded to τ_0 etc. As a rule a series consists of 4–6 transient tests of 30 s each. Only logarithmic relaxations are considered below. At low enough temperatures and for positive work-hardening coefficients, the following features are observed: (i)$\Delta\tau_j$ decreases as j, the relaxation number in

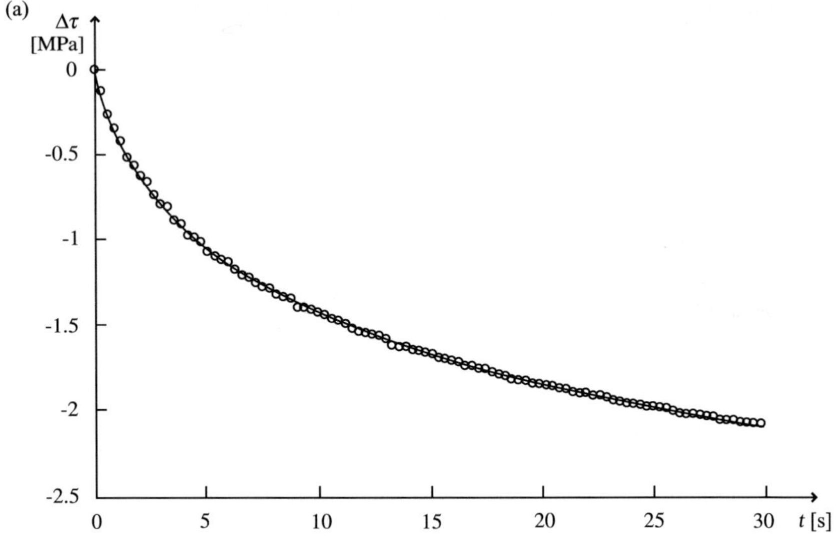

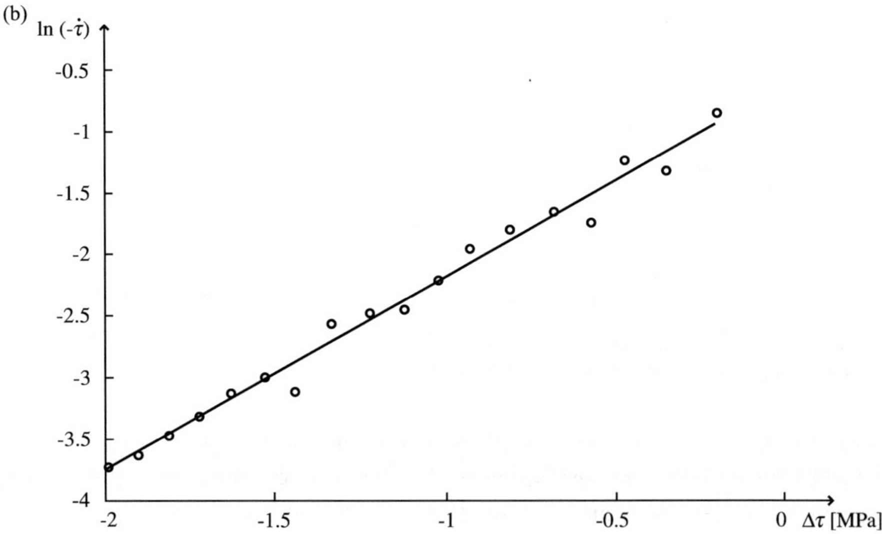

Figure 2.3. Two representations of the same logarithmic stress relaxation. [$\bar{1}$23] Ni_3(Al, Hf) single crystal. $T = 300$ K. $\gamma_p = 4\%$. (a) Experimental and fitted curves, yielding $V_r = 380b^3$ (b is the Burgers vector of a superpartial dislocation) and $c_r = 1.26$ s. (b) Linear dependence of $\ln(-\dot{\tau})$ as a function of $\Delta\tau$ yielding $V_r = 386b^3$ (Spätig, 1995).

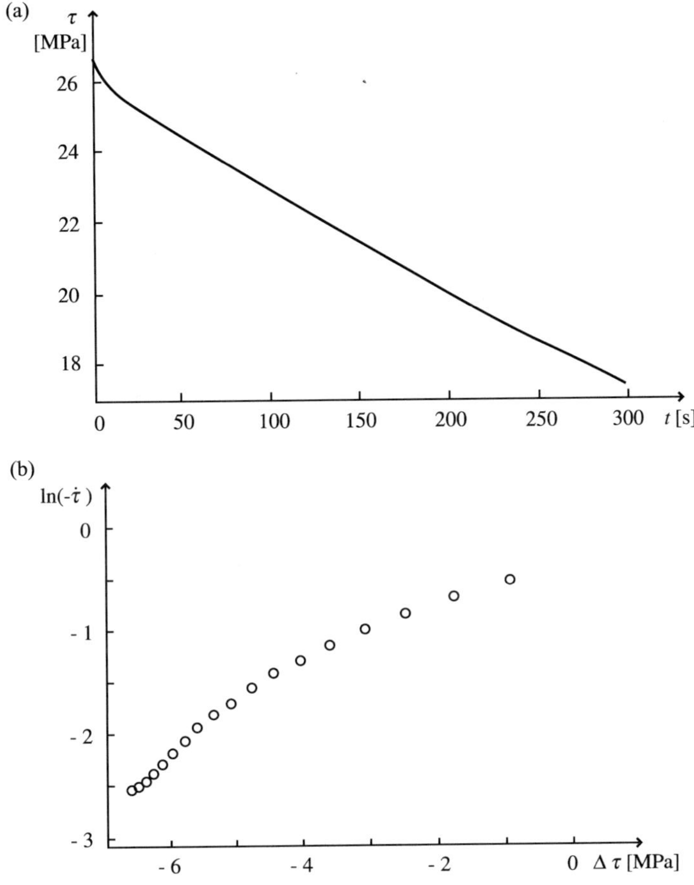

Figure 2.4. Non-logarithmic relaxations in a [$\bar{1}$23] Ge single crystal (a) $T = 750$ K, $\tau = 26$ MPa, before the upper yield point (b) $T = 700$ K, $\tau = 38$ MPa, after the lower yield point. (Charbonnier et al., 2001).

the series, increases; (ii) analysing the relaxation curves of the series in terms of relation (2.8), V_r is found to be constant, while the time constants c_{rj} depend on j. This accounts for the changes in relaxation curves along the series (see Section 2.1.6).

The slowing down of the relaxation along the series is due to hardening subsequent to plastic deformation and to the decrease in mobile dislocation density along the series. A complete description is provided in Section 2.1.4. When the monotonic test is resumed after the series, a small yield point of amplitude $\Delta\tau_R$ (Figure 2.5) may be observed. It is thought to correspond to dislocation multiplication which compensates for exhaustion during the transient.

The essential "trick" of the method is that the quasi elastic reloading prevents any substructural changes. As an example, $\dot{\gamma}_{f1}$, and $\dot{\gamma}_{i2}$ which correspond, respectively, to

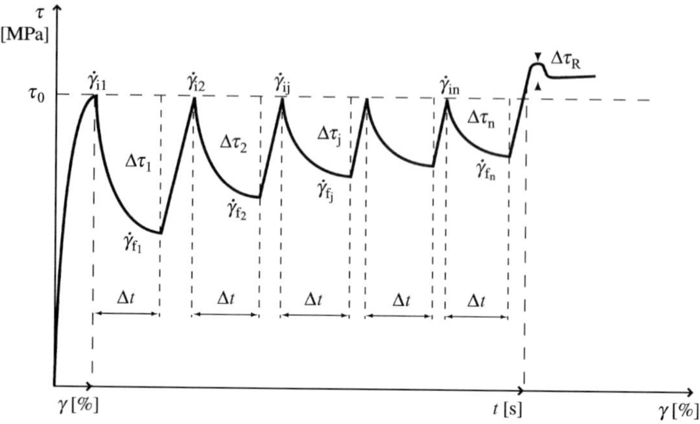

Figure 2.5. The procedure of successive stress relaxations (schematics). Definition of the parameters used. See text.

the end of stress relaxation 1 and the onset of stress relaxation 2 (Figure 2.5) can be compared. The same dislocation density is moving in both cases. Therefore, relation (2.4) indicates that the volume defined by

$$V = kT\frac{\ln(\dot{\gamma}_{i2}/\dot{\gamma}_{f1})}{\Delta\tau_1} \tag{2.12}$$

characterizes the stress dependence of the dislocation velocity (see Section 2.1.4.1).

2.1.3 Creep tests

At a given point of the stress–strain curve, the monotonic test is interrupted and the stress is kept constant. The strain is recorded as a function of time. Under such conditions relation (2.5) indicates that strain increments, $\Delta\gamma$, from the onset of the transient are plastic strain increments, $\Delta\gamma_p$.

An example of such a test is shown in Figure 2.6. The slope of the curve yields the plastic strain-rate, $\dot{\gamma}_p$.

A logarithmic increase of strain as a function of time can be observed (e.g. in Figure 2.6) depending on materials and deformation conditions. In such cases, the strain increase as a function of time follows the relation

$$\Delta\gamma_p = (kT/MV_c)\ln(1 + t/c_c) \tag{2.13}$$

where V_c has the dimension of a volume and c_c is a time constant. Fitting this relation with the creep curve provides V_c and c_c. In the test of Figure 2.6, $V_c = 30b^3$ and $c_c = 11$ s. Consequently, the creep rate is:

$$\dot{\gamma}_p = (kT/MV_c)[1/(c_c + t)] \tag{2.14}$$

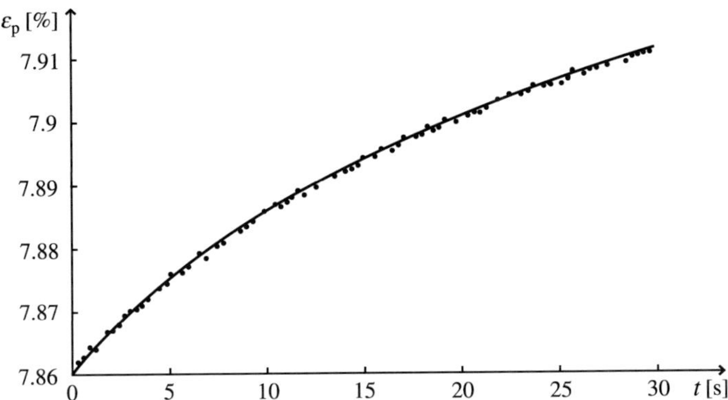

Figure 2.6. A transient creep test in a $Ni_{75}Al_{25}$ polycrystal. 300 K. $\varepsilon_p = 7.86\%$. Experimental and fitted curves according to relation (2.13). $V_c = 30b^3$, $t_c = 11$ s. (Lo Piccolo, 1999).

Relations (2.13) and (2.14) for creep correspond to Eqs. (2.8) and (2.9), respectively, for relaxation tests.

A repeated creep experiment was proposed by Orlova together with equations describing the material's response (see Orlova et al., 1995). Although it was applied successfully to the study of γ-TiAl polycrystals (Bonneville et al., 1997a) it could not be used for Cu or Ni_3Al single crystals, because of the creep strains being too low. A new procedure was proposed (Lo Piccolo et al., 2000; Martin et al., 2000) which is depicted in Figure 2.7. As the monotonic test is interrupted at stress τ_0 the specimen is allowed to creep during Δt. The stress is then increased quasi elastically by an amount, $\Delta\tau$, with a subsequent strain increment, $\Delta\gamma$, and the specimen creeps anew and so on. The same "trick" is used here: the change in creep rate immediately before and after the stress increment $\Delta\tau$ is the signature of a change of dislocation velocity. Similarly to repeated relaxation tests an activation volume of the dislocation velocity can be measured through the relation.

$$V = kT\frac{\ln(\dot{\gamma}_{i2}/\dot{\gamma}_{f1})}{\Delta\tau} \tag{2.15}$$

where creep test numbers 1 and 2 are considered as an example.

2.1.4 *Interpretation of repeated stress relaxation tests*

There are two ways of interpreting relations (2.8) and (2.13): (i) using Hart's equation (Hart, 1970), which was developed for a different purpose following a promising idea produced by Saada et al. (1997); (ii) in the framework of thermal activation theory. This latter approach is presented below since it has so far seen more development.

A complete interpretation of repeated relaxation tests can be found in Martin et al. (2002) and of creep tests in Lo Piccolo et al. (2000).

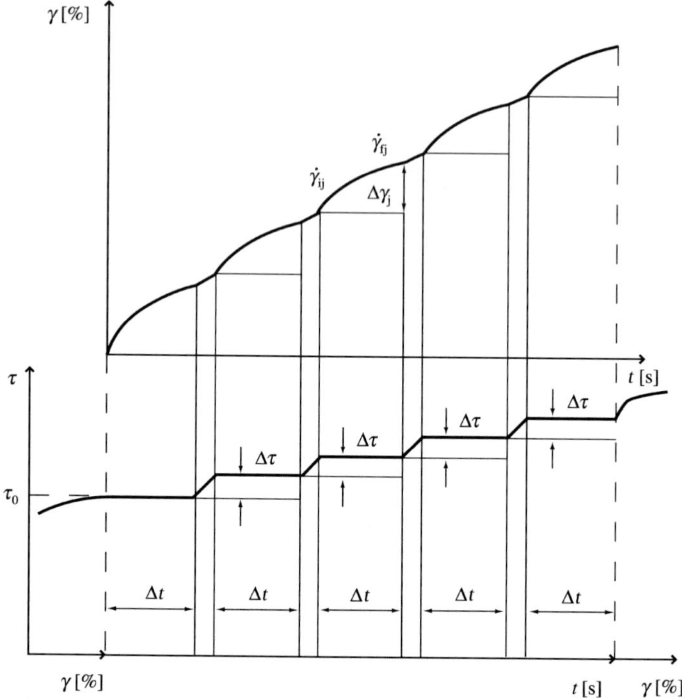

Figure 2.7. Schematic representation of a repeated creep test. The lower diagram represents the stress incurred to the specimen as a function of time during the transient. The upper diagram shows the corresponding creep strain.

2.1.4.1 General considerations. The following assumptions are made for short transient tests.

(i) *The stress τ can be decomposed into an athermal stress τ_μ and an effective stress τ^* (Seeger et al., 1957)*:

$$\tau = \tau_\mu + \tau^* \tag{2.16}$$

This relation expresses the necessity for the applied stress τ to counterbalance stresses due to the lattice resistance and the microstructure. τ^* is temperature-and strain-rate-dependent and corresponds to localized obstacles or energy barriers such as impurities, solute atoms, forest dislocations, etc. For those, thermal activation helps dislocation motion. τ_μ corresponds to athermal obstacles. Therefore, τ_μ should not depend on temperature significantly, but rather on strain. Relation (2.16) is strictly valid in the case of a homogeneous dislocation density.

For relaxation conditions, the change in internal stress is considered to be proportional to the change in strain:

$$\Delta\tau_\mu = K^{\text{r}}\Delta\gamma_{\text{p}} \tag{2.17}$$

Deriving $\Delta\gamma_p$ from relation (2.5) ($\Delta\gamma = 0$), and using Eqs. (2.16) and (2.17) yields:

$$\Delta\tau^* = (1 + K^r/M)\Delta\tau \tag{2.18a}$$

Similarly, for creep:

$$-\Delta\tau^* = \Delta\tau_\mu = K^c\Delta\gamma_p \tag{2.18b}$$

In these relations ΔX (with $X = \tau^*$, τ_μ, τ or γ) is the variation of parameter X during the transient. K^r and K^c are work-hardening coefficients during relaxation and creep, respectively.

(ii) *The dislocation velocity v is thermally activated*:

$$v = \nu d\exp(-\Delta G/kT) \tag{2.19}$$

where ΔG is here the activation energy of the mobility mechanism, ν the vibration frequency of the average dislocation segment, and d the distance over which this segment moves after a successful activation event. The activation volume of the dislocation velocity is defined here as:

$$V = -\partial\Delta G/\partial\tau^* \tag{2.20}$$

since τ^* is the stress component that acts on the dislocation. τ^* is responsible for the velocity v. V (a physical activation volume) is obviously different from V_r or V_c in relations (2.8) and (2.13), respectively. Therefore V cannot be determined via a single transient test.

2.1.4.2 Activation volume and microstructural parameters. During a short stress relaxation test, provided the change in $\Delta\tau^*$ is small enough and the pre-exponential factor in Eq. (2.19) is constant, the corresponding change in activation energy is $-\Delta\tau^* V$. The subsequent dislocation velocity is:

$$v = v_{or}\exp(V\Delta\tau^*/kT) \tag{2.21}$$

after considering relations (2.19) and (2.20) and v_{or} being the velocity at the onset of the transient ($\Delta\tau^* = 0$). Eq. (2.21) expresses the ratio v/v_{or} as a function of $\Delta\tau^*$. The mobile dislocation density is also a relevant parameter and its variation during the transient is of key importance. This was emphasized long ago (Rhode and Nordstrom, 1973). The following development aims at proving that a coupling of the dislocation velocity and the mobile density exists. It can be expressed by a power relation (exponent β):

$$\rho_m/\rho_{mo} = (v/v_{or})^\beta \tag{2.22}$$

where ρ_{mo} is the initial mobile dislocation density. Under stress relaxation and creep conditions, β has different values β^r and β^c, respectively.

This coupling is in fact a consequence of the Orowan equation, of the logarithmic relaxation and of the thermal activation of v. To show this, the former equation is written at time t and at the onset of relaxation, respectively. The two expressions are divided, which yields

$$(\dot{\gamma}/\dot{\gamma}_0) = (\rho_m/\rho_{mo})(v/v_{or}) \tag{2.23}$$

The strain-rate ratio can be expressed as a function of the time constant using relation (2.9):

$$(c_r/c_r + t) = (\rho_m/\rho_{mo})(v/v_{or}) \tag{2.24a}$$

The left hand side of this equation can be transformed due to Eq. (2.8):

$$\exp(V_r\Delta\tau/kT) = (\rho_m/\rho_{mo})(v/v_{or})$$

Then $\Delta\tau^*$ is used instead of $\Delta\tau$ (relation (2.18a))

$$\exp[(\Omega_r V/kT)(\Delta\tau^*/(1 + K^r/M))] = (\rho_m/\rho_{mo})(v/v_{or}) \tag{2.24b}$$

where a new parameter Ω_r is introduced:

$$\Omega_r = V_r/V \tag{2.25}$$

The ratio v/v_{or} (from Eq. (2.21)) can be evidenced in the left hand side of Eq. 2.24(b).

$$(v/v_{or})^{\Omega_r/(1+K^r/M)} = (\rho_m/\rho_{mo})(v/v_{or})$$

This relation is similar to (2.22) provided the following relation is fulfilled:

$$\Omega_r = (1 + \beta^r)(1 + K^r/M) \tag{2.26}$$

At this stage, Ω_r can be determined as follows (relation (2.25)):

In a repeated stress relaxation series V_r is obtained usually by a fit of the first relaxation curve (the longest one) with the logarithmic law (2.8). V can be computed using relation (2.15).

Ω_r being known, Eq. (2.26) shows that β^r and K^r cannot be determined independently. Though hardening under stress relaxation conditions is still an open problem, K^r can be considered to be not too different from θ, the work-hardening coefficient of the monotonic curve. Therefore, β^r can be deduced using relation (2.26). β^r provides useful information about the mobile dislocation density during the transient. In particular, it allows us to quantify mobile dislocation exhaustion during relaxation.

Indeed, a combination of Eqs. (2.21) and (2.22) yields ρ_m/ρ_{mo} as a function of $\Delta\tau^*$, then of $\Delta\tau$ using Eq. 2.18(a). Expressing $\Delta\tau$ via relations (2.8), (2.25) and (2.26) gives:

$$\rho_m/\rho_{mo} = \left(\frac{c_r}{c_r + t}\right)^{\beta^r/(1+\beta^r)} \tag{2.27}$$

which is the time decrease of the mobile dislocation density during a stress relaxation test.

At this stage the slowing down of the relaxations along a series (Figure 2.5) can be accounted for: during one of the stress relaxation tests, the plastic strain increases as well as the athermal stress (relation (2.17)). Although each relaxation test starts at the same stress τ_0, the corresponding effective stress τ^*, and therefore the dislocation velocity, is smaller as the relaxation number increases. The mobile dislocation density is also smaller under the same conditions. Therefore, the initial deformation rate decreases along a series for the two reasons given above.

Another way of determining parameter Ω_r (relation (2.25)) is now derived. For various quantities below, the subscripts i and f will be used, which refer to the onset and the end of the relaxation test, respectively. Subscript j refers to the relaxation number in the series. The strain-rates at the end of relaxation number j and the onset of relaxation number $(j + 1)$ are compared (same ρ_m):

$$\dot{\gamma}_{i,j+1}/\dot{\gamma}_{f,j} = v_{i,j+1}/v_{f,j} = \exp(-V\Delta\tau_j/kT) \tag{2.28}$$

where the Orowan equation and relation (2.21) have been used. Combining Eqs. (2.28) and (2.9) yields:

$$V = (kT/\Delta\tau_j)\ln\left(\frac{c_{r,j} + \Delta t}{c_{r,j+1}}\right) \tag{2.29}$$

Eq. (2.29) provides another way of determining V using the time constants of two successive relaxations in the series. To express Ω_r as a function of experimental data, the strain-rates are compared at the onset and the end of test number j, using Eq. (2.9) for $t = 0$ and $t = \Delta t$, respectively. This yields:

$$\dot{\gamma}_{i,j}/\dot{\gamma}_{f,j} = (c_{r,j} + \Delta t)/c_{r,j} = \exp(-\Omega_r V\Delta\tau_j/kT)$$

where relation (2.8) has been used for relaxation number j at $t = \Delta t$. Combining this expression with Eq. (2.28) gives:

$$\dot{\gamma}_{i,j+1}/\dot{\gamma}_{i,j} = \exp[(\Omega_r - 1)V\Delta\tau_j/kT] \tag{2.30}$$

which provides a recurrent relation between $\dot{\gamma}_{i,n}$ (n relaxations in the series) and $\dot{\gamma}_{i,1}$:

$$\dot{\gamma}_{i,n}/\dot{\gamma}_{i,1} = \exp[(\Omega_r - 1)V\sum_1^{n-1}(\Delta\tau_j/kT)]$$

The ratio $\dot{\gamma}_{i,n}/\dot{\gamma}_{i,1}$ equals c_{r1}/c_{rn} according to Eq. (2.9) with $t = 0$. Then the time constant ratio is expressed via Eqs. (2.8) and (2.25), which yields:

$$\dot{\gamma}_{i,n}/\dot{\gamma}_{i,1} = [\exp(-\Omega_r V\Delta\tau_n/kT) - 1]/[\exp(-\Omega_r V\Delta\tau_1/kT) - 1]$$

Eliminating $\dot{\gamma}_{i,n}/\dot{\gamma}_{i,1}$ between the two preceding relations yields:

$$\Omega_r^{-1} = 1 - \left(kT/\left(V_r \sum_1^{n-1} \Delta\tau_j\right)\right)\ln[(\exp(-V_r\Delta\tau_n/kT) - 1]/[\exp(-V_r\Delta\tau_1/kT) - 1)] \quad (2.31)$$

Relation (2.31) allows Ω_r to be determined as a function of experimental data such as V_r measured along the first relaxation of the series and the $\Delta\tau_j'$s. Then Eq. (2.25) is used to obtain V. This last method is preferred to determine V since V_r and the $\Delta\tau_j'$s are easily and safely measured.

2.1.5 *Interpretation of repeated creep tests*

Transient creep tests have so far been investigated less than stress relaxation experiments. The state-of-the-art is described below.

Under creep conditions, relation (2.6) yields:

$$\dot{\gamma} = \dot{\gamma}_p \quad (2.32)$$

Relations (2.19)–(2.21) related to dislocation velocities can be used. Combining Eqs. (2.18b) and (2.21) gives:

$$\mathrm{v} = \mathrm{v}_{or} \exp(-V\Delta\gamma_p K^c/kT) \quad (2.33)$$

Relation 2.24(a) is valid and, together with Eq. (2.22), gives:

$$(\mathrm{v}/\mathrm{v}_{or})^{\beta_c+1} = c_c/(c_c + t)$$

In this expression $\mathrm{v}/\mathrm{v}_{or}$ is expressed via Eq. (2.33):

$$(c_c/c_c + t) = \exp(-V\Delta\gamma_p K^c(1 + \beta_c)/kT)$$

Taking the logarithm of both sides and using Eq. (2.13) for $\Delta\gamma_p$ yields:

$$-\ln(1 + t/c_c) = -(V/MV_c)K^c(1 + \beta^c)\ln(1 + t/c_c)$$

or

$$\Omega c = (1 + \beta_c)(K^c/M) \quad (2.34)$$

with

$$\Omega_c = V_c/V \quad (2.35)$$

similar to Eq. (2.25) for relaxations. The volume V_c can be measured in a single creep test, while a creep series is necessary to determine Ω_c and therefore V. Similarly to stress relaxations, Ω_c also provides information about β_c.

The latter parameter provides useful information about the change in mobile dislocation density during the transient. It can be computed as for stress relaxations starting from Eqs. (2.21) and (2.22). $\Delta\tau^*$ is expressed via Eq. 2.18(b), $\Delta\gamma_p$ using Eq. (2.13). Expressing Ω_c with Eq. (2.34) yields:

$$\rho_m/\rho_{mo} = \left(\frac{c_c}{c_c + t}\right)^{\beta_c/1+\beta_c} \tag{2.36}$$

The evaluation of Ω_c from repeated creep experiments follows the above interpretation of repeated stress relaxation tests. Three methods are available to determine V. In the present case, a relation equivalent to Eq. (2.28) is:

$$\dot{\gamma}_{i,j+1}/\dot{\gamma}_{f,j} = v_{i,j+1}/v_{f,j} = \exp(V\Delta\tau/kT) = \exp(VM\Delta\gamma/kT) \tag{2.37}$$

where $\Delta\gamma$ is the strain increment during reloading which corresponds to $\Delta\tau$. Relation (2.37) indicates in particular that the above strain-rate or velocity ratios are independent of j if the dislocation mobility mechanism (i.e. V in relation (2.21)) is the same. Relation (2.37) provides one method of determining V, by measuring the $\dot{\gamma}$ values as the slope of the creep curves and deducing $\Delta\gamma$ from $\Delta\tau$. A second method provides a relation equivalent to (2.29):

$$V = (kT/\Delta\tau)\ln\left(\frac{c_{c,j} + \Delta t}{c_{c,j+1}} \frac{V_{c,j}}{V_{c,j+1}}\right) \tag{2.38}$$

In this case, V is computed from the parameters of Eq. (2.13) fitted along two successive creep tests in the series (usually the 1st and 2nd ones). The comparison between two successive creep tests can be pursued to estimate Ω_c (relation (2.34)). As above, the ratio $\dot{\gamma}_{i,j+1}/\dot{\gamma}_{i,j}$ is computed using two different procedures which yield

$$\dot{\gamma}_{i,j+1}/\dot{\gamma}_{i,j} = \exp((MV_{c,j}/kT)(\Delta\gamma/\Omega_{c,j} - \Delta\gamma_j))$$

which is equivalent to Eq. (2.30).

The second way of computing the above strain-rate ratio consists of using Eq. (2.14):

$$\dot{\gamma}_{i,j+1}/\dot{\gamma}_{i,j} = V_{c,j}c_{c,j}/(V_{c,j+1}c_{c,j+1})$$

The ratio of the time constants is calculated by considering relation (2.13) for $t = \Delta t$ for tests number j and $j + 1$, respectively, whence:

$$\dot{\gamma}_{i,j+1}/\dot{\gamma}_{i,j} = (V_{c,j}/V_{c,j+1})[\exp(MV_{c,j+1}\Delta\gamma_{j+1}/kT) - 1]/[\exp(MV_{c,j}\Delta\gamma_j/kT) - 1]$$

Eliminating $\dot{\gamma}_{i,j+1}/\dot{\gamma}_{i,j}$ between this last relation and the first one gives:

$$(\Omega_{c,j} - 1)^{-1} = \frac{1}{V_{c,j}\Delta\tau}\left[kT\ln\frac{V_{c,j}[\exp(V_{c,j+1}M\Delta\gamma_{j+1}/kT) - 1]}{V_{c,j+1}[\exp(V_{c,j}M\Delta\gamma/kT) - 1]}\right] + V_{c,j}M\Delta\gamma_j \tag{2.39}$$

This expression provides $\Omega_{c,j}$ as a function of the V_cs and the $\Delta\gamma$s, measured along two successive tests number j and $j+1$. $\Omega_{c,j}$ being known as well as $V_{c,j}$, V is computed with relation (2.34).

The respective advantages of relaxation and creep tests are discussed in Section 2.1.6.4.

2.1.6 *Experimental assessments*

Some experimental data are presented now, which support the assumptions made in Sections 2.1.2–2.1.5 for the interpretation of logarithmic stress relaxations and creep tests. The corresponding experiments are performed on different set-ups under an He or Ar atmosphere between 80 and 1300 K. The temperature stability achieved is close to ± 0.1 K. The resolution in stress is of 1 MPa and in strain of 10^{-4}. Computer programs have been designed for data acquisition and for controlling the machines under various operating modes such as constant strain-rate, constant strain, constant stress, etc. More details are given by Lo Piccolo (1999).

2.1.6.1 Transition between monotonic and transient tests. This is the first step toward verifying the consistency of the above description of the transients. Continuity of the plastic strain-rates, at the end of the monotonic test and at the onset of the transient is expected, as suggested by Saada et al. (1997). To test this point, single transients are performed as follows. Two samples of polycrystalline Ni_3Al are deformed at 300 K at two different strain-rates $\dot{\varepsilon}_p = 4.4 \times 10^{-4}\ s^{-1}$ and $3.6 \times 10^{-5}\ s^{-1}$. Both monotonic tests are interrupted at the same stress level, $\sigma = 360$ MPa, and the sample is allowed to relax (Figure 2.8(a)). The plastic strain-rate at the onset of relaxation $\dot{\varepsilon}_{po}$ is estimated using relation (2.9) for $t = 0$ and compared with the imposed plastic strain-rate $\dot{\varepsilon}_p$. The values of $\dot{\varepsilon}_p$ and $\dot{\varepsilon}_{po}$ shown in Figure 2.8(a) are in fair agreement at both applied strain-rates. The continuity observed for $\dot{\varepsilon}_p$ supports the assumptions made in the previous sections. In addition, the volume V_c is found to be $48b^3$ for the high strain-rate and $50b^3$ for the lower one. This suggests that this parameter is stress (and not strain) dependent.

The following experiment confirms the previous result over a range of stresses (Figure 2.9). Along a stress–strain curve at a constant strain-rate of $5 \times 10^{-5}\ s^{-1}$ ($Ni_{74}Al_{26}$ polycrystal at 300 K), stress relaxations are performed at increasing strains. At a stress of about 300 MPa, the applied strain-rate is multiplied by 11. As expected in this type of material (see e.g. Thornton et al. (1970)), the strain-rate sensitivity is close to zero and no corresponding change in stress is observed along the stress–strain curve. Values of $\dot{\varepsilon}_p$ and $\dot{\varepsilon}_{po}$ are compared for all relaxations. The agreement shown in Figure 2.9 between the two strain-rates within the experimental error confirms the expected continuity.

A similar type of experiment using creep transients is illustrated in Figure 2.8(b). Two monotonic tests with respective plastic strain-rates of $4.4 \times 10^{-4}\ s^{-1}$ and $3.6 \times 10^{-5}\ s^{-1}$,

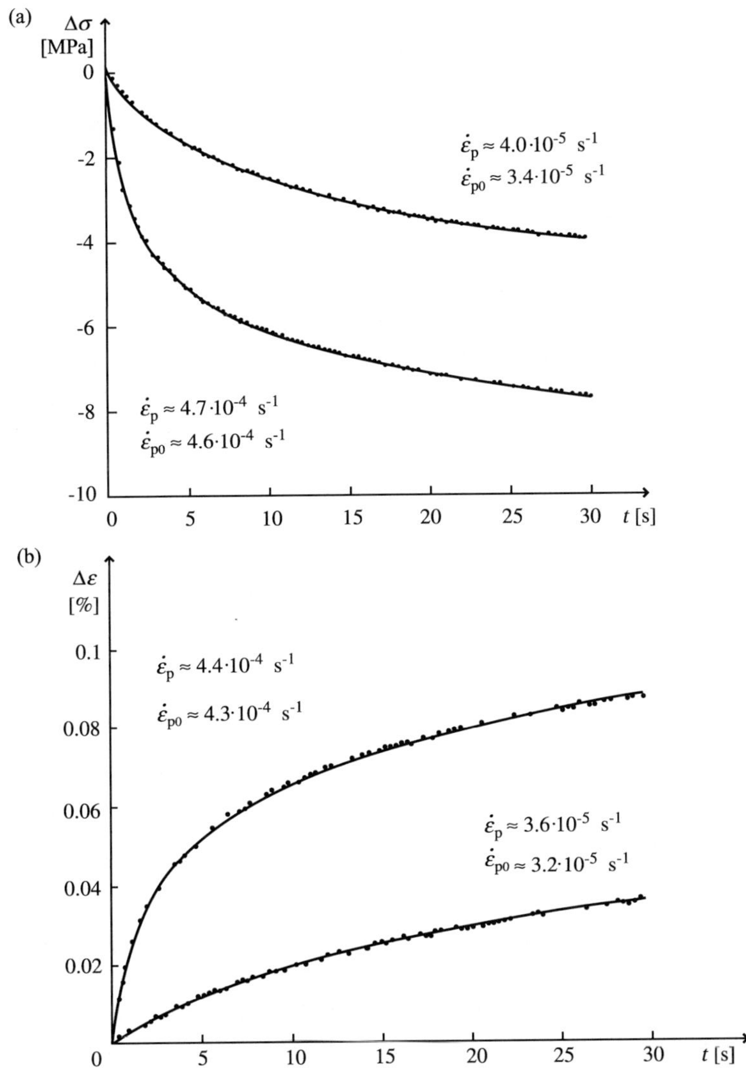

Figure 2.8. Transient tests corresponding to two monotonic tests at two different strain-rates $\dot{\varepsilon}_p$ interrupted at the same stress. $\dot{\varepsilon}_{po}$ is the initial rate of the transients: (a) stress relaxation test, $Ni_{74}Al_{26}$ polycrystal. $T = 300$ K, (b) creep test, $Ni_{75}Al_{25}$ polycrystal. $T = 300$ K.

on a polycrystal of $Ni_{75}Al_{25}$ are interrupted at a stress of 210 MPa. They are then allowed to creep. The plastic strain-rate values indicated in Figure 2.8(b) confirm the support given by the stress relaxation experiments. The strain-rate continuity is also tested along a stress–strain curve (Figure 2.10) interrupted by creep transients ($Ni_{75}Al_{25}$ polycrystal). Two strain-rate jumps are performed along the stress–strain curve: $\dot{\varepsilon}_p$ is increased from

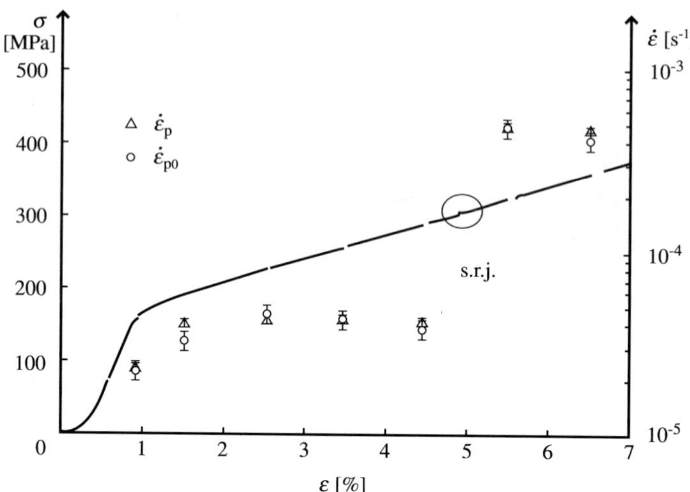

Figure 2.9. Stress–strain curve with a strain-rate jump (s.r.j.) in the circled region. $Ni_{74}Al_{26}$ polycrystal. $T = 300$ K. Stress relaxations are performed along the curve. $\dot{\varepsilon}_p$ and $\dot{\varepsilon}_{po}$ are the plastic strain-rates just before and at the onset of relaxation, respectively.

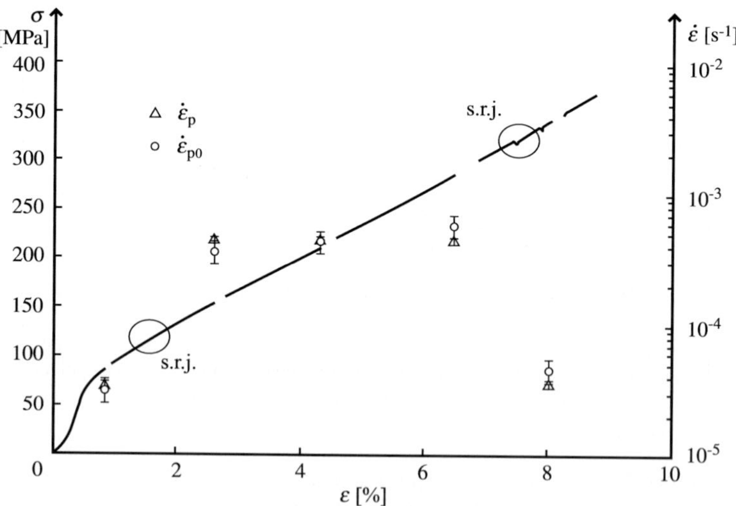

Figure 2.10. Transient creep tests along a stress–strain curve. $Ni_{75}Al_{25}$ polycrystals. 300 K. The monotonic test starts at a low strain-rate then a strain-rate jump (s.r.j.) is performed upwards, then downwards near the end of the test. Transients are performed where the stress–strain curve is interrupted. The plastic strain-rates $\dot{\varepsilon}_p$ and $\dot{\varepsilon}_{po}$ are indicated (see text).

$4.8 \times 10^{-5}\,s^{-1}$ to $5.3 \times 10^{-4}\,s^{-1}$ at a stress of about 110 MPa and then decreased to its former value at 325 MPa. $\dot{\varepsilon}_p$ and $\dot{\varepsilon}_{po}$ are equal within the experimental error at various stresses and strain-rates along the deformation curve. These results of relaxation and creep tests support the assumptions made for the interpretation of the transients. The continuity of the plastic deformation-rate at the transition between the monotonous test and the transient can be understood in terms of the Orowan relation: as the imposed strain-rate test is interrupted, all the dislocations moving at a given velocity also contribute to strain at the onset of the transient, therefore with the same net deformation-rate.

2.1.6.2 Examples of repeated creep tests. Such transients have been extensively used in one of our laboratories and constitute a well-established technique (see e.g. Bonneville and Martin, 1991; Martin et al., 1999; Spätig, 1995; Bonneville et al., 1997a; Lo Piccolo, 1999). They have provided a set of V values characteristic of various mobility mechanisms for different types of materials and conditions (see Martin et al., 1999). They also yield useful information about dislocation exhaustion during plastic deformation. Figure 2.11 illustrates an example of repeated creep test.

2.1.6.3 Results of stress relaxation series. Various aspects of stress relaxation series are now recalled.

Figure 2.12 shows the small stress drop achieved during stress relaxation, which should alter as little as possible the dislocation substructure at constant strain-rate.

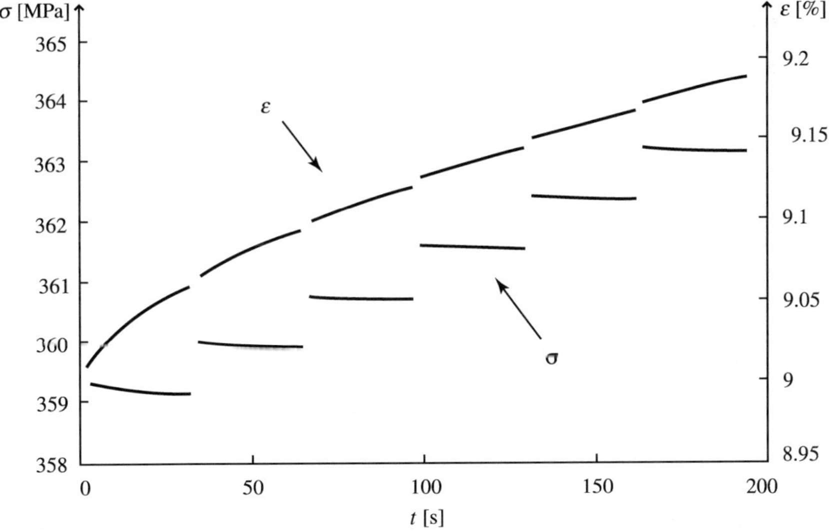

Figure 2.11. Experimental curve corresponding to a repeated creep test in a $Ni_{75}Al_{25}$ polycrystal. $T = 300$ K, $\sigma = 359$ MPa (Lo Piccolo, 1999).

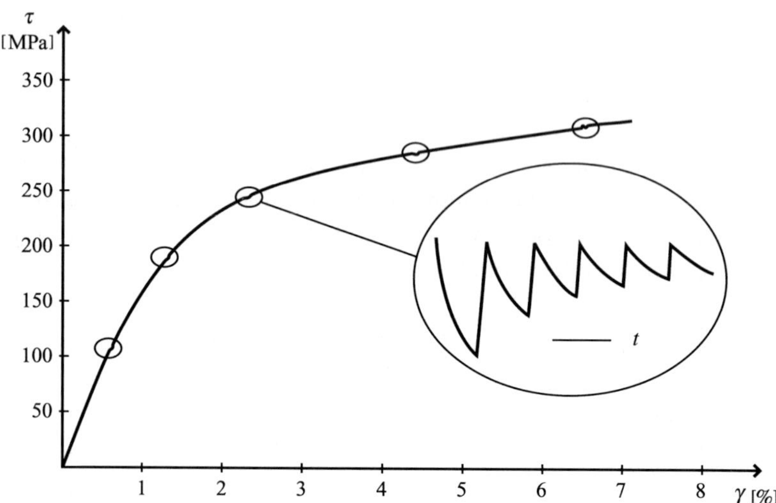

Figure 2.12. Stress–strain curve for a ⟨123⟩ $Ni_3(Al,Hf)$ single crystal. $T = 423$ K. Circled areas indicate repeated stress relaxation tests. The insert shows one of them. (Spätig, 1995).

The consistency of assumptions made for the determination of Ω_r (relation (2.31)) is tested in Figure 2.13. Ω_r values for a single crystal of Ni_3Al are found to be close to 1.9 and rather constant for an increasing number of relaxations (between 2 and 7). Ω_r is close to 1.6 for TiAl polycrystals and also constant between 2 and 6 relaxation tests in a series. These Ω_r values emphasize the difference between V_r and V (relation (2.25)).

The validity of the method used to determine V is tested in the experiment illustrated in Figure 2.14. A single crystal of $Ni_3(Al, Hf)$ is compression tested in two set-ups which

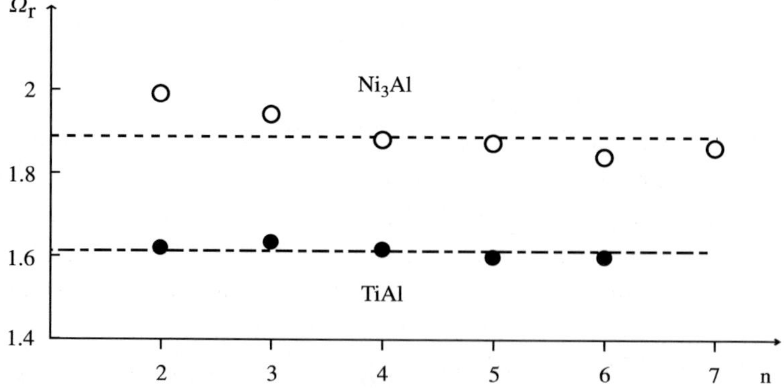

Figure 2.13. Values of the structural parameter Ω_r as a function of the relaxation number n in the series. $Ni_3(Al, Hf)$ single crystal. $T = 293$ K. $\gamma_p = 4\%$ (Spätig, 1995). γ-TiAl polycrystal. $T = 473$ K, $\sigma = 342$ MPa (Viguier et al., 1995).

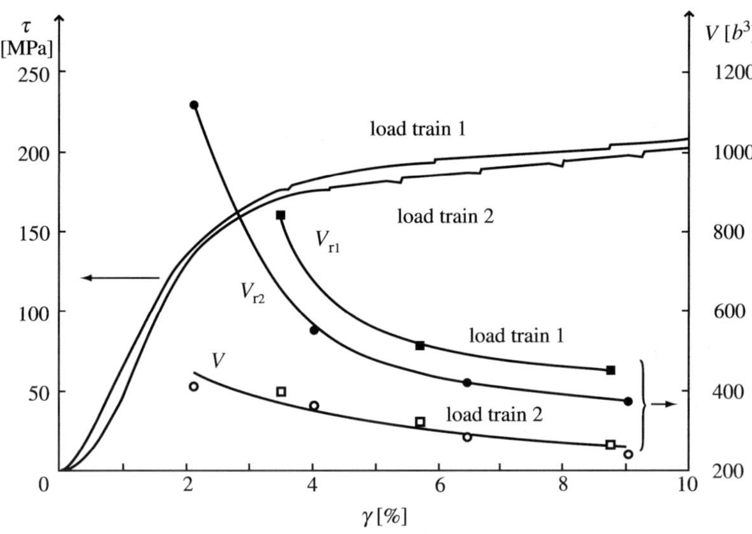

Figure 2.14. Comparison between activation volumes V_r and V measured using two different load trains. ⟨123⟩ $Ni_3(Al,Hf)$ single crystals. $T = 293$ K. V_{r1} and V_{r2} refer to load trains 1 and 2, respectively.

exhibit different stiffnesses (load trains 1 and 2, respectively). Along the stress–strain curves, repeated stress relaxation tests are performed. They yield activation volumes which are plotted as a function of strain. Figure 2.14 clearly shows that the apparent volumes V_{r1} and V_{r2}, respectively depend on the set-up, in agreement with the Ω dependence on M (relation (2.26)). V_{r1} is larger ($M_1 = 3250$ MPa) as compared to V_{r2} ($M_2 = 4000$ MPa). Nevertheless, the values of V (relation (2.20)) follow a single curve as a function of strain. This suggests that the dislocation mobility mechanism is the same in both experiments. The difference in V_r values indicates different substructural changes in the two samples (relation (2.4)).

At this stage, meaningful quantities such as the plastic strain, the dislocation velocity and the mobile density can be represented as a function of time during the relaxation series. These parameters are shown in Figure 2.15 for a relaxation series which consists of three successive tests of 30 s each, in Ni_3Al. Figure 2.15 shows that the dislocation velocity decreases as stress relaxes according to relations (2.21) and (2.18a). The plastic strain ($\Delta\gamma_p$) increases (as well as τ_μ—relation (2.17)). The mobile dislocation density decreases markedly (Eq. (2.22)), thanks to the efficient mobile dislocation exhaustion mechanism by Kear–Wilsdorf lock formation in these crystals (Kear and Wilsdorf, 1962). A detailed description of this process can be found in Section 10.1. At this stage, it is possible to introduce some information about the respective contributions to the strain-rate of the mobile dislocation density and velocity, e.g., during the first 30 s of relaxation.

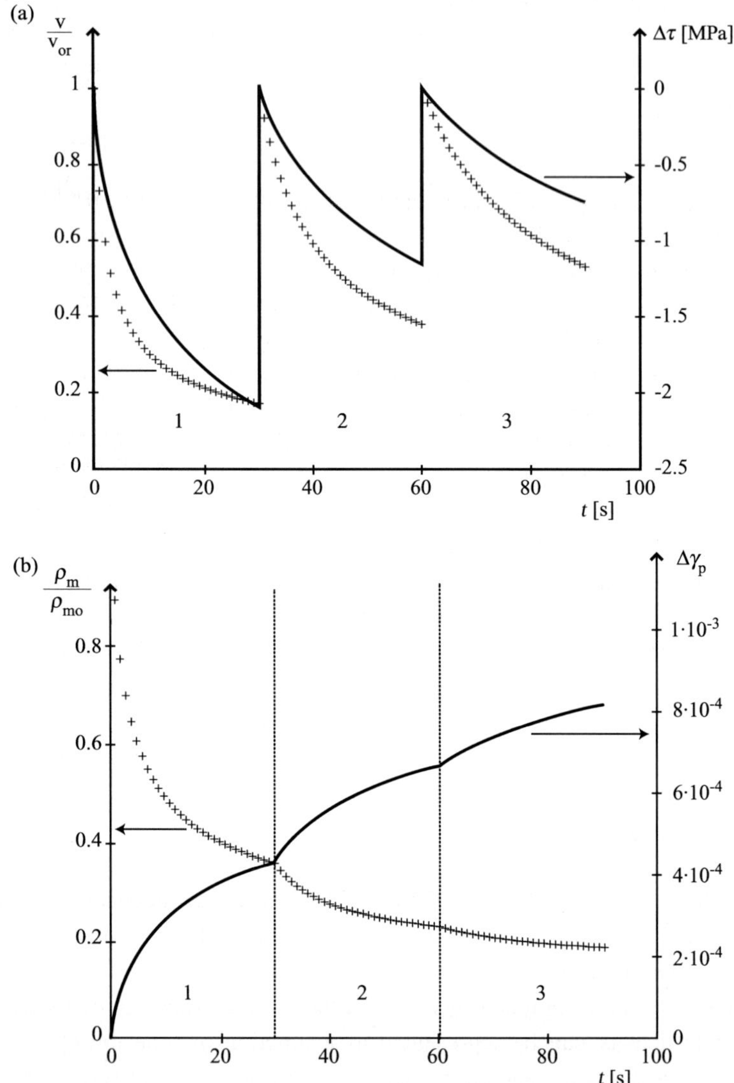

Figure 2.15. Time dependence of macroscopic and microstructural parameters during a repeated stress relaxation experiment. ⟨123⟩ $Ni_3(Al, Hf)$ single crystal. $T = 293$ K. $\gamma_p = 4\%$. The relaxation number is indicated. (a) Stress decrease $\Delta\tau$ and dislocation velocity v normalized for its onset value v_{or}. (b) Plastic strain increase $\Delta\gamma_p$ and mobile dislocation density ρ_m/ρ_{mo}.

The strain-rate retains about 4% of its initial value, as compared to 16% for the velocity and 24% for the mobile dislocation density. In this case, the decrease in strain-rate is accommodated by a decrease in dislocation density slightly lower than that of the average dislocation velocity.

2.1.6.4 Results of creep series and comparison with stress relaxations. We have verified that the creep curves of Figure 2.11 are logarithmic. The volumes, $V_{c,j}$, and time constants, $c_{c,j}$, have been fitted to these curves using relation (2.13). The dependence of $V_{c,j}$ on j is illustrated in Figure 2.16 for the same material at a lower strain.

Figure 2.16 shows that $V_{c,j}$ decreases along the series for $j = 1, 2, 3$. $\Omega_{c,j}$, obtained as explained above, is also found to decrease, while V appears to be constant. This is at variance from stress relaxation tests where V_r and Ω_r are constants as a function of j. The latter result suggests that the dislocation mobility mechanism is the same along the series, while the structural parameters β_c and K_c, which are included in $V_{c,j}$ (relation (2.34)), are changing with j. The values of parameter V, which is the signature of the dislocation mobility mechanism, are now compared when determined by relaxation and creep transients. V values along a stress–strain curve, determined by both techniques are shown in Figure 2.17. Within the experimental error, the data are along a single curve of variation as a function of strain, irrespective of the type of transient. Both techniques seem to

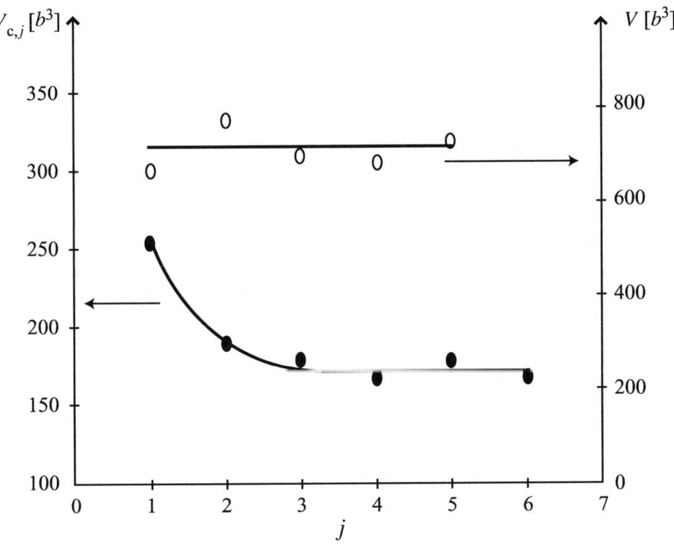

Figure 2.16. Variation of $V_{c,j}$ and V as a function of j. Repeated creep test in a $Ni_{75}Al_{25}$ polycrystal. $T = 300$ K. $\sigma = 78.5$ MPa. $\varepsilon_p = 0.3\%$.

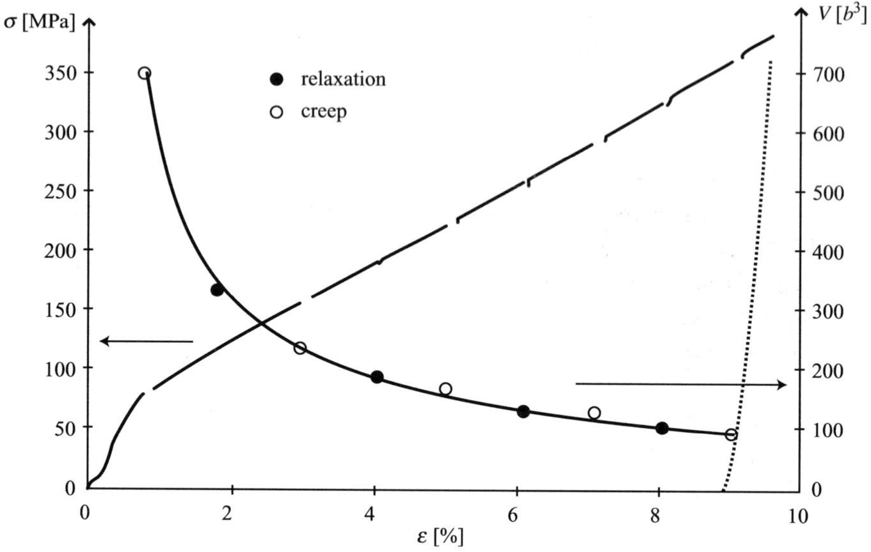

Figure 2.17. Activation volume V as a function of strain along a compression curve, measured by relaxation and creep tests, respectively. $Ni_{75}Al_{25}$ polycrystals. $T = 300$ K.

indicate the same mobility mechanism in the material investigated. Therefore, creep transients appear to be as reliable and useful as repeated relaxations.

Similar to the analysis of stress relaxation experiments, some microstructural parameters are estimated for creep transients as a function of time. The results are presented in Figure 2.18 for a polycrystal of $Ni_{75}Al_{25}$ at 300 K.

The thermal and athermal parts of the stress have been estimated (relation (2.18b)) and their evolution as a function of time is shown in Figure 2.18(a). The decrease in the dislocation velocity as well as the mobile dislocation density is illustrated in Figure 2.18(b). In this case too, the contribution of v and ρ_m to the creep rate can be estimated. Over the first time interval of the series the creep rate retains 28% of its onset value, v retains 51% and ρ_m 55%. Therefore, a significant decrease in the creep rate is accommodated in this case by equivalent reductions in mobile dislocation densities and velocities.

A further comparison of the methods of repeated relaxations and creep tests shows that both techniques differ in the plastic strains imparted to the specimen. In the above Ni_3Al polycrystal (Figure 2.11), the total plastic strain after six creep transients of 30 s each is close to 0.15%. In repeated relaxation tests of the same duration, the plastic strain is in the range of 0.02–0.08%. The sample is less deformed in the latter case. Consequently, repeated relaxations appear to be more attractive than the present procedure of creep transients. Therefore, the information gained about dislocation mobilities and

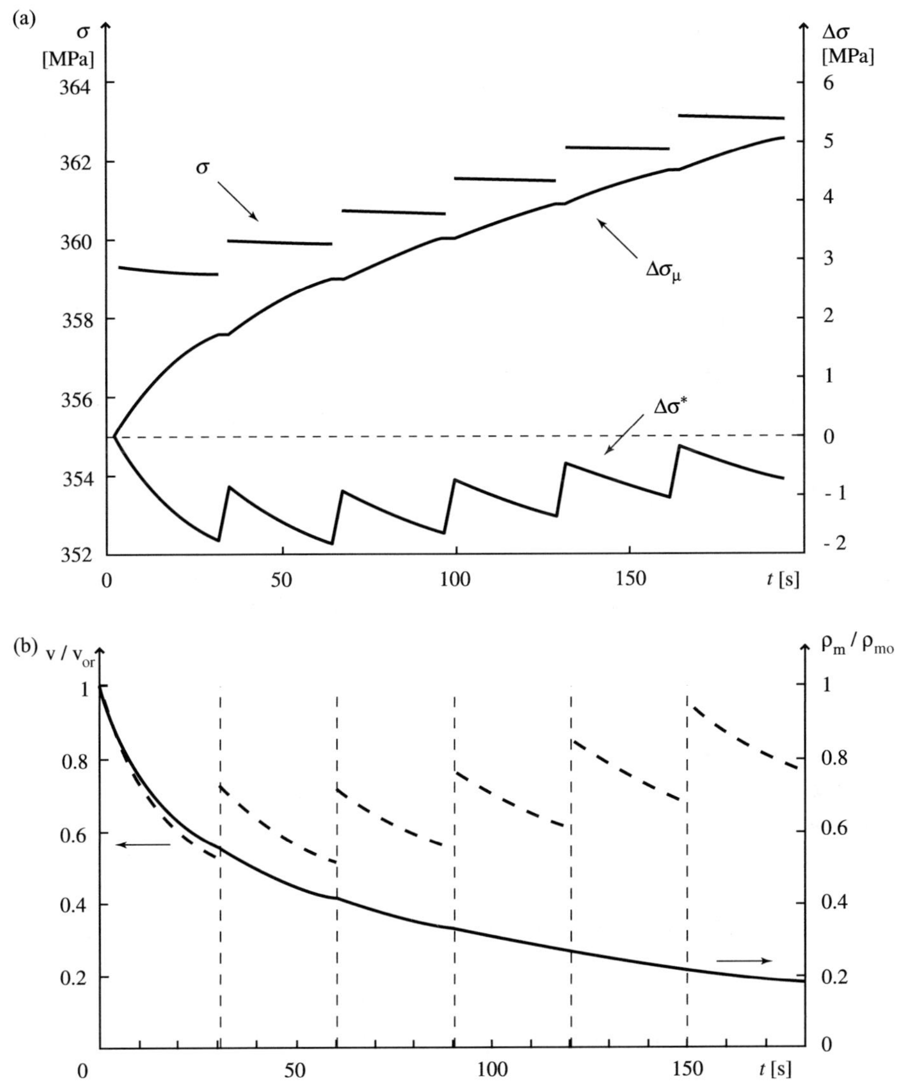

Figure 2.18. Microstructural parameters during a creep series, $Ni_{75}Al_{25}$ polycrystal, $T = 300$ K, $\sigma = 355$ MPa. (a) Imposed stress σ and variation of $\Delta\sigma^*$ and $\Delta\sigma_\mu$. (b) dislocation velocity normalized to v_{or} and mobile Dislocation density ρ_m/ρ_{mo}.

the microstructure evolution during plastic deformation is likely to be closer to the corresponding constant strain-rate mechanisms. However, the more recent technique of creep series may yield interesting results when applied to materials other than intermetallics.

2.1.7 Stress reduction experiments

To our knowledge stress reduction experiments were proposed for the first time by Gibbs (1966). His original idea is illustrated in Figure 2.19 for a monotonic test. Rapid reduction of applied stress either directly from σ_0 or in the early stages of relaxation is used. The crosshead of the machine is then stopped and the stress evolution as a function of time is studied. If the stress is reduced to σ_μ, a short period of constant stress is observed. If the stress is lowered below σ_μ, the stress exhibits an initial rise. The stress reduction leading to constant or rising stress corresponds to glissile dislocations encountering internal stress fields acting in the reverse direction to the applied stress. This is expressed by relation (2.16).

Such a procedure is nowadays called a stress-dip test, as opposed to a strain-dip test, which is generally performed under creep conditions. In the latter test, the stress is reduced and kept constant, while the subsequent evolution of strain with time is observed.

The simple interpretation proposed by Gibbs for the Mg data is certainly valid, as long as dislocations are homogeneously distributed. In addition, obstacles to dislocation motion as well as microstructures must not exhibit appreciable recovery during stress reduction.

For deformation conditions which yield a heterogeneous dislocation distribution at a mesoscopic scale, σ_μ is not constant through the crystal. For example, as cells form, hard

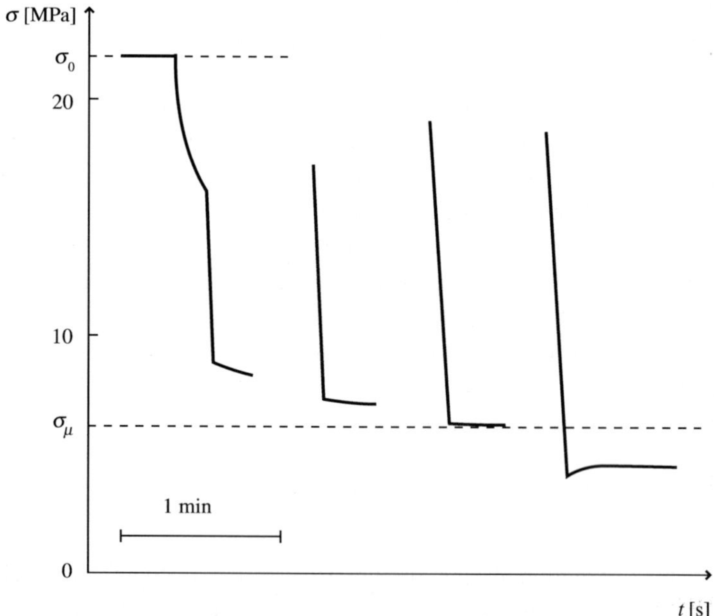

Figure 2.19. Stress reduction experiments along a constant strain-rate curve as proposed by Gibbs (1966). Mg polycrystals. $T = 565$ K. $\dot{\varepsilon} = 6.67 \times 10^{-4}$ s^{-1}.

and soft zones have to be considered that correspond, respectively, to dislocation walls where the local dislocation density is very high and cell interiors where it is very low. In the former zone, the internal and applied stresses add up to push dislocations forward, while in the latter ones σ_μ resists the applied stress (see the experimental facts in Mughrabi, 1983 and Ungar et al., 1984). More details about the corresponding "composite model" can be found in Mughrabi and Ungar (2002). In this model, the overall flow stress is a simple average of the two local flow stresses, weighted according to the volume fractions of hard and soft zones. Comparable results about stress heterogeneities have been produced for creep subgrains and subboundaries (Morris and Martin, 1984). In the latter case a model for mobile dislocations cutting through subboundaries was proposed (Caillard, 1985). It accounts for the build-up of high stresses at subboundaries.

Another case of heterogeneous deformation, which so far has not been modelled satisfactorily, is that of crystals containing dynamic pile-ups: covalent crystals in which the early stages of deformation correspond to rapid dislocation multiplication, solid solution alloys deformed at low temperatures, short range ordered alloys, etc.

Consequently, a variety of material responses to such tests is observed and specific interpretations have been proposed. For example, Mills et al. (1985) study anelastic back strains subsequent to unloading during creep of Al–5.5 at.% Mg alloys. The deformation conditions correspond to a viscous motion of dislocation loops as the rate controlling process. σ_μ is found to be smaller than or equal to 10% of the applied stress. Blum et al. (1989) study the unloading response of an Al–Zn alloy during creep. The observed transients are interpreted in terms of the rearrangement of the subgrain structure during unloading combined with a reduction of individual dislocation velocity. By performing unloading experiments during creep of an ordered $Cu_{50}Zn$ alloy, Milicka (1999) gives evidence of a linear relation between the effective stress and the applied stress. Nen et al. (2000) measure σ_μ during constant strain-rate tests and determine $\sigma_\mu \sim 0.75\sigma$ at intermediate and high homologous temperatures.

The impression is that the materials investigated so far behave differently and the interpretation of stress reduction tests has to be adapted in each case. Classes of materials according to their response have not been established yet. A wide field of investigation is therefore open. A rewarding attitude is to combine stress reduction tests and microstructural investigations (see e.g. Mills et al., 1985). Unfortunately several studies consider mechanical test data exclusively.

2.1.8 Conclusions about transient mechanical tests

The various experimental results given in Section 2.1.6 provide a body of evidence about the validity of the assumptions used for the interpretation of various transient mechanical tests (logarithmic transients). Repeated creep, as well as stress relaxation, experiments yield several unique kinds of information about the microscopic deformation mechanisms which operate. The activation volume of the average dislocation velocity can be unambiguously determined. This parameter is really the signature of a dislocation mobility mechanism. Indeed, although V corresponds to the stress dependence of the activation energy ΔG (relation (2.20)), it can change by orders of magnitude depending on the mechanisms, crystals and deformation conditions. However, under the same situation, ΔG does not change significantly over the temperature range commonly investigated. The determination of V is slightly more complicated than that of the apparent volume through single stress relaxation or creep transients. V can be up to two times smaller than V_a.

Examples of data about V can be found: screw dislocation cross-slip in Cu according to Bonneville et al., 1988 (see also Chapter 5), dislocation glide in single crystals of Ni_3(Al, Ta) (Spätig et al., 1995), Ni_3(Al, Hf) (Bonneville et al., 1995), binary Ni_3Al (Bonneville et al., 1997b), polycrystals of binary Ni_3Al of various compositions (Matterstock et al., 1999), of Fe_3(Al, Cr) (Kràl, 1996), and in superalloy single crystals (Nazar et al., 1993).

Another important kind of information gained from the same transient mechanical tests concerns the average mobile dislocation density. This parameter as a rule is poorly documented. In the case of logarithmic transients, a power law dependence of ρ_m on the average dislocation velocity has been evidenced. The exponent β of this relation can be determined, with reasonable assumptions on hardening during the transient. The mobile dislocation exhaustion rate can be measured with values that depend on the crystal type and deformation conditions. A fair correlation is found between this parameter and the work-hardening coefficient (see e.g. Kruml et al., 2002, for Ni_3Al). Since the former and the latter quantities are measured, respectively, during the transient and monotonic tests this proves again that the mechanisms operating under both conditions are similar.

It is worth noting that the above data illustrate the potential of the techniques of stress relaxation and creep series. These open a wide new field of investigation for a variety of mechanisms, lattices and imposed conditions, which includes the stress dependence of the average dislocation velocity, the coupling between this velocity and the mobile dislocation density and the mobile dislocation exhaustion rates in connection with work-hardening.

2.2. DEFORMATION EXPERIMENTS IN THE ELECTRON MICROSCOPE

In the history of electron microscopy, 1956 is a crucial date: the first foils were made, thin enough to yield images of crystalline defects. Hirsch et al. (1956) were the first ones to report dislocation movements under thermal stresses generated by the beam of 100 keV electrons. Two years later, Wilsdorf (1958) performed the first in situ tensile experiments in a 100 kV TEM. He identified several dislocation mechanisms that were proposed previously. After these pioneering works, such experiments marked time because surface forces were too disturbing in very thin foils. However, at the end of the 1960s, high-voltage electron microscopes (HVEMs) were built operating between 600 and 1000 kV. In situ experiments started anew in thicker foils, on deformation mechanisms in FCC and body-centred cubic (BCC) metals (see e.g. Fujita, 1966; Vesely, 1968; Furubayashi, 1969; Caillard and Martin, 1975). Since then this technique has been used continuously and successfully, as shown in various chapters of this book. It is worth noting that since the early 1980s, intermediate voltage microscopes operating between 200 and 400 kV took over. They provide access to reasonable foil thickness with no or limited radiation damage and they are easier to use than HVEMs. Some technical aspects of such experiments are being briefly reviewed together with their limitations and advantages.

2.2.1 Some key technical points

The microtensile specimens have to fulfil two requirements: be suitable for TEM observations and be adequately shaped for a tensile test. The hole with thin edges must be as circular as possible and at the centre of the specimen, to ensure a predictable stress distribution. The rim of the perforation must be free of cracks (otherwise straining will proceed by crack propagation at the expense of homogeneous deformation). The specimen area to be thinned is smaller than the conventional 3 mm diameter disc and this operation must be done with great care. An extensive description of the preparation procedure is given by Couret et al. (1993). For metallic specimens, 400 μm thick sheets are spark erosion machined, then thinned down to 60–70 μm using sand papers with finer and finer grains. The specimen faces have to be parallel within 2–3 μm to ensure successful chemical or electro-chemical thinning. To perform this latter procedure, dedicated holders have to be built, adapted to the small specimen size. As far as the chemical or electro-chemical solutions are concerned, their composition, temperature and, for the latter, current intensity and voltage have to be optimized.

The most frequent specimen shape is rectangular, of dimensions close to 3×1 mm^2. These are glued to the specimen holder. They correspond to experiments performed between 80 and 600 K. For crystals that exhibit high strength at high temperatures, e.g. ordered intermetallic alloys, the microsamples are mechanically clamped to the grips (see an example in Figure 2.20). The specimen tensile axis is chosen so as to favour a glide system of interest, according to the Schmid law. The choice of the foil plane results from a compromise between several considerations: diffracting planes should be reached easily, to allow for easy observation and characterization of dislocation lines and the active slip plane should not be end on, so that moving defects can be easily seen.

A variety of straining holders have been built, each one operating at a range of temperatures (Couret et al., 1993). A set-up built in house, and operating between room temperature and 1300 K is presented in Figure 2.21.

During the experiment, the dynamic sequences are recorded via a video system. The video images are analysed frame by frame. After the experiment, the microsample is often observed again "post mortem" using a conventional double tilt holder to perform a deeper analysis of the defects present in the foil.

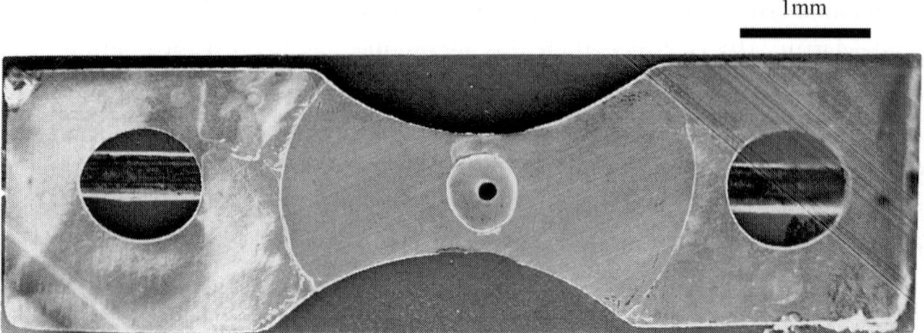

Figure 2.20. A microtensile superalloy specimen for a 1300 K in situ deformation experiment.

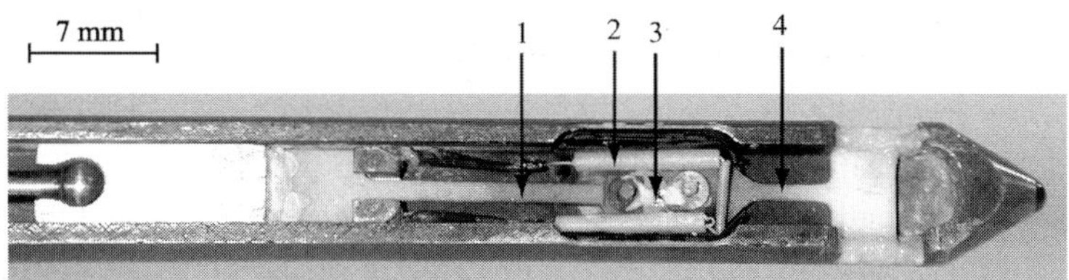

Figure 2.21. Tip of a high temperature (300–1300 K) straining holder (Couret and Caillard, 1998): 1 mobile jaw, 2 heating resistor, 3 microsample, 4 fixed jaw.

2.2.2 *Quantitative information provided by in situ experiments*

Slip traces at the foil surfaces allow the dislocation glide planes to be determined. This is of particular interest for the identification of non-closed-packed slip planes in close-packed structures (see Chapter 6). The Burger's vectors are usually determined after the dynamic observations.

An estimation of the local stress is made according to the following considerations. The local direction of the normal tensile stress is predicted using a finite element approach (circular hole in a rectangular sheet). Figure 2.22 shows how the normal stress direction changes as a function of the position around the hole. It shows in particular that the local tensile direction is parallel to the applied one near the upper and the lower hole rims on the figure. Higher normal stress values are found in these areas as well. Prediction of locally active slip systems can be made using these results.

Quantitative estimation of the effective stress τ^* (relation (2.16)) acting on a dislocation can be made by measuring its radius of curvature R. It can be very different from the applied stress. For an equilibrium configuration, the relation between τ^* and R is:

$$\tau^* = \hat{T}/bR$$

where $\hat{T}$ is the line tension. Evaluations of $\hat{T}$ can be found in Hirth and Lothe (1982, p. 174) in the frame of elastic isotropy and have also been attempted in anisotropic materials

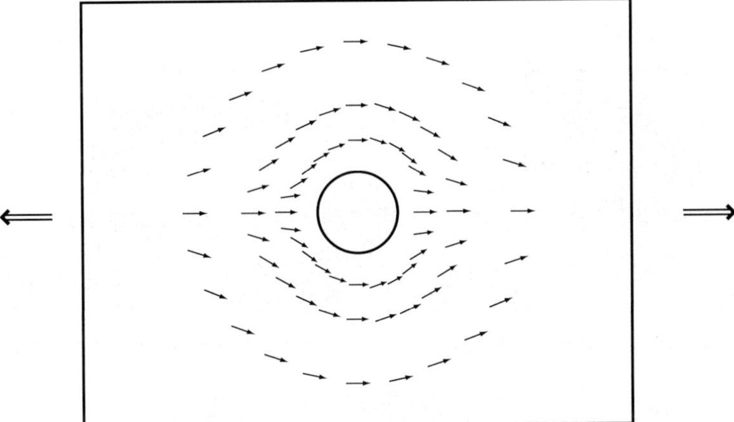

Figure 2.22. Finite element determination of the local normal stress direction around a circular hole in a rectangular plate. The tensile stress applied to the rectangle is also represented (from Couret et al., 1993).

(see e.g. Douin et al. (1986) for dislocations curved on the cube plane of Ni_3Al). Such an estimation does not yield accurate values of τ^* because of approximations in the expression of $\hat{T}$ and uncertainties in the measurement of R. However, it provides information about the variation of τ^* as a function of imposed conditions (e.g. the temperature).

As an example, dislocation unpinning is quantitatively studied for ordinary dislocations gliding in a γ-TiAl lattice. Figure 2.23(a) is the last video picture before unpinning. The radius of curvature is measured by superimposition of a calibrated elliptical loop, as well as the angles at P between the loop segments and the Burger's vector direction b (Figure 2.23(b)). In the present case, the force exerted by the dislocation on P is close to 2.7 nN.

2.2.3 *Reliability of in situ experiments in TEM*

In the early days of HVEM in situ experiments, a list of advantages and limitations of such tests was discussed (Martin and Kubin, 1978). Some additional information is offered by Clément et al. (1991).

The first type of questions refers to the stability of a dislocation line in a thin foil. Orders of magnitude of the practical observed thickness are: 100–500 nm for 200 kV microscopes and 500–1500 nm for 1 MV microscopes. For sake of comparison, "post mortem" observations are made in 5 nm foil thickness for atomic resolution experiments and 10–200 nm for weak-beam conditions. The observable foil thickness depends on the amount of resolution required, but also on the atomic number of the atoms.

The free surfaces are characterized by a zero stress field which results in forces acting on dislocations. These image forces have been evaluated for different dislocation

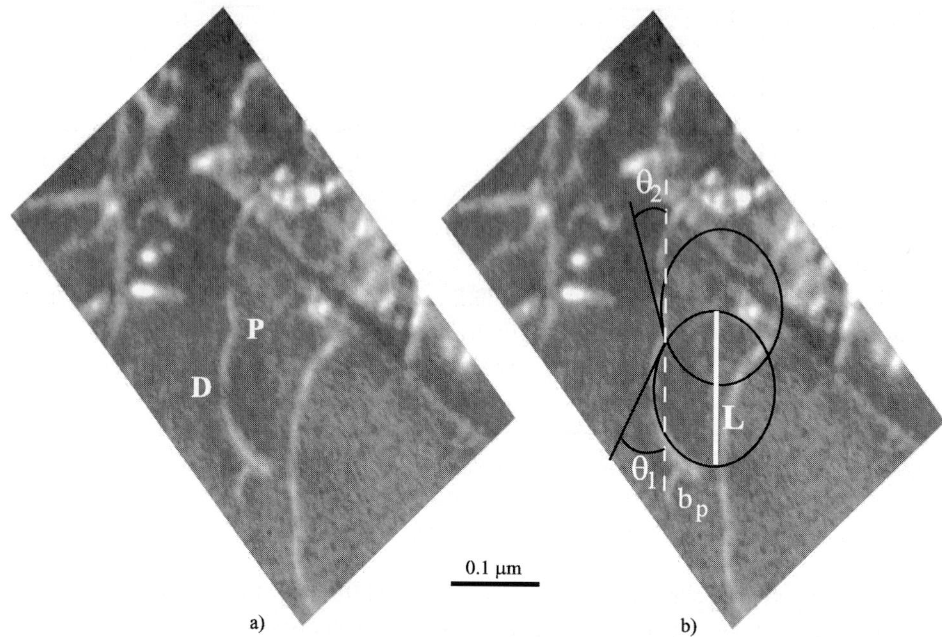

Figure 2.23. Video frame showing the critical configuration of an ordinary dislocation pinned at point P. γ-TiAl at 300 K. Weak beam micrograph, in (a). Determination on the same micrograph of the radius of curvature ($R = L/2$) and escaping angles θ_1 and θ_2 in (b). (from Pettinari et al., 2001).

geometries (see a review in Martin and Kubin, 1978). A screw dislocation lying parallel to the foil surface at distance d is submitted to a stress:

$$\mu b/4\pi d$$

These image forces can be neglected at distances d from the surface where they are substantially lower than the stress of other sources acting on the dislocation.

When the lattice friction is weak, dislocations in a foil may take peculiar orientations, different from those in the bulk material. This is the net result of lowering their energy by reducing their length, of image forces and applied stress. In this case too, dislocations may be attracted by the foil surfaces, thus leaving the crystal with a resultant density lower than in the bulk crystal. In practice, the sample thickness must be larger than a critical value, of the order of $\mu b/\sigma$, where σ is the flow stress.

Radiation damage due to the electron beam is another type of possible artefact (see a review by Martin and Kubin, 1979). For each type of crystal, a threshold voltage can be determined above which electron irradiation takes place. Corresponding values are reviewed by Urban (1976).

Given the above list of possible artefacts, the validity of the results of an in situ deformation experiment has to be assessed. This can be done by comparing the data with results of macroscopic deformation tests and post mortem TEM observations. Several examples of successful in situ experiments will be presented in the following chapters related to a variety of deformation mechanisms. Unique types of such information can be obtained. Qualitatively, the aspect of dislocation motion, jerky as opposed to viscous, provides a hint about the mobility mechanisms involved. It can be seen whether the crystal deforms homogeneously or heterogeneously, at a microscopic scale. Dislocation glide cross-slip or climb can be identified depending on the conditions. Glide planes can be unambiguously determined using the slip traces. Quantitative information about the effective stress acting on the dislocations and their velocities can be gained as a function of temperature.

2.3. IN SITU SYNCHROTRON X-RAY TOPOGRAPHY

The specimen is deformed in the intense photon beam of a synchrotron, while topographic dislocation images are recorded as a function of time. As for the experiments in Section 2.2, the idea is to identify directly the dislocation mechanisms which are responsible for strain. The main differences from the TEM conditions are: (i) the penetration of X-rays is high as compared to TEM electrons so that bulk specimens can be used; (ii) the dislocation image width is much larger with photons, i.e. there is a poor resolution in microstructural observations. Consequently, only crystals containing very low dislocation densities are good candidates for such experiments.

Pioneer work consisted of the observation of subgrain structure formation and evolution in $\langle 100 \rangle$ NaCl single crystals submitted to compression creep in the LURE synchrotron. The dedicated creep machine was designed for $2.5 \times 2.5 \times 6\ \mathrm{mm}^3$ specimens and allowed reflection topography. The creep conditions were successively 700°C–0.7 MPa, 580°C–1.5 MPa, 480°C–3 MPa. The resolution achieved allowed a safe observation of subgrain sizes of 500 μm or larger (Fries et al., 1983). Subgrain misorientation is observed to increase rapidly for strains lower than 4.10^{-2} and much more slowly afterwards. At the onset of creep, the as grown microstructure is replaced by a new one which results from subboundary formation (by dislocation accumulation). They are observed to migrate, which accounts for part of the creep strain in primary creep. The subgrain misorientation increases with time (increase of subboundary dislocation density) which slows down their motion in secondary creep.

Recently, new experiments were undertaken at the ID19 beam line at the European Synchrotron Radiation Facility (ESRF). The point was to characterize dislocation multiplication processes in perfect silicon single crystals during in situ creep.

The tensile specimens had a gauge section of 4×0.7 mm^2 and a length of 15 mm. The creep conditions corresponded to temperatures between 975 and 1075 K and loads from 22 to 44 MPa (resolved shear stresses of 10 and 20 MPa on primary systems). The $\langle 114 \rangle$ symmetrical orientation is combined with two different types of sample faces, which allows for a variety of diffraction conditions. Therefore, a better dislocation identification is possible using classical visibility criteria. X-ray topographs are made in transmission (Jacques et al., 2000; Vallino et al., 2000). In this preliminary work, a full description of multiplication processes is not available. However, some dislocation mechanisms have been identified which participate in the sequence of events leading to fresh dislocation generation as virgin crystals are deformed. The sample faces were polished carefully to avoid dislocation nucleation sites at surfaces. Vicker's microindents were done on the front and rear faces to create volume dislocation sources at well-defined spots. The experiment is interrupted when the dislocation density becomes too high. The specimen is then cooled under load and subsequent Burger's vector analysis performed post mortem, using a variety of diffraction conditions. Dislocation half loops form from the indents near the surfaces, lying on different $\{111\}$ planes. They consist of linear $\langle 110 \rangle$ segments. The different sets of half loops increase in size with time, each segment moving at a rate which depends on its Schmid factor and particular velocity law (see Chapter 7). No new sources appeared in the bulk or at the surfaces during the following observations, apart from those which form at the previously moving dislocations. An example of loop generation and evolution is shown in Figure 2.24. A model is proposed for such loop formation which involves dislocation cross-slip.

Other evidence of cross-slip in the bulk is provided by these observations. Because of dislocation splitting, cross-slip necessitates constrictions along the dislocation line. At that stage, the authors explain the observed frequency of cross-slip by the presence of dissociated jogs along dislocations (Hirsch, 1962) or of favourable junctions (Washburn mechanism, 1965).

This preliminary work establishes a detailed list of elementary dislocation mechanisms which participate in the multiplication process. It also shows the advantages of bulk investigations.

One has to bear in mind that the experiments in TEM (Section 2.2) and in the synchrotron (present section) can provide complementary information on a given process. Figure 2.25 illustrates a fair agreement between dislocation velocities measured respectively by in situ X-ray topography and in the HVEM over a range of temperatures and stresses. Along the same line, Figure 2.26 shows similar types of microstructures observed by the two techniques.

Figure 2.24. Half loop generation at the surface of a $\langle 114 \rangle$ Si single crystal. In situ creep experiment in the ESRF synchrotron. $T = 1075$ K, $\sigma = 22$ MPa. Pictures are taken after 11 min (a), 23 min 15 s (b), 25 min 15 s (c) 26 min 40 s (d) and 28 min (e) after the onset of loading. (from Jacques et al., 2000).

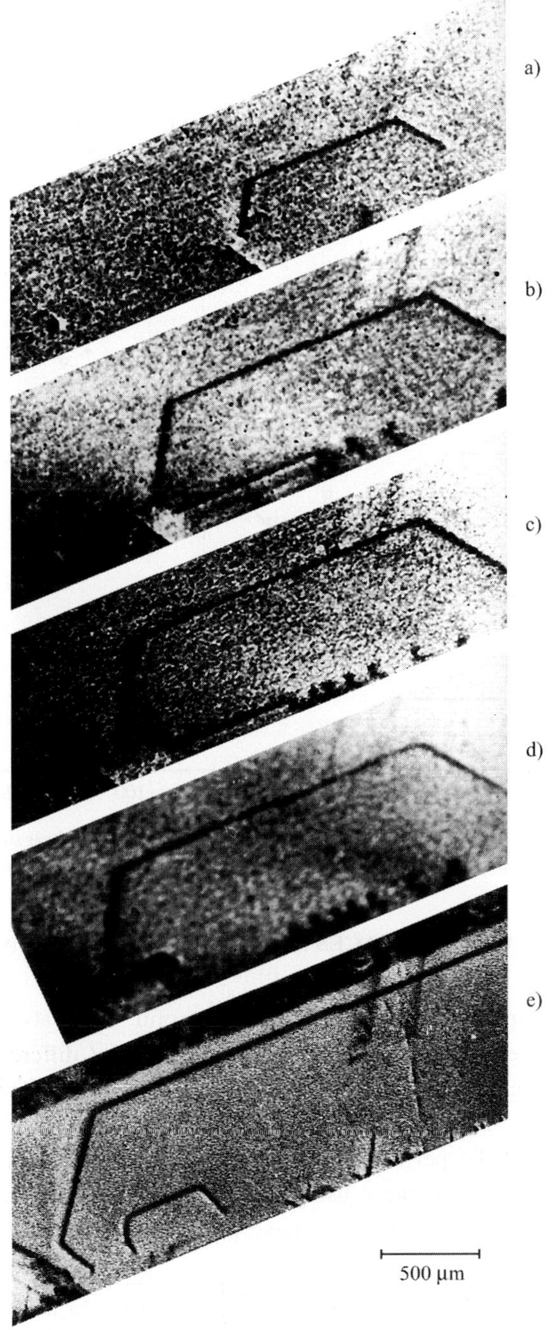
a)
b)
c)
d)
e)
500 μm

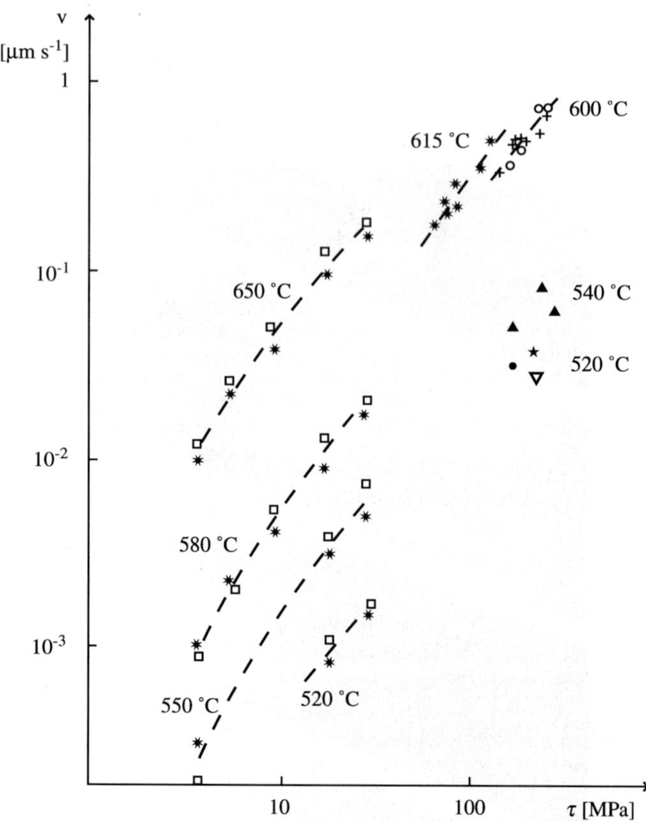

Figure 2.25. Dislocation velocities in Si single crystals as a function of resolved shear stress. Two sets of data are presented from X-ray topography and HVEM, below and above 50 MPa respectively (Louchet, 1981).

2.4. OBSERVATION OF SLIP TRACES AT THE SPECIMEN SURFACE

The geometry of slip traces can also provide useful hints about the deformation mechanisms. The observation of such traces can be made at different scales, using the optical microscope, the scanning electron microscope (SEM), the replica technique in TEM and the atomic force microscope (AFM), which can provide a high resolution. These observations can be performed post mortem or during straining.

Static observations of slip traces using the optical microscope can be found in the literature of the 1950s and before. They have provided information about the slip line

---

Figure 2.26. Dislocation structure ahead of crack tips. (a) X-ray topograph of a Si single crystal (Michot et al., 1994). (b) In situ TEM image in Ti_3Al. Loop planes are labelled according to their traces at foil surfaces. g is the diffraction vector. (After Legros, 1994). Note the difference in scales.

(a)

300 μm

(b)

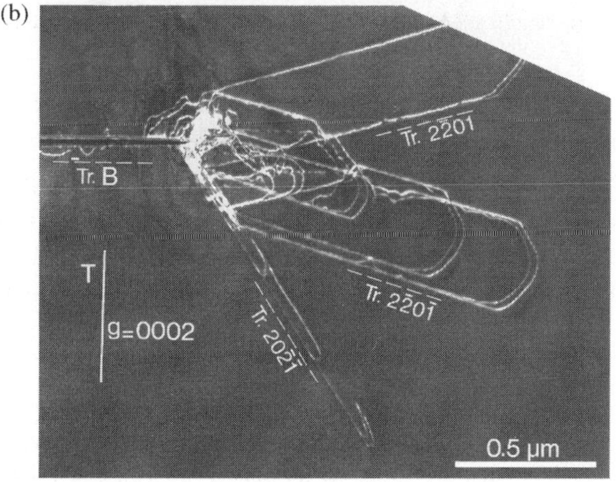
Tr. $2\bar{2}0\bar{1}$
Tr. B
T
g=0002
Tr. $2\bar{2}0\bar{1}$
Tr. $20\bar{2}\bar{1}$
0.5 μm

length, the heterogeneity of glide that corresponds to the various hardening stages of FCC single crystals (Seeger et al., 1957) and the wavy aspect of slip traces during low temperature glide of BCC metals.

In particular, it has been shown that a coarsening of slip patterns takes place by alloying FCC metals (see reviews by, e.g., Seeger, 1958; Haasen, 1983). In stage I hardening, faint and long slip lines, rather randomly distributed, are observed in copper single crystals and solid solutions with up to a few % of solute (see e.g. Neuhäuser and Schwink, 1993 for Cu–Be). For higher concentrations slip becomes heterogeneous. A complete description of this process can be found in Neuhäuser and Schwink (1993).

The dynamic recording of slip step formation during in situ straining in an optical microscope has brought unique information about slip band formation in concentrated solid solutions (see reviews by Neuhäuser, 1983, 1988). They appear as a succession of slip lines which can be formed in a very short time (10 μs). This corresponds to dislocation velocities up to the ms^{-1} range. This rate slows down by orders of magnitude when the number of lines per band increases. The recording of such events necessitates a high speed movie camera (Neuhäuser et al., 1975).

The observation of slip lines at a higher resolution using the AFM has brought information about the glide mode in Ni_3Al (Coupeau et al., 1999). Under conditions which correspond to the positive temperature dependence of the flow stress, it has been shown that the mean slip line length decreases as temperature increases. The detectable slip trace height is in the range of 0.1–0.2 nm under such conditions (Figure 2.27). This supports the idea of a thermally activated dislocation exhaustion mechanism. As the temperature is raised, dislocation locking is more frequent and the stress has to increase so as to activate

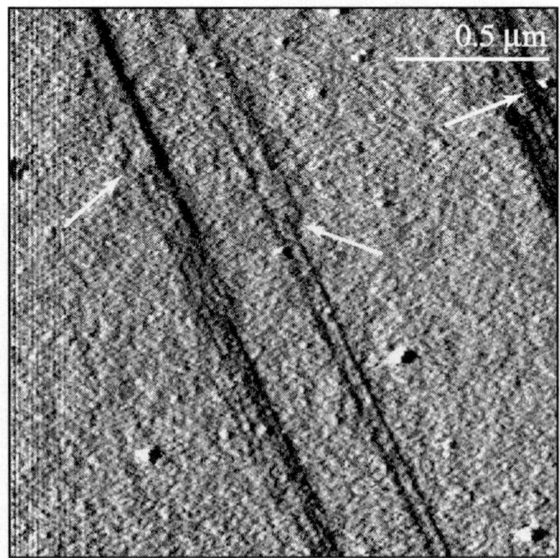

Figure 2.27. AFM high resolution image of slip traces in a $Ni_{75}Al_{24}Ta_1$ single crystal after room temperature deformation. The arrows indicate cross-slip events. From Coupeau et al. (1999).

shorter dislocation sources. This explains why the flow stress increases as temperature increases (see Chapter 10). An accurate imaging of the slip traces provides evidence of cross-slip events from {111} onto {001} planes over very short distances of the order of Burger's vector. This is illustrated in Figure 2.27. This suggests that the incomplete locks which are formed can also yield during plastic deformation (see Chapter 10). The authors also plan to perform such observations during straining according to the technique developed by Small, et al. (1995) and Coupeau et al. (1998).

2.5. CONCLUSION ABOUT THE CHARACTERIZATION OF DISLOCATION MECHANISMS

A few techniques have been reviewed in this chapter that allow us to identify dislocation mobility processes which underlie the mechanical behaviour of the material. Their potentialities have been illustrated by some examples but more of them are presented in the following chapters.

It is essential to combine information about the macroscopic behaviour of the crystal together with its microstructural evolution, observed at different scales, either "in situ" or "post mortem". Unfortunately too many publications are still available nowadays in which the authors care only about the mechanical properties or the microstructural features of the studied material. This leads, in most cases, to an erroneous description of the corresponding deformation mechanisms.

REFERENCES

Blum, W., Rosen, A., Cegielska, A. & Martin, J.L. (1989) *Acta Met.*, **37**, 2439.

Bonneville, J. & Martin, J.L. (1991) in *High Temperature Ordered Intermetallic Alloys*, IV, vol. **213**, Eds. Johnson, L.A., Pope, D.P. & Stiegler J.O., MRS, Warrendale, p. 629.

Bonneville, J., Escaig, B. & Martin, J.L. (1988) *Acta Met.*, **36**, 1989.

Bonneville, J., Spätig, P. & Martin, J.L. (1995) in *High Temperature Ordered Intermetallic Alloys*, vol. **364**, Eds. Horton, J.A., Baker, I., Hanada, S., Noebe, R.D. & Schwartz D.S., MRS, Warrendale, p. 369.

Bonneville, J., Martin, J.L., Spätig, P., Viguier, B. & Matterstock, B. (1997a) in *High Temperature Ordered Intermetallic Alloys,* vol. **460**, Eds. Coch, C.C., Liu, C.T., Stoloff, N.S. & Wanner, A., MRS, Warrendale, p. 419.

Bonneville, J., Viguier, B. & Spätig, P. (1997b) *Scripta Mater.*, **36**, 275.

Caillard, D. (1985) *Phil. Mag. A*, **51**, 157.

Caillard, D. & Martin, J.L. (1975) in *Proceedings of 4th Int. Congress on H.V.E.M*, Ed. Jouffrey, B., Soc. Fr. Micros. Elec., Paris, p. 305.

Charbonnier, C., Kruml, T. & Martin, J.L. (2001) in *Multiscale Modeling of Materials-2000*, vol. **653**, Eds. Kubin, L.P., Selinger, R.L., Bassani, J.L. & Cho K., MRS, Warrendale, p. Z5.7.1.

Clément, N., Couret, A. & Caillard, D. (1991) *Phil. Mag. A*, **64**, 669.

Coupeau, C., Bonneville, J., Matterstock, B., Grilhé, J. & Martin, J.L. (1999) *Scripta Mater.*, **41**, 945.

Coupeau, C., Givard, J.C. & Grilhé, J. (1998) *J. Vac. Sci. Technol. B*, **16**, 1964.

Couret, A. & Caillard, D. (1998) *Acta Met.*, **36**, 2515.

Couret, A., Crestou, J., Farenc, S., Molenat, G., Clément, N., Cougon, A. & Caillard, D. (1993) *Microsc. Microanal. Microstruct.*, **4**, 153.

Dotsenko, V.I. (1979) *Phys. Stat. Sol (b)*, **93**, 11.

Douin, J., Veyssière, P. & Beauchamp, P. (1986) *Phil. Mag. A*, **54**, 375.

Feltham, P. (1961) *Phil. Mag.*, **1**, 259.

Fries, E., Deschamps, J. & Castaing, J. (1983) *Rad. Eff.*, **74**, 329.

Fujita, H. (1966) in *Electron Microscopy*, vol. **1**, Ed. Uyeda, R., Maruzen, Tokyo, p. 289.

Furubayashi, E. (1969) *J. Phys. Soc. Jpn*, **27**, 130.

Gibbs, G.B. (1966) *Phil. Mag.*, **13**, 317.

Guiu, F. & Pratt, P.L. (1964) *Phys. Stat. Sol.*, **6**, 111.

Haasen, P. (1983), in *Physical Metallurgy*, Eds. Cahn, R.W. & Haasen P., Elsevier, Amsterdam, p. 1341.

Hamersky, M., Trojanova, Z. & Lukac, P. (1992) *Acta Techn. CSAV*, **37**, 263.

Hart, E.W. (1970) *Acta Met.,* **18**, 599.

Hirsch, P.B. (1962) *Phil. Mag.*, **7**, 67.

Hirsch, P.B., Horne, J.W. & Whelan, M.J. (1956) *Phil. Mag.*, **1**, 677.

Hirth, J.P. & Lothe, J. (1982) *Theory of Dislocations*, Wiley-Interscience publication, New York.

Jacques, A., Vallino, F., Serbena, F. & George, A. (2000) *J. Phys.: Condens. Matter*, **12**, 10045.

Johnston, W.G. & Gilman, J.J. (1959) *J. Appl. Phys.*, **30**, 129.

Kear, B.H. & Wilsdorf, H.G.F. (1962) *Trans. Metall. Soc. AIME*, **224**, 382.

Kràl, F (1996) PhD thesis no 11685, Swiss Federal Institute of technology, Zürich.

Kruml, T., Conforto, E., Lo Piccolo, B., Caillard, D. & Martin, J. L. (2002) *Acta Mater.*, **1**, 5091.

Kubin, L.P. (1974) *Phil. Mag.*, **30**, 705.

Legros, M. (1994) PhD thesis no 1739, Universitéte P-Sabatier (Toulouse, France).

Lo Piccolo, B. (1999) PhD thesis no 2044, Ecole Polytechnique Fédérale, Lausanne.

Lo Piccolo, B., Martin, J.L. & Bonneville, J. (2000), in *Thermomechanical Processing of Steels*, Eds. Yue, S. & Essadiqi E., Canadian Inst. of Mining, Metallurgy and Petroleum, Montreal, p. 183.

Louchet, F. (1981) *Phil. Mag.*, **43**, 1289.

Louchet, F. & Kubin, L.P. (1975) *Scripta Met.*, **9**, 111.

Makin, M.J. (1958) *Phil. Mag.*, **3**, 384.

Martin, J.L. & Kubin, L.P. (1978) *Ultramicroscopy*, **3**, 215.

Martin, J.L. & Kubin, L.P. (1979) *Phys. Stat. Sol.*, **56**, 487.

Martin, J.L., Matterstock, B., Spätig, P. & Bonneville, J. (1999) in *Proc. 20th Risø Int. Symp. on Mater. Sci.*, Eds. Bilde-Sørensen, J.B., Cartensen, J.L., Hansen, N., Juul Jensen, D., Leffers, T., Pantleon, W., Pedersen, O.B. & Winther G., Risø Nat. Lab, Roskilde, p. 103.

Martin, J.L., Lo Piccolo, B. & Bonneville, J. (2000) *Intermetallics*, **8**, 1013.

Martin, J.L., Lo Piccolo, B., Kruml, T. & Bonneville, J. (2002) *Mater. Sci. Eng. A*, **322**, 118.

Matterstock, B., Conforto, E., Kruml, T., Bonneville, J. & Martin, J.L. (1999) in *High Temperature Ordered Intermetallic Alloys*, vol. **552**, MRS, Warrendale, p. 781.

Michot, G., Loyola de Olivera, M.A. & George, A. (1994) *Mater. Sci. Eng. A*, **176**, 99.

Milicka, K. (1999) *Acta Met.*, **47**, 1831.

Mills, M.J., Gibeling, J.C. & Nix, W.D. (1985) *Acta Met.*, **33**, 1503.

Morris, M. & Martin, J.L. (1984) *Acta Met.*, **32**, 549.

Mughrabi, H. (1983) *Acta Met.*, **31**, 1367.

Mughrabi, H. & Ungar, T. (2002) in *Dislocations in Solids*, vol. **11**, Ed. Nabarro, F.R.N. & Duesberry, M.S., Elsevier Science, Amsterdam, p. 343.

Nazar, A., Bonneville, J. & Saucy, C. (1993) *High Temperature Materials and Processes*, **12**, 97.

Nen, R.W., Scott, D.T. & Woodmanse, M.W. (2000) *Int. J. Plasticity*, **16**, 283.

Neuhäuser, H. (1983) in *Dislocations in Solids*, vol. **6**, Ed. Nabarro, F.R.N., North Holland, Amsterdam, p. 319.

Neuhäuser, H. (1988) *Res. Mechanica*, **23**, 113.

Neuhäuser, H. & Schwink, C. (1993) in *Materials Science and Technology*, vol. **6**, Eds. Cahn, R.W., Haasen, P. & Kramer E.J., VCH, Basel, p. 211.

Neuhäuser, H., Koropp, J. & Heege, R. (1975) *Acta Met.*, **23**, 441.

Orlova, A., Bonneville, J. & Spätig, P. (1995) *Mater. Sci. Eng. A*, **191**, 85.

Pettinari, F., Couret, A., Caillard, D., Molénat, G., Clément, N. & Coujon, A. (2001) *J. Microsc.*, **203**, 247.

Rhode, R.W. & Nordstrom, T.V. (1973) *Scripta Met.*, **7**, 317.

Saada, G., Bonneville, J. & Spätig, P. (1997) *Mater. Sci. Eng. A*, **234–236**, 263.

Sargent, G., Jones, G. & Conrad, H. (1969) *Scripta Met.*, **3**, 481.

Seeger, A. (1958) in *Handbuch der Physik*, vol. **VII/2**, Ed. Flügge, S., Springer, Berlin, p. 1.

Seeger, A., Diehl, J., Mader, S. & Rebstock, H. (1957) *Phil. Mag.*, **2**, 323.

Small, M.K., Coupeau, C. & Grilhé, J. (1995) *Scripta Met. Mater.*, **32**, 1573.

Spätig, P (1995) PhD thesis no 1407, Ecole Polytechnique Fédérale, Lausanne.

Spätig, P., Bonneville, J. & Martin, J.L. (1993) *Mater. Sci. Eng. A*, **167**, 73.

Spätig, P., Bonneville, J. & Martin, J.L. (1995) in *High Temperature Ordered Intermetallic Alloys*, vol. **364**, Eds. Horton, J.A., Baker, I., Hanada, S., Noebe, R.D. & Schwartz D.S., MRS, Warrendale, p. 713.

Thornton, P.H., Davies, R.G. & Johnston, T.L. (1970) *Met. Trans.*, **1**, 207.

Ungar, T., Mughrabi, H., Rönnpagel, D. & Wilkens, M. (1984) *Acta Met.*, **32**, 333.

Urban, K. (1976), in *Microscopie Electronique à Haute Tension 1975*, Eds. Jouffrey, B. & Favard P., Soc. Fr. Micros. Elec., Paris, p. 159.

Vallino, F., Jacques, A. & George, A. (2000) *Phys. Stat. Sol (b)*, **222**, 51.

Vesely, D. (1968) *Phys. Stat. Sol.*, **28**, 675.

Viguier, B., Bonneville, J. & Martin, J.L. (1996) *Acta Mater.*, **44**, 4403.

Washburn, J. (1965) *Appl. Phys. Lett.*, **7**, 183.

Wilsdorf, H.G.F. (1958) *ASTM Spec. Tech. Publ.*, **245**, 43.

Chapter 3

Interactions Between Dislocations and Small-size Obstacles

Chapter 3
Interactions Between Dislocations and Small-size Obstacles

This chapter describes the main properties of dislocations interacting with "small-size" obstacles such as "forest" dislocations cutting the slip plane or individual solute atoms or small clusters. For more details, especially concerning interactions with solute atoms, the reader can refer to Yoshinaga and Morozumi (1971), Hirth and Lothe (1982), Hirth (1983), Wille et al. (1987) and Neuhauser and Schwink (1993).

3.1. THERMALLY ACTIVATED GLIDE ACROSS FIXED SMALL-SIZE OBSTACLES

The glide movement of dislocations across small-size obstacles, assumed here to be fixed, can be enhanced by thermal activation. The main equations describing this interaction are described below for several short-range phenomenological potentials.

When a dislocation is pinned, the force exerted on the pinning point is, in the isotropic approximation (Figure 3.1):

$$F = 2\hat{T} \sin \alpha \approx \mu b^2 \sin \alpha \tag{3.1}$$

where $\hat{T}$ is the line tension.

(In order to take into account differences between edge and screw segments, the effective line tension defined by Kocks et al., 1975, must be used.)

In the case of a low density of weak obstacles (small α) this force can also be written:

$$F = \tau b L_{\mathrm{F}} \tag{3.2}$$

where τ is the effective stress and L_{F} the distance between obstacles along the dislocation line. In this approximation by Fleischer (1961), the net force acting on L_{F} is transferred to

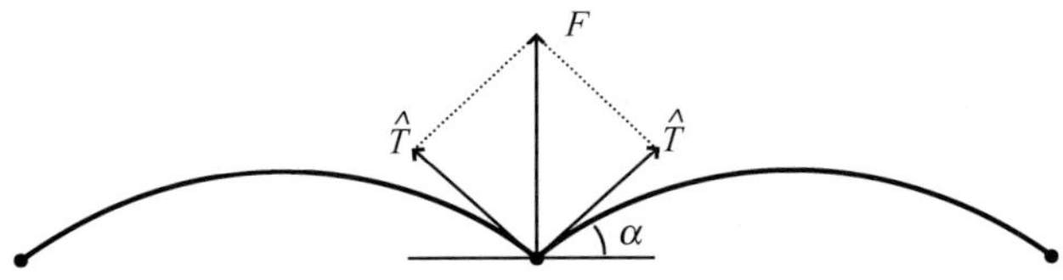

Figure 3.1. Forces exerted by a dislocation on a pinning point.

the obstacle point. Combining Eqs. (3.1) and (3.2) yields:

$$\tau = \frac{\mu b \sin \alpha}{L_{\mathrm{F}}} \tag{3.3}$$

This length L_{F} has been estimated by Friedel (1964, p. 224) in the case of weak dislocation–obstacle interactions. It is defined as follows (see Figure 3.2): in a steady-state regime, each time one dislocation crosses a pinning point, another one is met. The area swept during the process is $\mathscr{A} = L_{\mathrm{F}}\lambda$, where $\lambda = R(1 - \cos \alpha) \approx (R\alpha^2/2)$, R is the radius of curvature and $\alpha \approx (L_{\mathrm{F}}/R)$.

Accordingly:

$$\mathscr{A} = \frac{1}{2}\frac{L_{\mathrm{F}}^3}{R} \approx L_{\mathrm{F}}^3 \frac{\tau}{\mu b} \tag{3.4}$$

The steady-state condition also implies:

$$\mathscr{A} = d^2 = b^2/c_b \tag{3.5}$$

where d is the in-plane average distance between obstacles and c_b is their atomic concentration (c_b can be much smaller than the average concentration of solute atoms if the efficient obstacles are clusters of two or three atoms).

Combining Eqs. (3.4) and (3.5) then yields the Friedel length:

$$L_{\mathrm{F}} = d\left(\frac{\mu b}{\tau d}\right)^{1/3} = b\left(\frac{\mu}{\tau c_b}\right)^{1/3} \tag{3.6}$$

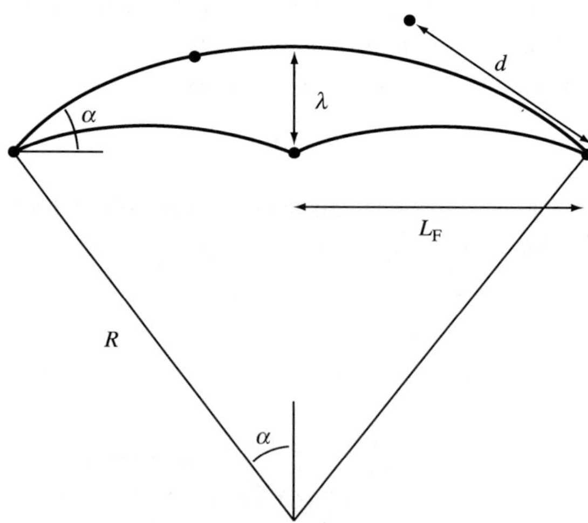

Figure 3.2. Dislocation escaping a pinning point in the Friedel approximation.

L_F is found to be stress dependent because more strongly curved dislocations interact with a higher density of obstacles.

Combining Eqs. (3.3) and (3.6), the effective stress becomes:

$$\tau = \mu \frac{b}{d} (\sin \alpha)^{3/2} \tag{3.7a}$$

Or, using Eq. (3.1):

$$\tau = \mu \frac{b}{d} \left(\frac{F}{\mu b^2} \right)^{3/2} \tag{3.7b}$$

In the case of a high density of stronger and more diffuse obstacles, the Mott–Labusch theory yields another stress–force relation (see, e.g. Labusch, 1970; Haasen, 1979, 1983; Neuhauser and Schwink, 1993):

$$\tau \approx \mu \left(\frac{b}{d} \right)^{4/3} \left(\frac{F}{\mu b^2} \right)^{4/3} \left(\frac{w}{2b} \right)^{1/3} \tag{3.8}$$

where w is the width of the obstacles.

Following Mott and Labusch, the validity limit between both the approximations is determined by the parameter:

$$\beta = \frac{w}{d} \left(\frac{2\hat{T}}{F} \right)^{1/2} \tag{3.9}$$

The Fleischer–Friedel approximation is valid for $\beta \ll 1$ ($w \sim b$) and the Mott–Labusch approximation must be used for $\beta > 1$.

To predict the effect of temperature several approximations have been made for the short-range energy–distance and force–distance profiles. As discussed by Kocks et al. (1975, p. 141), the energy profile can be expressed as

$$U(\tau) = E_{max} \left[1 - \left(\frac{\tau}{\tau_{max}} \right)^p \right]^q$$

where $0 < p < 1$ and $1 < q < 2$. τ_{max} is the maximum value of the effective stress τ at which the obstacle is crossed without thermal activation.

Several types of energy or force profiles have been proposed that correspond to specific values of p and q. They are described in the following. In each case, the stress versus temperature and activation volume versus temperature relations are established.

3.1.1 The rectangular force–distance profile

The force of the obstacle is constant and equal to F_{max} over distance w (Figure 3.3a). The corresponding energy–distance profile is $E(y) = F_{max} y$, and its maximum value is $E_{max} = F_{max} w$ (see Figure 3.3b).

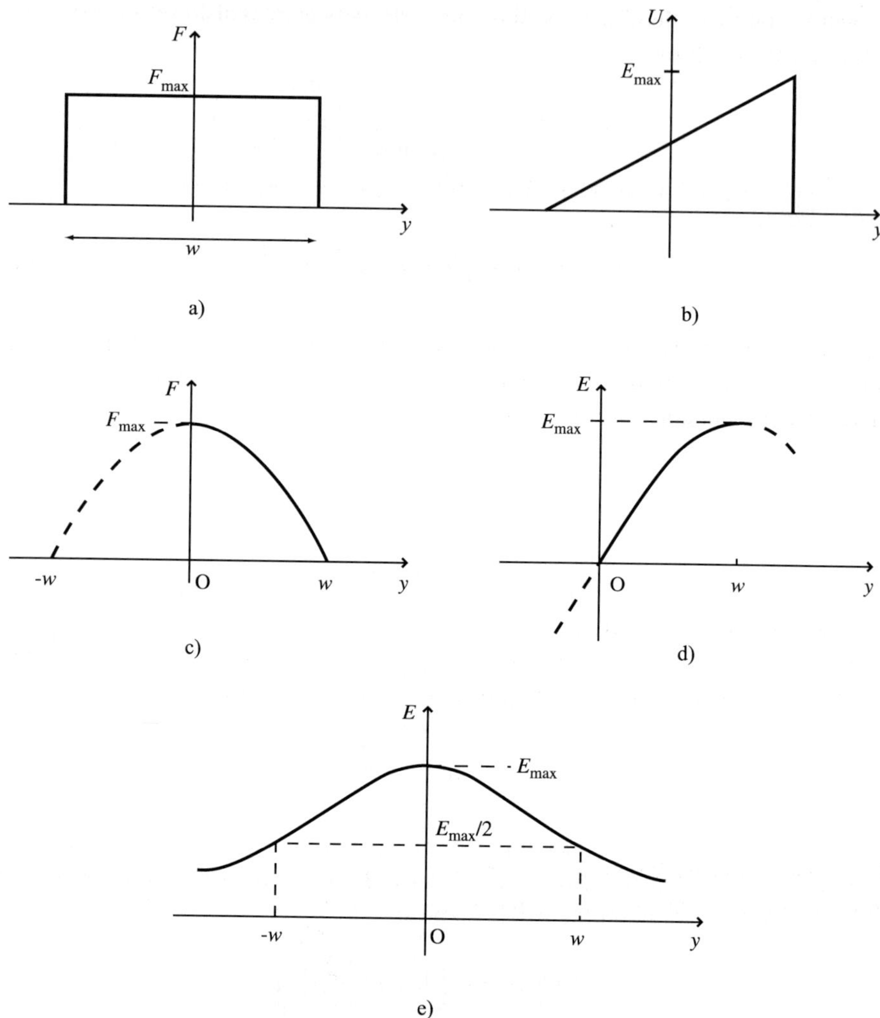

Figure 3.3. Different obstacle profiles: (a) rectangular force–distance profile; (b) corresponding energy; (c) and (d) parabolic approximation; and (e) Cottrell–Bilby potential.

This profile is used to describe the forest mechanism when gliding dislocations have no long-range elastic interaction with the intersecting trees (Friedel, 1964, p. 221). In this case, E_{max} is the energy of the two jogs created on the two intersecting dislocations and w is the width of the dislocation cores. For attractive junctions close to their unzipping configuration, E_{max} may contain an additional term of dislocation line energy.

Under an applied force F, the height of the barrier is decreased by the quantity Fw, and the activation energy is:

$$U(F) = E_{\max} - Fw = E_{\max}\left(1 - \frac{F}{F_{\max}}\right) \tag{3.10a}$$

In the Friedel approximation (Eq. (3.7b)) the activation energy becomes:

$$U(\tau) = E_{\max}\left[1 - \left(\frac{\tau}{\tau_{\max}}\right)^{2/3}\right] \tag{3.10b}$$

where $\tau_{\max} = \mu(b/d)(F_{\max}/\mu b^2)^{3/2}$. Eq. (3.10b) provides an energy of the type proposed by Kocks et al. (1975) with $p = 2/3$ and $q = 1$.

The corresponding activation volume is:

$$V = -\frac{\partial U}{\partial \tau} = \frac{E_{\max}}{\tau_{\max}}\left(\frac{\tau}{\tau_{\max}}\right)^{-1/3} \tag{3.11a}$$

$$V = L_F bw \tag{3.11b}$$

where L_F is given by Eq. (3.6).

Writing $U = kT\ln(\dot{\gamma}_o/\dot{\gamma})$ and $E_{\max} = kT_o \ln(\dot{\gamma}_o/\dot{\gamma})$, where T_o is the athermal temperature, the stress can be expressed as a function of temperature as:

$$\tau = \tau_{\max}\left(1 - \frac{T}{T_o}\right)^{3/2} \tag{3.12}$$

3.1.2 The parabolic force–distance profile

The force–distance profile in the direction of motion is given by (Figure 3.3c):

$$F(y) = F_{\max}\left(1 - \left(\frac{y}{w}\right)^2\right) \tag{3.13a}$$

The corresponding energy–distance profile is (Figure 3.3d):

$$E(y) = wF_{\max}\left(\frac{y}{w} - \frac{1}{3}\left(\frac{y}{w}\right)^3\right) \tag{3.13b}$$

Its maximum value is $E_{\max} = \frac{2}{3}wF_{\max}$, for $y = w$, size of the obstacle.

This profile is often used to describe interactions between dislocations and solute atoms. It is, however, thought to be valid only for $F \sim F_{\max}$, i.e. at high stresses and low temperatures (Wille et al., 1987). In the particular case of the size–effect interaction, which is discussed in Section 3.2.1, $F_{\max}$ is approximately given by Eq. (3.25).

Under an applied force F, the saddle position is reached when the net force on the dislocation is zero, namely when $(\partial E(y)/\partial y) - F = 0$. Using Eq. (3.13) this condition is satisfied for the critical value $y = y_c$ given by:

$$\frac{y_c}{w} = \left(1 - \frac{F}{F_{max}}\right)^{1/2} \tag{3.14}$$

The corresponding activation energy is $U(F) = E(y_c) - Fy_c$, i.e.

$$U(F) = \frac{2}{3} w F_{max}\left(1 - \frac{F}{F_{max}}\right)^{3/2} = E_{max}\left(1 - \frac{F}{F_{max}}\right)^{3/2} \tag{3.15a}$$

Using Eq. (3.7b), the activation energy can be written as:

$$U(\tau) = E_{max}\left(1 - \left(\frac{\tau}{\tau_{max}}\right)^{2/3}\right)^{3/2} \tag{3.15b}$$

where τ_{max} has the same value as for Eq. (3.10b). Again, $U(\tau)$ is of the type proposed by Kocks et al. (1975) with $p = 2/3$ and $q = 3/2$.

The activation volume dependence on stress is:

$$V = \frac{\partial U}{\partial \tau} = \frac{E_{max}}{\tau_{max}}\left(\frac{\tau}{\tau_{max}}\right)^{-1/3}\left(1 - \left(\frac{\tau}{\tau_{max}}\right)^{2/3}\right)^{1/2} \tag{3.16a}$$

As in the preceding case, the stress–temperature dependence is given by:

$$\tau = \tau_{max}\left(1 - \left(\frac{T}{T_o}\right)^{2/3}\right)^{3/2} \tag{3.17}$$

and the activation volume dependence on temperature is:

$$V = \frac{E_{max}}{\tau_{max}}\left[\frac{T}{T_o}\right]^{1/3}\left[1 - \left(\frac{T}{T_o}\right)^{2/3}\right]^{-1/2} \tag{3.16b}$$

3.1.3 The Cottrell–Bilby potential (Cottrell and Bilby, 1949)

The energy–distance profile is (Figure 3.3e):

$$E(y) = \frac{E_{max}}{1 + \left(\frac{y}{w}\right)^2} \tag{3.18}$$

This profile is used to describe dislocation–solute interactions, with less restrictions than the parabolic one (Wille et al., 1987). Here again, F_{max} is approximately given by Eq. (3.25) below in the particular case of size–effect interaction.

Using the same procedure as above analytical solutions can be found. Following Wille et al. (1987), they can be approximated by:

$$U(F) = E_{max}\left(1 - \left(\frac{F}{F_{max}}\right)^{0.69}\right)^{3/2} \tag{3.19a}$$

with $E_{max} = (8/3\sqrt{3})wF_{max}$.

This expression is similar to those given by Eqs. (3.10a) and (3.15a). In the Friedel approximation (Eq. (3.7b)), it can be expressed as:

$$U(\tau) = E_{max}\left(1 - \left(\frac{\tau}{\tau_{max}}\right)^{0.46}\right)^{3/2} \tag{3.19b}$$

namely

$$\tau = \tau_{max}\left(1 - \left(\frac{T}{T_o}\right)^{2/3}\right)^{2.17} \tag{3.20}$$

where τ_{max} has the same value as for Eqs. (3.10b) and (3.15b). Eq. (3.20) can be compared with Eqs. (3.12) and (3.17) with $p = 2/3$ and $q = 2.17$.

The activation volume is:

$$V = 0.69\frac{E_{max}}{\tau_{max}}\left(\frac{\tau}{\tau_{max}}\right)^{-0.54}\left(1 - \left(\frac{\tau}{\tau_{max}}\right)^{0.46}\right)^{1/2} \tag{3.21a}$$

which is not too different from Eqs. (3.11a) and (3.16a), or

$$V = 0.69\frac{E_{max}}{\tau_{max}}\left(\frac{T}{T_o}\right)^{1/3}\left(1 - \left(\frac{T}{T_o}\right)^{2/3}\right)^{-1.17} \tag{3.21b}$$

3.2. DISLOCATIONS INTERACTING WITH MOBILE SOLUTE ATOMS

As the temperature rises solute atoms become sufficiently mobile to diffuse towards dislocations. The driving forces for this process are long-range interactions, different from the short-range ones described in the preceding section. They are estimated below and then used to compute dislocation mobility.

3.2.1 Long-range elastic interactions

The most important interaction is the paraelastic one or size effect. In cylindrical coordinates, at a distance r from an edge dislocation and along a direction at an angle θ

from the Burgers vector, the hydrostatic pressure is $p = \frac{1}{3}(\sigma_{rr} + \sigma_{\theta\theta} + \sigma_{zz})$, where

$$\sigma_{rr} = \sigma_{\theta\theta} = -\frac{\mu b \sin\theta}{2\pi(1-\nu)r} \quad \text{and} \quad \sigma_{zz} = \nu(\sigma_{rr} + \sigma_{\theta\theta})$$

whence:

$$p = \frac{\mu b \sin\theta}{3\pi r}\frac{1+\nu}{1-\nu} \tag{3.22}$$

The interaction energy with a solute atom which induces a local change of atomic volume $\Delta\Omega$ is accordingly (Haasen, 1979):

$$U_{\text{int}} = 3\frac{1-\nu}{1+\nu}p\Delta\Omega = \frac{1}{\pi}\mu b\Delta\Omega\frac{\sin\theta}{r} \tag{3.23}$$

At given r and θ, Eq. (3.23) shows that the sign of U_{int} is directly connected to that of $\Delta\Omega$.

The corresponding radial interaction force is:

$$F_{\text{int}} \approx \frac{1}{\pi}\mu b\frac{\Delta\Omega \sin\theta}{r^2} \tag{3.24}$$

The maximum value of F is obtained when the slip plane is one interatomic distance from the obstacle, i.e. $z \approx b$ (see Figure 3.4), whence:

$$F_{\max} \approx \frac{1}{\pi}\frac{\mu\Delta\Omega}{b} \approx \frac{1}{3}\mu b^2\frac{\Delta\Omega}{\Omega} \tag{3.25}$$

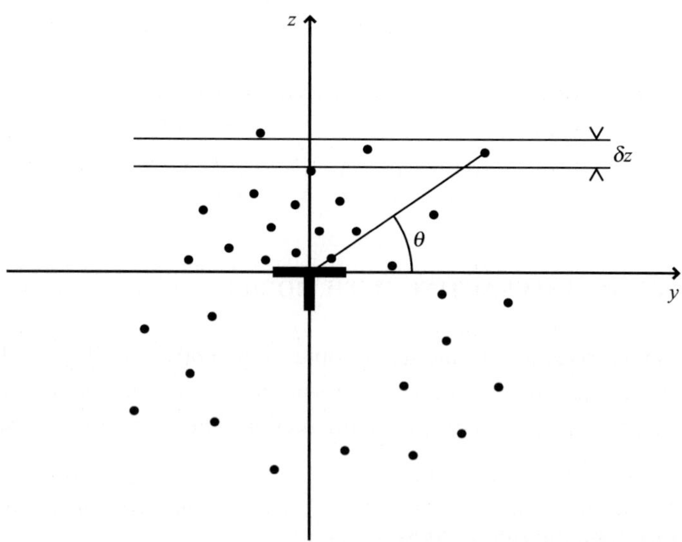

Figure 3.4. Schematic description of the solute atom concentration around an edge dislocation. The size effect here corresponds to $\Delta\Omega > 0$.

U_{int} and F are often expressed in terms of the change of lattice parameter a with solute concentration: $\delta = \mathrm{d}\ln a/\mathrm{d}c$, taking into account $\Delta\Omega/\Omega = 3\delta$.

Screw dislocations are also subjected to the paraelastic interaction, provided they are dissociated in several mixed partials.

The dielastic interaction, based on the so-called modulus effect, is weaker.

Several other interactions can be considered, assuming that short-range atomic movements of solute atoms can take place in the vicinity of dislocation cores, e.g. short-range ordering or local disordering in intermetallics (Haasen, 1983; Neuhauser and Schwink, 1993). The small-size obstacles considered in this chapter can be small aggregates of several atoms.

Another important effect must be mentioned here, that is discussed in Chapter 6: Peierls forces in metals and alloys can substantially increase when the local concentration of solute atoms increases (e.g. oxygen in titanium). This effect is not a local pinning but it can also induce stress instabilities.

3.2.2 Static ageing, dynamic strain ageing and the Portevin–Lechâtelier effect

This section introduces the main features of these complex phenomena.

The static equilibrium concentration of solute atoms around a dislocation is

$$c = c_o \exp\left(-\frac{U_{int}}{kT}\right) \tag{3.26}$$

where c_o is the average solute concentration and U_{int} is the dislocation–solute interaction energy. For a size–effect interaction, U_{int} depends on r and θ, according to Eq. (3.23). The corresponding concentration is described schematically in Figure 3.4. Moving dislocations tend to drag their atmosphere of solute atoms and a dynamic equilibrium is established which depends on temperature and dislocation velocity. If all solute atoms are assumed to move only along the direction of dislocation motion ($y > 0$ in Figures 3.4 and 3.5) then their concentration obeys:

$$-D\frac{\partial c}{\partial y} - \frac{Dc}{kT}\frac{\partial U_{int}}{\partial y} = (c - c_o)\mathrm{v} \tag{3.27}$$

where D is the solute-diffusion coefficient, given by an expression similar to Eq. (8.5).

The first left-hand side term is the diffusion flux due to the concentration gradient $\partial c/\partial y$ and the second one is the transport flux under the driving force $\partial U_{int}/\partial y$. At steady state, the whole concentration profile around the dislocation is assumed to move at the dislocation velocity v. Consequently, the right-hand-side term of Eq. (3.27) expresses the solute flux through the crystal. The resolution of this equation allows the dynamic equilibrium profiles around moving dislocations to be determined.

Figure 3.5 shows the dynamic solute concentration along planes at various distances z from the dislocation core.

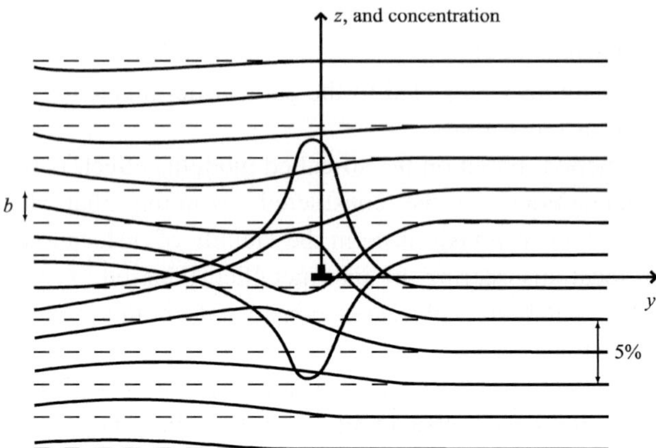

Figure 3.5. Computed solute atmosphere around an edge dislocation, moving to the right, along different planes above and below the slip plane. From Sakamoto (1981).

Figure 3.6a shows the dynamic concentration along the plane located at one interatomic distance from the slip plane for various dislocation velocities. The general shape and the maximum concentration value remain close to their equilibrium values at rest up to a velocity of 10^2 nm/s. In addition, a depletion is formed just ahead of the moving dislocation. At higher velocities the maximum concentration decreases and a tail develops.

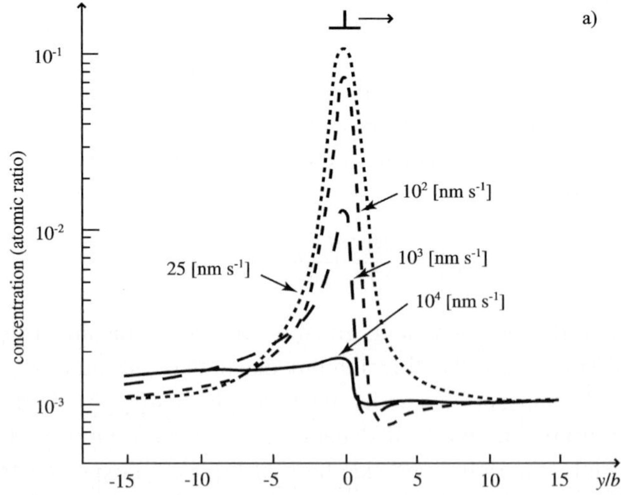

Figure 3.6. Computed solute atmosphere around a moving edge dislocation (from Yoshinaga and Morozumi, 1971): (a) in a plane located one interatomic distance from the slip plane (the dislocation moves to the right); (b) corresponding friction stress as a function of dislocation velocity; and (c) same as (b), for different increasing temperatures $T_1 < T_2 < T_3$.

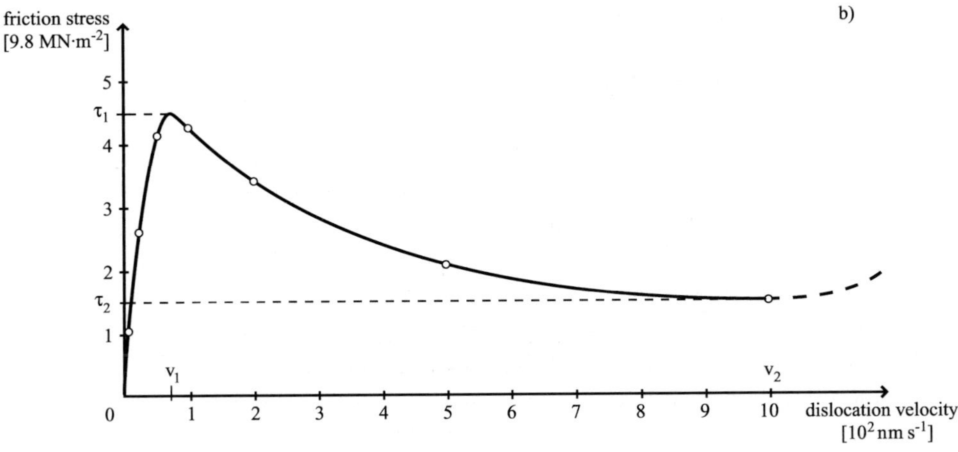

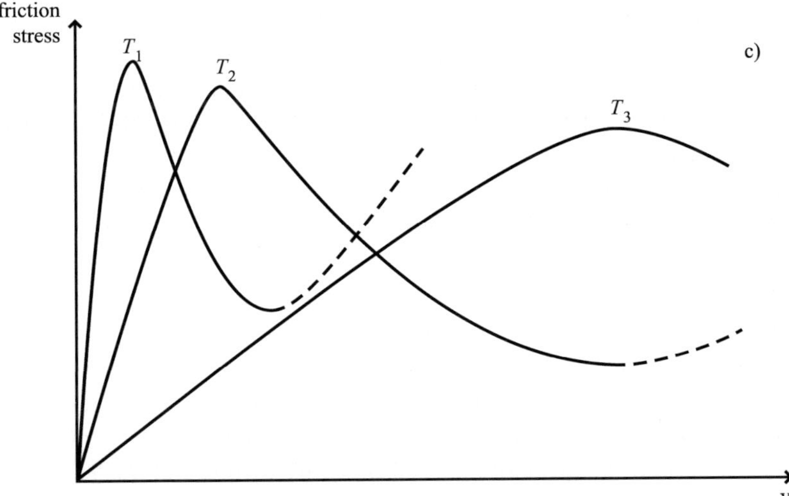

Figure 3.6. (*continued*)

The corresponding stress–velocity dependence is shown in Figure 3.6b. The stress increases up to τ_1 as the velocity increases to $v_1 = 70$ nm/s. Cottrell and Jaswon (1949) showed that this critical velocity is of the order of 12 $kTD/mb\Delta\Omega$, where D is the solute-diffusion coefficient. This regime corresponds approximately to the translation of the equilibrium cloud discussed above, considering Figure 3.6a. Above v_1 the friction stress decreases as the maximum concentration decreases. Above a critical velocity v_2 the stress increases again because solute atoms can be considered as immobile with respect to the moving dislocation, and the conditions of Section 3.1 are satisfied. Similar stress–velocity curves are shown at different temperatures in Figure 3.6c. It shows in particular that, as the temperature increases, v_1 and v_2 increase as well and τ_1 and τ_2 decrease.

From the curves in Figure 3.6b and c, stress instabilities can be anticipated for stresses and velocities that correspond to $d\tau/dv < 0$. The origin and the properties of stress instabilities (or Portevin–Lechâtelier (PLC) effect) have been discussed in several articles (see, e.g. MacCormick, 1972; Van den Beukel, 1975; Mulford and Kocks, 1979; Strudel, 1980; Estrin and Kubin, 1989; Kubin and Estrin, 1990). Let us consider a sample containing a density of mobile dislocations, ρ, deformed at an imposed strain-rate, $\dot{\varepsilon}$, ranging between $\rho b v_1$ and $\rho b v_2$. Figure 3.7a shows that when the stress increases to the critical value τ_1 dislocations suddenly accelerate from v_1 to $v(\tau_1) > v_2$. Since $\rho b v(\tau_1) > \dot{\varepsilon}$, the tensile machine relaxes and the applied stress decreases to the second critical value τ_2. The dislocation velocity then decreases instantaneously to $v(\tau_2) < v_1$. Since $\rho b v(\tau_2) < \dot{\varepsilon}$, the applied stress increases again and another cycle starts. This behaviour leads to stress instabilities or the PLC effect. These considerations show that the part of the curve of Figure 3.7a between v_1 and v_2 has no physical meaning. Increasing the applied strain-rate increases the total time spent by the dislocations in the high-velocity regime but the average flow stress keeps oscillating between τ_1 and τ_2. This corresponds to a zero stress–strain rate sensitivity. However, Figure 3.6c shows that this average flow stress decreases with increasing temperature.

A slightly different behaviour is expected at low temperatures (lower than T_1 in Figure 3.6c). Then, v_1 is so small that dislocations moving at $v < v_1$ can be considered as immobile (Figure 3.7b). Solute atoms then start to move to the dislocation core (static ageing). This results in an unpinning stress increasing with increasing waiting time (or decreasing strain-rate) $\tau_1' > \tau$. Figure 3.7b shows that the average flow stress increases with decreasing strain-rate, which results in a negative stress–strain rate sensitivity. With increasing temperature, the average flow stress can also increase, which results in a yield-stress anomaly. The same behaviour is expected for larger values of v_1, provided mobile dislocations are momentarily slowed down by extrinsic obstacles such as forest dislocations (Estrin and Kubin, 1989).

Let us note that, in reality, the situation is still more complex because of strain localization.

3.2.3 *Diffusion-controlled glide*

Inspection of Figure 3.6b,c shows that, at a given dislocation velocity (given applied strain-rate), deformation can take place in the low-velocity regime ($v < v_1$) provided the temperature is high enough ($T \geq T_3$). The aim of this section is to derive analytical expressions for this low-velocity/high-temperature regime.

Consider the dislocation described schematically in Figure 3.4. When it moves along the y-direction, the solute atoms are assumed to move only in this direction.

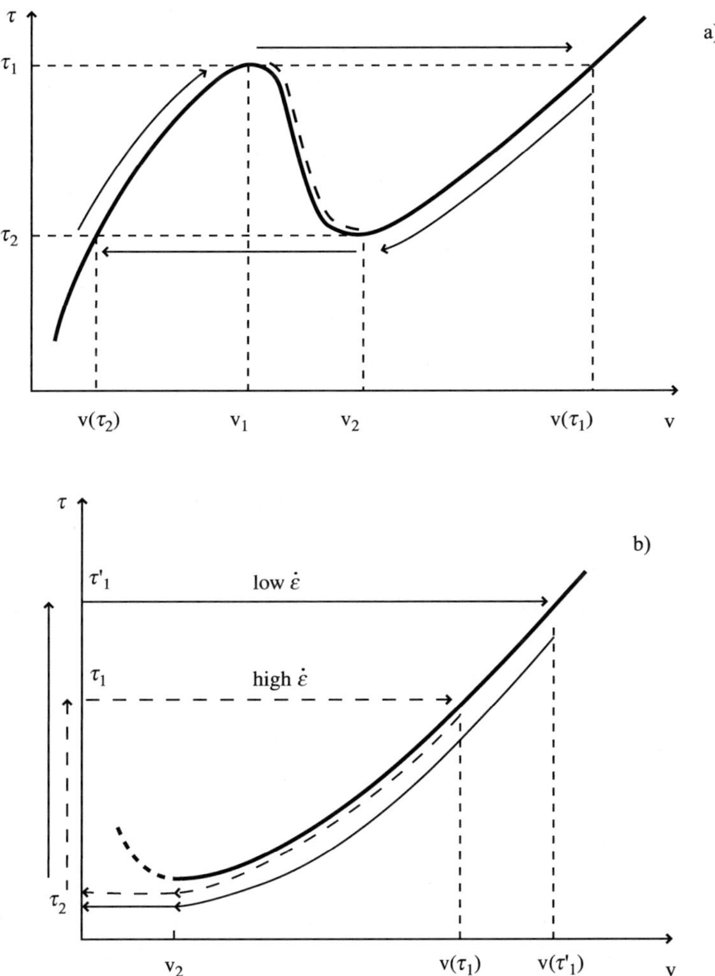

Figure 3.7. Origin of stress instabilities: (a) on the basis of Figure 3.6b and c; and (b) in the case $v_1 \approx 0$ (low temperature).

The dislocation velocity is estimated in the frame of two different approximations.

– For $U_{int} \gg kT$, Eqs. (3.23) and (3.26) show that the cloud is highly asymmetrical, with a high concentration of solute atoms in the region corresponding to $U_{int} < 0$ and a weak depletion in the opposite one ($U_{int} > 0$). This property results from the asymmetrical shape of the exponential function. According to Friedel (1964, p. 410), in the limit of unsaturated clouds the solute atoms are all very close to the dislocation core, forming a row of pinning points with average distance λ (Figure 3.8). The work done by the applied stress during the diffusion of one solute atom over one interatomic distance is $\tau b^2 \lambda$.

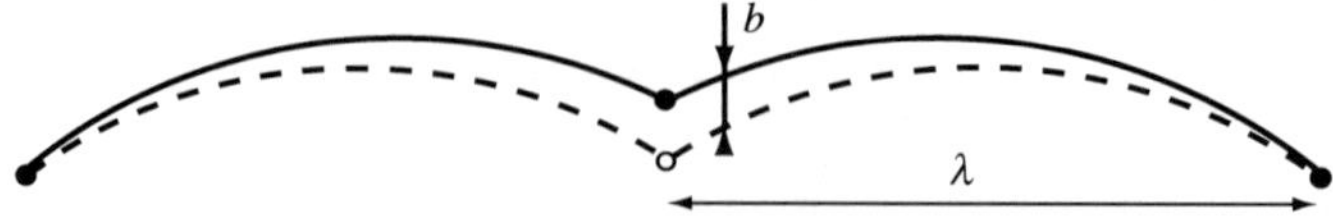

Figure 3.8. Schematic description of the diffusion-controlled glide of a dislocation pinned by a row of solute atoms.

The frequency of this event is

$$\frac{1}{2}\nu_{\rm D}\frac{b}{\lambda}\exp\left(-\frac{U_{\rm d}-\tau b^2\lambda}{T}\right)$$

in the forward direction, and

$$\frac{1}{2}\nu_{\rm D}\frac{b}{\lambda}\exp\left(-\frac{U_{\rm d}+\tau b^2\lambda}{kT}\right)$$

in the backward direction (cf. Section 7.2.2.2) ($\nu_{\rm D}$ is the Debye frequency and $U_{\rm d}$ is the solute-diffusion activation energy).

The dislocation velocity is accordingly:

$$\mathrm{v}=\frac{1}{2}\nu_{\rm D}\frac{b^2}{\lambda}\left[\exp\left(-\frac{U_{\rm d}-\tau b^2\lambda}{kT}\right)-\exp\left(-\frac{U_{\rm d}+\tau b^2\lambda}{kT}\right)\right]$$

or, assuming that $\tau b^2\lambda \ll kT$

$$\mathrm{v}=\nu_{\rm D}\frac{\tau b^4}{kT}\exp\left(-\frac{U_{\rm d}}{kT}\right)=D\frac{\tau b^2}{kT}$$

This yields

$$\tau=\frac{1}{b^2}\frac{kT}{D}\mathrm{v} \tag{3.28}$$

Under these circumstances, the friction force is independent of the solute concentration in the core.

– For $U_{\rm int} < kT$, the cloud is symmetrical. The local concentration (Eq. (3.26)) can indeed be approximated by $c - c_{\rm o} \approx c_{\rm o}(U_{\rm int}/kT)$ so that $c - c_{\rm o}$ has opposite values above and below the slip plane (points y, z and y, $-z$ in Figure 3.4). The average concentration around the dislocation is thus $c_{\rm o}$, which a priori should yield no frictional force. However, a detailed description shows that this statement is erroneous. It is based on the works of Cottrell and Jaswon (1949) and Fuentes-Samaniego (1979), summarized by Hirth and Lothe (1982). It takes into account the energy dissipated by the solute atoms moving either in the direction of the dislocation or in the opposite one. The dynamic concentration profile is as in Figure 3.5. For the sake of simplicity, the dislocation can be considered as surrounded by an excess of solute atoms on one side and by an excess of holes on the opposite one. The motion of a hole in one direction corresponds to the motion of a solute atom in the opposite one.

All atoms and holes are assumed to move along the y-axis in such a way that Eq. (3.27) applies. The energy dissipated corresponds to the average work done by the solute atoms and holes in excess under the driving forces $F = -(\partial U_{\mathrm{int}}/\partial y)$ and $-F$, respectively. The elementary friction force in a strip of width δz is thus equal to the sum of these forces, namely:

$$\sum F(\delta z) = \frac{\delta z}{\Omega}\int_{-\infty}^{+\infty} -(c - c_{\mathrm{o}})\frac{\partial U_{\mathrm{int}}}{\partial y}\mathrm{d}y \tag{3.29}$$

For $z > 0$, the expression in the integral is the work done by the solute atoms in excess. It is positive for $y < 0$ $(\partial U_{\mathrm{int}}/\partial y < 0)$ and negative for $y > 0$ $(\partial U_{\mathrm{int}}/\partial y > 0)$. The integral is, however, strictly positive because of the asymmetrical shape of $c(y)$ shown in Figures 3.5 and 3.6a.

For $z < 0$, the expression in the integral can be understood as the work done by the holes (concentration $c - c_{\mathrm{o}}$) under the driving force $(\partial U_{\mathrm{int}}/\partial y)$. The integral is also positive.

Replacing $(\partial U_{\mathrm{int}}/\partial y)$ by its value deduced from Eq. (3.27) yields:

$$\sum F(\delta z) = \frac{\delta z}{\Omega}kT\left[\frac{\mathrm{v}}{D}\int_{-\infty}^{+\infty}\frac{(c - c_{\mathrm{o}})^2}{c}\mathrm{d}y - \int_{-\infty}^{+\infty}\frac{c - c_{\mathrm{o}}}{c}\frac{\partial c}{\partial y}\right]$$

The second integral in the brackets is $|c - c_{\mathrm{o}} \ln c|_{-\infty}^{+\infty} = 0$. Since $U_{\mathrm{int}} < kT$ we have $c - c_{\mathrm{o}} \approx c_{\mathrm{o}}U_{\mathrm{int}}/kT$ and $c \approx c_{\mathrm{o}}$, whence:

$$\sum F(\delta z) = \frac{c_{\mathrm{o}}}{DkT\Omega}\mathrm{v}\delta z\int_{-\infty}^{+\infty} U_{\mathrm{int}}^2\,\mathrm{d}y$$

and, using Eq. (3.23)[1]

$$\sum F(\delta z) = \frac{kT}{D\Omega}\mathrm{v}\delta z\, c_{\mathrm{o}}\left(\frac{\mu b\Delta\Omega}{\pi kT}\right)^2\int_{-\infty}^{+\infty}\frac{z^2}{(z^2 + y^2)^2}\mathrm{d}y$$

Then,

$$\sum F(\delta z) = \frac{kT}{D\Omega}\mathrm{v}\delta z\, c_{\mathrm{o}}\left(\frac{\mu b\Delta\Omega}{3kT}\right)^2\frac{\pi}{2z} \tag{3.30}$$

---

[1] A typing error in Eq. (18.53) of Hirth and Lothe (1982, 1992) has been corrected.

After integration along the z-direction, between a minimum value z_o equal to a few interatomic distances and a maximum value estimated as D/v, the resulting friction stress is:

$$\tau = \frac{1}{2\pi} c_o \mu \frac{\mathrm{v}b}{D} \frac{\mu \Delta \Omega}{kT} \frac{\Delta \Omega}{\Omega} \ln \frac{D}{\mathrm{v} z_o}$$

or

$$\tau \approx c_o \frac{\mu b}{DkT} \frac{\mu \Delta \Omega^2}{\Omega} \mathrm{v} \tag{3.31}$$

The friction stress is thus similar to that given by Eq. (3.28), however, with a more complex constant factor.

Taking $c_o \sim 10^{-2}$, $\mu b^3/kT \sim 10^2$ and $\Delta\Omega/\Omega \sim 10^{-1}$ this factor has the same value as the one in the first approximation $((1/b^2)(kT/D)$ in Eq. (3.28)).

Another estimation proposed by Hirth and Lothe is based on a simplified interaction energy, varying as $1/r$ as in Eq. (3.2), but independent of θ. The result is twice the above estimation. Cottrell and Jaswon (1949) showed that taking into account diffusion along the z-direction introduces only a factor of 1/2 in Eq. (3.31). Therefore, it seems that Eq. (3.28) provides a satisfactory approximation, valid in the whole temperature range.

3.3. COMPARISON WITH EXPERIMENTS

3.3.1 The forest mechanism

Experimental evidence of the forest mechanism comes from the "Cottrell–Stokes" experiments in single and polycrystalline FCC metal (Cottrell and Stokes, 1955; Thornton et al., 1962), recently discussed by Nabarro (1990) and Saada (1999). The deformation curves exhibit strong linear and parabolic hardening stages (called stages II and III, respectively) which result from the storage of forest sessile dislocations. The flow stress is usually decomposed into an athermal or internal stress, τ_μ, and a temperature and strain-rate dependent effective stress, τ^* (Section 2.1.4.1). The internal stress is irreversible upon strain-rate and temperature changes. It is due to the long-range stress field of forest dislocations and to the formation of attractive junctions. The effective stress is reversible upon strain-rate and temperature changes. Figure 3.9 shows that the ratio of flow stresses at high and low temperatures is almost independent of strain or total stress. This Cottrell–Stokes law indicates that the effective stress remains approximately proportional to the internal stress during deformation. This shows that both stresses are related to the same obstacles, namely the forest dislocations. In addition, the strong decrease of the activation volumes with increasing strain and stress that they observe shows that the density of thermally activated obstacles increases with strain (Eqs. (3.6) and (3.11b)) as expected for a forest mechanism.

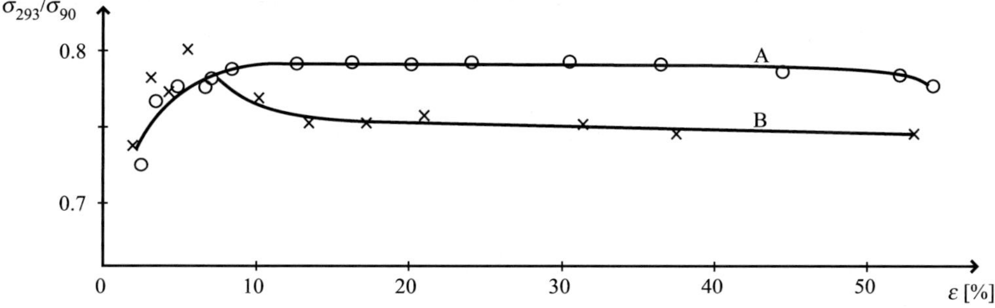

Figure 3.9. Ratio of flow stresses at room temperature and in liquid air, of an Al single crystal. Curve A was obtained by transitions from 293 to 90 K, and curve B by the reverse transitions. From Cottrell and Stokes (1955).

More quantitatively, the average distance between forest dislocations is related to the internal stress by the usual relation: $d = \mu b/\tau_\mu$.

Using Eqs. (3.6) and (3.11b) then yields the activation volume:

$$V = \mu b^2 w/\tau_\mu$$

Since for high strains we can assume $\tau_\mu \approx \tau$, we obtain:

$$V\tau \approx \mu b^2 w$$

which is a constant. This property is indeed satisfied, at least at large strains, as shown in Figure 3.10. Saada (1999), however, pointed out that this analysis does not take into account cross-slip that is also active in stage III.

The exact origin of thermal activation can be the formation of jogs on intersecting dislocations or the recombination of short attractive junctions.

For a very low forest density, the activation volumes are no longer proportional to the distance between adjacent trees. This behaviour has been interpreted as a breakdown of thermal activation considerations based on fully relaxed dislocation configurations when the mean spacings in the glide plane become very large (Argon and East, 1979).

3.3.2 Dislocations–solute atoms interactions

The different situations described in Sections 3.1 and 3.2 are observed experimentally in many alloys as a function of temperature. Figure 3.11 shows the temperature dependence of the CRSS and of the stress–strain rate sensitivity in several CuAl alloys. Four domains can be identified.

– In domain 1 (very low temperatures) the phonon frictional force becomes so low that dislocation velocities can reach very high values. The inertial effects are

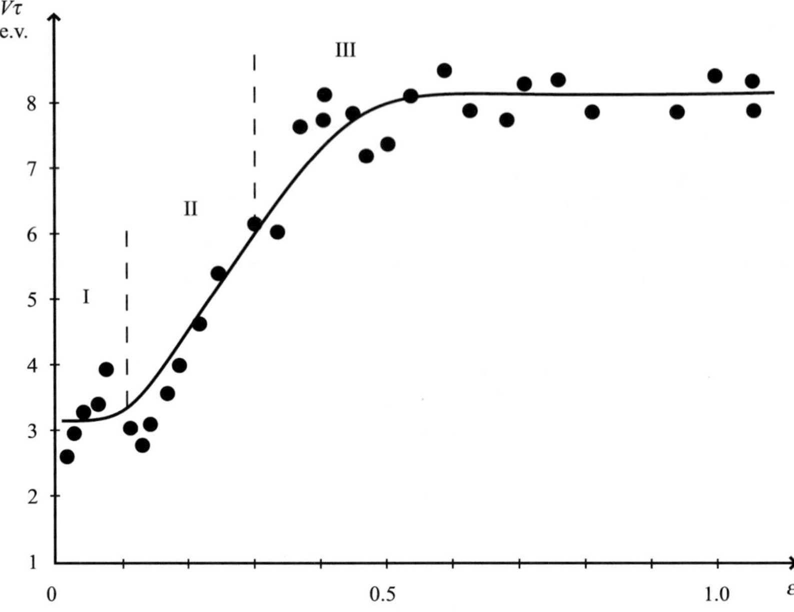

Figure 3.10. Curve $V\tau$ as a function of strain in a copper single crystal strained at 473 K. I, II, III refer to the classical hardening stages. From Thornton et al. (1962).

then sufficiently important to help dislocations passing through solute atoms ("underdamped" dislocation motion, see Figure 3.12). This phenomenon has been described by Granato (1971).

- Domain 2 corresponds to the "overdamped" thermally activated motion of dislocations across fixed obstacles described in Section 3.1. Sometimes solute atoms can already start to move to dislocations (Flor and Neuhauser, 1980).
- Domain 3, in which stress instabilities are associated to a small yield-stress anomaly and to a negative stress–strain rate sensitivity, corresponds to the situation described in Section 3.2.2.
- In domain 4, dislocation motion is controlled by the drag of solute atmosphere (Section 3.2.3).

The corresponding dislocation kinetics have been clearly identified in the in situ TEM experiments of Monchoux and Neuhauser (1987) in CuGe alloys. In particular, in the domain of stress instabilities, sudden vigorous dislocation movements shook the specimen and made observations impossible.

Domains 2–4 are described in more detail in what follows.

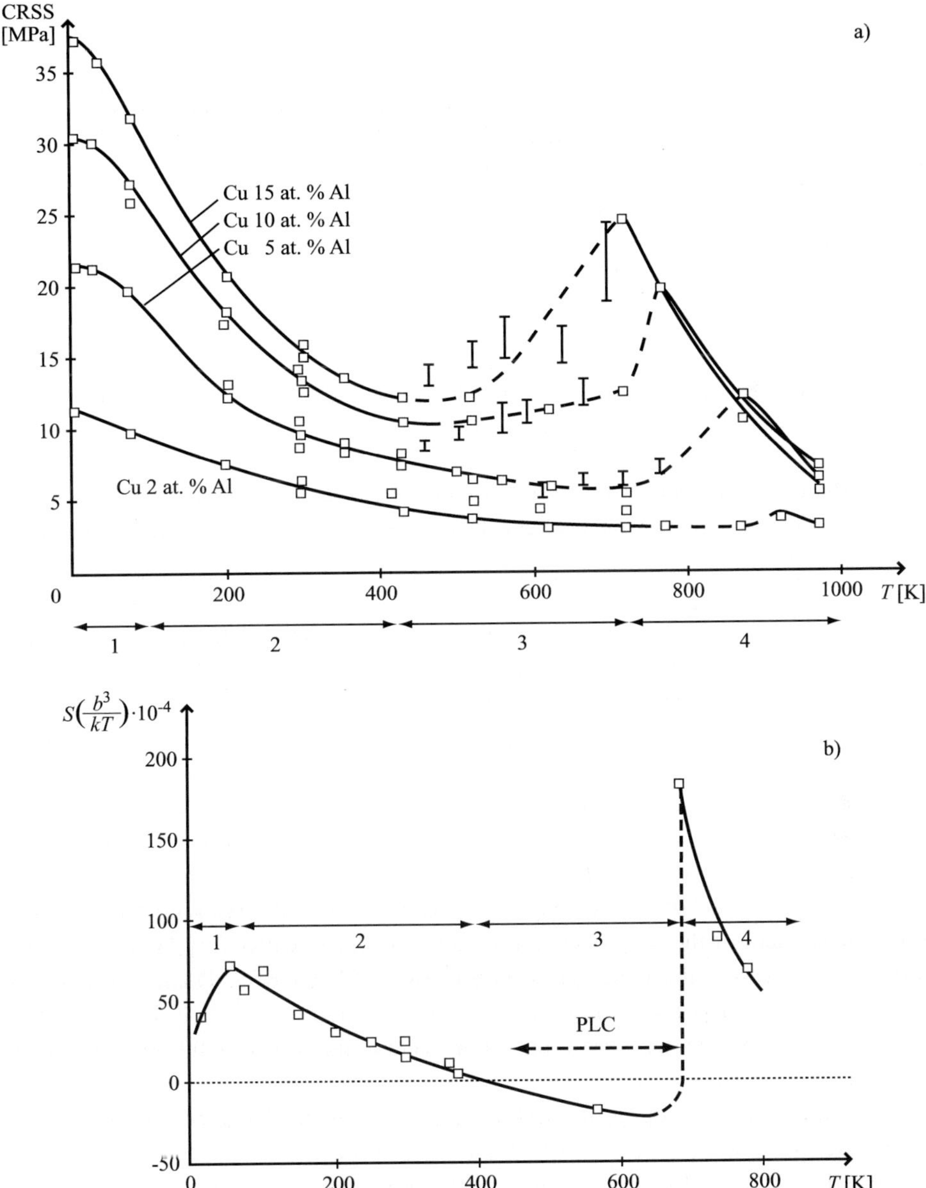

Figure 3.11. Mechanical properties of CuAl alloys (from a review by Neuhauser and Schwink, 1993). Temperature domains 1–4 are described in the text. (a) CRSS as a function of temperature. Data from Suzuki and Kuramoto (1968), Startsev et al. (1979), Nixon and Mitchell (1981) and Neuhauser et al. (1990). The PLC regime is indicated by dotted lines and bars that refer to the amplitudes of the stress instabilities. (b) Stress–strain rate sensitivity of Cu–15% Al as a function of temperature and PLC domain. Data from Kopenaal and Fine (1962), Komnik and Demiskii (1981) and Neuhauser et al. (1990).

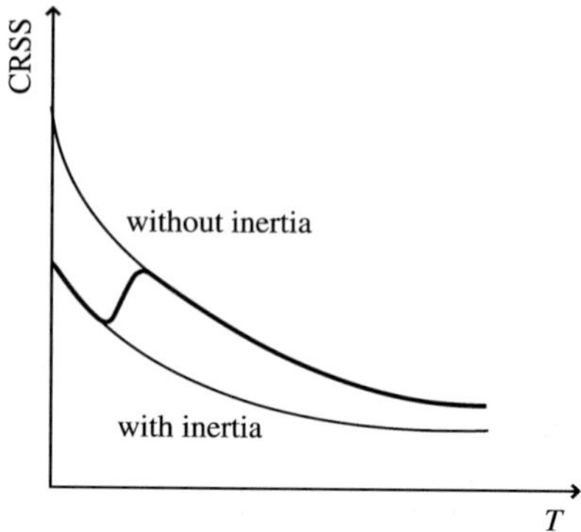

Figure 3.12. Schematic description of the influence of inertia effects at low temperature.

3.3.2.1 Domain 2: thermally activated motion across fixed obstacles. As discussed by Wille et al. (1987), comparison between theory and experiment is complicated by the following points:

(i) Results in domain 2 must be extrapolated to 0 K in order to determine E_{max} and τ_{max}. This is rather questionable when the extension of domain 1 is large.
(ii) There is usually a spectrum of obstacles of different strengths.
(iii) The solute distribution is often not uniform and segregation can take place at stacking faults. Short-range ordering or clustering may also alter the mobility of dislocations.
(iv) Deformation tends to be heterogeneous.

In order to avoid the above effects, Wille et al. (1987) have selected the CuMn system, where the tendency to form short-range order is negligible, the stacking fault energy is almost independent of the Mn concentration and the size effect is very large.

Figure 3.13a shows the temperature dependence of the CRSS. Domain 1 is smaller than in Figure 3.11a, which allows for an easier extrapolation to 0 K. Figure 3.13b shows that this stress varies with temperature according to Eq. (3.20) (Cottrell–Bilby potential).

The activation volume in Figure 3.14a shows that domain 1 extends from 0 to 25 K. Above 25 K its stress dependence obeys Eq. (3.21a) if the variation of $(1 - (\tau/\tau_{max})^{0.46})^{1/2}$ is neglected (Figure 3.14b). It also varies with temperature according to Eq. (3.21b), as shown in Figure 3.14c.

The values of τ_{max} and E_{max} corresponding to the best fits are shown in Table 3.1. E_{max} ranges between 1.2 and 1.4 eV and τ_{max} increases with the solute concentration. The internal stress τ_{μ} is assumed to be the high-temperature stress in Figure 3.13a. F_{max} is deduced from E_{max} included in Eq. (3.19a), assuming that $w = 2.5b$. Then, Eq. (3.16b) for

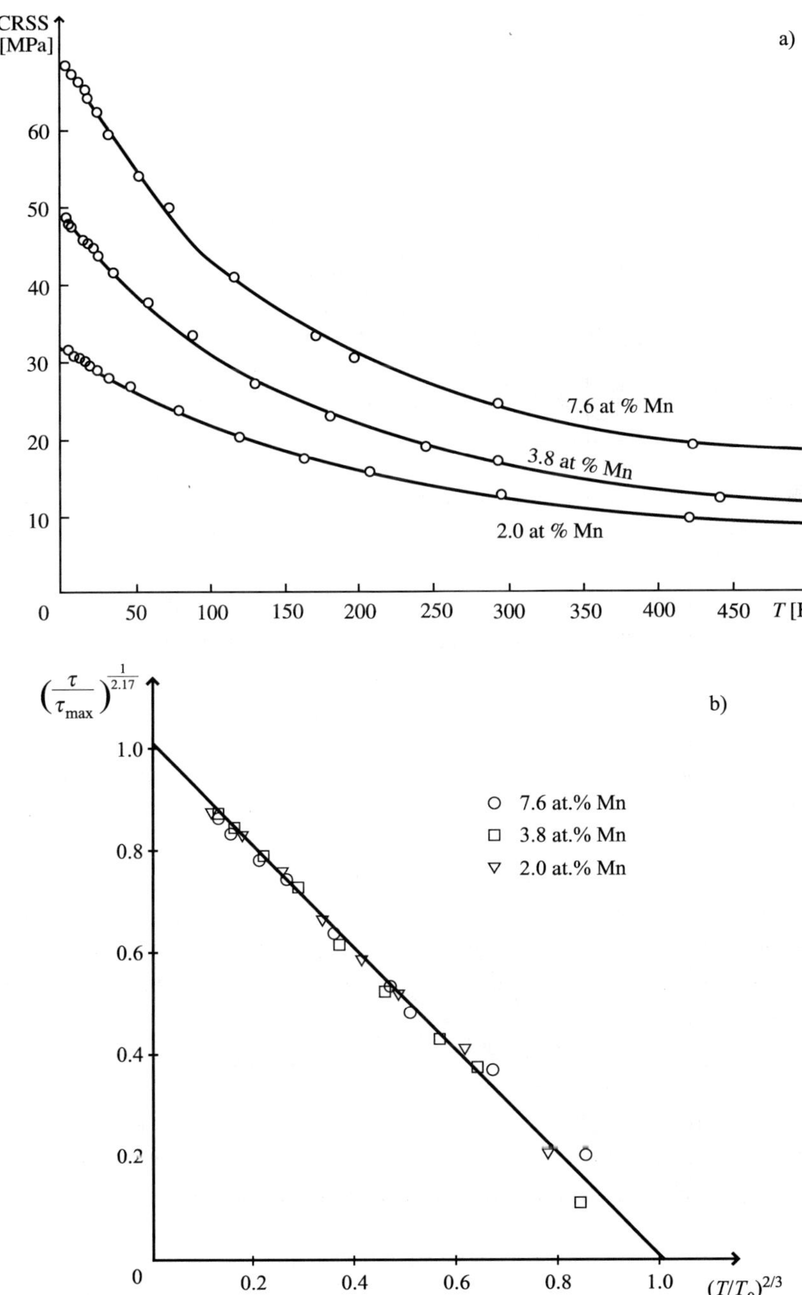

Figure 3.13. CRSS in CuMn alloys (from Wille et al., 1987): (a) as a function of temperature; and (b) plotted so as to check the validity of Eq. (3.20).

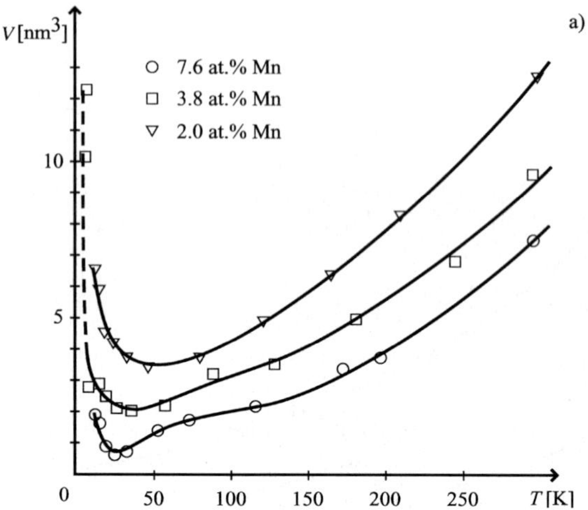

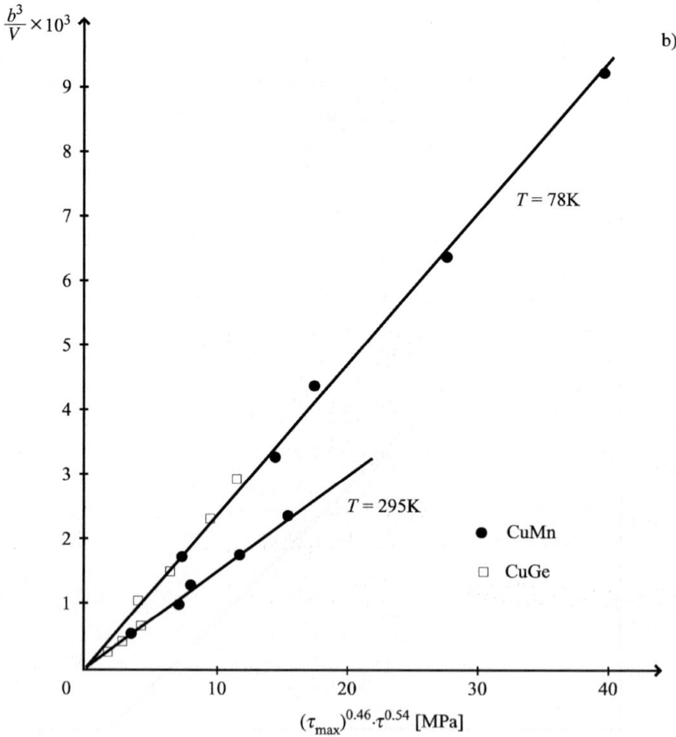

Figure 3.14. Activation volumes in CuMn alloys (from Wille et al., 1987): (a) as a function of temperature; (b) plotted so as to check the validity of Eq. (3.21a); and (c) plotted so as to check the validity of Eq. (3.21b).

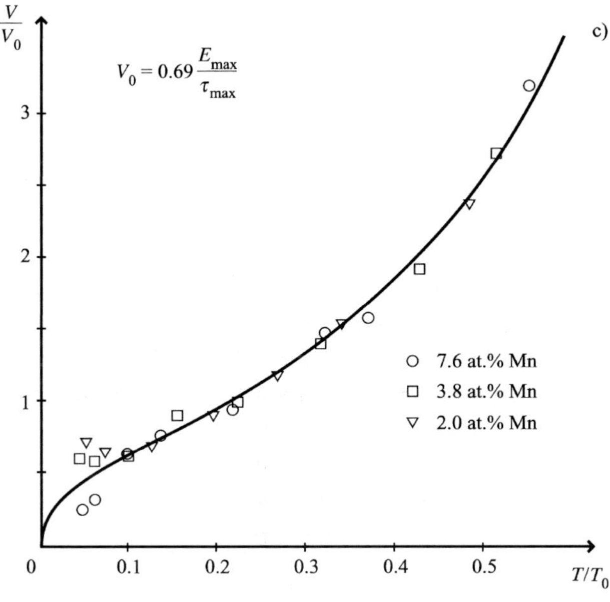

Figure 3.14. (*continued*)

$F = F_{max}$ and $\tau = \tau_{max}$ yields the atomic concentration of obstacles $c_b = (b/d)^{1/2}$. All these parameters are given in Table 3.1.

This table shows that:

- The estimation of β (Eq. (3.9)) ensures that the Fleischer–Friedel approximation is valid.
- The concentration of obstacles c_b is 20 times smaller than the average solute concentration c. In addition, the interaction energy E_{max} is found to be about 1.3 eV.

Table 3.1. Mechanical parameters of CuMn alloys (from Wille et al., 1987).

c (at.%)	0.4 Mn	1.2 Mn	2.0 Mn	3.8 Mn	7.6 Mn
$\dot{\gamma}_o/\dot{\gamma}$	23.5	25	23.5	25.8	26.5
τ_{max} (MPa)	10.3	20.7	25.0	40.2	58.2
T_o (K)	658	631	610	569	533
E_{max} (eV)	1.34	1.36	1.21	1.28	1.23
τ_μ (MPa)	2.0 ± 0.2	5.7 ± 0.3	10.3 ± 0.6	14 ± 1	20.4 ± 1.5
w/b	2.5	2.5	2.5	2.5	2.5
$c_b \times 10^{-4}$	1.33	5.1	10.8	23.5	53.8
$L_F(\tau_{max})/b$	324	164	120	79	53
$\beta \times 10^{-2}$	2.9	5.6	8.1	12.0	18.3

Since that expected for individual solute atoms is predicted to be smaller than 0.4 eV, thermal activation is thought to correspond to the crossing of doublets or triplets of solute atoms.

- The athermal stress τ_μ increases with the solute concentration. Since it has been observed that τ_μ is independent of the dislocation density, which can be changed by annealing, it seems accordingly that clusters involving more than three solute atoms, which cannot be overcome by thermal activation, are at the origin of this stress.

This analysis shows that calculations in Section 3.1 are reliable, at least as long as solute atoms are really immobile. In CuZn, on the contrary, Flor and Neuhauser (1980) measured non-logarithmic relaxations for which they assumed an increase of the activation energy with time due to solute segregation. In addition, $E_{\max}$ is observed to increase with increasing temperature, which indicates that the yield-stress is controlled by obstacles of different strengths at different temperatures. These results are of course much more difficult to compare with theoretical estimates.

3.3.2.2 Domain 3: stress instabilities and PLC effect. For a review of the experimental results in this temperature domain, the reader can refer to Strudel (1980) or Kubin and Estrin (1991).

More recently, an exhaustive study of stress instabilities was made in binary CuMn and CuAl alloys by Schwink and Nortmann (1997). The activation energies of the onset and of the end of the PLC effect are well below the volume-diffusion energy of Mn and Al atoms in a Cu lattice, which suggests the occurrence of pipe diffusion in the core of dislocations.

Surprisingly, plasticity in this domain may often be controlled by the movement of screw dislocations which are less subjected to pinning than edge ones. The release of screw segments generates fresh edge segments that can form dislocation sources (Neuhauser and Schwink, 1993; Suzuki, 1985). Since screw segments are also often subjected to Peierls-type friction forces, a combination of these two mechanisms (Peierls friction and solute effect) is expected in many cases. Note that solute atoms can also modify the core structure of sessile screw dislocations, e.g. in titanium and zirconium (Chapter 6). Such an interaction is different from that treated in this chapter. It may, however, contribute to enhance static and dynamic strain ageing when a Peierls mechanism acting on screw dislocations is rate controlling. The anomalous behaviour of several intermetallics, in which straight screw dislocations move in bursts, may be explained in this way (see Chapter 10).

3.3.2.3 Domain 4: glide controlled by solute-diffusion. Dislocation glide controlled by diffusion of solute atoms is often considered to explain the creep properties of “class I” alloys (see, e.g. Takeuchi and Argon, 1976). Taking a dislocation density proportional to the stress squared, the resulting creep rate is indeed expected to vary as the third power of stress and the activation energy is expected to be that of solute-diffusion in agreement with some experimental results. In many cases, however, the problem is considered to be too complex to have simple solutions (see, e.g. Poirier, 1976).

REFERENCES

Argon, A.S. & East, G.H. (1979) in *Strength of Metals and Alloys*, Eds. Haasen, P., Gerold, V. & Kostorz G., Pergamon Press, Oxford, p. 9.

Cottrell, A.H. & Bilby, B.A. (1949) *Proc. Phys. Soc. London A*, **62**, 49.

Cottrell, A.H. & Jaswon, M.A. (1949) *Proc. Roy. Soc. A*, **199**, 104.

Cottrell, A.H. & Stokes, R.J. (1955) *Proc. Roy. Soc. A*, **233**, 17.

Estrin, Y. & Kubin, L.-P. (1989) *J. Mech. Behav. Mater.*, **2**, 255.

Fleischer, R.L. (1961) *Acta Met.*, **9**, 996.

Flor, H. & Neuhauser, H. (1980) *Acta Met.*, **28**, 939.

Friedel, J. (1964) *Dislocations*, Pergamon Press, Oxford.

Fuentes-Samaniego, R. (1979) PhD thesis, Stanford University, California.

Granato, A.V. (1971) *Phys. Rev.* **B4**, p. 2196; *Phys. Rev. Lett.* **27**, p. 660.

Haasen, P. (1979) in *Dislocations in Solids*, vol. **4**, Ed. Nabarro, F.R.N., North Holland, Amsterdam, Chap. 15.

Haasen, P. (1983) in *Physical Metallurgy*, Part II, vol. **8**, Eds. Cahn, R.W. & Haasen P., 3rd Edition, North Holland Physics Publishing, Amsterdam, p. 1341.

Hirth, J.P. (1983) in *Physical Metallurgy*, Part II, vol. **8**, Eds. Cahn, R.W. & Haasen P., 3rd Edition, North Holland Physics Publishing, Amsterdam, p. 1223.

Hirth, J.P. & Lothe, J. (1982) *Theory of Dislocations*, 2nd Edition, Wiley Interscience, New York; (1992) second reprint edition, Krieger Pub. Comp., Malabar, Florida.

Kocks, U.F., Argon, A.S. & Ashby, M.F. (1975) *Thermodynamics and Kinetics of Slip*, Pergamon Press, Oxford.

Komnik, S.N. & Demiskii, V.V. (1981) *Czech. J. Phys. B*, **31**, 187.

Kopenaal, T.J. & Fine, M.E. (1962) *Trans AIME*, **224**, 347.

Kubin, L.P. & Estrin, Y. (1990) *Acta Met. Mat.*, **38**, 697.

Kubin, L.P. & Estrin, Y. (1991) *J. Phys. III*, **1**, 929.

Labusch, R. (1970) *Phys Stat. Sol.*, **41**, 659.

MacCormick, P.G. (1972) *Acta Met.*, **20**, 351.

Monchoux, F. & Neuhauser, H. (1987) *J. Mater. Sci.*, **22**, 1443.

Mulford, R.A. & Kocks, U.F. (1979) *Acta Met.*, **27**, 1125.

Nabarro, F.R.N. (1990) *Acta Metall. Mater.*, **38**, 161.

Neuhauser, H. & Schwink, C. (1993) in *Material Science and Technology*, vol. **6**, Eds. Cahn, R.W., Haasen, P. & Kramer E.J., VCH Verlag, Weinheim, p. 191.

Neuhauser, H., Plessing, J. & Schülke, M. (1990) *J. Mech. Behav. Met.*, **2**, 231.

Nixon, W.E. & Mitchell, J.W. (1981) *Proc. Roy. Soc. London A*, **376**, 343.

Poirier, J.P. (1976) *Plasticité à Haute Température des Solides Cristallins*, Eyrolles, Paris.

Saada, G. (1999) *Deformation-Induced Microstructures: Analysis and Relation to Properties*, Proceedings of 20th Risø International Symposium on Materials Science, Eds. Bilde-sorensen, J.B., Cartensen, J.V., Hansen, N., Jensen, D.J., Leffers, T., Pantleon, W., Pedersen, O.B. & Winther G., Risø National Laboratory, Roskilde, Denmark, p. 147.

Sakamoto, M. (1981) *Bull. Jpn. Inst. Met.*, **20**, 912.

Schwink, Ch. & Nortmann, A. (1997) *Mat. Sci. Eng. A*, **234–236**, 1.

Startsev, V.I., Demirskii, V.V. & Komnik, S.N. (1979) in *Strength of Metals and Alloys*, Eds. Haasen, P., Gerold, V. & Kostorz G., Pergamon Press, Oxford, p. 265.

Strudel, J.L. (1980) in *Dislocations et Déformation Plastique*, Eds. Groh, P., Kubin, L.P. & Martin J.L., Les Editions de Physique, Les Ulis, p. 199.

Suzuki, H. (1985) in *Strength of Metals and Alloys*, Eds. Mc Queen, H.J., Bailon, J.P., Dickson, J.L., Jonas, J.J. & Akben M.G., Pergamon, Toronto, p. 1727.

Suzuki, H. & Kuramoto, E. (1968) *Trans. JIM*, **9**(suppl.), 697.

Takeuchi, S. & Argon, A.S. (1976) *Acta Met.*, **24**, 883.

Thornton, P.R., Mitchell, T.E. & Hirsch, P.B. (1962) *Phil. Mag.*, **7**, 337.

Van den Beukel, A. (1975) *Phys. Stat. Sol. A*, **30**, 197.

Wille, T.H., Gieseke, W. & Schwink, C.H. (1987) *Acta Met.*, **35**, 2679.

Yoshinaga, H. & Morozumi, S. (1971) *Phil. Mag.*, **23**, 1367.

Chapter 4

Frictional Forces in Metals

Chapter 4
Frictional Forces in Metals

It is well known that dislocation glide in metals is easy along close-packed planes. This is the case for octahedral glide in FCC metals and basal glide in HCP metals. Indeed dislocations can dissociate along these planes in a glissile configuration. However, such a situation is an exception rather than the rule. In a number of deformation conditions, dislocations exhibit non-planar core configurations which makes glide difficult.

In the present chapter, the various approaches that describe the corresponding friction are revisited and treated using a unified formalism. These include: (i) the general treatment by Dorn and co-workers which considers several types of Peierls potentials; (ii) the description by Hirsch and Escaig of the moving dissociated cores in various crystal structures and (iii) the "locking–unlocking" mechanism more recently proposed.

The corresponding experimental situations are reviewed in Chapter 6.

4.1. DISLOCATION CORE STRUCTURES AND PEIERLS POTENTIALS

Peierls (1940) and Nabarro (1947) were the first ones to remark that the energy (E) of gliding dislocations has necessarily the same periodicity as the crystal lattice. This variation is described by the Peierls potential as represented schematically in Figure 4.1. Its amplitude ΔE depends on the change in the dislocation core structure as a function of its displacement, y.

The dislocation core structure is determined by local atomic displacements that are too large to be described by linear elasticity. They depend on the anisotropy of the crystal

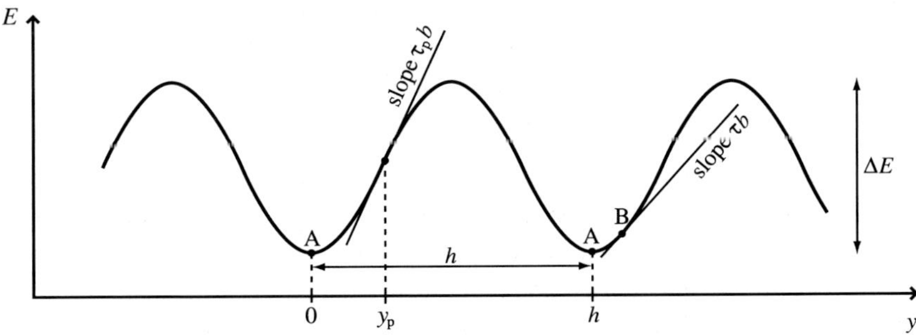

Figure 4.1. Schematic description of the variation of the dislocation energy, E, as a function of its displacement, y, in a periodic Peierls potential.

lattice and corresponding atomic bonds. Dislocation cores thus tend to extend along the directions of easier shear displacements.

In atomistic calculations, core structures are obtained by finding minimum values of the configuration energy. Accordingly, calculations under a zero applied stress yield structures corresponding to the bottom of Peierls valleys (Figure 4.1, in A). Calculations under an applied stress τ yield a more energetic structure corresponding to the minimum in the total energy $E - \tau by$, where y is the displacement of the dislocation with respect to the bottom of the Peierls valley (Figure 4.1, in B). The Peierls stress, τ_p, which allows dislocation movements across Peierls valleys without any thermal activation, corresponds to the maximum slope of the Peierls potential for $y = y_p$. In atomistic calculations it is the stress for which no energy minima of $E - \tau by$ can be obtained. They show that τ_p and ΔE strongly depend on the dislocation core structure. These structures are usually described in the perfect lattice by arrows connecting adjacent atoms, the length of which are proportional to their relative displacement when introducing the dislocation. It is important to note that each diagram corresponds to displacements along a given direction that is specified in each case and that can be either in or outside the diagram plane (the directions of atomic displacements do not correspond to the directions of arrows). Equi-displacement contours materialize the extension and the shape of the core and allow the estimation of the dislocation mobility.

- When the core is spread in the slip plane (Figure 4.2(a)), ΔE and τ_p are low and the dislocation is fairly glissile. The core indeed contains regions in all possible states of distortion, in such a way that the sum of these distortions do not vary significantly during the movement. In other words, the core smoothes the lattice periodicity. This description applies to dislocations dissociated in their slip plane.
- When the core is compact, ΔE and τ_p are necessarily larger.
- When the core is non-planar, ΔE and τ_p are high and the dislocation is at least difficult to move or definitely sessile. An example is shown in Figure 4.2(b). Figure 4.2(a) is also a non-planar core with respect to a vertical displacement in the prismatic plane. In those cases, dislocation movements can take place only by very energetic collective atomic displacements or by a transformation into a more compact core structure. An example of this situation is the glide of a screw dislocation in a plane different from its dissociation plane.

Dislocations often have several possible core structures. In the case of screw dislocations, different equivalent core structures result from the symmetry of the lattice, e.g. $\frac{1}{2}\langle 110\rangle$ dislocations can spread in two possible intersecting $\{111\}$ planes in the FCC structure and two equivalent dissociation modes can exist for $\langle 111\rangle$ dislocations in the BCC structure, as shown in Figure 4.2(b). Non-equivalent core structures can, however, be found as well, the one of lowest energy being stable and the others being metastable. This idea first proposed by Régnier and Dupouy (1968, 1970) to explain some aspects of

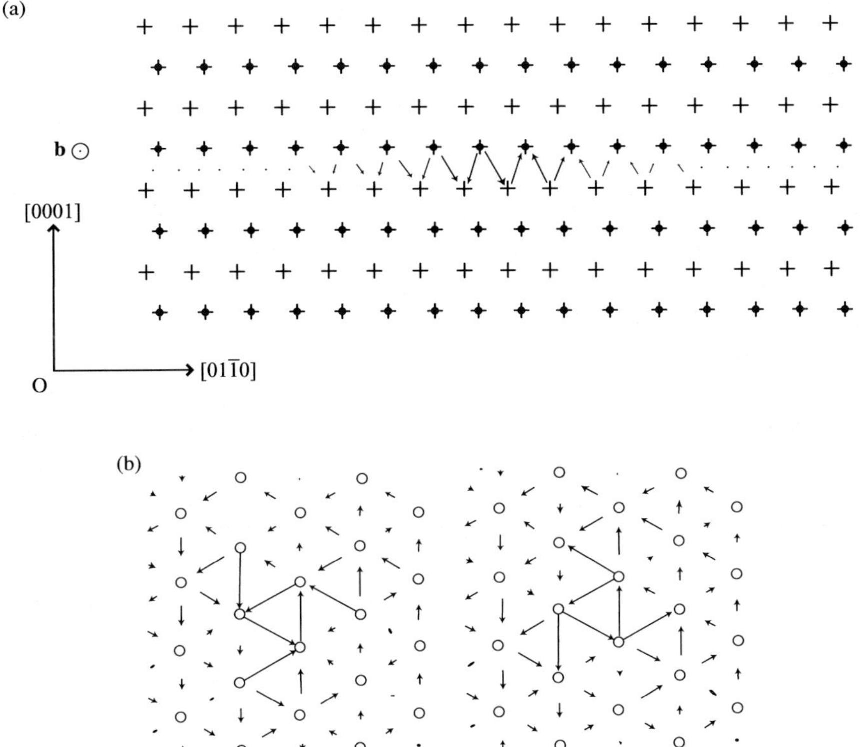

Figure 4.2. Results of atomistic core structure calculations of end on screw dislocations extended along: (a) the basal plane of an HCP lattice (Bacon and Martin, 1981) and (b) several planes in a BCC lattice (Takeuchi, 1981). Two possible configurations of same energy are shown.

prismatic slip in close-packed HCP metals (beryllium, magnesium). Screw dislocations were assumed to have two possible core structures: a stable one of lower energy dissociated in the basal plane and a metastable one of higher energy dissociated in the prismatic plane. The first one allows for easy glide in the basal plane but entails difficult glide in the prismatic plane, whereas the second one allows for easy glide in the prismatic plane. The two types of dislocation core structures have been obtained by atomistic calculations by Vitek and Igarashi (1991). Similar calculations in Ti_3Al (HCP-type ordered structure) have been carried out by Cserti et al. (1992). Ordinary screw dislocations in TiAl can also take either a planar or a non-planar core structure according to Mahapatra et al. (1995) and Simmons et al. (1997) (see Figure 4.3).

As a conclusion, dislocations can often take several core configurations which induce different Peierls potentials in the different possible slip planes. The mobility of dislocations can thus be changed drastically by a change in their core structure.

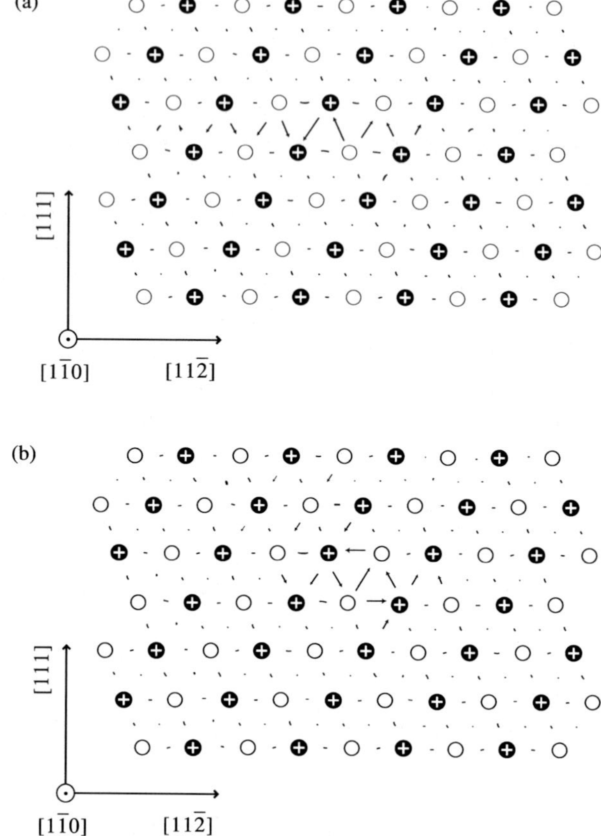

Figure 4.3. Two possible core structures for screw ordinary dislocations (seen end on) in TiAl (from Simmons et al., 1997): (a) planar and glissile in the {111} plane; (b) non-planar and subjected to a high friction force.

Two different glide modes that take into account the dislocation core structure and its evolution under stress are described in what follows. These are i) the kink-pair mechanism that controls dislocation glide across Peierls valleys (Section 4.2), ii) the thermally activated core transformation that changes the Peierls potentials felt by gliding dislocations (Section 4.3).

4.2. KINK-PAIR MECHANISM

This mechanism describes the lattice friction acting on a dislocation. In what follows the main assumptions about this mechanism are presented, the various periodic potentials along the slip plane are exposed, followed by energy estimations.

4.2.1 Principles

Dislocations glide between adjacent Peierls valleys by nucleation of kink-pairs which subsequently glide apart rapidly. The kinks are here assumed to be fairly mobile, namely they are not slowed down by another Peierls potential. The alternative case, typical of covalent materials, will be discussed in Chapter 7. The different steps of the process are described in Figure 4.4(a).

A bulge is formed first on the straight dislocation in the bottom of the Peierls valley (step 1). Then two kinks of opposite signs are formed when the leading part of the dislocation reaches the neighbouring valley (step 2). The two kinks interact elastically as long as they remain close to each other, until they become independent (step 3). Let Δx be the bulge width or the kink separation. For a zero stress, the energy of the configuration shown in (a), $U_{\mathrm{kp}}^{(0)}(\Delta x)$, is equal to the bulge energy $U_{\mathrm{b}}^{(0)}(\Delta x)$ in domain 1 and to the energy of two interacting kinks, $U_{\mathrm{ik}}^{(0)}(\Delta x)$, in domain 2. It tends asymptotically to be twice the energy of an isolated kink, $2U_{\mathrm{k}}$, when Δx tends to infinity. The work done by the stress, $W(\tau, \Delta x)$, must be subtracted from this energy. It is equal to $\tau bh\Delta x$ when the kinks are well formed (domains 2 and 3) and it varies non-linearly in domain 1. The total energy $\mathrm{U_{kp}}\langle\tau, \Delta \mathrm{x}\rangle$ is equal to $\mathrm{U_b}(\tau, \Delta\mathrm{x}) = \mathrm{U_b^{(0)}} - \tau \mathrm{bh}\Delta x$ in domain 1, and to $\mathrm{U_{ik}}(\tau, \Delta \mathrm{x}) = \mathrm{U_{ik}^{(0)}},\ \tau\mathrm{bh}\Delta\mathrm{x}$ in domain 2. It goes through a maximum, the activation energy of the process, $U_{\mathrm{kp}}^{(\mathrm{c})}$, for $\Delta x = x_{\mathrm{c}}$. This critical value falls in domain 1 if τ is large (case of Figure 4.4(b)) and then $U_{\mathrm{kp}}^{(\mathrm{c})} = U_{\mathrm{b}}^{(\mathrm{c})}$. It falls in domain 2 if τ is small (case of Figure 4.4(c)) and then $U_{\mathrm{kp}}^{(\mathrm{c})} = U_{\mathrm{ik}}^{(\mathrm{c})}$.

The nucleation rate of kink-pairs per unit dislocation length is (see Guyot and Dorn, 1967)[1]

$$P_{\mathrm{kp}} = \nu_{\mathrm{D}} \frac{b}{x_{\mathrm{c}}^2} \exp\left(- \frac{U_{\mathrm{kp}}^{(\mathrm{c})}(\tau)}{kT} \right) \tag{4.1}$$

where $U_{\mathrm{kp}}^{(\mathrm{c})} = U_{\mathrm{b}}^{(\mathrm{c})}$ or $U_{\mathrm{ik}}^{(\mathrm{c})}$, depending on the stress, $\nu_{\mathrm{D}} b/x_{\mathrm{c}}$ is the dislocation vibration frequency of wavelength x_{c} and $1/x_{\mathrm{c}}$ is the density of antinodes of vibration (i.e. the number of possible nucleation sites of kink-pairs).

The velocity of the dislocation of length L is accordingly:

$$\mathrm{v_{kp}} = \nu_{\mathrm{D}} \frac{bL}{x_{\mathrm{c}}^2} h \exp\left(- \frac{U_{\mathrm{kp}}^{(\mathrm{c})}(\tau)}{kT} \right) \tag{4.2}$$

4.2.2 Several Peierls potentials and associated Peierls stresses

Several functions can be used to describe the periodic variation of the total dislocation energy along the slip plane. The first three below are smooth potentials; the fourth one is an

[1] The more simple pre-exponential term $\nu_{\mathrm{D}} L/x_{\mathrm{c}}$ has been used by Dorn and Rajnak (1964).

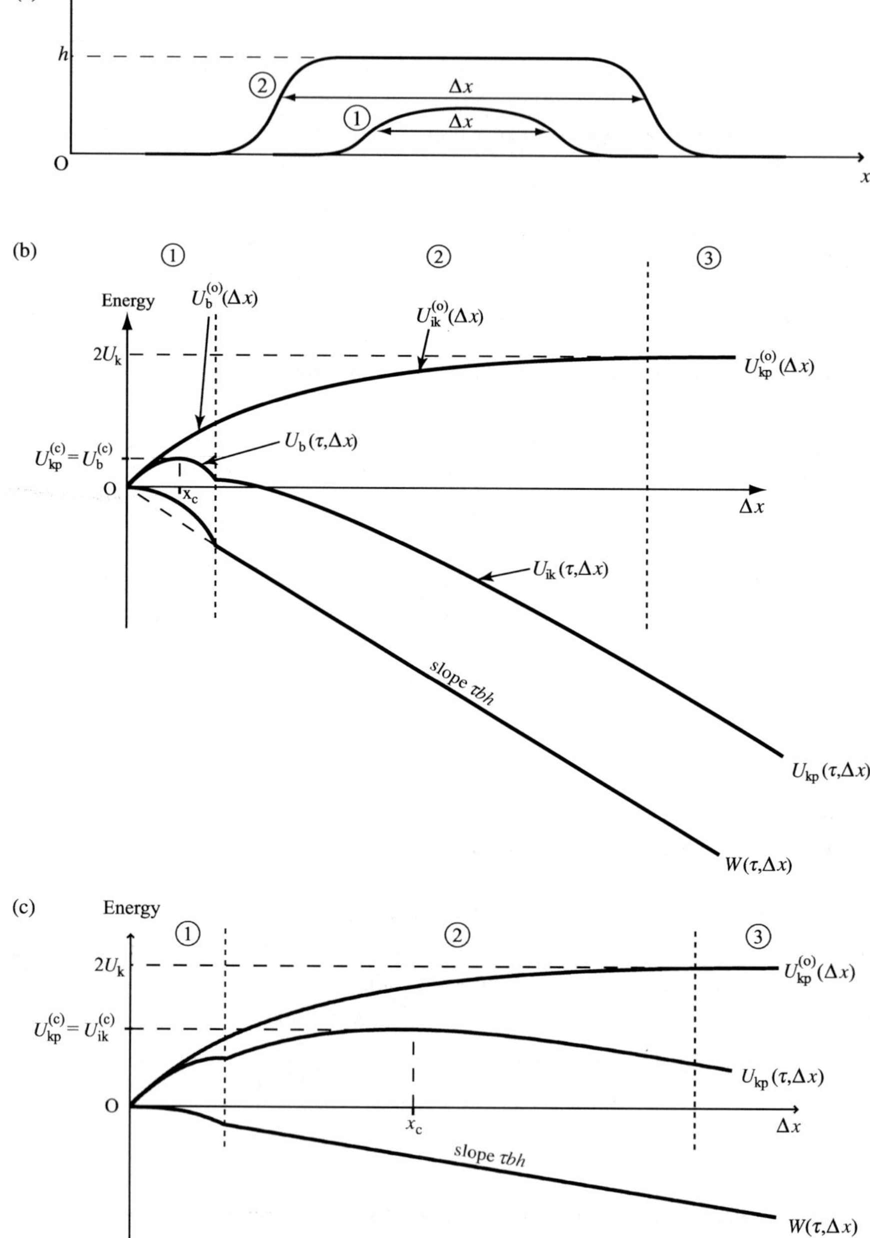

Figure 4.4. Dislocation moving in a Peierls potential by nucleation and glide of a kink-pair. The dislocation is along the x direction and moves along y. (a) Bulge in 1 and interacting kinks in 2. (b) Energy diagram (high stress). (c) Energy diagram (low stress).

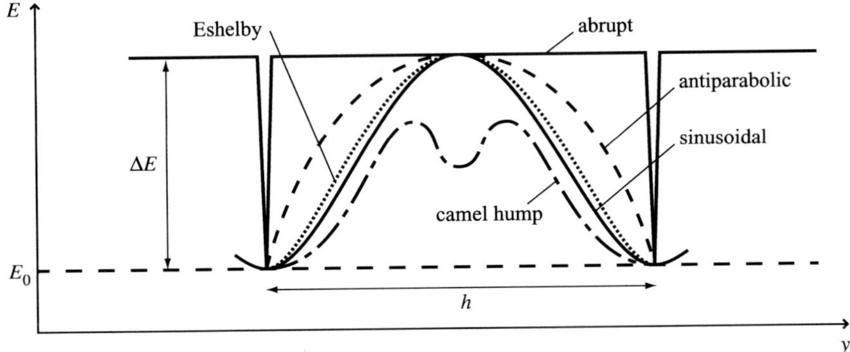

Figure 4.5. Schematic description of the different Peierls potentials used in the following.

abrupt potential (Figure 4.5). The wall-like potential for covalent materials will be discussed in Chapter 7. It induces very high values of τ_p (Suzuki et al., 1995).

- Sinusoidal potential

$$E = E_0 + \frac{\Delta E}{2}\left(1 - \cos\frac{2\pi y}{h}\right) \tag{4.3a}$$

$$\tau_p = \frac{\pi \Delta E}{hb} \tag{4.3b}$$

Several variants have been used by Dorn and Rajnak (1964) and by Koizumi et al. (1993). They can have intermediate minima as the "camel-hump" potential shown in Figure 4.5.

- Eshelby potential (Eshelby, 1962)

$$E = E_0 + 16\Delta E\left(\frac{y}{h}\right)^2\left(1 - \frac{y}{h}\right)^2 \tag{4.4a}$$

$$\tau_p = 3.08\frac{\Delta E}{hb} \tag{4.4b}$$

E being non-periodic, it is necessary to take its value over the interval $[0 + \varepsilon, h - \varepsilon]$ and to repeat it periodically.

- Antiparabolic potential

$$E = E_0 + 4\Delta E\frac{y}{h}\left(1 - \frac{y}{h}\right) \tag{4.5a}$$

$$\tau_p = 4\frac{\Delta E}{hb} \tag{4.5b}$$

(Same remark as for the Eshelby potential.)

- Abrupt potential
  The dislocation energy E is assumed to be constant, except in sharp valleys, distant from h, where it is decreased by ΔE. The core is described in terms of alternate planar and non-planar dissociations of respective energies $E_0 + \Delta E$ and E_0. The corresponding Peierls stress, τ_p, is infinite. This potential is fairly simple to use, but it is necessary to introduce constrictions to smooth the sharp transition between the two core structures. It has been used by Vitek (1966), Escaig (1968a,b) and Duesberry and Hirsch (1968).

4.2.3 *Energy of an isolated kink*

The kink energy is the difference between the energy of the dislocation shown in Figure 4.6 and the energy of a straight dislocation in a single Peierls valley, for $y = 0$, with E_0 as energy per unit length (Figure 4.5).

E_0 is the part of the dislocation energy per unit length that is involved in the formation of kinks. It is assumed to be independent of the character (edge, screw, mixed). It is thus similar to the line tension $\hat{T}$ defined by Friedel (1964):

$$\hat{T} \sim \frac{\mu b^2}{4\pi} \ln \frac{x}{b}$$

where x is a length equal to either the kink-length (Section 4.2.6) or the critical kink-pair separation x_c (Section 4.2.5). Taking an orientation-dependent energy E_0 would lead to more complex calculations.

4.2.3.1 *Dorn and Rajnak calculation (smooth potentials).* The energy of the kinked dislocation is the integral of the energies of elementary segments of length $\sqrt{dx^2 + dy^2}$, with energy per unit length $E(y)$. E(y) varies between E_0 and $E_0 + \Delta$E. Accordingly, the kink energy can be written as:

$$U_k = \int_{-\infty}^{+\infty} \left[E(y)\sqrt{1 + \left(\frac{dy}{dx}\right)^2} - E_0 \right] dx \tag{4.6}$$

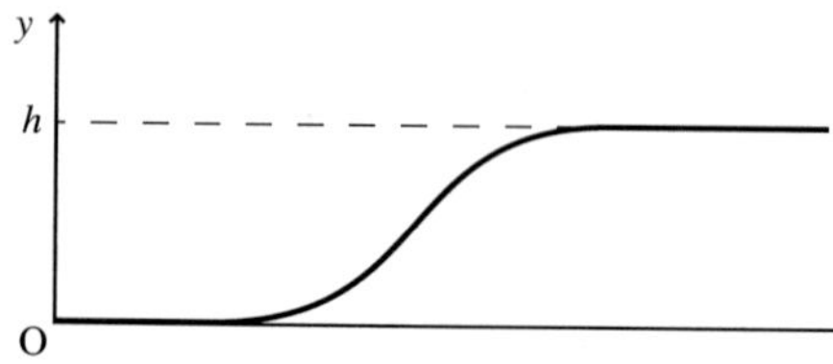

Figure 4.6. Equilibrium shape of an isolated kink (smooth potential).

The kink energy is at a minimum for

$$\frac{d}{dy}\left(E(y)\Big/\sqrt{1+\left(\frac{dy}{dx}\right)^2}\right)=0$$

(Euler condition), which yields after integration

$$\frac{E(y)}{\sqrt{1+\left(\frac{dy}{dx}\right)^2}}=E_0$$

and

$$\left(\frac{dy}{dx}\right)^2=\left(\frac{E(y)}{E_0}\right)^2-1 \qquad (4.7)$$

This condition gives the equilibrium shape of the kink as well. After combining Eqs. (4.6) and (4.7) and changing the variable, the kink energy becomes:

$$U_k=\int_0^h\sqrt{E^2(y)-E_0^2}\,dy \qquad (4.8)$$

4.2.3.2 Line tension approximation. The above relations can be simplified assuming that $dy/dx \ll 1$, which is always verified in metals. The detailed calculations below have been carried out by Eshelby (1962), Seeger (1984) and Suzuki et al. (1991).

Eq. (4.6) becomes:

$$U_k=\int_{-\infty}^{\infty}\left[E(y)\left(1+\frac{1}{2}\left(\frac{dy}{dx}\right)^2\right)-E_0\right]dx \qquad (4.9)$$

The energy can be minimized using the line tension, $\hat{T}$, to express the dislocation equilibrium condition. The relation between the local curvature, d^2y/dx^2, and the stress dE/dy due to the potential variation, is:

$$\hat{T}\frac{d^2y}{dx^2}=\frac{dE}{dy} \qquad (4.10)$$

which yields after integration:

$$\left(\frac{dy}{dx}\right)^2=\frac{2}{\hat{T}}(E(y)-E_0) \qquad (4.11)$$

Assuming $E(y) - E_0 \ll E_0$ and $\hat{T} \approx E_0$ (see §4.2.3.1), this relation is equivalent to Eq. (4.7). Combining Eqs. (4.9) and (4.11) and changing the variable then yields:

$$U_k = \sqrt{2}E_0 \int_0^h \sqrt{\frac{E(y)}{E_0} - 1}\, dy \tag{4.12}$$

This relation is equivalent to Eq. (4.8).

Relation (4.8) or (4.12) yield simple values of U_k for different potentials, which are very close to each other:

- Sinusoidal (Dorn and Rajnak, 1964):

$$U_k = \frac{2^{3/2}}{\pi} h(E_0 \Delta E)^{1/2} \tag{4.13}$$

- Antiparabolic (Guyot and Dorn, 1967):

$$U_k = \frac{\pi}{2^{3/2}} h(E_0 \Delta E)^{1/2} \tag{4.14}$$

- Eshelby (Eshelby, 1962):

$$U_k = 0.943h(E_0 \Delta E)^{1/2} \tag{4.15}$$

4.2.3.3 Abrupt potential. The kink energy can be expressed as (see Figure 4.7):

$$U_k = (E_0 + \Delta E)\frac{h}{\sin\theta} - E_0 \frac{h}{tg\theta}$$

Since θ is small, this value can be simplified as:

$$U_k = h\left(\frac{\Delta E}{\theta} + E_0 \frac{\theta}{2}\right)$$

The kink energy is at a minimum for $\partial U_k / \partial \theta = 0$, namely $\theta_c = \sqrt{2\Delta E / E_0}$, whence

$$U_k = h\theta_c E_0 = h\sqrt{2}(E_0 \Delta E)^{1/2}$$

The dislocation has a non-planar dissociation in the bottom of the valley (energy E_0) and either a planar dissociation or a compact core at the kink. In order to smooth the transition

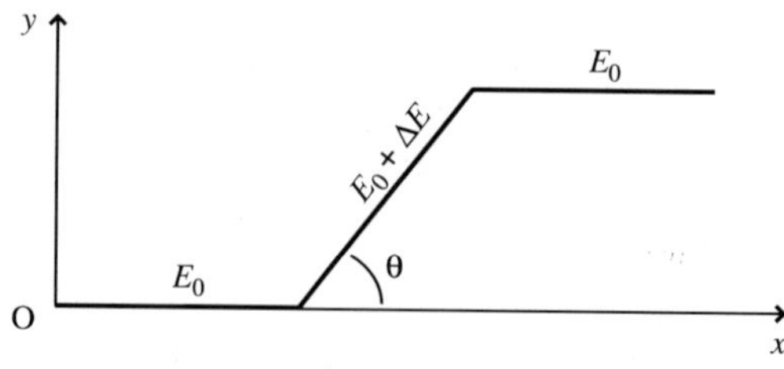

Figure 4.7. Equilibrium shape of an isolated kink (abrupt potential).

between the two core structures it is necessary to add a constriction energy which will be discussed in Section 4.3.2. The true kink energy is then:

$$U_k = h\sqrt{2}(E_0\Delta E)^{1/2} + U_{constr} \tag{4.16}$$

Even with this very crude potential, the kink energy is close to those corresponding to smooth potentials.

4.2.4 *Energy of a critical bulge (high stress approximation)*

This approximation is valid in the case of Figure 4.4(b), that is when the maximum of $U_{kp}(\tau, \Delta x)$ lies in region 1. The basic simplifying hypothesis is that the variation in the line energy during the bulge formation is much more important than the variation of the elastic interaction between the different parts of the bulge.

For each value of the stress τ, there is an equilibrium position for which the dislocation remains straight and moves in its original Peierls valley, from $y = 0$ to y_0 such that $\tau b = \partial E/\partial y$ (see Figure 4.8). There is also a bulged metastable equilibrium position with energy U_b. The following calculation follows the same procedure as for the isolated kink.

4.2.4.1 *Dorn and Rajnak calculation (1964).* The energy of the mechanical equilibrium configuration is:

$$U_b = \int_{-\infty}^{\infty} \left[E(y)\sqrt{1 + \left(\frac{dy}{dx}\right)^2} - E(y_0) - \tau b(y - y_0) \right] dx \tag{4.17}$$

This relation corresponds to Eq. (4.6) where the work done by the stress has been added and E_0 has been replaced by $E(y_0)$.

The bulge energy is at a minimum for

$$\frac{d}{dy}\left(E(y) \Big/ \sqrt{1 + \left(\frac{dy}{dx}\right)^2} \right) = \tau b$$

(Euler condition), which, after integration, yields,

$$\frac{E(y)}{\sqrt{1 + \left(\frac{dy}{dx}\right)^2}} = E(y_0) + \tau b(y - y_0)$$

and

$$\left(\frac{dy}{dx}\right)^2 = \left(\frac{E(y)}{E(y_0) + \tau b(y - y_0)}\right)^2 - 1 \tag{4.18}$$

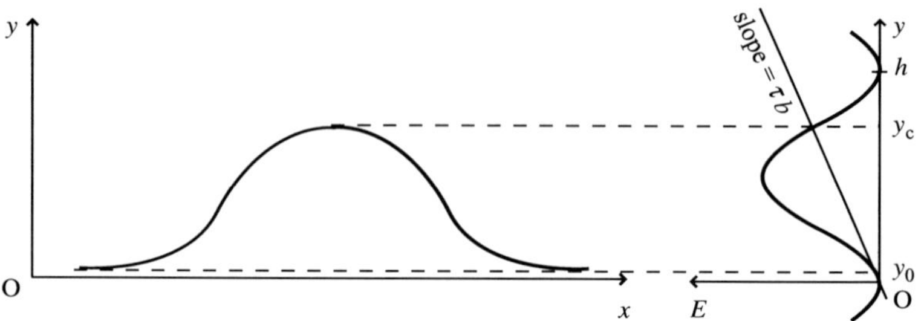

Figure 4.8. Bulged metastable equilibrium configuration and corresponding energy diagram.

The critical extension of the bulge, y_c, is such that $dy/dx = 0$, or

$$E(y_c) - E(y_0) = \tau b(y_c - y_0) \tag{4.19}$$

(see Figure 4.8) and, after combining Eqs. (4.17) and (4.18) and changing the variable, the activation energy is:

$$U_b^{(c)} = 2\int_{y_0}^{y_c} \sqrt{E^2(y) - [\tau b(y - y_0) + E(y_0)]^2} dy \tag{4.20}$$

where the factor 2 comes from the two sides of the bulge.

The activation energy, $U_b^{(c)}$, and the corresponding activation area, $A_b = (1/b)(\partial U_b^{(c)}/\partial \tau)$, have been calculated numerically for a sinusoidal potential (Figure 4.9).

At high stresses ($\tau \rightarrow \tau_p$) the activation energy tends to zero. The critical bulge has a small amplitude and the dislocation is near the position $y = y_p$ of maximum force (see Figure 4.10, for $\tau/\tau_p = 0.9$). At low stress ($\tau \rightarrow 0$) the activation energy tends to $2U_k$. In that case, however, the critical dislocation shape is a kink-pair and neglecting the elastic interaction between the two kinks is a very bad approximation, as discussed in Section 4.2.5. For the same reason, in this approximation, the activation volume tends to a finite value when the stress tends to zero, whereas it should tend to infinity.

4.2.4.2 Line tension approximation. The same procedure as for an isolated kink has been used: (Celli et al., 1963; Seeger, 1984; Suzuki et al., 1991).

Eq. (4.17) can be developed to the first order approximation:

$$U_b = \int_{-\infty}^{\infty} \left[E(y)\left(1 + \frac{1}{2}\left(\frac{dy}{dx}\right)^2\right) - E(y_0) - \tau b(y - y_0) \right] dx \tag{4.21}$$

The minimum value of U_b is obtained using the line tension equilibrium condition

$$\hat{T}\frac{d^2y}{dx^2} = \frac{dE}{dy} - \tau b$$

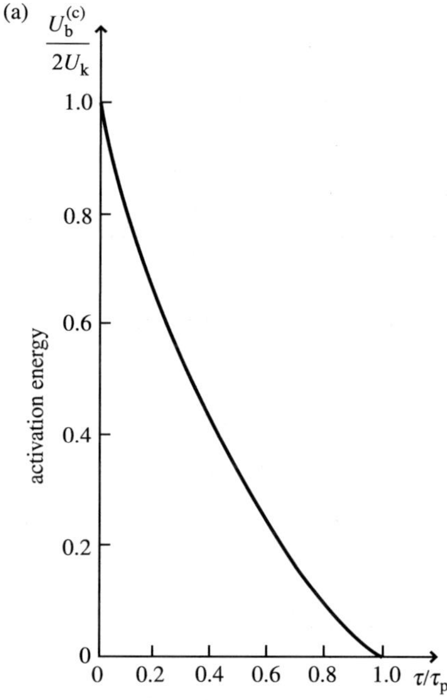

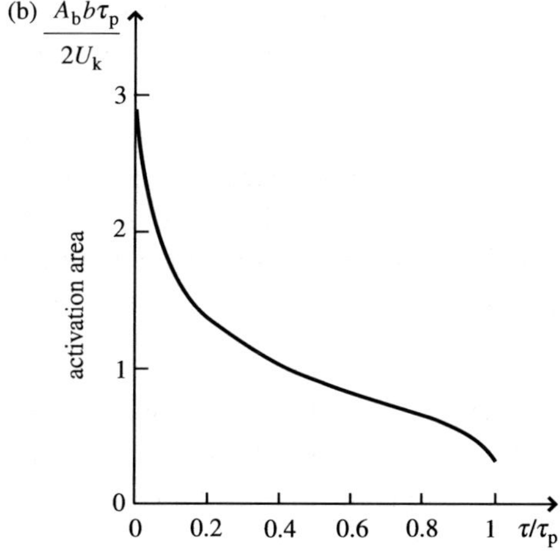

Figure 4.9. Activation energy $U_b^{(c)}$(a) and activation area A_b (b) of the critical bulge configuration for a sinusoidal potential (Dorn and Rajnak, 1964).

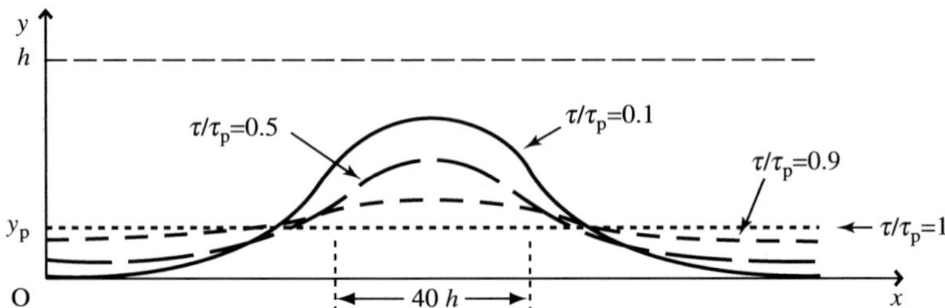

Figure 4.10. Critical bulge shapes at various stresses, for a sinusoidal potential. From Kocks et al. (1975).

which, after integration, yields

$$\hat{T}\left(\frac{\mathrm{d}y}{\mathrm{d}x}\right)^2 = 2[E(y) - E(y_0) - \tau b(y - y_0)] \tag{4.22}$$

Assuming $E(y) - E(y_0) \ll E_0$ and $\hat{T} \approx E_0$, Eq. (4.22) yields Eq. (4.19) for the condition $\mathrm{d}y/\mathrm{d}x = 0$ $(y = y_c)$. Combining Eqs. (4.21) and (4.22) and changing the variable then yields the activation energy of the bulge:

$$U_b^{(c)} = 2\sqrt{2E_0}\int_{y_0}^{y_c}\sqrt{E(y) - E(y_0) - \tau b(y - y_0)}\,\mathrm{d}y \tag{4.23}$$

This relation in the line tension approximation is equivalent to relation (4.20). For the antiparabolic potential, the activation energy becomes (Kocks et al., 1975):

$$U_b^{(c)} = 2U_k\left(1 - \frac{\tau}{\tau_p}\right)^2 \tag{4.24}$$

It is compared with the activation energy given by the sinusoidal potential in Figure 4.11. The two variations are close to each other.

The general expression of the activation area deduced from Eq. (4.23) is:

$$A_b = \sqrt{2E_0}\int_{y_0}^{y_c}(y - y_0)[E(y) - \tau b(y - y_0)]^{-1/2}\mathrm{d}y \tag{4.25}$$

The activation parameters Eqs. (4.23) and (4.25) have been estimated within various stress limits.

For the lowest stress values compatible with the high stress approximation, Seeger (1984) and Holzwarth and Seeger (1991) derive approximate solutions of Eq. (4.25). The work done by the stress is neglected and the only stress-dependence is that of the integration limit y_c in Eq. (4.20). In addition, the following two approximations are made:

$$E(y_0) \approx E_0 \approx E(h)$$

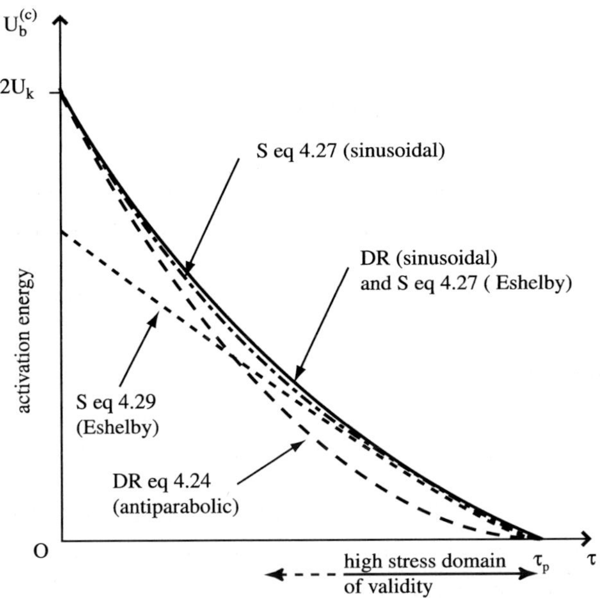

Figure 4.11. Activation energy of bulge nucleation versus stress, using the Dorn and Rajnak (DR) and Seeger (S) approximations for two different potentials. From Kocks et al (1975) and Brunner and Diehl (1991a,b).

$$E(y) - E(h) \approx \frac{E''(h)}{2}(h - y)^2$$

The second approximation is only valid for $y \approx h$, namely near the bulge extremity ($y = y_c$). It is expected, however, to yield a reasonable order of magnitude of the activation area because the area swept by the bulge is mainly determined by the radius of curvature near the bulge extremity (see, e.g. Figure 4.8). The activation area is then:

$$A_b = h\sqrt{\frac{\hat{T}}{E''(h)}}\left(1 + \ln\frac{\hat{\tau}}{\tau}\right) \tag{4.26}$$

(Holzwarth and Seeger, 1991), where $\hat{\tau}$ is a constant, whence, after integration:

$$U_b^{(c)} = 2U_k - h\sqrt{\frac{\hat{T}}{E''(h)}}\tau\left(1 + \ln\frac{\hat{\tau}}{\tau}\right) \tag{4.27}$$

The term $E''(h) = E''(0)$ is the curvature of the Peierls potential near its minimum value, that is:

$$E''(h) = \frac{2\pi^2}{h^2}\Delta E \quad \text{for a sinusoidal potential, and}$$

$$E''(h) = \frac{32}{h^2}\Delta E \quad \text{for an Eshelby potential.}$$

The stress $\hat{\tau}$ is, for an Eshelby potential:

$$\hat{\tau} = 12\sqrt{3}\tau_p = 64\frac{\Delta E}{hb} \qquad \text{(Werner and Seeger, 1988).}$$

Another approximation is valid for very high stresses close to τ_p, namely when the bulge amplitude is small and y remains close to y_p. A third order polynomial expression of $E(y) - \tau by$ can be used in relation (4.23), which yields (Mori and Kato, 1981):

$$U_b^{(c)} = \frac{96}{5} b^2 E_0^{1/2} [-2bE'''(y_p)]^{-3/4} (\tau_p - \tau)^{5/4} \tag{4.28}$$

Using an Eshelby potential, this expression reduces to:

$$U_b^{(c)} = \frac{12}{5} U_k \left[\frac{2}{3} \left(1 - \frac{\tau}{\tau_p} \right) \right]^{5/4} \tag{4.29}$$

Several values calculated by Brunner and Diehl (1991a,b) using relations (4.27) and (4.29) are plotted in Figure 4.11. It shows that $U_b^{(c)}$ depends more on the chosen potential rather than the theoretical approach.

4.2.4.3 Abrupt potential. The following calculations have been proposed by Friedel (1959) for describing the basal-prismatic cross-slip in HCP metals. Since the potential is flat between two Peierls valleys, the critical bulge shape is an arc of circle of radius $R = \hat{T}/\tau b$, provided its maximum extension, y_c, is smaller than the distance between Peierls valleys, h (Figure 4.12).

The bulge energy has various contributions: (i) extraction of segment AB from the Peierls valley; (ii) bending of this segment; (iii) work done by the applied stress and

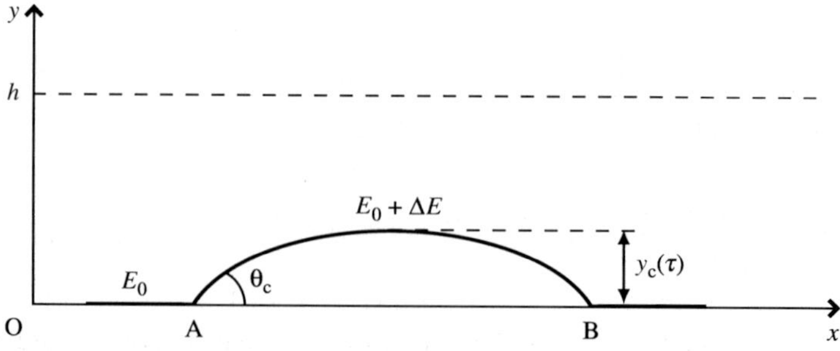

Figure 4.12. Critical bulge shape for an abrupt potential.

(iv) two half constrictions at A and B. It is accordingly:

$$U_b = 2R\sin\theta\Delta E + (E_0 + \Delta E)(2R\theta - 2R\sin\theta) - \tau b R^2(\theta - \sin\theta\cos\theta) + U_{constr}$$

After a Taylor expansion at (small θ), U_b goes through a maximum for the critical angle

$$\theta_c = \left(\frac{2\Delta E}{2\hat{T} - E_0}\right)^{1/2} \approx \left(\frac{2\Delta E}{E_0}\right)^{1/2}$$

which is the same θ_c as for an isolated kink (see §4.2.3.3). The condition of line tension equilibrium, $\hat{T} = (\hat{T} + \Delta E)\cos\theta_c$, yields the same result, assuming $\hat{T} \approx E_0$. The corresponding value of U_b is the activation energy:

$$U_b^{(c)} = \frac{2^{5/2}}{3}\frac{E_0^{1/2}\Delta E^{3/2}}{\tau b} + U_{constr} \tag{4.30a}$$

Escaig (1967) showed that U_{constr} depends on τ and that this term can be preponderant at high stresses. For BCC metals, it is

$$U_{constr} \approx 6.7\mu b^3\left(\frac{\tau_p - \tau}{\mu}\right)^2 \tag{4.30b}$$

The maximum extension of the bulge is

$$y_c = \frac{R\theta_c^2}{2} \approx \frac{\Delta E}{\tau b} \tag{4.31}$$

The calculation is valid as long as $y_c \leq h$, namely for

$$\tau > \tau_{min}^{(b)} = \frac{\Delta E}{hb} \tag{4.32}$$

The corresponding activation area is:

$$A_b = \frac{2^{5/2}}{3}\frac{E_0^{1/2}\Delta E^{3/2}}{\tau^2 b^2} = \frac{U_b^{(c)} - U_{constr}}{\tau b} \tag{4.33}$$

It is equal to the area swept by the dislocation between the segment AB and the critical curved configuration of extension y_c (Figure 4.12).

4.2.5 Energy of a critical kink-pair (low stress approximation: Coulomb elastic interaction)

This approximation is valid in the case of Figure 4.4(c), i.e. when the maximum of $U_{kp}(\tau, \Delta x)$ lies in region 2. The critical configuration is a pair of well separated kinks and the dislocation segments parallel to the Peierls valleys verify $y_0 \approx 0$ and $y_c \approx h$.

When the kinks are distant enough ($\Delta x \gg h$) their elastic interaction does not depend on their exact shape, i.e. the shape of the Peierls potential. According to Eshelby (1962), the elastic interaction potential is $-\mu h^2 b^2 / 8\pi\Delta x$. Detailed calculations can be found in Hirth and Lothe (1982, 1992) and Seeger (1984). The energy of two interacting kinks is accordingly:

$$U_{\mathrm{ik}}(\Delta x) = 2U_{\mathrm{k}} - \frac{\mu h^2 b^2}{8\pi\Delta x} - hb\Delta x\tau \tag{4.34}$$

This energy is at a maximum (saddle position) for

$$\Delta x = x_{\mathrm{c}} = \left(\frac{hb}{8\pi}\frac{\mu}{\tau}\right)^{1/2} \tag{4.35}$$

The activation energy is thus

$$U_{\mathrm{ik}}^{(\mathrm{c})} = 2U_{\mathrm{k}} - (hb)^{3/2}\left(\frac{\mu\tau}{2\pi}\right)^{1/2} \tag{4.36}$$

and the corresponding activation area is

$$A_{\mathrm{ik}} = h\left(\frac{hb}{8\pi}\frac{\mu}{\tau}\right)^{1/2} = hx_{\mathrm{c}} \tag{4.37}$$

4.2.6 Transition between high stress and low stress regimes

The preceding estimations of the activation parameters are different in the two stress regimes. The transition is now considered.

Starting from the lower stresses, the elastic interaction approximation becomes questionable when the critical kink separation x_{c} is too small, i.e. equivalent to the kink-width. For the abrupt potential, the kink-width is well defined and equal to $(h/\theta_{\mathrm{c}}) = h\sqrt{E_0/2\Delta E}$ (see §4.2.3.3). In the same approximation, x_{c} is given by relation (4.35).

Equating these two parameters yields the maximum stress allowed by the interacting kink approximation $\tau_{\max}^{(\mathrm{ik})} = (\mu b/4\pi h)(\Delta E/E_0)$ or, with $E_0 = 1/2\mu b^2$:

$$\tau_{\max}^{(\mathrm{ik})} = \frac{\Delta E}{2\pi hb} \tag{4.38}$$

Starting from the higher stresses, the critical bulge approximation becomes invalid when the leading part of the critical bulge approaches the bottom of the neighbouring valley and tends to straighten. For the abrupt potential, this corresponds to the stress $\tau_{\min}^{(\mathrm{b})}$ given by Eq. (4.32).

In the interval between $\tau_{\max}^{(\mathrm{ik})} = \Delta E/2\pi hb$ and $\tau_{\min}^{(\mathrm{b})} = \Delta E/hb$ neither of these two approximations is really valid.

Different approaches have been made to this transition domain. Seeger (1984) extrapolates the results obtained in the low stress and high stress approximations and concludes that a discontinuity should appear, making a hump on the activation area versus

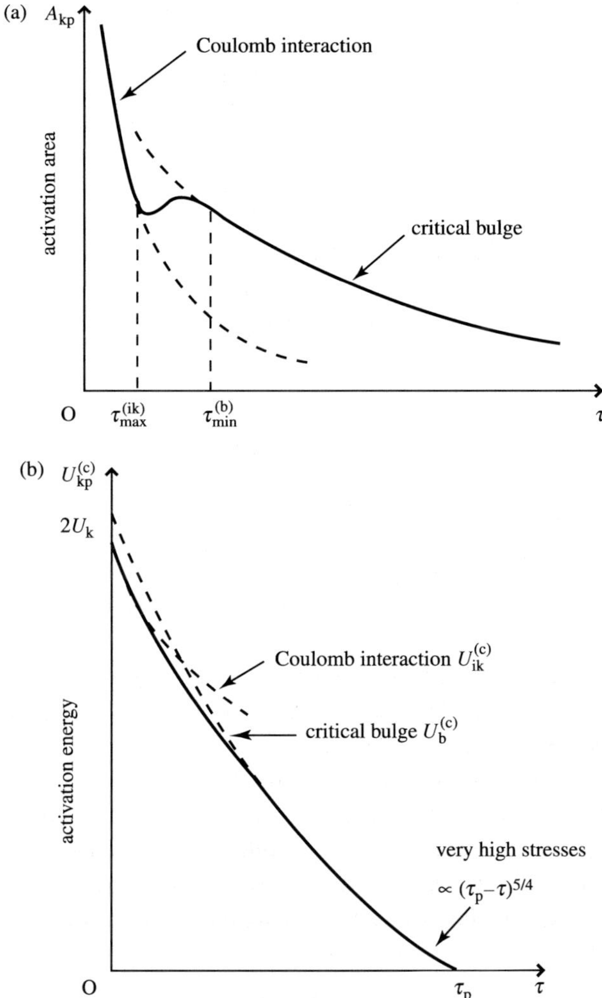

Figure 4.13. Transition between the low and high stress regimes of the kink-pair mechanism. (a) Discontinuous transition, from Seeger (1981). (b) and (c) Continuous transition, from Suzuki et al. (1991).

stress curve (Figure 4.13(a)). On the contrary, Suzuki et al. (1991) believe that the transition is smooth, as illustrated in Figure 4.13(b) and (c). It is not possible to decide which description is the most reliable.

A model has been proposed by Escaig (1968a) for the transition in the frame of the abrupt potential. The kinks are assumed to be well formed but very close to each other.

Different line tensions are attributed to the different dislocation segments (Figure 4.14):

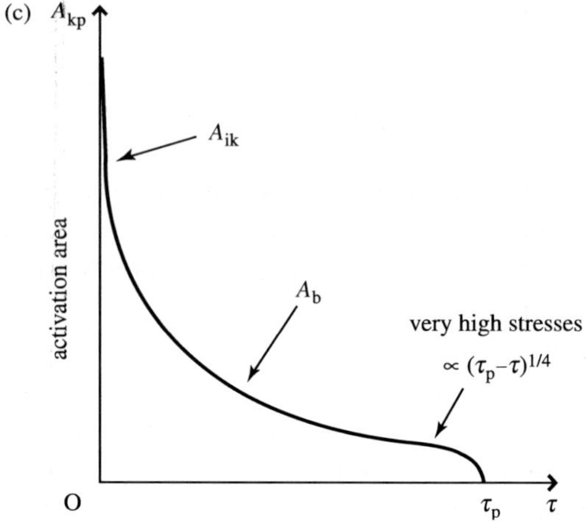

Figure 4.13. (*continued*)

- $E_0 \approx \mu b^2/2$ for the starting dislocation.
- $\hat{T}(\Delta x) = (\mu b^2/4\pi)\ln(\Delta x/b)$ for the kinks PP' and QQ', separated by the distance Δx. $\hat{T}(\Delta x)$ describes the elastic interaction between the three segments PP', $P'Q'$ and QQ'. Then, $\hat{T}(\Delta x)$ tends to E_0 for large separation distances Δx.
- $\hat{T}(\Delta x) - f(\Delta x)\Delta E$ for the segment PQ, with $f(0) = 0$ and $f(\Delta x) \to 1$ when Δx is large. The term $f(\Delta x)\Delta E$ describes the decrease in the line tension $\hat{T}(\Delta x)$ for segment $P'Q'$, due to its increasing dissociation at increasing separation distance Δx. Indeed, when $P'Q'$ is small, the two half constrictions in P' and Q' overlap.

At the saddle position, the line tension equilibrium conditions are different in P and P' (respectively, Q and Q'), which introduces different angles θ_c and θ'_c in keeping with the curvature of the kinks (Figure 4.14). The corresponding activation energy is then:

$$U_E = 2U_{\text{constr.}} + 2\mu b^3\left(\frac{\Delta E}{\mu b^2}\right)^{1/2}\left(\frac{1}{2\pi}\ln\frac{(\mu\Delta E)^{1/2}}{2\pi\tau b}\right)^{1/2}\left(1+\frac{\tau b^2}{4\Delta E}\right) \tag{4.39}$$

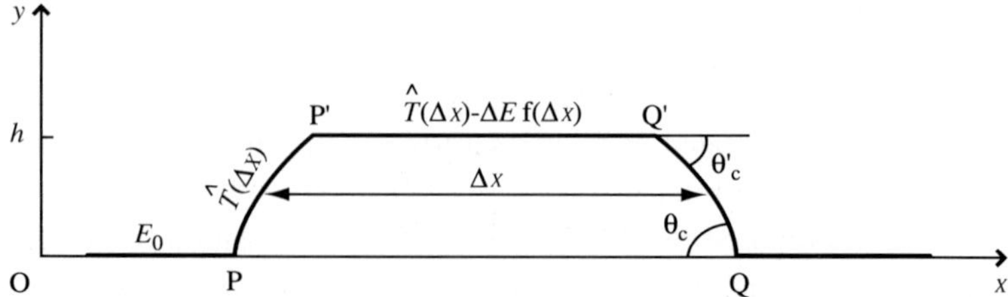

Figure 4.14. Critical kink-pair shape, in the variable line tension approximation (Escaig, 1968a,b). The energies per unit length are indicated for each segment.

and the activation area is

$$A_{\mathrm{E}} = \frac{b^2}{\pi}\left(\frac{\mu b^2}{\Delta E}\right)^{1/2}\left[\frac{\Delta E}{\tau b^2} - \pi\right] \tag{4.40}$$

At the highest stress level compatible with this approximation, $\tau_{\min}^{(b)}$, the distance $P'Q'$ is zero and the saddle position is an arc of circle in continuity with the high stress description using the abrupt potential (Section 4.2.4). The transition obtained at $\tau_{\min}^{(b)}$ in the stress dependence of the activation volume (Figure 4.15) is thus significant. The amplitude of this discontinuity is, however, small.

When Δx increases to large values (i.e. τ decreases to zero), since $\hat{T}$ cannot exceed E_0, the activation energy tends to twice the kink energy given by Eq. (4.16), with $h = b$. The elastic interaction approximation is, however, better justified when $\tau < \tau_{\max}^{(ik)}$.

Atomistic calculations carried out by Duesberry (1983) in a BCC lattice (Figure 4.16), where dislocations have a non-planar core structure, show that the kink-pair energy at zero stress, $U_{\mathrm{kp}}^{(0)}$, varies smoothly across the transition domain, as in Figure 4.4(b,c). As shown by Farenc et al. (1995), this indicates that the activation area $A_{\mathrm{kp}}(\tau)$ varies monotonically. These atomistic calculations are also in good agreement with the previous calculations because the critical bulge shapes shown in Figure 4.16(a) are close to those shown in Figure 4.10.

Finally, atomistic calculations show that straight dislocations under stress do not move much with respect to the bottom of Peierls valleys, which indicates that the Peierls potential has acute minima. Since the bulge shapes are different from arcs of circle, the potential is not constant between adjacent valleys unlike the abrupt potential. The antiparabolic potential may be the best approximation in that case.

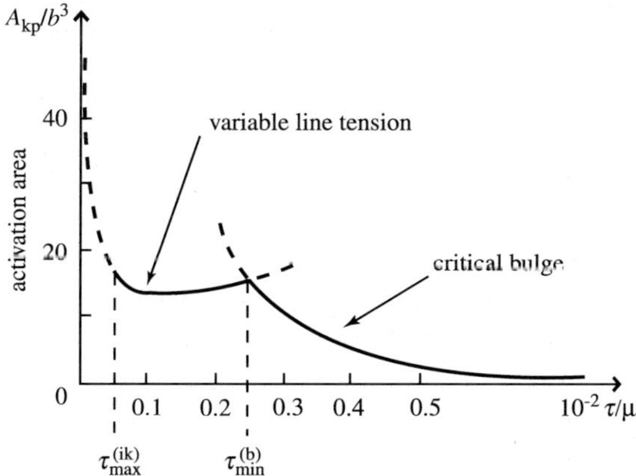

Figure 4.15. Discontinuous transition in the stress variation of the activation area, in the variable line tension approximation (Escaig, 1968a,b).

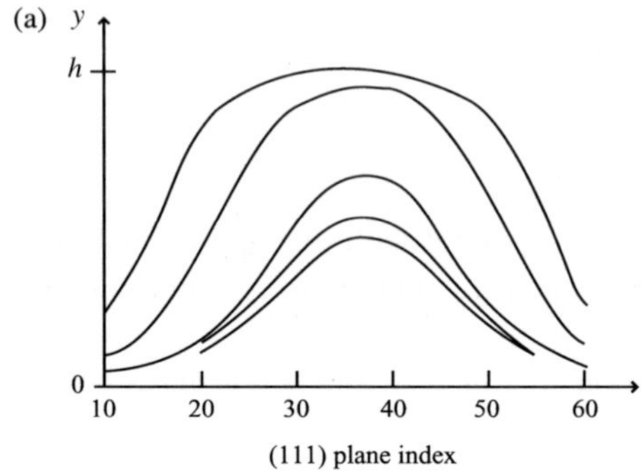

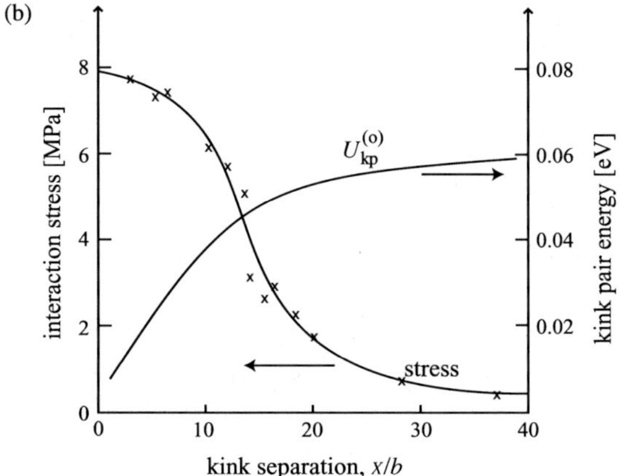

Figure 4.16. Results of atomistic calculations in BCC potassium, from Duesberry (1983): (a) critical bulge shapes for different stresses (a small extension in the y direction corresponds to a large stress); (b) kink pair energy and interaction stress as a function of kink separation (measured at $y = h/2$).

As a conclusion, there is no indication of a strong discontinuity for the activation parameters at the transition between the low stress and high stress regimes for the potentials considered so far. As shown in Figure 4.17, the activation energy is expected to decrease monotonically with increasing stress. In all cases the activation energy is $2U_k$ at low stresses and decreases to zero at $\tau = \tau_p$. Smooth potentials yield low τ_p values and moderate increase of stress at decreasing temperatures whereas potentials with more acute minima (i.e. the abrupt potential or the wall potential of Suzuki et al., 1995) yield high τ_p and a steep increase of stress. The various approximations valid in the different stress ranges are summarized in Table 4.1.

The situation may, however, be different if the Peierls potential exhibits intermediate minima between the main valleys (camel-hump potential, see Figure 4.5). In that case, Guyot and Dorn (1967) showed that the activation energy as a function of stress

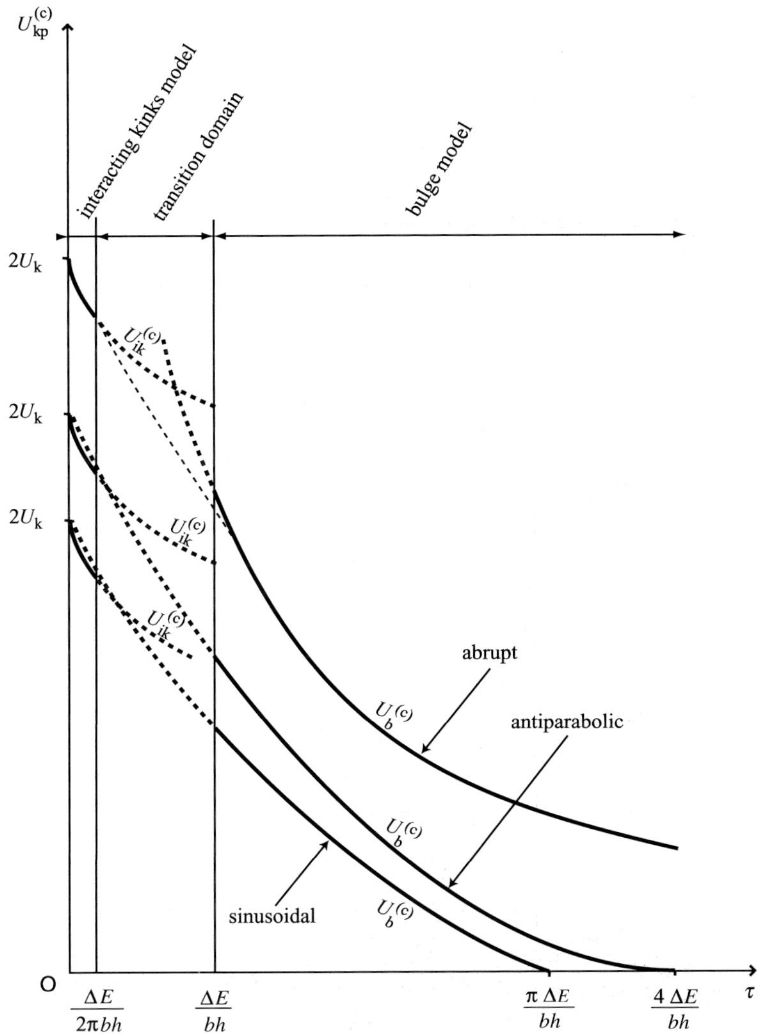

Figure 4.17. Activation energies versus stress for the kink-pair mechanism.

calculated in the bulge approximation exhibits inflexion points (Figure 4.18(a)) and that the activation area versus stress exhibits a peak (Figure 4.18(b)). The same result has been obtained by Koizumi et al. (1993), assuming that the threshold configuration is a pair of kinks of varying height. The lowering of the activation energy at low stresses corresponds to a crossing of Peierls hills in two steps separated by a metastable position in the intermediate valley. The more difficult first step is then rate controlling according to Koizumi et al. (1993).

Table 4.1. Summary of the approximations used to estimate the activation energy of the kink-pair mechanism in various stress intervals.

Stress range	Low $\tau_{max}^{(ik)} \approx \frac{\Delta E}{2\pi bh}$	transition $\tau_{min}^{(b)} \approx \frac{\Delta E}{hb}$	High	Very high	
Approximation	Elastic interaction between kinks	Varying line tension (Escaig)	Critical bulge		
Activation energy	$U_{ik}^{(c)}$ Eq. (4.36) (any potential)	U_E Eq. (4.39) (abrupt potential)	$U_b^{(c)}$ Eq. (4.27)	$U_b^{(c)}$ Eq. (4.24) (antiparabolic potential) $U_b^{(c)}$ Eq. (4.30a) (abrupt potential)	$U_b^{(c)}$ Eq. (4.29) (Eshelby potential) $U_b^{(c)} \approx U_{constr.}$ Eq. (4.30b) (abrupt potential)

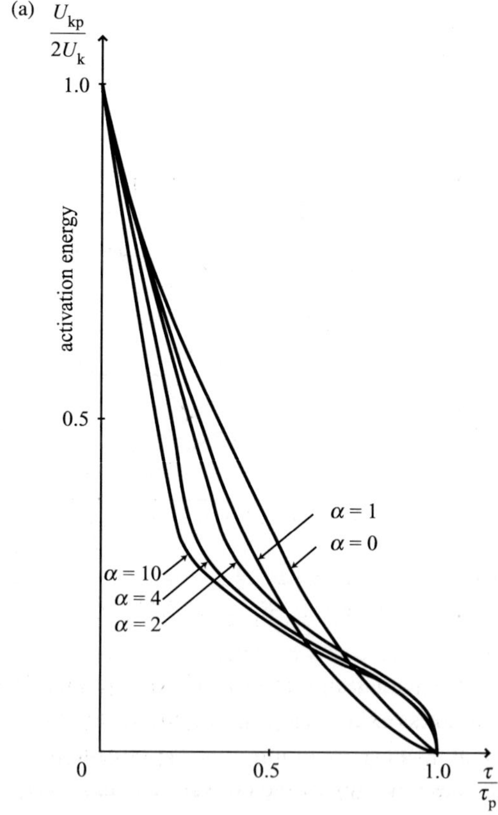

Figure 4.18. Activation parameters for a camel-hump potential. α determines the depth of the intermediate minimum (Figure 4.5). (a) Activation energy versus stress. (b) Activation area versus stress. After Guyot and Dorn (1967).

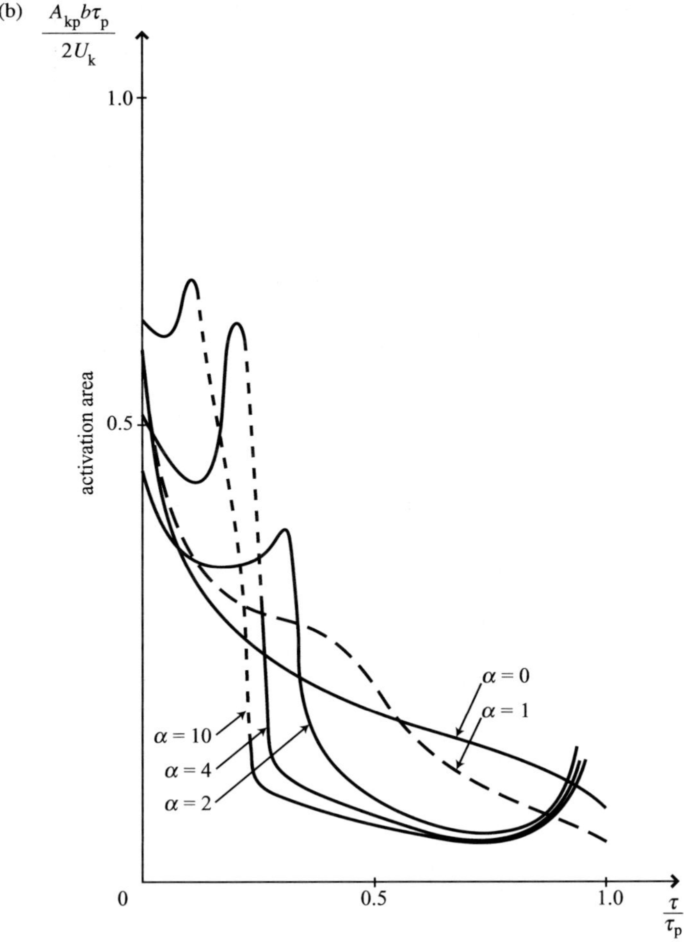

Figure 4.18. (*continued*)

4.2.7 Properties of dislocations gliding by the kink-pair mechanism

- Mobile dislocations are usually rectilinear along the direction of Peierls valleys. In the stress and temperature domain, where the kink-pair mechanism is thermally activated, the time for nucleating a kink-pair is indeed much larger than the time for moving them to the dislocation extremities. As a result the average kink density on the moving dislocation is very low. Kinks cannot coalesce at dislocation extremities because of their elastic repulsion. They pile up and yield an average radius of curvature $R = \hat{T}/\tau b$ (Figure 4.19).
- Since the dislocation velocity is described by Eq. (4.2), it is proportional to the length L of its rectilinear part.

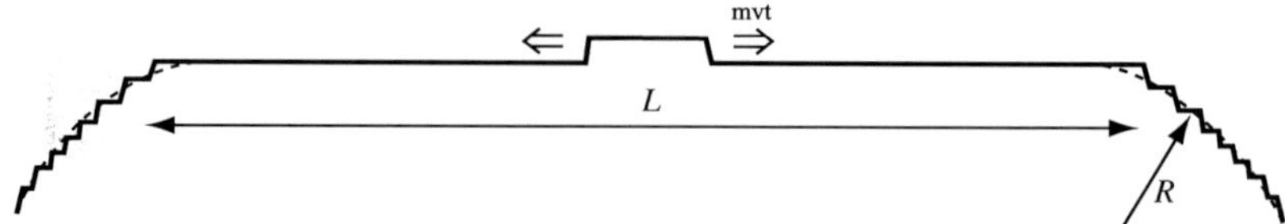

Figure 4.19. Schematic description of the nucleation and pile up of kinks on a dislocation gliding in a Peierls potential.

- The glide process, which is a series of jumps over the distance h, appears to be smooth and continuous at the scale of in situ observations in the transmission electron microscope. Since the activation volumes are small, dislocation velocities are not very sensitive to variations of the local stress and thus dislocation movements look fairly homogeneous and uniform.
- If dislocation motion is hindered by both extrinsic local obstacles and Peierls friction forces, the situation can be described as follows (Figure 4.20):
  Kinks pile up against the pinning points and the line tension, $\hat{T}$, exerts on them the net force, $F = 2\hat{T}\cos\theta$. Rectilinear dislocation segments go on moving by kink-pair nucleations, at a decreasing velocity as their length decreases for geometrical reasons. The angle θ decreases until it reaches the critical value θ_c at which F is large enough for unpinning. Therefore, pinning points slow down the dislocation

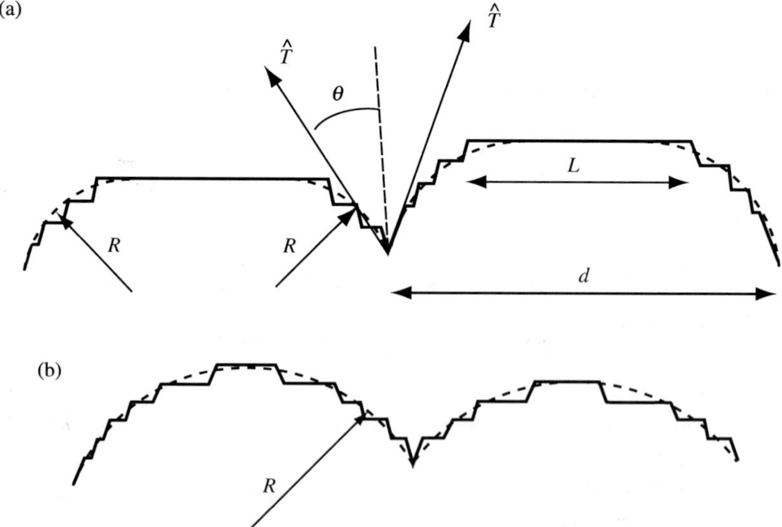

Figure 4.20. Combination of a Peierls friction force and pinning points. (a) High stress or large distance between pinning points and (b) low stress or small distance between pinning points.

but never stop it as long as L does not decrease to zero. On the contrary, if L tends to zero before θ and F reach the critical values for unpinning, the dislocation bends along two arcs of circle with radius R and stops (Figure 4.20(b)). It can only glide further if the stress is increased. The transition between these two behaviours takes place for:

$$R = \frac{d}{2\cos\theta_c}, \quad \text{i.e. } \tau = \frac{2\hat{T}}{bd}\cos\theta_c \tag{4.41}$$

where d is the distance between pinning points along the Peierls valleys.

4.3 THERMALLY ACTIVATED CORE TRANSFORMATIONS

As discussed in Section 4.1, dislocations may have different possible core configurations. Here dislocations are assumed to have at least two distinct core structures with different energies and different mobilities. The lower and higher energies correspond to stable and metastable configurations, respectively. These two states are separated by an energy threshold and the core transformation in both directions requires some amount of thermal activation.

As a rule, glissile–glissile transitions correspond to a change in slip plane of screw dislocations, namely to a cross-slip process (see Chapter 5) . Glissile–sessile transitions correspond to a locking process and sessile–glissile transitions correspond to the reverse unlocking process. Series of sessile–glissile transitions is also named the locking–unlocking mechanism.

Two types of transitions are considered in what follows, respectively, to a state of higher energy (stable–metastable transitions) and to a state of lower energy (metastable–stable transitions).

4.3.1 Transformation into a higher energy core structure

This core transformation can take place only with the help of applied stress. It corresponds to a sessile–glissile transition, namely to an unlocking process. The core structure of the sessile dislocation is extended out of the slip plane, whereas the core structure of the mobile curved segment is more compact and slightly extended in the slip plane. The formalism of Section 4.2.4, developed for the critical bulge in the abrupt potential (Figure 4.12), can be used for the present situation.

The configuration of maximum energy is shown in Figure 4.21 and the corresponding threshold energy is given by relation (4.30):

$$U_b^{(c)} = \frac{2^{5/2}}{3}\frac{E_0^{1/2}\Delta E^{3/2}}{\tau b} + U_{constr}$$

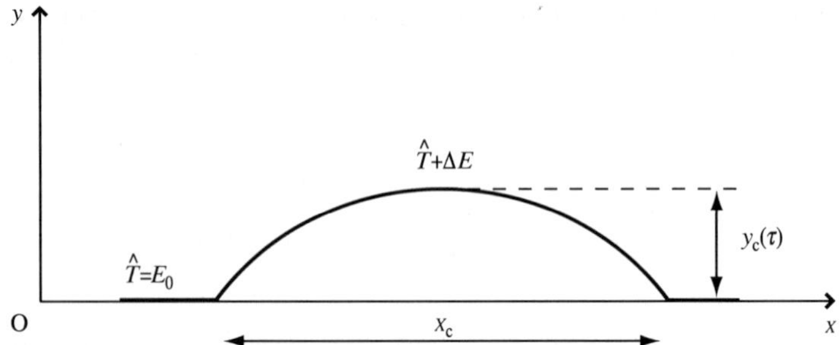

Figure 4.21. Critical bulge configuration for the sessile–glissile transformation (same as Figure 4.12, but y_c can be larger than h).

The corresponding activation area is Eq. (4.33):

$$A_b = \frac{2^{5/2}}{3} \frac{E_0^{1/2} \Delta E^{3/2}}{\tau^2 b^2}$$

However, the difference with respect to the calculations of Section 4.2.4 is that the bulge height, y_c, can be larger than h. ΔE is the difference between the core energies of glissile and sessile structures. In the original treatment of Friedel (1959), in order to describe the basal-prismatic cross-slip in the HCP structure, dislocations were assumed to glide over large distances along the prism plane after cross-slip. Although this glissile structure is metastable its lifetime is, however, generally short and the dislocation goes back to its initial sessile structure after some limited amount of glide (Section 4.3.3).

The probability of the unlocking process per unit time and unit dislocation length can be expressed in the same way as Relation 4.1:

$$P_{ul} = \nu_D \frac{b}{x_c^2} \exp\left(-\frac{U_b^{(c)}}{kT} \right) \tag{4.42}$$

where x_c is the width of the critical arc.

4.3.2 Transformation into a lower energy core structure

This core transformation has been discussed by Escaig (1968b, 1974) in the particular case of cross-slip in the FCC structure (Chapter 5), but the results can be easily transposed to other situations. When the core structure can be described by a dissociation in two partials, the threshold configuration is the intermediate configuration shown in Figure 4.22, where the dislocation is incompletely dissociated over a short length in the plane of lower energy.

The corresponding activation energy equals the energy of a constriction in the initial high energy plane, plus the energy of an incomplete constriction in the final low energy plane. Accordingly, it lies between one and two constriction energies. In a more general

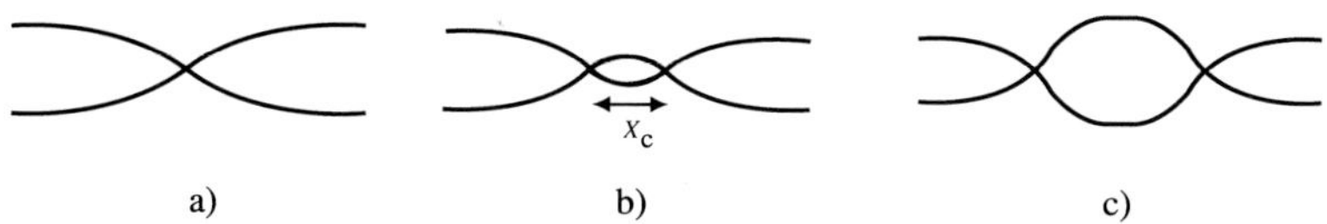

Figure 4.22. Transformation of a dissociated dislocation into a configuration of lower energy by dissociation in another plane. (a) Constriction in the plane of high energy; (b) threshold configuration and (c) lateral movement of the two constrictions. From Escaig (1968a, 1974).

description in terms of extended dislocation cores, constriction means transition between two different core structures along the same dislocation line.

Transformations into lower energy core structures may correspond to a cross-slip process, when the final structure is glissile (case of cross-slip in the FCC structure, see Chapter 5), or to a locking process, when the final structure is sessile. Therefore, the same relations apply for the probabilities of cross-slip and locking per unit time and unit length, respectively:

$$P_{\mathrm{cs}} = \nu_{\mathrm{D}} \frac{b}{x_{\mathrm{c}}^2} \exp\left(-\frac{U_{\mathrm{cs}}}{kT}\right) \tag{4.43a}$$

and

$$P_{\mathrm{l}} = \nu_{\mathrm{D}} \frac{b}{x_{\mathrm{c}}^2} \exp\left(-\frac{U_{\mathrm{l}}}{kT}\right) \tag{4.43b}$$

where x_c is the width of the critical configuration for both situations, x_c is of the order of magnitude of several times the core size. The activation energies U_{cs} and U_l are small. The activation area A_l is of the order of the size of the threshold configuration described schematically in Figure 4.22 and is therefore small (Escaig, 1968b). The activation parameters are described in more detail in Chapter 5.

4.3.3 Sessile–glissile transformations in series (locking–unlocking mechanism)

This process corresponds to the alternation of the two mechanisms described in Sections 4.3.1 and 4.3.2, namely to series of transformations between a low-energy-stable-sessile configuration, and a high-energy-metastable-glissile one. This is described schematically in Figure 4.23.

The low energy of the sessile configuration corresponds to the trough of the Peierls potential shown in Figure 4.23. Instead of moving through this potential by a kink-pair mechanism, the dislocation takes the metastable glissile configuration of higher energy, which allows fast glide at the speed v_g. After moving over the distance y_g, the dislocation recovers its initial configuration in another Peierls valley.

Vitek (1966) was the first to remark that the kink-pair mechanism in BCC metals corresponds to the transformation of a non-planar core into a planar core in the slip plane

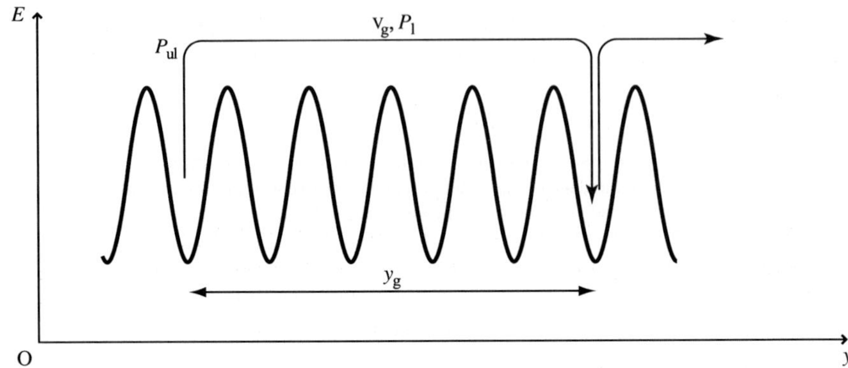

Figure 4.23. Energy-distance diagram for sessile–glissile transformations in series (locking–unlocking).

and that accordingly the reverse transformation can also be thermally activated. Duesberry and Hirsch (1968) have subsequently proposed this mechanism to explain prismatic slip in HCP metals. The detailed mechanism has been described by Couret and Caillard (1989). The outlines of their model are recalled below.

The average dislocation velocity $\bar{v}$ is determined by the probabilities per unit time and unit length of unlocking, P_{ul} (relation (4.42)), and locking, P_l (relation (4.43b)), whence:

$$\bar{v} = \bar{v}_g \frac{P_{ul}}{P_l} = P_{ul} L \bar{y}_g \tag{4.44}$$

with

$$\bar{y}_g = \frac{v_g}{P_l L} \propto \exp\left(\frac{U_l}{kT}\right) \tag{4.45}$$

v_g is the dislocation free glide velocity expected to be controlled by phonon scattering.

Combining relations (4.42), (4.43b) and (4.44) yields:

$$\bar{v} = v_g \left(\frac{x_c^{(l)}}{x_c^{(ul)}}\right)^2 \exp\left(-\frac{U_b^{(c)} - U_l}{kT}\right) \tag{4.46}$$

The global activation energy is $U_b^{(c)} - U_l$. It is positive because order-of-magnitude estimates yield $U_b^{(c)} > U_l$. The corresponding activation area is $A_b - A_l \approx A_b$ (see Eq. (4.33)).

The pre-exponential term of relation (4.46) has an unusual form because the frequencies of vibrations contained in P_{ul} and P_l disappear. This term should be smaller than for the kink-pair mechanism.

Increasing the stress increases P_{ul} but does not substantially increase P_l because A_l is considered to be small. The dislocation velocity thus increases with stress, in agreement with a strictly positive total activation area $A_{ul} - A_l$. In the same way, increasing the temperature increases P_{ul} more than P_l because $U_b^{(c)} > U_l$. This increases the dislocation velocity, in agreement with a strictly positive activation energy $U_b^{(c)} - U_l$. An experimental study of this mechanism is described in Chapter 6.

4.3.4 Transition between the locking–unlocking and the kink-pair mechanism

For the locking–unlocking mechanism, the average free-glide distance, $\bar{y}_g$, is given by Eq. (4.45).

When the temperature increases, $\bar{y}_g$ decreases and eventually reaches its minimum value, h. The locking process then becomes athermal and the dislocation moves by the kink-pair mechanism.

In order to fully understand this transition it is also necessary to examine the variation of the critical arc height, y_c. As the stress increases (and as the temperature usually decreases), y_c decreases according to Eq. (4.31):

$$y_c(\tau) = \frac{\Delta E}{\tau b}$$

The relative variations of $\bar{y}_g(T)$ and $y_c(\tau)$ determine the range of existence of different mechanisms (see Figure 4.24). Several situations can be anticipated:

- The locking–unlocking mechanism requires that the mean jump length $\bar{y}_g$ is larger than the distance between Peierls valleys, h. This happens below the temperature T_1 i.e. above the stress τ_1.
- The bulge mechanism requires the development of the critical arc of height y_c, i.e. $y_c < \bar{y}_g$. This happens above the stress $\tau_2 = \Delta E/\bar{y}_g b$. In the kink-pair regime, ($\tau_2 < \tau_1$) this reduces to $\tau_2 = \tau_{min}^{(b)} = \Delta E/hb$ (Eq. (4.32)).
  The above conditions are sufficient to explain the domains in the case of Figure 4.24(a)–(d).
- In Figure 4.24(a), where y_g remains equal to h, the two well-known modes of the Peierls mechanism are present. The stress τ_2 at which $y_c = \bar{y}_g$ coincides with $\tau_{min}^{(b)}$ at which $y_c = h$. For $\tau < \tau_2$, we have $y_c > \bar{y}_g$, which corresponds to either the elastic approximation or its transitions to the bulge approximation, whereas for $\tau > \tau_2$ we have $y_c < \bar{y}_g$ and the bulge mechanism takes over.
- In Figure 4.24(b), the mean jump length $\bar{y}_g$ becomes larger than h below the temperature T_1 (or above the stress τ_1), which results in a locking–unlocking mechanism. τ_1 is assumed to be larger than τ_2 so that $\tau_2 = \tau_{min}^{(b)}$.
- In Figure 4.24(c), τ_1 is assumed to be smaller than τ_2 so that $\tau_2 < \tau_{min}^{(b)}$. In addition to the kink-pair mechanism below τ_1 and to the locking–unlocking mechanism above

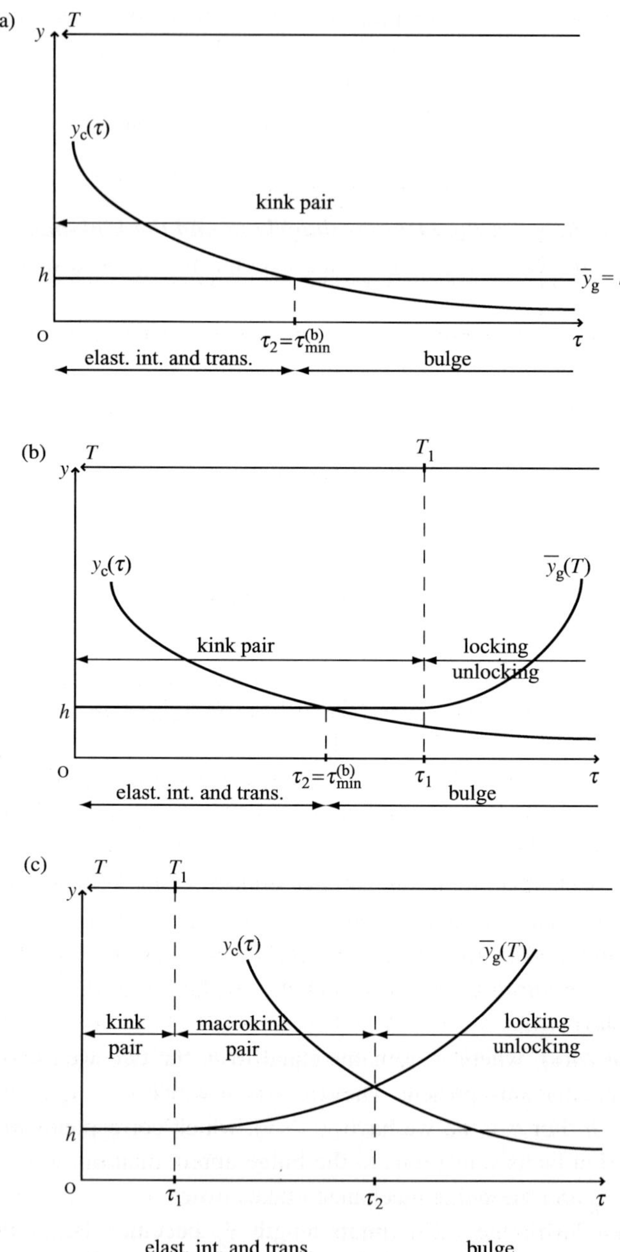

Figure 4.24. Schematic representation of the domains of existence of the different mechanisms involved in the motion of dislocations with non-planar cores. T_1 is the minimum temperature for the kink-pair (corresponding stress τ_1) and τ_2 is the minimum stress for the critical bulge configuration ($\tau_2 = \tau_{min}^{(b)}$ in the case of a kink-pair mechanism). (a) to (d) refer to different relative values of τ_1, τ_2, and $\tau_{min}^{(b)}$. See text.

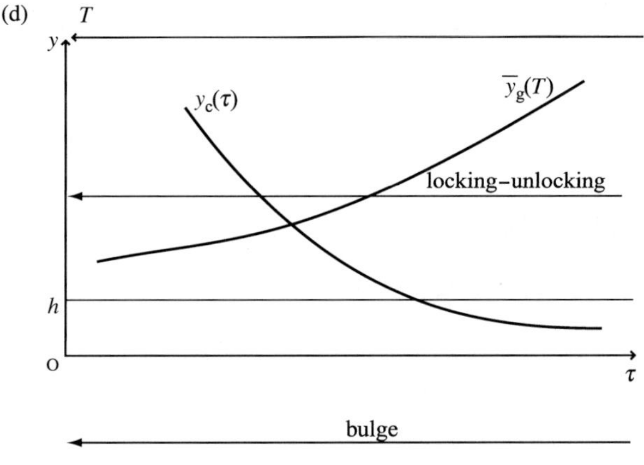

Figure 4.24. (*continued*)

τ_2, the thermally activated nucleation of macrokink pairs of height $\bar{y}_g$ takes place between τ_1 and τ_2. Locking and unlocking become coupled processes between τ_1 and τ_2.

– In Figure 4.24(d), $\bar{y}_g$ is always larger than h, so that locking–unlocking takes place in the whole stress and temperature range.

The stress dependence of the global activation energy and the corresponding threshold configurations are depicted in Figure 4.25, for the situation described in Figure 4.24(c). It is $U_{ik}^{(c)}$ for $\tau < \tau_1$ and $U_b^{(c)}$ for $\tau > \tau_2$. The anomalous variation between τ_1 and τ_2 corresponds to nucleations of macrokink pairs with varying height y_g.

The corresponding activation area versus stress is described schematically in Figure 4.26. It is A_b above τ_2 and A_{ik} below τ_1 and it follows different curves, noted as A_{Mkp}, corresponding to the nucleation of macrokink pairs with height varying from $2h$ to $4h$. This results in a hump on the A(τ) curve at the stress τ_2.

In the situations described in Figure 4.24(a) and (b), since $\tau_2 = \tau_{min}^{(b)}$, the activation energy does not exhibit any discontinuity but follows the dotted line below the hump in Figure 4.25. The activation area also varies monotonically.

The expected variations of yield stress and activation area are described schematically in Figure 4.27. Figure 4.27(a) has been deduced from Figure 4.25, assuming that the activation energy U is approximately proportional to temperature. Figure 4.27(b) is deduced from Figure 4.26.

The simplifying assumption $y_g = \bar{y}_g$ for all dislocations is, however, rather crude in some cases. In reality, the probability for a given glissile dislocation to jump over a

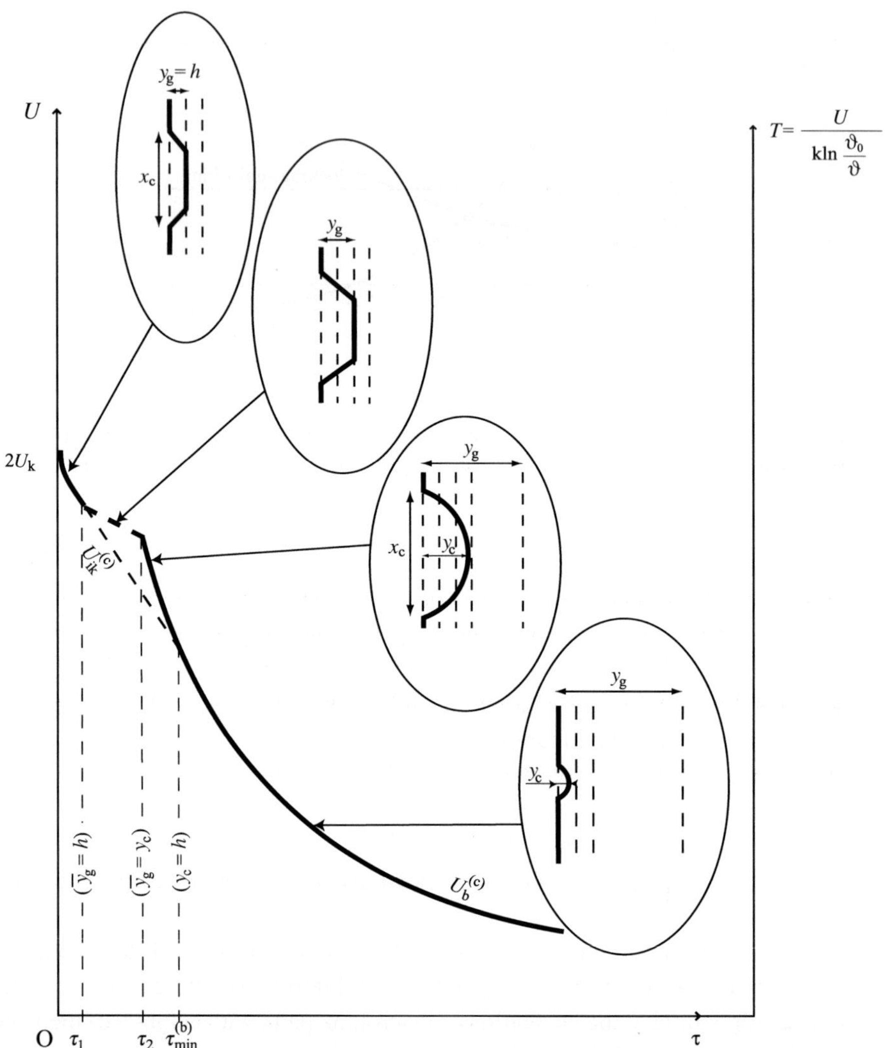

Figure 4.25. Activation energy versus stress corresponding to Relations 4.30 a and 4.36 showing the transition between the kink-pair and the locking–unlocking mechanism (in the case of Figure 4.24(c)). From Caillard and Couret (2000).

particular distance y_g is:

$$P(y_g) = \frac{h}{\bar{y}_g} \exp\left(-\frac{y_g}{\bar{y}_g} \right) \tag{4.47}$$

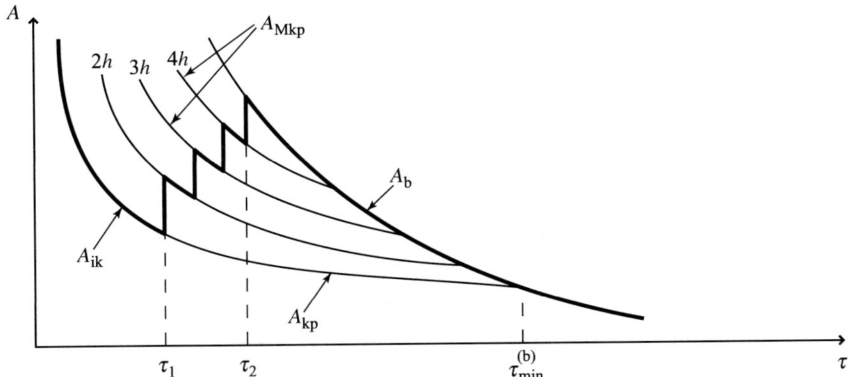

Figure 4.26. Schematic representation of the activation area versus stress showing the transition between kink-pair and locking–unlocking mechanisms (in the case of Figure 4.24(c)). From Farenc et al. (1995).

– For $\tau > \tau_{min}^{(b)}$, the critical configuration is a bulge of height $y_c < h$ whatever the jump length y_g. Then, the frequency of jumps over a given length y_g is $N(y_g) = P_{ul}LP(y_g)$, where P_{ul} is given by Eq. (4.42), and $P(y_g)$ is given by Eq. (4.47). The jump distances are expected to follow an exponential distribution and the relative frequency of jumps over a single interatomic distance h is expected to be $h/\bar{y}_g$.

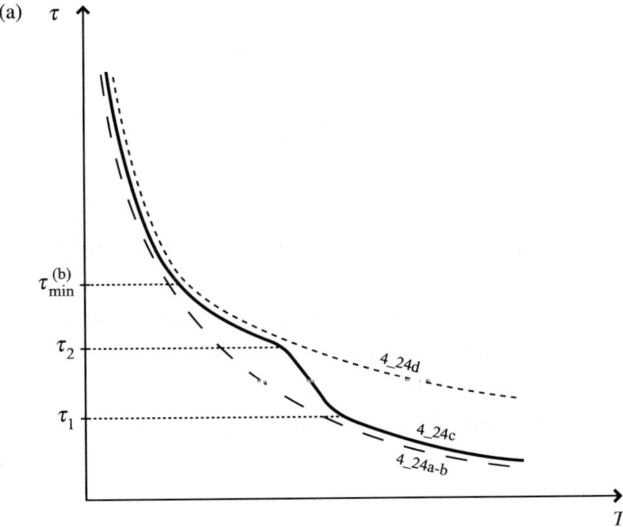

Figure 4.27. Expected variations of (a) stress as a function of temperature (or activation energy U) and (b and c) activation area A as a function of stress and temperature, for the various situations described in Figure 4.24 (a–d).

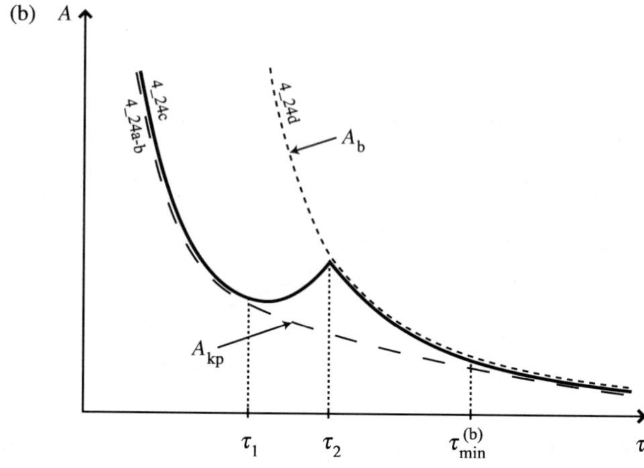

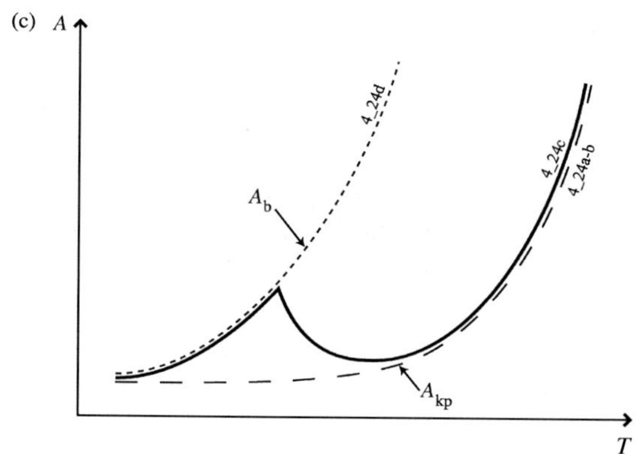

Figure 4.27. (*continued*)

- For $\tau_2 < \tau < \tau_{min}^{(b)}$, however, the situation is more complex because the critical configuration is a bulge for $y_g > y_c$, a kink-pair for $y_g = h$ and a pair of macrokinks for $h < y_g < y_c$. Then the frequency of jumps over a given length $y_g > y_c$ is $N(y_g) = P_{ul}LP(y_g)$ as above, whereas the frequency of jumps over a length h is $N(h) = P_{kp}LP(h)$, where P_{kp} is given by Eq. (4.1). Since $U_b^{(c)} > U_{kp}^{(c)}$ in this stress domain, the relative frequency of jumps over a single interatomic distance h is expected to be much higher than $h/\bar{y}_g$. This remark is important to explain several properties in Ni_3Al (see Section 10.1.4.3).

Taking these results into account would not, however, change the main conclusions concerning the transition between the Peierls and the locking–unlocking mechanisms discussed above.

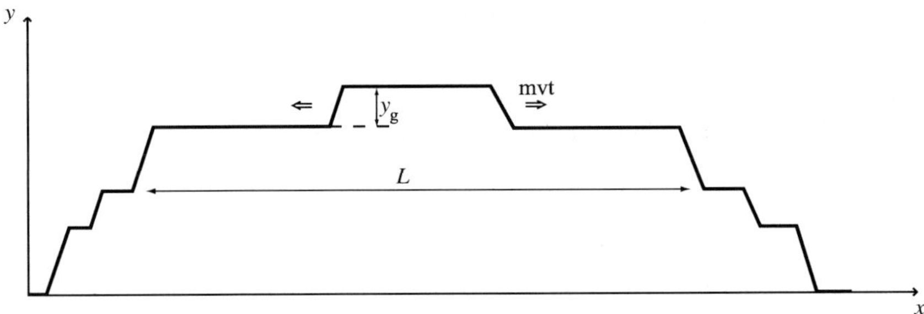

Figure 4.28. Dislocation gliding by a locking–unlocking mechanism. The macrokinks here can be several thousands h in height. Compare with Figure 4.19.

4.3.5 Properties of dislocations gliding by the locking–unlocking mechanism

This mechanism exhibits several common properties with the kink-pair mechanism, especially when the free-glide distance, y_g, is small, as well as important differences:

- Dislocations have a jerky movement with rectilinear locked configurations.
- Each jump forms a pair of macrokinks of height y_g, provided y_g is smaller than the dislocation length, L (Figure 4.28). Dislocation extremities are thus not smoothly curved, unlike Figure 4.19, but made of pile ups of macrokinks.
- The frequency of jumps over a given distance y_g (or the frequency of macrokink of height y_g) is proportional to $\exp(-y_g/\bar{y}_g)$, provided $y_g > y_c$. The average value of y_g is $\bar{y}_g = v_g/P_l L$ (Eq. 4.45).
- The frequency of a given waiting time t_l before unlocking is proportional to $\exp(-P_{ul} L t_l)$, provided the length of the jump that follows is larger than y_c. The average value of t_l is

$$\bar{t}_l = \frac{1}{P_{ul} L} \tag{4.48}$$

- The total activation energy is close to $U_b^{(c)}$ (relation (4.30)). When U_{constr} can be neglected, it varies as $1/\tau$.
- The total activation area is close to A_b (relation (4.33)). When U_{constr} can be neglected, it varies as $1/\tau^2$.
- The pre-exponential term is expected to be smaller than for the kink-pair mechanism.
- The probability of unlocking per unit time, $P_{ul} L$, is proportional to the dislocation length, L (relation (4.42)), but the dislocation velocity, v, is independent of L (relation (4.46)).

4.4 CONCLUSIONS

The various descriptions of the friction forces on dislocations gliding in metals appear to be fairly coherent. The general approach of Dorn and co-workers and the description by

Hirsch and Escaig in terms of dissociated dislocations yield similar expressions for dislocation velocities. The locking–unlocking mechanism describes how dislocations can jump along the glide plane over several Peierls hills in a row at sufficiently low temperatures. It degenerates into the classical kink-pair mechanism when the temperature increases. The transition between both mechanisms corresponds to discontinuities of the activation parameters as a function of stress or temperature.

The various expressions obtained for the dislocation glide velocities will be compared with experimental data in Chapter 6.

REFERENCES

Bacon, D.J. & Martin, J.W. (1981) *Phil. Mag. A*, **43**, 883.
Brunner, D. & Diehl, J. (1991a) *Phys. Stat. Sol (a)*, **124**, 455.
Brunner, D. & Diehl, J. (1991b) *Phys. Stat. Sol (a)*, **125**, 203.
Caillard, D. & Couret, A. (2000) *Mater. Sci. and Eng. A*, **322**, 108.
Celli, V., Kabler, M., Ninomiya, T. & Thomson, R. (1963) *Phys. Rev.*, **131**, 58.
Couret, A. & Caillard, D. (1989) *Phil. Mag. A*, **59**, 783.
Cserti, J., Kantha, M., Vitek, V. & Pope, D.P. (1992) *Mater. Sci. Eng. A*, **152**, 95.
Dorn, J.E. & Rajnak, S. (1964) *Trans. Met. Soc. AIME*, **230**, 1052.
Duesberry, M.S. (1983) *Acta Metall.*, **31**, 1759.
Duesberry, M.S. & Hirsch, P.B. (1968) in *Dislocation Dynamics*, Eds. Rosenfield, A.R., Hahn, G.T., Bement, A.L. & Jaffee R.I., Mc Graw-Hill Book Company, New York, p. 57.
Escaig, B. (1967) *J. Phys.*, **28**, 171.
Escaig, B. (1968a) *Phys. Stat. Sol.*, **28**, 463.
Escaig, B. (1968b) *J. Phys.*, **29**, 225.
Escaig, B. (1974) in *Journal de Physique*, 35, **C7**, 151.
Eshelby, J.D. (1962) *Proc. Roy. Soc. London*, **A266**, 222.
Farenc, S., Caillard, D. & Couret, A. (1995) *Acta Metall. Mater.*, **43**, 3669.
Friedel, J. (1959) *Intern. Stresses and Fatigue in Metals*, Elsevier P.C., Amsterdam, p. 220.
Friedel, J. (1964) *Dislocations*, Pergamon, Oxford, p. 31.
Guyot, P. & Dorn, J.E. (1967) *Can. J. Phys.*, **45**, 983.
Hirth, J.P. & Lothe, J. (1982) in *Theory of Dislocations*, 2nd Edition, Wiley-Interscience Publication, New York, (1992) 2nd reprint edition, Krieger Pub. Comp., Malabar, Florida.
Holzwarth, U. & Seeger, A. (1991) in *Strength of Metals and Alloys*, Eds. Brandon, D.G., Chaim, R. & Rosen A., Freund Publish. Company Ltd, London, p. 577.
Kocks, U.F., Argon, A.S. & Ashby, M.F. (1975) *Thermodynamics and Kinetics of Slip*, Pergamon, Oxford.
Koizumi, H., Kirchner, H.O.K. & Suzuki, T. (1993) *Acta Metall. Mater.*, **41**, 3483.
Mahapatra, R., Girshick, A., Pope, D.P. & Vitek, V. (1995) *Scripta Met. Mater.*, **33**, 1921.
Mori, T. & Kato, M. (1981) *Phil. Mag. A*, **43**, 1315.
Nabarro, F.R.N. (1947) *Proc. Phys. Soc.*, **59**, 256.
Peierls, R.E. (1940) *Proc. Phys. Soc.*, **52**, 34.
Régnier, P. & Dupouy, J.M. (1968) *Phys. Stat. Sol.*, **28**, 55.
Régnier, P. & Dupouy, J.M. (1970) *Phys. Stat. Sol.*, **39**, 79.

Seeger, A. (1981) *Z. Metallkde*, **72**, 369.
Seeger, A. (1984) in *Dislocations 1984*, Eds. Veyssière, P., Kubin, L. & Castaing J., CNRS, Paris, p. 141.
Simmons, J.P., Rao, S.I. & Dimiduk, D.M. (1997) *Phil. Mag. A*, **75**, 1299.
Suzuki, T., Takeuchi, S. & Yoshinaga, H. (1991) *Dislocation Dynamics and Plasticity*, Springer-Verlag, Berlin.
Suzuki, T., Koizumi, H. & Kirchner, H.O.K. (1995) *Phil. Mag. A*, **71**, 389.
Takeuchi, S. (1981) in *Interatomic Potentials and Crystalline Defects*, Ed. Lee, J.K., The Metals Society AIME, Warrendale, PA, p. 201.
Vitek, V. (1966) *Phys. Stat. Sol.*, **18**, 687.
Vitek, V. & Igarashi, M. (1991) *Phil. Mag. A*, **63**, 1059.
Werner, M. & Seeger, A. (1988) in *Strength of Metals and Alloys*, Eds. Kettunen, P.O., Lepistö, T.K. & Lehtonen M.E., Pergamon Press, Oxford, p. 173.

Chapter 5

Dislocation Cross-slip

Chapter 5
Dislocation Cross-slip

Cross-slip is a key process in crystalline plasticity. Jackson (1985) emphasized that it is the mechanism through which screw dislocations annihilate, thus forming low-energy structures in deformation processes such as work hardening, creep or fatigue. It is associated with the presence of stage III on the monotonic curves of FCC metals (see the review by Saada and Veyssière, 2002), with prismatic slip in HCP metals (Section 6.1), slip on non-close-packed planes in FCC metals (Section 6.2) and low temperature deformation of BCC crystals (Section 6.3). Cross-slip on various planes is claimed to lead to the formation of barriers opposing dislocation glide in $L1_2$, B2 and $L1_0$ intermetallic compounds (Chapter 10). It is also involved in several multiplication processes of Section 9.1.1. It can also soften a material by allowing screw dislocations to bypass obstacles of the primary glide plane as, for example, in dispersion hardened alloys as advocated by Humphreys and Hirsch (1970).

Let us note that two types of dislocation cross-slip have been introduced in the preceding chapter. One involves a core transformation towards a higher energy state (Section 4.3.1) and the other towards a lower energy state (Section 4.3.2).

The material in this chapter concerns the second situation and is organized as follows: Various models describing cross-slip are presented first, with emphasis on FCC metals. The available experimental observations about this mechanism are then reviewed and, finally, atomistic simulations of the process are exposed.

5.1. MODELLING CROSS-SLIP

For the past 45 years, different elementary mechanisms have been imagined which are presented now. Their relevance will be discussed in Section 5.4. Emphasis is given to the FCC structure in which dislocations lower their energy by dissociating along the close-packed glide plane into two Shockley partials that bind a stacking fault. Dissociations in other crystalline structures can be more complicated, involving more than two partials. Some examples of the latter are given by Escaig (1974).

5.1.1 Elementary mechanisms

Considering a dislocation on the primary glide plane, the local stresses may be more favourable for the dislocation to glide on another plane, thus inducing its cross-slip.

One can anticipate that a perfect screw dislocation cross-slips easily, as compared to a dissociated one.

5.1.1.1 The Fleischer model (1959). This model is illustrated in Figure 5.1. The primary and cross-slip planes are $(11\bar{1})$ and $(1\bar{1}\bar{1})$, respectively, and the Burgers vector of the screw is $a/2$ [101]. It dissociates into two Shockley dislocations as follows:

$a/2[101] \rightarrow a/6[112] + a/6[2\bar{1}1]$ on the primary plane

$a/2[101] \rightarrow a/6[1\bar{1}2] + a/6[211]$ on the cross-slip plane.

In the case of an obtuse dihedron, the author assumes that cross-slip takes place through the dissociation of the leading partial of the primary plane according to the scheme:

$a/6[2\bar{1}1] \rightarrow a/6[211] + a/3[0\bar{1}0]$

The leading partial in the cross-slip plane is a glissile Shockley, while the trailing one is a stair rod which is sessile.

In the case of an acute dihedron, which is energetically more favourable, the leading Shockley decomposes:

$a/6[2\bar{1}1] \rightarrow a/6[10\bar{1}] + a/6[1\bar{1}2]$

The second partial on the cross-slip plane, $a/6[1\bar{1}2]$, is a glissile Shockley. The trailing partial of the primary plane is attracted by the sessile partial and reacts with it according to

$a/6[112] + a/6[10\bar{1}] \rightarrow a/6[211]$

The resulting Shockley is glissile on the cross-slip plane. It is the second partial requested on this plane for the completion of cross-slip.

This model has been confirmed experimentally by Clarebrough and Forwood (1975) in a Cu–8 at.% Si alloy. It has also been claimed to play a role in the FCC → HCP martensitic transformation (Fujita and Veda, 1975), in deformation twin intersection in

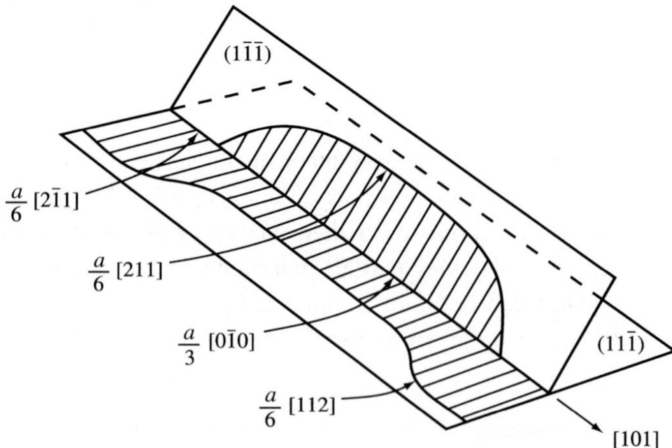

Figure 5.1. Dislocation cross-slip according to Fleischer (1959). The Burgers vectors of the partial dislocations are indicated. Case of an obtuse dihedron.

FCC alloys (Mori and Fujita, 1980; Coujou, 1981) and microtwin propagation in ordered structures such as CuAu (Pashley et al., 1969) and Ni_3V (Vanderschaeve, 1981).

However, energy considerations in the framework of anisotropic elasticity by Foreman (1955) and Stroh (1954) show that, as a rule, the generation of an additional stair rod is much more costly than recombining the screw. This confirms early considerations by Mott (1952) who predicted that the fault ribbon has to constrict at the intersection of the primary and cross-slip planes so as to avoid a too high joining energy for the two intersecting stacking faults.

The model of Fleischer is the only one in which the dislocation remains dissociated during cross-slip. All others assume either a pinching or a recombination of the dislocation. As an example, Hirsch (1962) considered sessile dissociated jogs along the screw as privileged sites for a deviation initiation. The activation energy to form the initial constriction is expected to be smaller at such spots. However, the jog dissociations postulated by the author cannot be assessed by any experimental evidence, as pointed out by Friedel (1964). Indeed, the jog is a localized core defect which cannot be investigated by TEM. Atomistic simulations may shed some light on the jog structure (see some recent attempts by Vegge and Jacobsen, 2002). On the other hand, Brown (2002) considers that Hirsch's approach accounts for several features of plasticity dominated by cross-slip.

5.1.1.2 The Washburn model (1965). The author proposes that cross-slip starts at an attractive junction reaction, as illustrated in Figure 5.2. The junction results from the intersection of a primary dislocation with a forest dislocation labelled M_1M_2 and F_1F_2, respectively, in Figure 5.2. The junction lies along the planes' intersection. All the possible junctions which can form have been listed by Whelan (1958) while interpreting TEM images of hexagonal dislocation networks. One of these junctions is split along the cross-slip plane. Under the line tension forces, this junction (OO′ in Figure 5.2) can

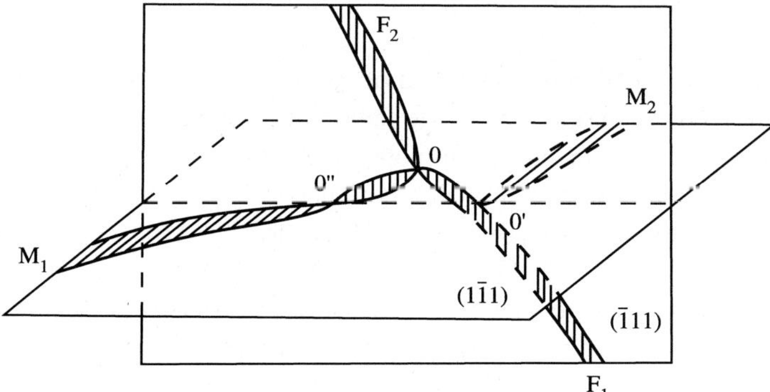

Figure 5.2. Dislocation cross-slip at a junction. $(1\bar{1}1)$ and $(\bar{1}11)$ are the primary and cross-slip planes, respectively. After Washburn (1965).

therefore glide in this plane, thus driving on its segment OM_1. Pulling the gliding dislocation along the cross-slip plane requires the nucleation of one constriction only in O''.

This mechanism has been observed to operate during TEM in situ straining of Cu foils at 300 K, hardened by alumina particles (Caillard and Martin, 1976). It is also likely to play a role in sub-boundary formation (Friedel, 1977). It can operate at relatively low temperatures, line tension forces compensating for the restricted thermal activation.

5.1.1.3 The Schoeck, Seeger, Wolf model. In the initial model by Schoeck and Seeger (1955), cross-slip implies the recombination of the screw, along some length. The recombined segment has to be long enough to become unstable in the cross-slip plane where it bulges under the shear stresses. This implies a high activation energy, particularly at low stresses. For example, in Cu (core splitting of about $5b$) this energy is close to 8 eV for stresses of the order of 10^{-4} μ.

Wolf (1960) improved the model as follows. After the initial recombination over a critical length, he considers the subsequent splitting of the recombined segment along the cross-slip plane. The process is sketched in Figure 5.3. In addition, to reach the proper stress level for recombination to occur, he postulates the existence of dislocation pile-ups to concentrate the stress.

This latter point does not seem to have been confirmed experimentally. The sophisticated observations of Mughrabi (1968) of dislocations pinned under load (through proper irradiation) in stage II Cu, revealed pile-ups. However, these were too short (15–20 dislocations) as compared with those postulated by Seeger et al. (1959) (20–100 dislocations). Moreover, the observed pile-ups were held up at Lomer–Cottrell locks which are not in screw orientation.

The experimental assessment of the model was undertaken considering that the stress τ_{III} corresponded to the onset of stage III. It was assumed to be the critical stress for cross-slip. The activation energy for cross-slip in this model is stress dependent and includes two parameters, namely the number n of dislocations per pile-up and the stacking fault energy γ.

To reproduce the experimental variation of τ_{III} with temperature and strain rate, large values of n are required and a rather large fault energy as compared to the one deduced from core splitting measurements. Let us note that the theory predicts a linear variation of $\ln \tau_{III}$ as a function of temperature which is rather well verified. Finally, the activation volume is expected to vary as the reciprocal of the stress.

5.1.1.4 The Friedel–Escaig cross-slip mechanism. Friedel (1957) suggested that cross-slip could occur more easily and more frequently than predicted by the preceding models. He considers that constrictions pre-exist along dissociated screw dislocations. These are present at jogs that result from dislocation intersections, or nodes at junction reactions. They act as nuclei for cross-slip. However, deviation will

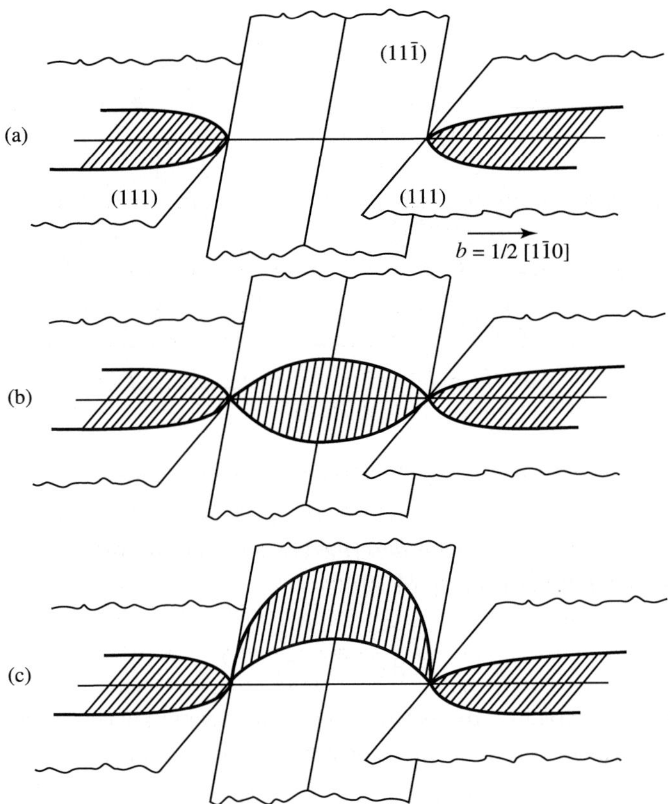

Figure 5.3. Schematics of dislocation cross-slip according to Wolf, Schoeck and Seeger.

also take place provided that the stresses constrict the two partial dislocations along the primary plane and move them away from each other along the cross-slip plane. This is illustrated in Figure 5.4. Since this description reproduces several experimental aspects of cross-slip (see Section 5.2) it will be presented in more detail.

Constriction formation being the key point of the cross-slip process, an estimation of the constriction energy is presented first. This is a prerequisite for Escaig's model which comes next.

5.1.2 Constriction energyxs

The treatment of Stroh (1954) is presented here with reference to equations introduced in Chapter 4. A constriction on a $1/2\langle 110\rangle$ screw dislocation dissociated into two Shockley partials is represented schematically in Figure 5.5. Each dislocation segment is subjected to forces that are assumed to be parallel to the y-axis. This approximation is valid as long as the constriction size along the x-axis is much larger than the dissociation width, d_0.

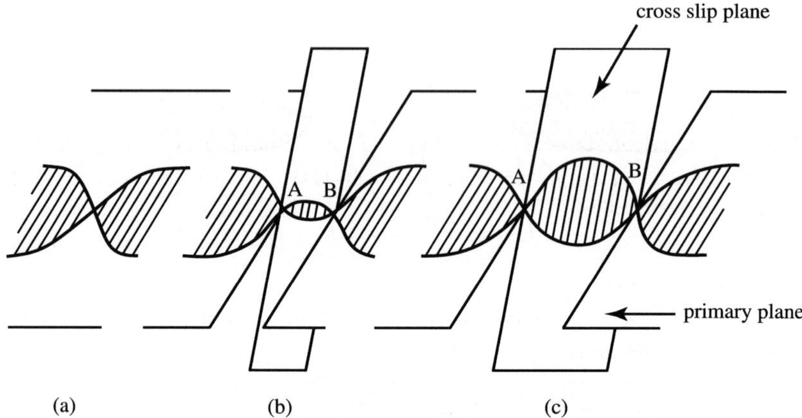

Figure 5.4. Schematics of the successive stages of cross-slip according to Friedel and Escaig.

The equilibrium of each segment is described by an equation similar to that already used to compute kink configurations (Eq. (4.10)):

$$\hat{T}\frac{d^2y}{dx^2} = -\frac{A}{2y} + \gamma \tag{5.1}$$

where $\hat{T}$ is the line tension, d^2y/dx^2 is the dislocation curvature and γ is the surface energy of the fault. $-A/2y$ is the elastic interaction between the two partials, with:

$$A = \frac{\mu b_p^2}{2\pi}\left(\cos^2 30° - \frac{\cos^2 60°}{1-\upsilon}\right) \approx \frac{\mu b^2}{16\pi}$$

In the absence of any external stress, the dissociation width is related to the fault energy by the relation $d_0 = A/\gamma$.

After integration, Eq. (5.1) becomes:

$$\left(\frac{dy}{dx}\right)^2 = -\frac{A}{\hat{T}}\ln y + \frac{2\gamma}{\hat{T}}y + constant$$

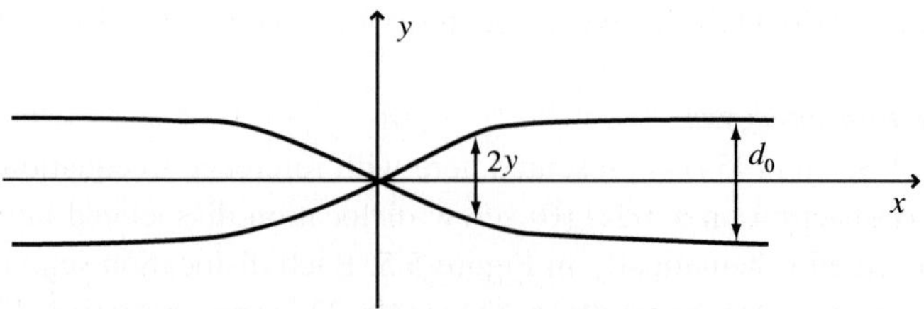

Figure 5.5. Constriction on a dissociated 1/2 ⟨110⟩ screw dislocation. Stroh (1954).

Since $dy/dx = 0$ for $y = d_0/2$ we obtain:

$$\left(\frac{dy}{dx}\right)^2 = -\frac{A}{\hat{T}}\ln\frac{2y}{d_0} + \frac{\gamma}{\hat{T}}(2y - d_0) \tag{5.2}$$

This differential equation describes the shape of one Shockley dislocation in Figure 5.5.

The constriction energy involves several components related successively to the elastic interaction energy between the two partials, their line energy and the fault energy. They are now calculated.

Elastic interaction energy:

The increase of the elastic energy per unit length between two parallel segments when their distance decreases from d_0 to $2y$ is:

$$\int_{2y}^{d_0} \frac{A}{y}\,dy = A\ln\frac{d_0}{2y} \tag{5.3}$$

The elastic interaction energy is then obtained by integration along the x direction:

$$U_1 = A\int_{-\infty}^{+\infty} \ln\frac{d_0}{2y}\,dx$$

Line energy:

For a segment of length dx, the increase in dislocation length as the constriction is created is proportional to

$$\left(\sqrt{1 + \left(\frac{dy}{dx}\right)^2} - 1\right) \approx \frac{1}{2}\left(\frac{dy}{dx}\right)^2$$

After integration along the x direction, it yields, for the two partials:

$$U_2 = E\int_{-\infty}^{+\infty} \left(\frac{dy}{dx}\right)^2 dx$$

where E is the line energy of Shockley partials.

Fault energy:

Since the surface of the fault decreases, the total fault energy decreases by the quantity

$$U_3 = -\int_{-\infty}^{+\infty} \gamma(d_0 - 2y)\,dx$$

The constriction energy is accordingly:

$$U_{\text{constr}} = U_1 + U_2 + U_3 = \int_{-\infty}^{+\infty}\left[A\ln\frac{d_0}{2y} - \gamma(d_0 - 2y) + E\left(\frac{dy}{dx}\right)^2\right]dx$$

After a change of variable it becomes:

$$U_{\text{constr}} = 2\int_{b/2}^{d_0/2}\left[A\ln\frac{d_0}{2y} - \gamma(d_0 - 2y) + E\left(\frac{dy}{dx}\right)^2\right]\frac{dx}{dy}dy \tag{5.4}$$

The cut-off radius, i.e. the minimum value of y, is assumed to be $b/2$.

Using Eq. (5.2) it becomes:

$$U_{\text{constr}} = 2(E+\hat{T})\int_{b/2}^{d_0/2}\frac{dy}{dx}dy, \text{ or}$$

$$U_{\text{constr}} = 2(E+\hat{T})\int_{b/2}^{d_0/2}\left(-\frac{A}{\hat{T}}\ln\frac{2y}{d_0} + \frac{\gamma}{\hat{T}}(2y - d_0)\right)^{1/2}dy \tag{5.5a}$$

Using $d_0 = A/\gamma$ it reduces to:

$$U_{\text{constr}} = 2(A\hat{T})^{1/2}\left(1+\frac{E}{\hat{T}}\right)\int_{b/2}^{d_0/2}\left(-\ln\frac{2y}{d_0} + \frac{2y}{d_0} - 1\right)^{1/2}dy \tag{5.5b}$$

Orders of magnitude estimated by Stroh (1954) are of several electron volts in the absence of stress. They will be further discussed in Section 5.1.3.4.

5.1.3 *Escaig's description of cross-slip (1968)*

A cross-slipping dislocation is described schematically in Figure 5.6. Escaig (1968) has estimated its energy.

5.1.3.1 *The activation energy for cross-slip.* The equilibrium dissociation widths in the primary and cross-slip planes are d' and d, respectively. They are a priori different because they depend on the components of the applied stress which tend to constrict or to widen the stacking fault ribbons in either plane (τ'_d and τ_d, respectively).

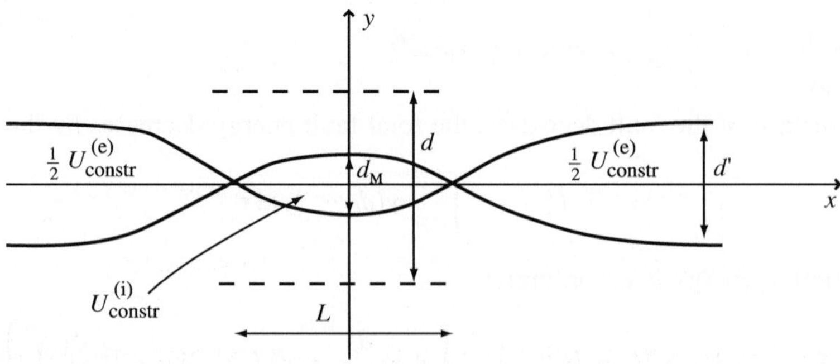

Figure 5.6. Definition of constriction energies for a cross-slipping screw dislocation. After Escaig (1968).

The equilibrium dissociation widths under stress can be expressed as $d = A/\gamma_e$ and $d' = A/\gamma'_e$, where γ_e and γ'_e are effective stacking fault energies, respectively, $\gamma_e = \gamma - \tau_d b/(2\sqrt{3})$ and $\gamma'_e = \gamma + \tau'_d b/(2\sqrt{3})$.

For a finite distance, L, between constrictions, the maximum dissociation width in the cross-slip plane is d_M. d_M is smaller than d but it tends to d as L tends to infinity, i.e. when cross-slip is completed.

Figure 5.6 shows that the two constrictions are different from the one described in Section 5.1.2 which is symmetrical with all segments lying in the same plane. Therefore, the constriction energies include $U^{(e)}_{\text{constr}}$ and $U^{(i)}_{\text{constr}}$ which correspond to two half constrictions on the primary and the cross-slip planes, external and internal, respectively.

Under these conditions, the energy of the cross-slipping dislocation can be decomposed into several parts, following the successive steps of Figure 5.7:

- the external constriction energy, $U^{(e)}_{\text{constr}}$ (Figure 5.7a). It is given by Eqs. (5.5a) or (5.5b), where d_0 is replaced by d' and γ is replaced by γ_e.
- the energy to recombine a dislocation segment of length L dissociated in the primary plane (dissociation width d') and to dissociate it in the cross-slip plane (dissociation width $d_M < d$) (Figure 5.7(b)).
- the internal constriction energy $U^{(i)}_{\text{constr}}$ (Figure 5.7(c)).
- the energy gained by the bowing out of the segment L in the cross-slip plane. This term will not be estimated because Escaig showed that it is negligible in all cases.

These various components are now estimated.

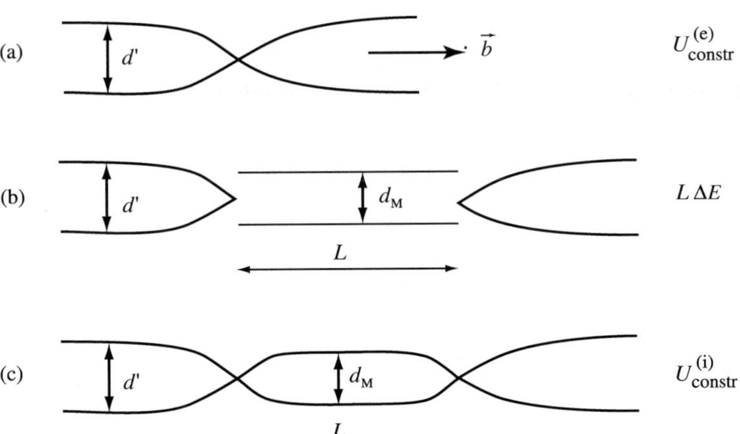

Figure 5.7. The successive steps of Escaig's cross-slip. The energy corresponding to each step is indicated (see text).

Internal constriction energy:
It is given by Eq. (5.5a) where d_0 is replaced by d_M and γ by $\gamma_e = A/d$:

$$U^{(i)}_{constr} = 2(A\hat{T})^{1/2}\left(1 + \frac{E}{\hat{T}}\right)\int_{b/2}^{d_M/2}\left(-\ln\frac{2y}{d_M} + \frac{d_M}{d}\left(\frac{2y}{d_M} - 1\right)\right)^{1/2} dy \tag{5.6}$$

Variation of dissociation energy:
The difference in dissociation energy per unit length, ΔE, has two components: the variation of the elastic interaction energy, $A \ln d'/d_M$ (cf. Eq. (5.3)), and the variation of the total fault energy plus the work done by the stress, $\gamma_e d_M - \gamma'_e d'$.

This yields:

$$\Delta E = A\left(-\ln\frac{d_M}{d} + \frac{d_M}{d} - 1 + \ln\frac{d'}{d}\right) \tag{5.7}$$

The length of the cross-slipped segment is

$$L = 2\int_{b/2}^{d_M/2}\frac{dx}{dy}dy$$

Using Eq. (5.2), where d_0 is replaced by d_M and γ by $\gamma_e = A/d$ it becomes:

$$L = 2\left(\frac{\hat{T}}{A}\right)^{1/2}\int_{b/2}^{d_M/2}\left[-\ln\frac{2y}{d_M} + \frac{d_M}{d}\left(\frac{2y}{d_M} - 1\right)^{-1/2}\right]dy \tag{5.8}$$

The total variation of dissociation energy is $L\Delta E$.
Total energy of the cross-slipping dislocation:
This energy is

$$U_{cs} = U^{(i)}_{constr} + L\Delta E + U^{(e)}_{constr} \tag{5.9}$$

Considering that $U^{(e)}_{constr}$ is a constant, the saddle configuration is obtained for $dU_{cs} = 0$, namely

$$dU^{(i)}_{constr} + Ld(\Delta E) + \Delta E d(L) = 0 \tag{5.10}$$

Setting

$$f(x) = -\ln\frac{x}{d} + \frac{x}{d} - 1 + \ln\frac{d'}{d} \tag{5.11}$$

the different terms involved in the total energy can be written as:

$$U^{(i)}_{constr} = 2(A\hat{T})^{1/2}\left(1 + \frac{E}{\hat{T}}\right)\int_{b/2}^{d_M/2}[f(2y) - f(d_M)]^{1/2}dy \tag{5.12}$$

$$L = 2\left(\frac{\hat{T}}{A}\right)^{1/2}\int_{b/2}^{d_M/2}[f(2y) - f(d_M)]^{-1/2}dy \tag{5.13}$$

and

$$\Delta E = Af(d_{\mathrm{M}}) \tag{5.14}$$

Combining Eqs. (5.11–5.14), yields:

$$\mathrm{d}U^{(\mathrm{i})}_{\mathrm{constr}} = \frac{\partial U^{(\mathrm{i})}_{\mathrm{constr}}}{\partial \Delta E}\mathrm{d}(\Delta E) = \frac{1}{A}\frac{\partial U^{(\mathrm{i})}_{\mathrm{constr}}}{\partial f(d_{\mathrm{M}})}\mathrm{d}(\Delta E) = -\frac{1}{2}\left(1 + \frac{E}{\hat{T}}\right)L\mathrm{d}(\Delta E)$$

Then, Eq. (5.10) reduces to

$$\frac{1}{2}\left(1 - \frac{E}{\hat{T}}\right)L\mathrm{d}(\Delta E) + \Delta E\mathrm{d}(L) = 0$$

or, with $E/\hat{T} \sim 0.7$

$$0.15L\mathrm{d}(\Delta E) + \Delta E\mathrm{d}(L) = 0 \tag{5.15a}$$

where $\mathrm{d}(\Delta E) < 0$, ΔE and $\mathrm{d}(L) > 0$.
Escaig defines a negative function δ such that $L\mathrm{d}(\Delta E) = \delta A\mathrm{d}(L)$

This yields

$$0.15\delta A + \Delta E = 0 \tag{5.15b}$$

This function is important only when $E \neq \hat{T}$, otherwise Eq. (5.15a) reduces to $\Delta E = 0$. Computed values of δ are between 0 and -1 and are small at low stresses ($d \sim d'$) (cf. Table 5.1).

Using Eqs. (5.11), (5.14) and (15.15b) allows one to write:

$$\ln\frac{d'}{d} = -\ln\frac{d_{\mathrm{M}}}{d} + \frac{d_{\mathrm{M}}}{d} - 1 + 0.15\delta \tag{5.16}$$

This equation yields values of d_{M} at the threshold position, as a function of the stress-dependent ratio d/d'.

At zero stress ($d = d'$) and for $E = \hat{T}$ (Eq. (5.16) independent of δ) the threshold position would be attained for $d_{\mathrm{M}} = d$, namely when the distance between constrictions, L, is infinite. For the more realistic case where $E \neq \hat{T}$, Escaig obtains the threshold position at $L \approx 10d$. Under an applied stress, L is always of the order of d′. Figure 5.8 shows the variation of $L/(d' \ln(d'/b))$, which is close to

$$\frac{L}{2d'(\hat{T}/A)^{1/2}}$$

Table 5.1. Computed values of δ (from Escaig, 1968)

d/d'	1	1.001	1.01	1.03	1.11	1.28	1.71	2.53	3.57	6.73
δ	−0.008	0.028	−0.087	−0.167	−0.311	−0.479	−0.663	−0.794	−0.861	−0.930

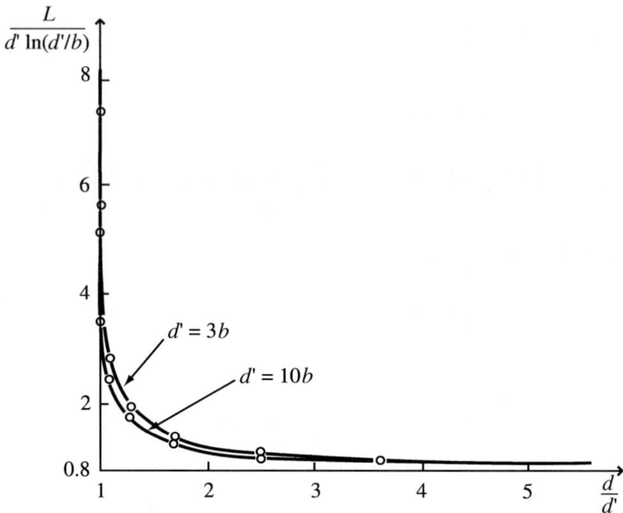

Figure 5.8. Critical distance L between constrictions as a function of stress expressed via d/d'. Two values of dissociation width d' are considered. After Escaig (1968).

and which is of the order of L/d' to $L/3d'$. Since the two curves for $d' = 3b$ and $d' = 10b$ are close to each other, the critical constriction separation distance L for a given stress (constant value of d/d') is almost proportional to d'.

With a pre-existing constriction, the corresponding activation energy $U_{cs} - U_{constr}^{(e)}$ is given by Eqs. (5.9) and (5.11–5.14), where d_M is given by Eq. (5.16). It is proportional to $d(A\hat{T})^{1/2}$. It is plotted on Figure 5.9 as a function of the stress-dependent ratio d/d'. Using a line tension:

$$\hat{T} = (\mu b_p^2/4\pi)\ln(d/b_p) = (\mu b^2/12\pi)\ln(d\sqrt{3}/b)$$

we obtain:

$$d(A\hat{T})^{1/2} = (\mu b^3/8\sqrt{3})(d/b)[\ln(d\sqrt{3}/b)]^{1/2}$$

This activation energy must be complemented by the external constriction energy, except if cross-slip takes place at a pre-existing constriction. This situation is actually the most likely one, because of a lower activation energy.

This activation energy can be estimated in the case of Cu using measurements of the width of splitting of screw dislocations by weak beam electron microscopy. Stobbs and Sworn (1971) provided a value of 1.8 ± 0.6 nm (or $d \approx 7b$). The corresponding energy is plotted on Figure 5.12 as a function of stress.

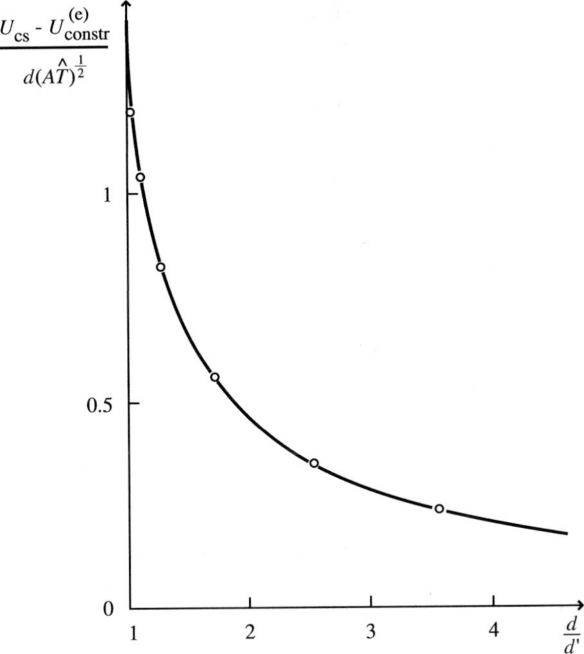

Figure 5.9. Cross-slip activation energy as a function of stress expressed via d/d'. After Escaig (1968).

5.1.3.2 The activation volume. A constriction is assumed to pre-exist on the cross-slipping dislocation.

Before the saddle position is reached, the work done by the stresses τ'_d and τ_d is proportional to the corresponding areas swept in the primary and cross-slip planes, respectively. Since the area swept in the cross-slip plane is between $Ld_M/2$ and Ld_M (see Figure 5.10) it has been taken as being equal to $2Ld_M/3$. Consequently, this work is:

$$W = \left(\tau'_d L d_M + \frac{2}{3}\tau_d L d_M\right)\frac{b_p}{2}$$

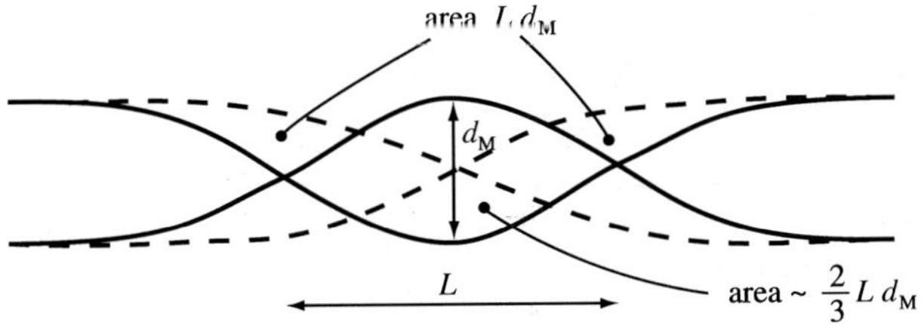

Figure 5.10. Estimation of the activation area for cross-slip.

where $b_p/2$ is the component of the Burgers vector of Shockley partials in a direction perpendicular to the dislocation line, sensitive to τ_d and τ'_d.

With $\tau_{eq} = \tau'_d + (2/3)\tau_d$, $b_p = b/\sqrt{3}$, and τ being the stress resolved in the primary system, the activation volume is:

$$V_{cs} = \frac{\partial U_{cs}}{\partial \tau} = \frac{\partial W}{\partial \tau_{eq}}\frac{\partial \tau_{eq}}{\partial \tau} = \frac{1}{2\sqrt{3}} Lbd_M \frac{\partial \tau_{eq}}{\partial \tau} \tag{5.17a}$$

or

$$V_{cs} = \frac{Lbd_M}{2\sqrt{3}} \frac{\frac{2}{3}\tau_d + \tau'_d}{\tau} \tag{5.17b}$$

As τ decreases, L and V_{cs} increase at the saddle position.

At very low stresses, we have $d_M \approx d_0$ and with $d/d' \approx 1$, fig. 5.8 yields $L \approx 10d_0$, whence:

$$V_{cs} \approx 3d^2 b \frac{\frac{2}{3}\tau_d + \tau'_d}{\tau} \tag{5.18}$$

These expressions show that, in all cases, V_{cs} is of the order of $d_0^2 b$ to a few times $d_0^2 b$.

5.1.3.3 Orientation effects. The activation area is very sensitive to the direction of the applied stress, because of the orientation factor $(\frac{2}{3}\tau_d + \tau'_d)/\tau$. In other words, in a constant strain rate test, the critical stress for cross-slip is expected to be orientation dependent. Indeed for a given activation energy, i.e. a given τV_{cs}, τ is smaller (easier cross-slip) for a larger volume. For the same reason, tension compression asymmetries are predicted by the model.

According to Figure 5.11, the above orientation factor is high and positive in area *B* of the stereographic triangle in compression. This situation is the most favourable to cross-slip (large V_{cs}, small τ). It is the opposite for areas *A* and *C*. The same factor is high but negative in area *B* in tension. This yields a "negative activation volume" which a priori

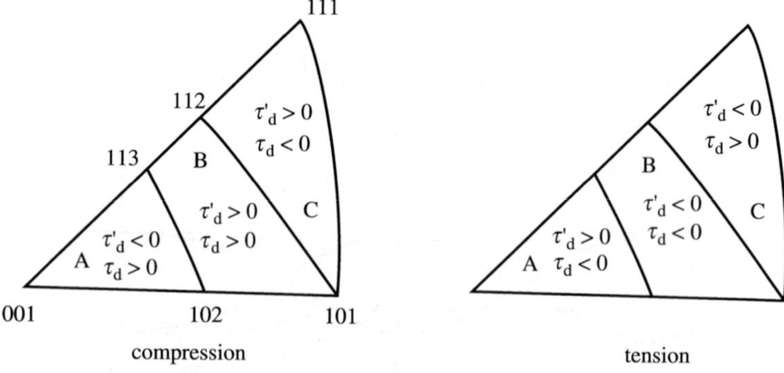

Figure 5.11. The signs of various stresses according to the single crystal orientation in compression and tension. (See text).

inhibits cross-slip. Note, however, that cross-slip can take place in front of a fixed obstacle in the primary plane. Then, the stress τ'_d has a different value.

In the most favourable orientations (area B in Figure 5.11, in compression) at very low stresses and assuming $\tau_\mathrm{d} \approx \tau'_\mathrm{d} \approx \tau$ the cross-slip activation volume is:

$$V_\mathrm{cs} \approx 6d_0^2 b \tag{5.19}$$

V_cs has been estimated in copper, for the most favourable orientation. It is plotted on Figure 5.12 as a function of stress.

5.1.3.4 Refinements in the activation energy estimation. Various attempts at improving the above estimation of the constriction energy are summarized below. (See an extensive review by Püschl (2002)).

After revisiting the early calculations by Stroh (1954) and Escaig (1968), Saada (1991) emphasizes the drastic influence of the cut-off radius, defined in Eq. (5.4), on the constriction energy, and consequently on the cross-slip activation energy. His results are presented in Figure 5.13 (see curves 5–7). The constriction energy is doubled as the core radius r_c decreases from $2b$ to $b/2$.

In the approach of Duesbery et al. (1992), each Shockley partial is decomposed into small adjacent straight segments. The elastic interactions of each of them with all the others are considered. The total energy of a constriction pair is minimized as a function of its geometry. This energy increases with the constriction distance, L, and reaches a constant value equal to two constriction energies ($2U_\mathrm{constr}$) for $L > 50b$. In particular, they

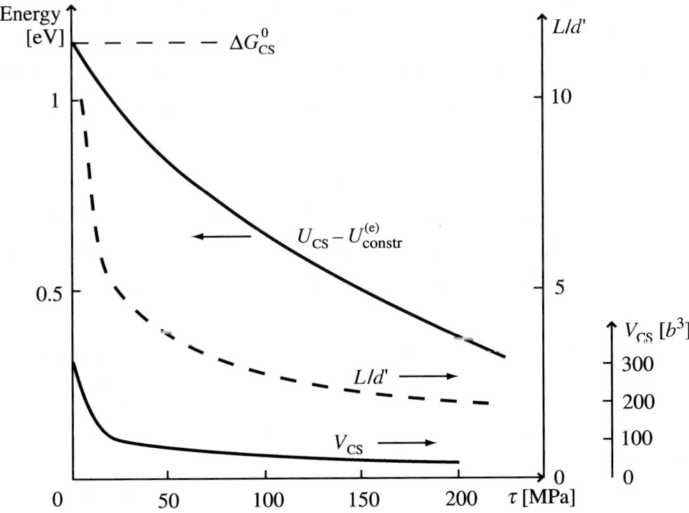

Figure 5.12. Cross-slip activation parameters in Cu as a function of stress, estimated in the framework of Escaig's theory.

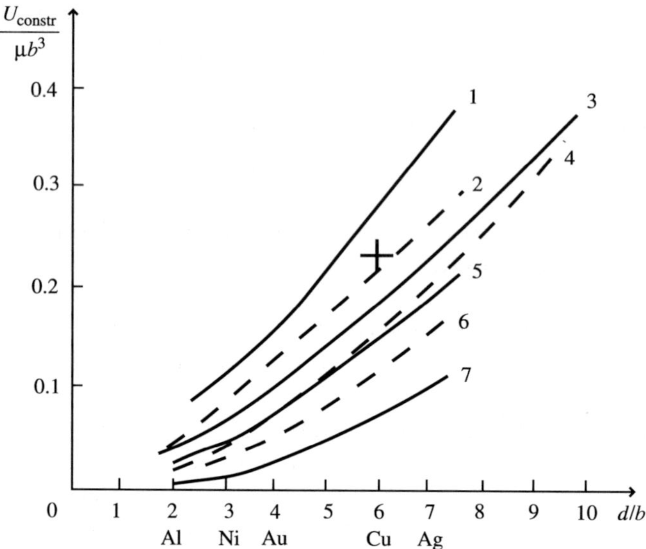

Figure 5.13. Constriction energy in units of μb^3 as a function of the dislocation splitting in units of b. Curve 1 is from Duesbery et al. (1992), 2 from Escaig (1968) (deduced from Figure 5.9 for low stresses), 3 from Püschl and Schoeck (1993), 4 from Bonneville and Escaig (1979) in the low stress approximation, 5–7 from Saada (1991) with successive cut-off radii $r_c = b/2$, b and $2b$. + is the experimental value of the cross-slip activation energy at zero stress in Cu determined by Bonneville et al. (1988) in Section 5.2.3.

predict that $2U_{\text{constr}} \approx 3.7$ eV in Cu. The constriction energy is the largest one among those plotted in Figure 5.13 (curve 1). According to the authors, such a large value is due to the role of the self-stress dipole force which is neglected in the line tension approximation. According to Püschl and Schoeck (1993), the approach of Duesbery et al. (1992) is better justified in the case of large dissociation widths. In particular, it is not appropriate for Cu.

Püschl and Schoeck (1993) avoid the problem of choosing the inner cut-off radius by introducing explicitly the core energy of the Peierls model. Their results correspond to curve 3 on Figure 5.13. They are in fair agreement with the estimations by Escaig (curves 2 and 4) and close to those of Saada (curves 5 and 6). The truth seems to lie between curves 2 and 5.

The most sophisticated developments based on atomistic calculations are described in Section 5.3.

5.2. EXPERIMENTAL ASSESSMENTS OF ESCAIG'S MODELLING

A special technique has been developed by Bonneville and Escaig (1979) in which an avalanche of cross-slip events is produced at yield that is important enough to overtake other thermally activated events such as tree cutting by gliding dislocations. Thus, it allows one to consider the yield stress as the critical stress for cross-slip. This technique is first presented, then applied to the experimental determination of the cross-slip activation parameters and then used to test the orientation effects.

5.2.1 *The Bonneville–Escaig technique*

A large [110] single crystal is predeformed in compression up to the end of stage II, i.e. the test is interrupted just before the triggering of cross-slip. The stress–strain curves exhibit a marked stage II, the corresponding activation volume being in the range of several thousand b^3. Such values are the signature of a forest mechanism. This multiple slip orientation is stable against lattice rotations and induces a homogeneous dislocation distribution. Indeed, four slip systems are simultaneously activated on two glide planes, along four directions: $(111)[01\bar{1}]$, $(111)[10\bar{1}]$, $(11\bar{1})[011]$ and $(11\bar{1})[101]$. This creates four families of forest dislocations of equal densities. New samples are then cut out of the predeformed single crystals and deformed again in tension or compression. The new deformation axis, $[\bar{4}21]$, is chosen so that the new primary system is $(1\bar{1}1)[011]$. $(1\bar{1}1)$ was not activated during predeformation because of a zero Schmid factor. Therefore, the [011] dislocations which were severely constrained by the forest are ready to escape by cross-slip from $(11\bar{1})$ onto $(1\bar{1}1)$ at the onset of deformation. On this plane, the resistance to slip should be comparable to that during predeformation, i.e. there is no latent hardening. An advantage of such a procedure is that all the samples have the same initial microstructural state, at least when deformed under conditions for which the predeformation substructure is stable. Thus, in each $[\bar{4}21]$ sample, cross-slip takes place in the presence of quite comparable dislocation substructures.

Typical dimensions for the $\langle 110 \rangle$ single crystals are 16 mm in diameter and 50 mm in length. The deformation samples are about $4 \times 4 \times 10$ mm^3. The sequence of operations that lead to the $[\bar{4}21]$ deformation samples is illustrated in Figure 5.14. Spark machining of the single crystals is followed by chemical polishing of the sample faces so as to remove any damage due to cutting.

Results are presented below in copper and aluminium.

5.2.2 *Experimental observations of cross-slip*

Cross-slip in Cu has been extensively studied and most data in this section refer to this element, unless otherwise specified. Deformation of the $[\bar{4}21]$ samples is performed at various temperatures, following predeformation at 473 K. Several converging features reveal that cross-slip takes place at yield.

5.2.2.1 TEM observations. These observations show that the initial microstructure of the $[\bar{4}21]$ samples consists of cells bounded by thick walls, an example of which is shown in Chapter 9 (Figure 9.23). An in situ TEM experiment has been performed at 293 K after

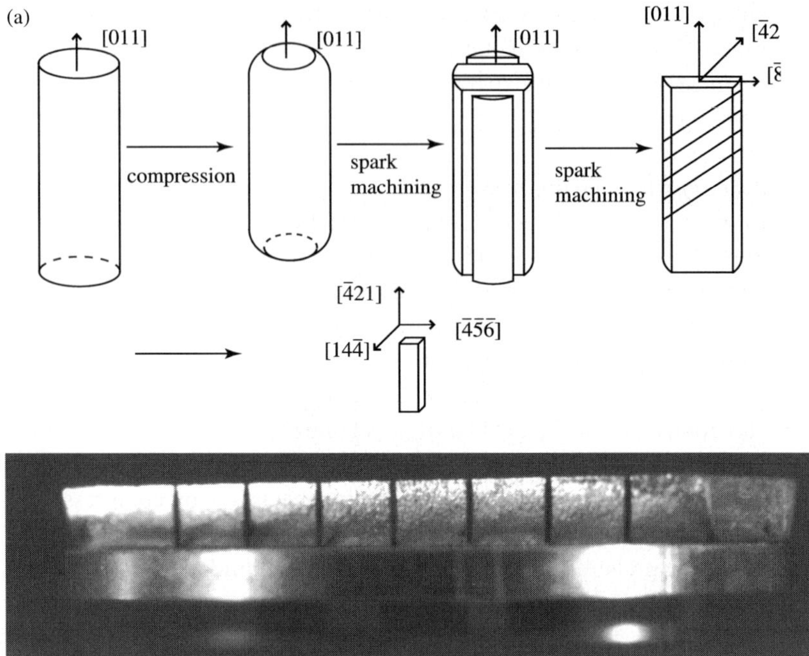

Figure 5.14. The preparation procedure of the [$\bar{4}$21] deformation samples (see text). (a) Schematics of the successive steps. (b) Cu samples seen end on. (Courtesy of J. Bonneville).

predeformation at 473 K. The point of which was to observe the substructure reorganization during deformation. In thin parts of the foil the cell structure is very unstable due to image forces as soon as the load is applied. However, in thicker parts, cross-slip events have frequently been observed. An example of such a deviation between two {111} planes is shown in Figure 5.15. Cross-slip of the dislocation segment AB is detectable because of its change of curvature, as sketched in Figure 5.15(c). This event is a representative of bulk cross-slip since it occurs here in the absence of any constriction at the surface. It is also worth noting that it takes place here in the absence of pile-ups, unlike the requirements of the Schoeck–Seeger–Wolf model, to ensure high enough stresses (Section 5.1.1.3).

5.2.2.2 Optical slip trace observations. Figure 5.16 clearly shows evidence for cross-slip along the expected planes for a test interrupted shortly after yield.

5.2.2.3 Peculiar features of the deformation curves. These are illustrated in Figure 5.17. For this 293 K test, the deformation curve exhibits three distinct sections. The first one

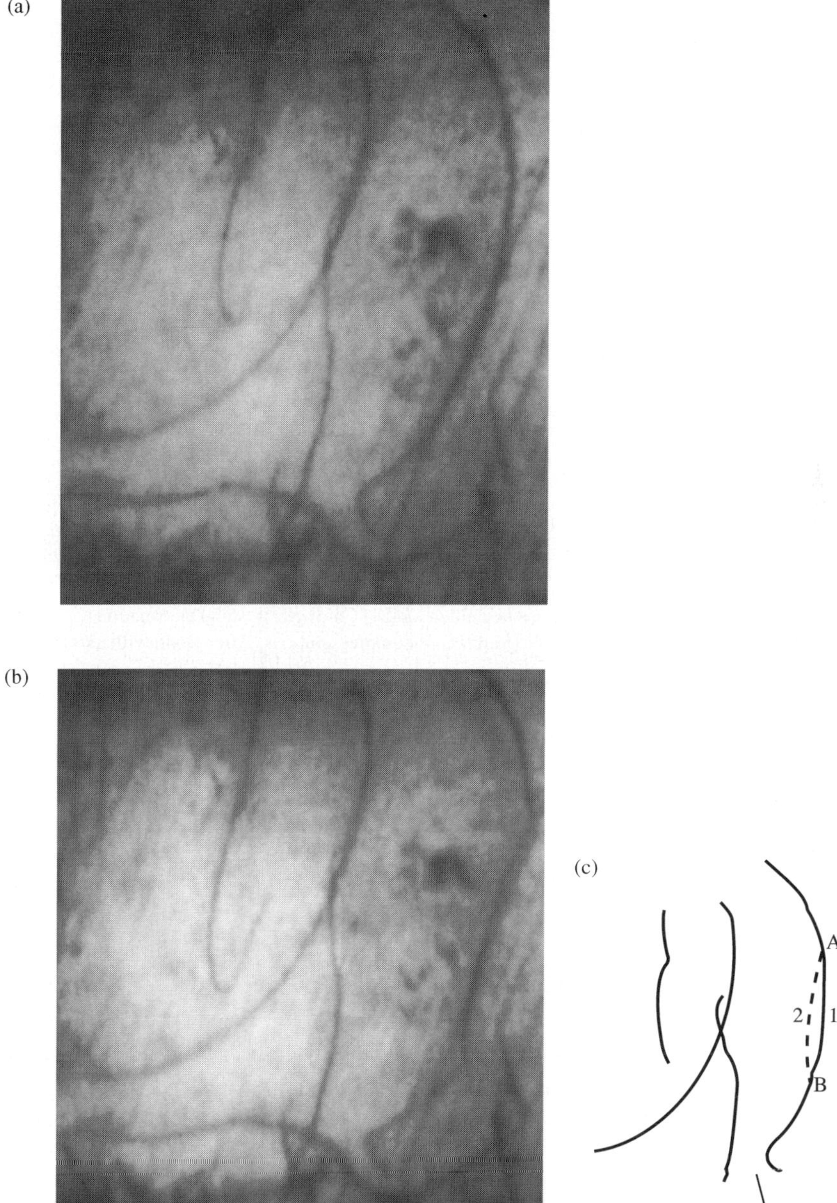

Figure 5.15. TEM in situ deformation experiment of a predeformed Cu single crystal at 293 K. 200 kV. Foil plane $(56\bar{4})$: (a) and (b) two successive positions of the cross-slipping dislocation (labelled 1 and 2, respectively); (c) corresponding schematics. T is the tensile axis. After Bonneville et al. (1988).

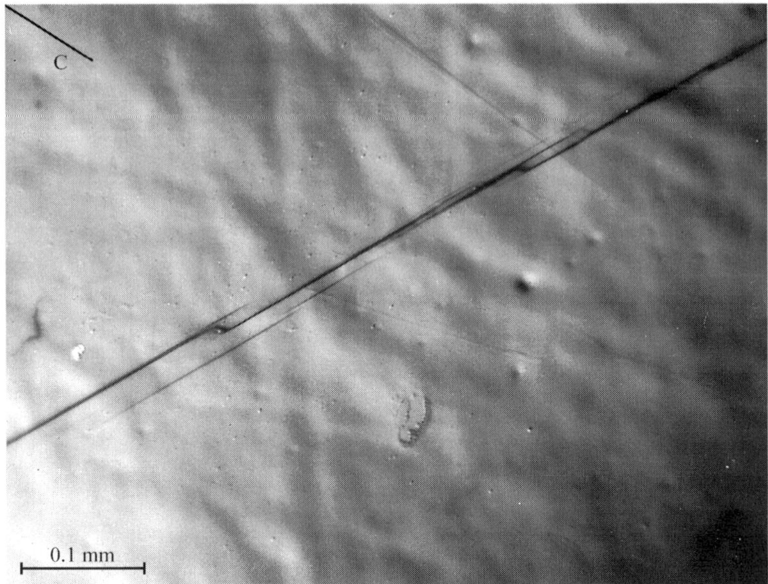

Figure 5.16. Slip traces of screw dislocations at yield in a [$\bar{4}$21] Cu single crystal. Deformation test at 293 K after predeformation at 473 K. Optical micrograph. C = compression axis. After Bonneville et al. (1988).

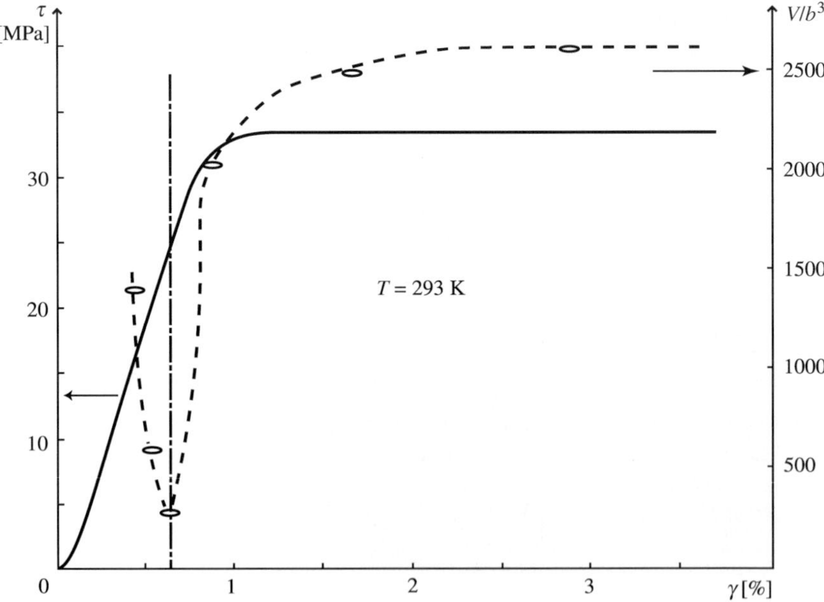

Figure 5.17. Stress and microscopic activation volume as a function of strain. [$\bar{4}$21] Cu single crystal. Deformation test at 293 K after predeformation at 473 K. After Bonneville et al (1988).

corresponds to a high work hardening rate ($\theta/\mu \cong 10^{-1}$), while the activation volume decreases as strain increases. This is typical of a preplastic stage in which shorter and shorter dislocation segments are activated as stress increases. In the second region, the work hardening rate declines while the activation volume increases rapidly. At the yield stress (24.5 ± 1.5 MPa or $5.10^{-4}\mu$) the activation volume is at a minimum and close to $280 \pm 65b^3$. In the third region, the hardening rate is negligible while the stress is close to 33.9 ± 0.5 MPa. The corresponding activation volume shows a moderate increase with strain with a value close to $2600b^3$.

The minimum activation volume at yield exhibits a value close to that predicted for cross-slip in Cu, according to the estimations of Section 5.1.3.3 for the most favourable orientation (area *B* of Figure 5.11 in compression). The third region of the stress–strain curve sets in all of the predeformation dislocations have completed cross-slip. The activation volume values are the signature of a forest mechanism.

A similar test in Al single crystals by Bonneville and Vanderschaeve (1985) reveals the same type of curves as in Cu (Figure 5.18). The minimum in activation volume is close to $43b^3$. Although the description of the core extension in Al in terms of splitting is not appropriate, estimations by Mills and Stadelmann (1989) using atomic resolution electron microscopy yielded a slight dissociation close to 0.55 ± 0.15 nm (about 2*b*) for 60° dislocations. It should be smaller for screws.

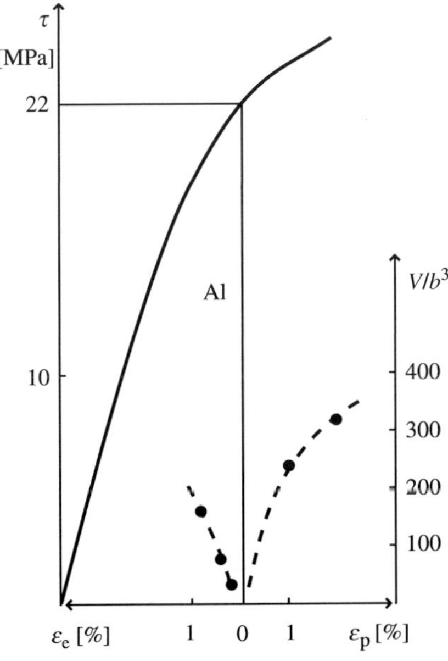

Figure 5.18. Stress and microscopic activation volume as a function of strain. [$\bar{4}$21] Al single crystal. Deformation and predeformation tests at 77 K. After Bonneville and Vanderschaeve (1985).

According to the estimations in Section 5.1.3.3, this would lead to V_{cs} of the order of $25b^3$ at very low stresses. Therefore, the minimum activation volume as a function of strain in Figure 5.18 corresponds to abundant cross-slip practically at yield ($\tau = 22$ MPa) in the case of Al.

It is also worth noting that the experimental results regarding orientation effects in Section 5.2.4 provide additional evidence of cross-slip operating at yield during deformation of the small samples extracted from the predeformed crystals.

To conclude this section, the present procedure allows one to determine the stress and activation volume that correspond to cross-slip.

5.2.3 The activation parameters

As shown above, the minimum activation volume observed at yield in the deformation experiment is a representative of dislocation cross-slip. The study of Cu was pursued at several temperatures between 150 and 473 K, following predeformation at 473 K. The yield stress together with the corresponding activation volume are represented in Figure 5.19 as a function of temperature.

Three temperature domains can be distinguished in this figure:

- Below 250 K: the yield stress is almost temperature independent. The activation volume measured at 230 K as a function of strain still exhibits a minimum at yield, but with a high value of $950b^3 \pm 180b^3$. This is interpreted as cross-slip taking place mixed together with other processes such as forest cutting. The almost athermal character of stress is understood in terms of the rearrangement of the predeformation

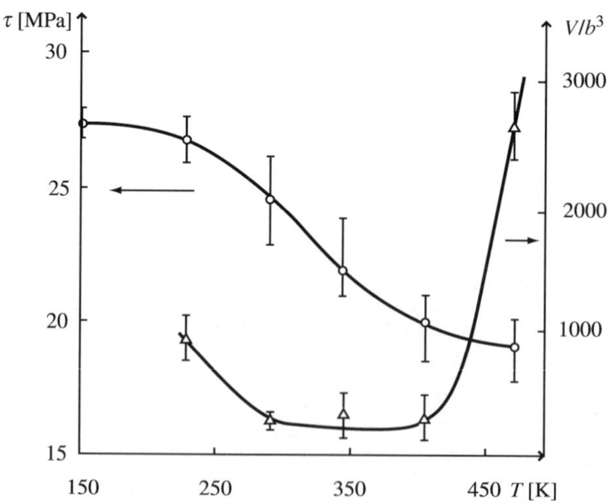

Figure 5.19. Critical stress for cross-slip and corresponding microscopic activation volume as a function of temperature. $[\bar{4}21]$ Cu single crystals. Predeformation tests at 473 K. After Bonneville et al. (1988).

forest under high stress values at low temperatures. Indeed, as the crystal yields at 230 K, the shear stress along the (111) [01$\bar{1}$] slip system is of the order of 22.5 MPa, quite close to the stress of about 24 MPa reached during predeformation. This system was one of the primary ones during predeformation.

- Above 410 K: the yield stress reaches a plateau in Figure 5.19, while the activation volume increases to high values ($2550b^3 \pm 250b^3$). It is thought that as the predeformation temperature is approached the predeformation forest is no longer stable. Due to the decrease in obstacle density the burst of cross-slip is no longer observed at yield.
- Between 250 and 410 K: the observed decrease in yield stress as temperature increases is indicative of a thermally activated process. The activation volume is constant within the experimental scatter and keeps a low value of the order of $300b^3$. This is close to the value predicted for Cu in Section 5.1.3.3. Due to the scatter, it is not possible to check accurately relations (5.17) and (5.18).

The more accurate 300 K data yield $V = 280b^3 \pm 65b^3$ and the slope $d\tau/dT \approx -5.3 \pm 0.9 \times 10^{-2}$ MPa/K can be measured on Figure 5.19. Using relation (1.6), an activation enthalpy $\Delta H = 0.47 \pm 0.16$ eV is obtained. Taking into account the entropy term (relation (1.5)), the activation energy is found to be $\Delta G = 0.42 \pm 0.16$ eV. Finally, adding the work τV done by the applied stress, the energy barrier associated with the cross-slip mechanism in Cu is found to be $\Delta G^0_{cs} = 1.15 \pm 0.37$ eV. In spite of the uncertainty, this value is, to our knowledge, the only one measured experimentally so far. It can be compared with the theoretical estimations in Section 5.1.3.1. A set of values is available, depending on the following approximations:

- In the low stress approximation, Escaig (1968) predicts $U_{cs} - U^{(e)}_{constr} = 0.86$ eV and $U_{cs} = 1.58$ eV, respectively, with and without a pre-existing constriction.
- According to the estimations of Section 5.1.3.1, Figure 5.12 shows that $U_{cs} - U^{(e)}_{constr} = 1.10$ eV at zero stress (with a pre-existing constriction).

The experimental value above is close to this latter value (Figure 5.13).

With regard to the measured activation volumes, the comparison of Cu and Al reveals drastically different values, in spite of the scatter. Relation (5.19) tells us that the activation volumes should compare as $d_0^2 b$. The above data yield $V_{cs}(\mathrm{Cu})/V_{cs}(\mathrm{Al}) \approx 7$ while $(d_0^2 b)_{\mathrm{Cu}}/(d_0^2 b)_{\mathrm{Al}} \approx 9.8$. Given the scatter on V_{cs} values in Cu and the few data available for Al the agreement is considered as satisfactory.

Therefore, the above features support Escaig's estimations in Section 5.1.3 and assess the pertinence of the technique set out in Section 5.2.1. Let us note that the latter deserves some improvements so as to observe cross-slip over the largest possible temperature interval. This should reduce the error bars on the activation parameter values. Other crystals should also be investigated using the same technique.

5.2.4 Experimental study of orientation effects

Using the same method as in Section 5.2.1, Bonneville and Escaig (1979) tested the orientation effects as follows. The predeformation and deformation tests were both performed at room temperature. However, the deformation tests were either in tension or compression along successive orientations in areas *A*, *B* or *C* of Figure 5.11. The yield stress was determined by inspection of the deformation curves, but also by determining the minimum activation volume as a function of strain as in Section 5.2.2. Most importantly, the minimum activation volume exhibits a low value, comparable with that found in Section 5.1.2 above, with some scatter. This precludes an experimental check of the orientation dependence of V. However, the measured values are in the range predicted by Eqs. (5.17) and (5.18) (see Figure 5.12), which confirms that cross-slip operates practically at yield. The critical stresses for cross-slip are indicated in Figure 5.20. A clear tension–compression asymmetry is observed: these critical stresses are above 40 MPa in tension and between 30 and 40 MPa in compression. In tension, the critical stress is lower for orientations A and B as compared to C, while the reverse is observed in compression. This is in perfect agreement with the predictions of Section 5.1.3.3. In addition, for the tensile data, cross-slip appears more difficult in orientation B (45 MPa) and easier for A and C (40 MPa). The reverse is observed for compression data (28 MPa for orientation B

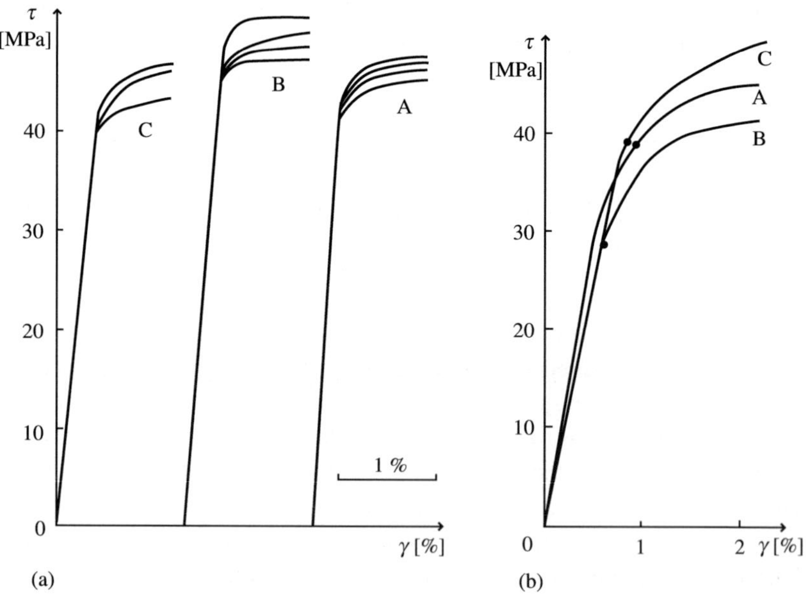

Figure 5.20. Stress–strain curves at 300 K of Cu single crystals of various orientations, predeformed at 300 K. (After Bonneville and Escaig, 1979). Orientations A, B and C refer to Figure 5.11. (a) Tension. (b) Compression. Critical stresses for cross-slip correspond to the yield stress in (a) and to the black dots in (b).

as compared to 38 MPa for orientations A and C). Such a trend was also predicted in Section 5.1.1.3.

These results assess not only the description proposed by Escaig, but also the relevance of the technique of Section 5.2.1 to separate cross-slip from the following dynamic recovery stage starting at τ_{III}.

5.3. ATOMISTIC MODELLING OF DISLOCATION CROSS-SLIP

This section focuses on the most recent three-dimensional calculations. These require a large number of atoms and consequently are very demanding in computer time. Several tricks have been used to reduce this time.

Rao et al. (1999) simulate cross-slipping core structures in model FCC crystals in the framework of the embedded atom method. The potentials are fitted to the elastic and structural properties of Ni. Green's function techniques are used to relax the boundary forces in the calculations. The authors study the core structure and energy of the constrictions during cross-slip, assumed to be of the Friedel–Escaig type. Two types of constrictions form edge and screw types, respectively, as illustrated in the schematics of Figure 5.21a. Conversely, a constriction on a single plane is of one type only (Figure 5.21b). The simulations show that the constrictions exhibit a diffuse core structure, as opposed to the point constriction considered so far. This core geometry is illustrated in Figure 5.22. At and near the constriction, atom displacements are along both the primary and the cross-slip plane. As one moves away from the constriction, they become localized along the cross-slip plane (above) or the primary plane (below).

The Danish group (Rasmussen et al., 1997a,b; Rasmussen, 2000) uses molecular dynamics and the "nudged elastic band" method. Its advantage is to avoid any assumption on the cross-slip process, since only the starting and final positions are imposed. An incremental

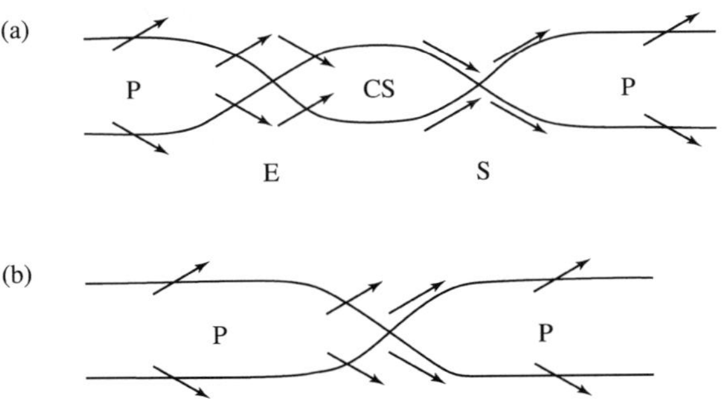

Figure 5.21. Schematics of constrictions on a cross-slipping dislocation. Arrows indicate Burgers vectors. (a) An edge (E) and a screw (S) constriction connect the dislocation segments on the primary (P) and the cross-slip (CS) planes. (b) The Stroh constriction is shown for comparison (the two partials are re-united).

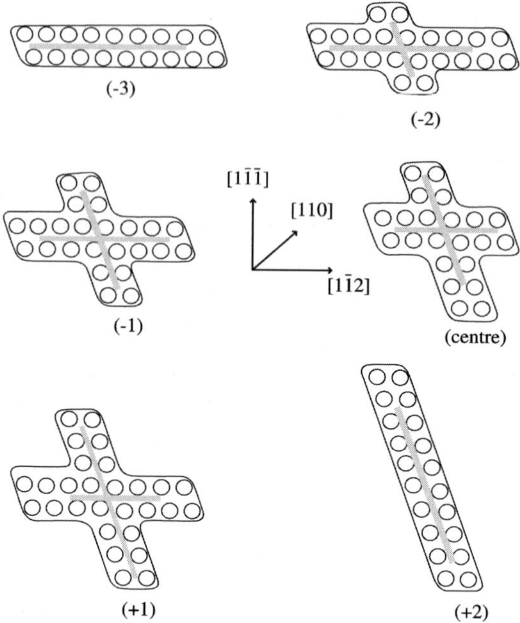

Figure 5.22. Atomic displacements in the dislocation core, according to Rao et al. (1999). Regions of significant displacements are highlighted. The various sections are perpendicular to the end on screw. Atom positions are represented at the constriction (centre), above it (+1, + 2) and below it (− 1 to − 3). Courtesy of Phil. Mag.

stress is applied, the system finds its equilibrium configuration and the corresponding energy is calculated. The activation energy is the saddle point energy. The resulting process is observed to be of the Escaig type and is illustrated in Figure 5.23.

Both simulations show that the two constrictions are of edge and screw types, respectively, as in Figure 5.21. The E and S constrictions have significantly different energies.

Activation energies and volumes have been computed for $\tau/\mu \approx 10^{-3}$, i.e. $\tau \approx 50$ MPa in Cu. Rao et al. (1999) find an activation energy proportional to $(d/b)[\ln(d/b)]^{1/2}$, in agreement with Escaig's model (see end of Section 5.1.3.1). It amounts to 1.2 eV without any pre-existing constriction. Rasmussen (2000) finds larger values of 2.7 eV in the same conditions while Vegge et al. (2001) obtain 1 eV with a pre-existing constriction at an elementary jog. This latter value fits rather well with the experimental one ($\Delta G^0_{cs} = 1.15 \pm 0.37$ eV) by Bonneville et al. (1988) in Section 5.2.3.

Under the same stress, activation volumes are equal to $20b^3$ in Cu, i.e. smaller than in the continuum theory. They are, however, of the same order of magnitude as in Saada's estimations ($40b^3$ in Cu). Rasmussen shows that activation volumes tend to infinity at

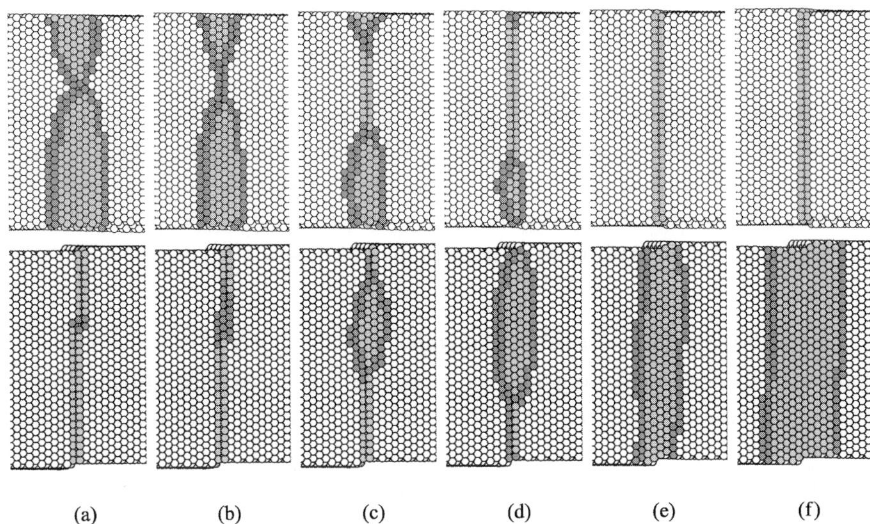

(a) (b) (c) (d) (e) (f)

Figure 5.23. Successive configurations of minimum energy from (a) to (e) during cross-slip. The screw dislocation has a pre-existing constriction at a jog in (a). The figures on the top are in the primary plane, those at the bottom are in the cross-slip plane. Note that in the initial step (b), redissociation has taken place on the cross-slip plane at the position of the jog. (Vegge and Jacobsen, 2002). Courtesy of J. Phys.

vanishing stress. This suggests that the δ function introduced by Escaig to avoid this trend (Section 5.1.3.1) may not be useful.

5.4. DISCUSSION AND CONCLUSIONS

The number of models in Section 5.1 that aim to describe the cross-slip process reflects the importance of this mechanism with regard to crystal plasticity over a range of temperatures too low for climb to be activated. We address below the question of which model is the most relevant and whether the study of τ_{III} can be of any use in learning about cross-slip.

5.4.1 Who is closer to the truth?

While presenting the various models in Section 5.1.1 we have set out a few criticisms specific to some of the attempts. In the present section, the models are compared with each other in terms of their relevance with respect to recent more sophisticated investigations (either theoretical or experimental).

It is remarkable to note how the early analytical estimation of the constriction energy in Stroh's (1954) enlightening paper, has survived more sophisticated descriptions which came out recently, as exemplified in Figure 5.13.

Among all the available models, the one from Escaig (1968), using Stroh derivations, has been confirmed by the atomistic simulations of the Danish group made without any assumptions about the cross-slip path. It has also been assessed by a variety of experimental observations reported in Section 5.2. In particular, the Bonneville–Escaig technique, which implies two successive deformations along two crystal orientations, allows one to determine the critical stress for cross-slip over a range of temperatures, the corresponding activation volume, and to estimate the activation energy. The values found are consistent with the atomistic simulations of Section 5.3. They suggest that cross-slip is nucleated at pre-existing constrictions. The present theory is not only a descriptive one, it also predicts orientation effects which have been confirmed experimentally (see Section 5.2.4).

5.4.2 Cross-slip and stage III in FCC metals

The decrease in work hardening rate, as stage III sets in, is undoubtedly due to the activation of recovery mechanisms that include cross-slip. Therefore, before the experiments described in Section 5.2, the characteristics of τ_{III} were considered as representative of those for the cross-slip critical stress. However, this statement was undermined by some observations that are summarized here.

Basinski (1968) presented evidence for the operation of cross-slip from half-way in stage II to half-way in stage III. Moreover, Nabarro (1986) examined the role of the stacking fault energy with respect to the presence of various hardening stages. He quoted that the $\tau(\gamma)$ curves for very pure BCC metals sometimes exhibit a form very similar to those of FCC crystals. This is observed at temperatures high enough for friction forces to be negligible but low enough to preclude climb. This suggests that, under the above conditions, dislocation interactions are responsible for the form of the $\tau(\gamma)$ curves. However, the stacking fault energy, which appears as a relevant parameter for FCC crystals, is not important for BCC crystals. Thus, the questions regarding the similarity of the flow curves for FCC and BCC crystals, as well as the exact role of cross-slip at τ_{III} are still open ones.

Furthermore, Nabarro (1986) examined the parameter $\gamma/\mu b$ (see Section 5.1.2) in connection with the form of the stress–strain curves at different temperatures. He drew sets of curves of stress, normalized to the shear modulus, as a function of strain at different homologous temperatures for five FCC metals. A convincing trend is observed. For high $\gamma/\mu b$ materials (e.g. lead) the deformation curve starts in stage III at low temperatures ($T/T_M = 0.07$). Conversely, at higher temperatures ($T/T_M = 0.13$), stages I and II are present in the $\tau(\gamma)$ curves for Cu and Ag (low values of $\gamma/\mu b$), while they are not seen in the curves for Ni and Pb. Therefore, the parameters $\gamma/\mu b$ and T/T_M seem to influence the stress–strain curves. However, Al constitutes an exception to the rule. This again questions the traditional interpretation of τ_{III}.

The results of Section 5.2 in Cu show that τ_{III} is not representative of cross-slip alone. Indeed, the shear stresses are different: e.g. 40 MPa for τ_{III} vs 29 MPa for cross-slip at 300 K according to Bonneville and Escaig (1979). The activation volumes are also different: e.g. larger than 1000 b^3 for τ_{III} (Section 5.2.2.1) as compared to 280 b^3 for cross-slip (Section 5.2.1).

Taking into account all the remarks of this section, the change in hardening rate which is observed at τ_{III} is likely to correspond to substructure rearrangements in which cross-slip is involved together with other processes such as, for example, the evolution of junction reactions.

REFERENCES

Basinski, Z. (1968), in *Dislocation Dynamics*, Eds. Rosenfeld, A.R., Hahn, G.T., Bement, A.L. & Jaffee R.I., McGraw-Hill, New York, p. 674.

Bonneville, J. & Escaig, B. (1979) *Acta Met.*, **27**, 1477.

Bonneville, J. & Vanderschaeve, G. (1985) Strength of Metals and Alloys, vol. **1**, Eds. McQueen, H.J., Bailon, J.P., Dickson, J.I., Jonas, J.J. & Akben M.G., p. 9.

Bonneville, J., Escaig, B. & Martin, J.L. (1988) *Acta Met.*, **36**, 1989.

Brown, L.M. (2002), in *Understanding Materials*, Ed. Humphreys, C.J., Maney, London, p. 135.

Caillard, D. & Martin, J.L. (1976), in *Electron Microscopy 1976*, vol. **1**, Ed. Brandon, D.G., TAL Int. Pub. Company, Jerusalem, p. 587.

Clarebrough, L.M. & Forwood, T.C. (1975) *Phys. Stat. Sol.(a)*, **32**, K15.

Coujou, A (1981) PhD thesis no 982, Toulouse.

Duesbery, M.S., Louat, N.P. & Sadananda, K. (1992) *Acta Met. Mater.*, **40**, 149.

Escaig, B. (1968) *Le Journal de Physique*, **29**, 225.

Escaig, B. (1974) *Le Journal de Physique*, **35-C7**, 151.

Fleischer, R.L. (1959) *Acta Met.*, **7**, 134.

Foreman, A.J.E. (1955) *Acta Met.*, **3**, 322.

Friedel, J. (1957) *Dislocations and Mechanical Properties of Crystals*, Wiley, New York, p. 330.

Friedel, J. (1964) *Dislocations*, Pergamon, New York, p. 223.

Friedel, J. (1977) *Rev. Phys. Appl.*, **12**, 1649.

Fujita, H. & Veda, S. (1975) *Acta Met.*, **9**, 631.

Hirsch, P.B. (1962) *Phil. Mag.*, **7**, 67.

Humphreys, F.J. & Hirsch, P.B. (1970) *Proc. Roy. Soc. A*, **318**, 73.

Jackson, P.J. (1985) *Prog. Mater. Sci.*, **29**, 139.

Mills, M.J. & Stadelmann, P. (1989) *Phil. Mag.*, **60**, 355.

Mori, T. & Fujita, H. (1980) *Acta Met.*, **28**, 771.

Mott, N.F. (1952) *Phil. Mag.*, **43**, 1151.

Mughrabi, H. (1968) *Phil. Mag.*, **18**, 1211.

Nabarro, F.R.N. (1986), in *Strength of Metals and Alloys*, vol. **3**, Eds. McQueen, H.J., Bailon, J.P., Dickson, J.I., Jonas, J.J. & Akben M.G., Pergamon, New York, p. 1667.

Pashley, D.W., Robertson, J.L. & Stowell, M.J. (1969) *Phil. Mag.*, **19**, 83.

Püschl, W. (2002) *Prog. Mater. Sci.*, **47**, 415.

Püschl, W. & Schoeck, G. (1993) *Mater. Sci. Eng. A*, **164**, 286.

Rao, S., Parthasarathy, T.A. & Woodward, C. (1999) *Phil. Mag.*, **79**, 1167.

Rasmussen, T., Jacobsen, K.W., Leffers, T. & Pedersen, O.B. (1997a) *Phys. Rev.*, **56**, 2977.

Rasmussen, T., Jacobsen, K.W., Leffers, T., Pedersen, O.B., Srinavasan, S.G. & Jonsson, H. (1997b) *Phys. Rev. Lett.*, **79**, 3676.
Rasmussen, T. (2000) In *Multiscale Phenomena in Plasticity*, NATO Science Series, Eds. Lépinoux, J., Mazière, D., Pontikis, V. & Saada G., Kluwer Acad. Publ., London, p. 281.
Saada, G. (1991) *Mater. Sci. Eng. A*, **137**, 177.
Saada, G. & Veyssière, P. (2002), in *Dislocations in Solids*, vol. **11**, Eds. Nabarro, F.R.N., Duesbery, M.S. & Hirth J., Elsevier, Amsterdam, p. 413.
Schoeck, G. & Seeger, A. (1955) *Report on the Conf. "Defects in Crystalline Solids"*, The Physical Society, London.
Seeger, A., Berner, R. & Wolf, H. (1959) *Z. Naturforsch.*, **155**, 249.
Stobbs, W.M. & Sworn, C.H. (1971) *Phil. Mag.*, **24**, 1365.
Stroh, A.N. (1954) *Proc. Phys. Soc.*, **B67**, 427.
Vanderschaeve, G (1981) PhD thesis no 509, Lille.
Vegge, T. & Jacobsen, K.W. (2002) *J. Phys.: Condens. Matter*, **14**, 2929.
Vegge, T., Rasmussen, T., Leffers, T., Pedersen, O.B. & Jacobsen, K.W. (2001) *Phil. Mag. Lett.*, **81**, 137.
Washburn, J. (1965) *Appl. Phys. Lett.*, **7**, 183.
Whelan, J. (1958) *Proc. Roy. Soc.*, **A249**, 114.
Wolf, H. (1960) *Z. Naturforsch.*, **15A**, 180.

Chapter 6

Experimental Studies of Peierls–Nabarro-type Friction Forces in Metals and Alloys

Chapter 6

Experimental Studies of Peierls–Nabarro-type Friction Forces in Metals and Alloys

This chapter reviews experimental data related to glide mechanisms in metals and alloys for which dislocation cores lie outside the slip plane. The resulting Peierls–Nabarro-type friction has been theoretically examined in Chapter 4. The data refer first to close-packed crystals gliding along non-close-packed planes. These include hexagonal close-packed (HCP) lattices deformed by prismatic slip and then FCC crystals gliding along non-octahedral planes. In these two cases, dislocations have a natural tendency to split along close-packed planes. Therefore, their core is sessile with respect to glide on other types of planes. Then data about slip in BCC metals will be analysed. The cores of screw dislocations in these crystals are split along three intersecting planes.

6.1. PRISMATIC SLIP IN HCP METALS

Metals with a hexagonal close-packed (HCP) structure deform by glide of $a = \frac{1}{3}\langle 11\bar{2}0\rangle$ dislocations in basal (0001) and prism $(1\bar{1}00)$ planes, and by the more difficult glide of $c + a = \frac{1}{3}\langle 11\bar{2}3\rangle$ dislocations in several types of pyramidal planes. (Note that they can also deform by climb of $c = \langle 0001\rangle$ dislocations, cf. Section 8.4). The relative ease of basal, as compared to prismatic, slip depends on the ratio of the corresponding stacking fault energies, as shown using atomistic calculations by Legrand (1984). For instance, the easiest slip system is prismatic in Ti and Zr, whereas it is basal in Mg and Be. Prismatic slip has been extensively studied in both types of materials and some important results are summarized in this section.

6.1.1 Prismatic slip in titanium

The CRSS of prismatic slip in single crystals is shown as a function of temperature in Figure 6.1. It increases rapidly with decreasing temperature and increasing oxygen content. A small hump can be detected at 400 K in the experiments of Akhtar and Teghtsoonian (1975) and at 500 K in the experiments of Naka et al. (1988). To this hump corresponds a peak in the temperature dependence of the activation area, A, (Figure 6.2) which varies as the reciprocal of $\partial\tau/\partial T$ (Relation 1.8). A similar peak has been evidenced in Ti and in several Ti–Al single crystal alloys by Sakai and Fine (1974; Figure 6.2(c)) and in Ti polycrystals by Tung and Sommer (1970).

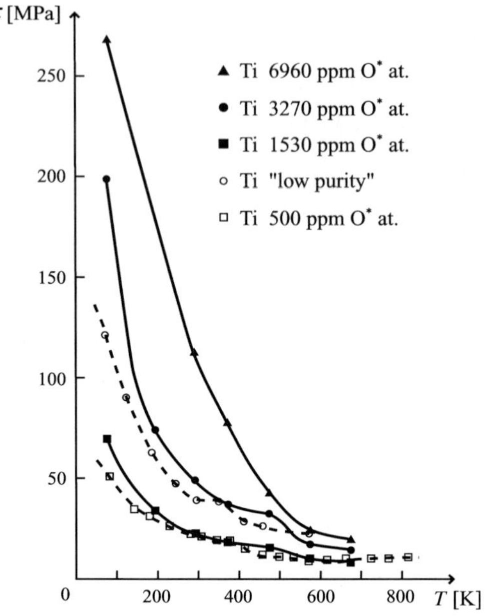

Figure 6.1. Critical resolved shear stress for prismatic slip in titanium of different purities as a function of temperature. Full symbols from Naka et al. (1988) and open symbols from Akhtar and Teghtsoonian (1975). (O^* is the equivalent oxygen concentration defined by $O^* = O + 2N + C$).

The activation enthalpy of Relation 1.6 can be written

$$\Delta H_a = -AbT \frac{\partial \tau}{\partial T}\bigg|_{\dot{\varepsilon}} \tag{6.1}$$

also exhibits a discontinuity, which indicates that two different thermally activated mechanisms may operate, respectively, below and above the discontinuity.

Biget and Saada (1989) have shown that the activation areas measured below 300 K in high-purity polycrystals vary as $A \propto \tau^{*^{-2}}$ where τ^* is the effective stress. The same variation can be deduced from the measurements of Levine (1966) in high-purity single crystals (see Figure 6.3).

A second maximum in the temperature dependence of the activation area is present at 600 K, above the thermally activated domain just described. It also corresponds to a hump on the stress versus temperature curve. It is usually attributed to dynamic strain ageing (Naka et al., 1991; Trojanova et al., 1991).

Microscopic observations show rectilinear screw dislocations in Ti deformed at 77 and 300 K (Naka et al., 1988) and in Ti–5.2 at.% Al deformed at 300 K (Sakai and Fine, 1974).

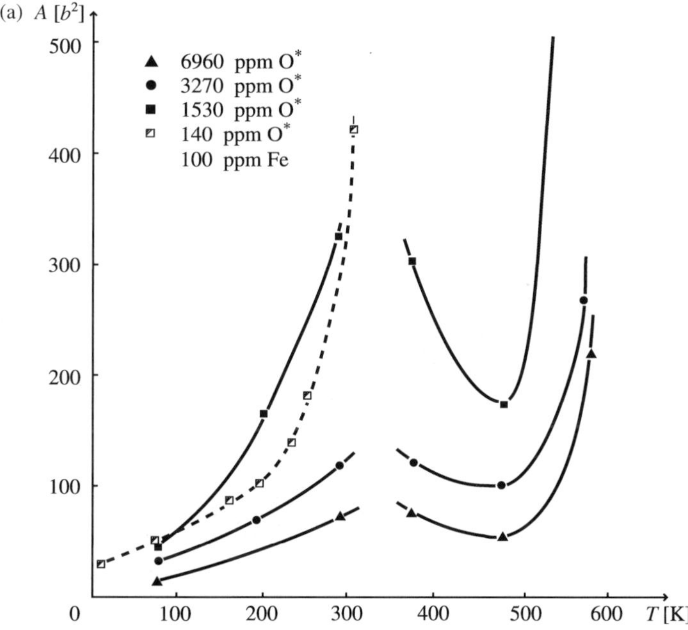

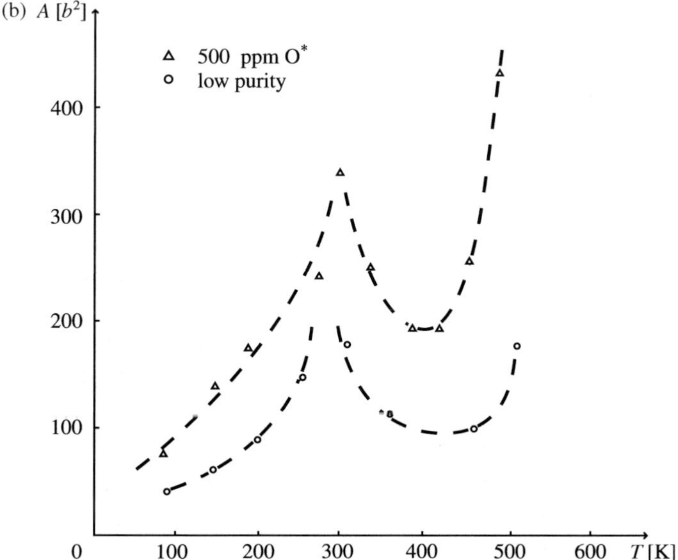

Figure 6.2. Activation area of prismatic slip as a function of temperature. (a) Ti of different purities. Full symbols from Naka et al. (1988) and mixed symbols from Levine (1966). (b) Ti of different purities from Akhtar and Teghtsoonian (1975). (c) Ti and Ti–Al alloys from Sakai and Fine (1974).

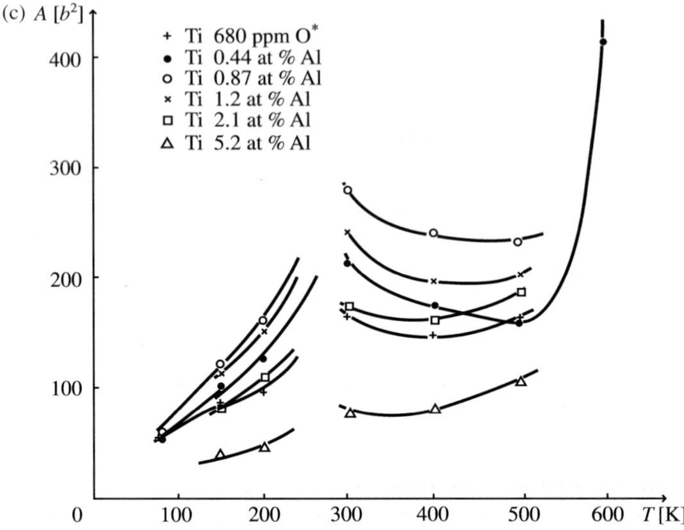

Figure 6.2. (*continued*)

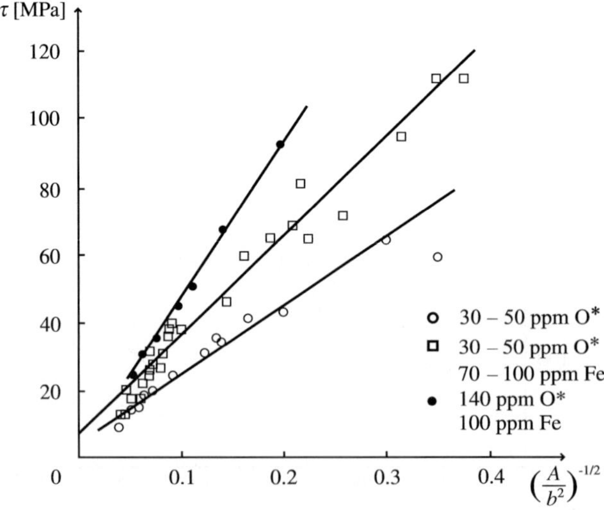

Figure 6.3. Stress-dependence of the activation area of prismatic slip in Ti in the low-temperature range ($T < 300$ K). Open symbols are from Biget and Saada (1989), and full symbols are from Levine (1966).

In situ experiments have been made by Naka et al. (1988), at 373 K, and by Farenc et al. (1993), between 100 and 473 K. They both reveal that the slowest dislocations are long and rectilinear screws. The local stress acting on individual dislocations has been plotted as a function of temperature for two alloys: the fairly good correspondence between microscopic and the corresponding macroscopic stresses ensures that the same mechanisms take place both in the thin foils and in the bulk material (Figure 6.4).

Above 373 K straight screw dislocations move steadily with an average velocity depending more on their intrinsic mobility between visible obstacles than on their interactions with these obstacles. This behaviour is exactly what is expected for a kink-pair mechanism (Section 4.2.7).

At temperatures below 300 K, however, the same dislocations move jerkily. They stay immobile and then jump very quickly to the next position. The flight time is usually so short that the start and final positions can be seen on the same frame (Figure 6.5). The temperature dependence of the mean jump length, $\bar{y}_g$, is plotted in Figure 6.6 and compared with the corresponding variation of the activation area in the same material. The jump length decreases to non measurable values above the peak temperature of the activation area (373 K).

Different interpretations have been proposed for the mechanical properties of titanium when prismatic slip is activated. Because of the very strong impurity effects a Fleisher-type interaction between dislocations and interstitial solute atoms has been considered

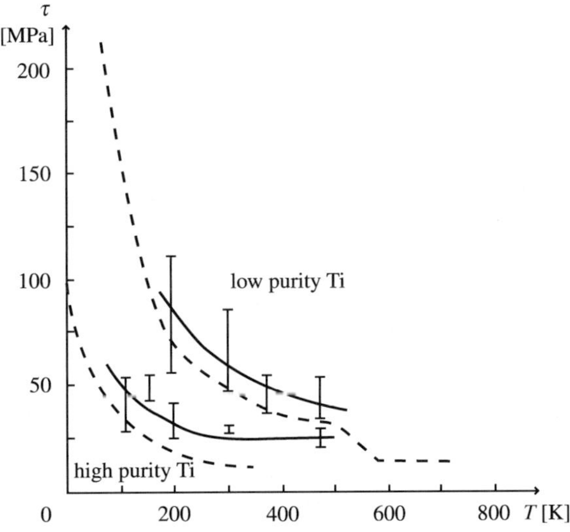

Figure 6.4. Local stress measurements on dislocations gliding in prismatic planes of Ti and comparison with the corresponding macroscopic CRSS. In situ measurements (bars) from Farenc et al. (1993); macroscopic data on high-purity Ti (upper dotted line) from Biget and Saada (1989); and macroscopic data on low purity Ti (lower dotted line) from Naka et al. (1988).

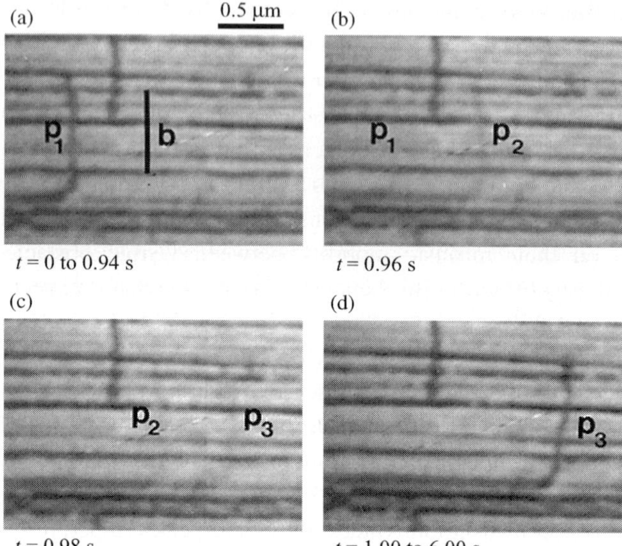

Figure 6.5. Jerky movement of a screw dislocation in a prismatic plane of a low-purity Ti ($O^* = 3270$ ppm) at 150 K. The locking positions are denoted P_i and b is the projection of the Burger's vector direction. From Farenc et al. (1995).

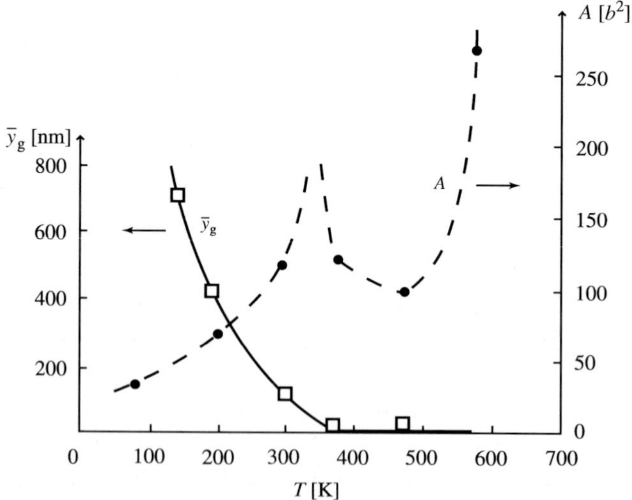

Figure 6.6. Temperature dependence of the mean jump length $\bar{y}_g$ in a low purity Ti (3270 ppm O^*). In situ measurement by Farenc et al. (1995). The macroscopic activation area in the same material is shown for comparison. (Naka et al., 1988).

(Section 3.1). It takes place in the whole thermally activated temperature range according to Tanaka and Conrad (1972) and below 250 K according to Akhtar and Teghtsoonian (1975). The kink-pair mechanism has also been proposed, in the whole temperature range by Sastry and Vasu (1972) and Naka et al. (1988) and below 200 K by Levine (1966) and Sakai and Fine (1974). Unknown mechanisms were assumed to take place above 200 K by Levine (1966), Sakai and Fine (1974) and Akhtar and Teghtsoonian (1975). Lastly, Biget and Saada (1989) pointed out that the stress-dependence of the activation area below 300 K shown in Figure 6.3 is typical of the Friedel cross-slip mechanism (Eq. (4.33)).

On the basis of electron microscopy observations, it is now obvious that Peierls-type friction forces on screw dislocations are rate controlling. This conclusion is consistent with violations of the Schmid law observed by Naka et al. (1988) because the core structure responds to various components of the stress tensor. These authors accordingly interpreted the hump in the $\tau(T)$ curves as a hardening induced by cross-slip and dipole formation. The hump may also be accounted for by the camel-hump potential (see Section 4.2.6). In both cases, however, the movement of screws should remain steady, in agreement with the kink-pair mechanism described in Section 4.2.7. This is not the case, as shown by in situ straining experiments.

In fact, all observations exactly correspond to what is expected from the transition between the locking–unlocking and kink-pair mechanisms, described in Section 4.3.4:

- Screw dislocations move jerkily at low temperatures (Section 4.3.5) and steadily above the transition. The jump length decreases with increasing temperature and reaches low values compatible with the distance between Peierls valleys at the transition temperature (Figure 6.6).
- The stress versus temperature curves exhibit the same hump as in Figure 4.27(a) and the activation area varies as in Figure 4.27(b) and (c).
- In the low-temperature range, the activation area varies as the inverse of the stress squared (Figure 6.3), in accordance with the Friedel cross-slip involved in the locking–unlocking mechanism (Section 4.3.5). The fit is less satisfactory, although still compatible, for the other—less accurate—available data.

A difficult problem is to account for the strong hardening effect of impurities seen in e.g. Figure 6.1. Naka et al. (1988) showed that the elastic interaction energy between edge dislocations and interstitial atoms cannot exceed 0.3 eV, a value well below the activation energies measured. They correctly pointed out that impurity effects necessarily concern the rate controlling screws and not the edges. Accordingly, they proposed that interstitial atoms modify the core structure of screw dislocations resulting in a more difficult sessile–glissile transition. The dislocation core structure has been computed by Legrand (1985) in the frame of the tight-binding approximation. It is definitely non-planar and difficult to move in both prism and basal planes (Figure 6.7). Recombination energies of screw dislocations, ΔE, can be deduced from the slopes of the curves shown in Figure 6.3, using

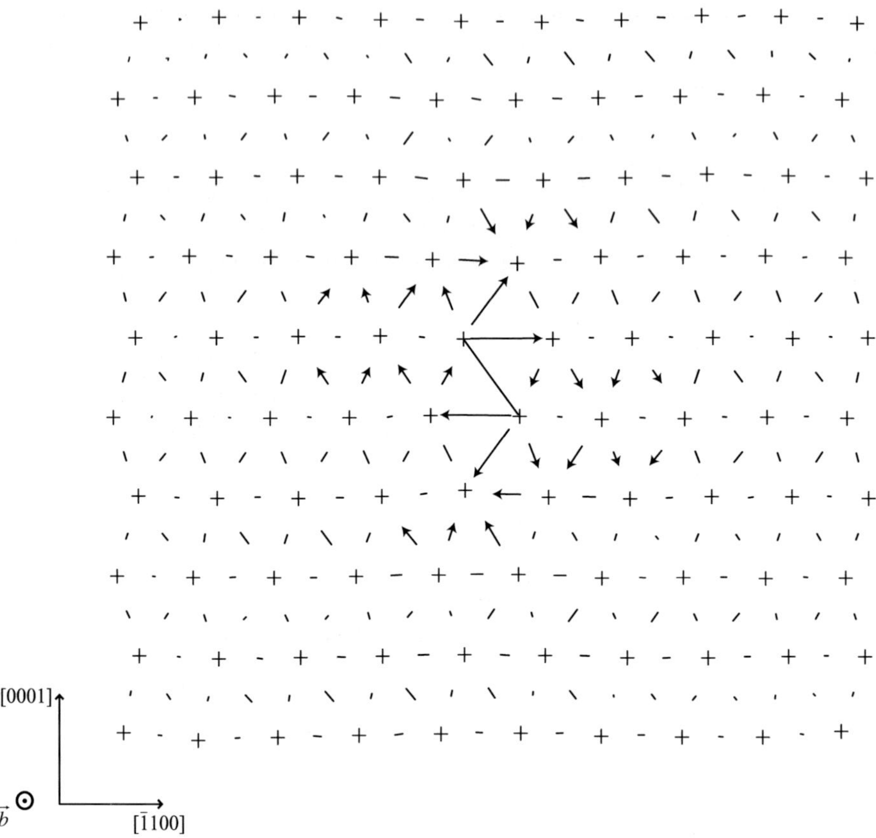

Figure 6.7. Core structure of a screw dislocation in Ti. Atomistic calculation by Legrand (1985).

Table 6.1. Experimental values of the recombination energy ΔE of screw dislocations in Ti with different amounts of impurities. From Farenc et al. (1995).

Purity	Reference	$\Delta E(\mu b^2) \sim$
30–50 ppm O*	Biget and Saada (1989)	1.6×10^{-3}
30–50 ppm O*; 70–100 ppm Fe	Biget and Saada (1989)	2.1×10^{-3}
140 ppm O*; 100 ppm Fe	Levine (1966)	2.3×10^{-3}
500 ppm O*	Akhtar and Teghtsoonian (1975)	3.5×10^{-3}
1530 ppm O*	Naka et al. (1988)	3.5×10^{-3}
3270 ppm O*	Naka et al. (1988)	6×10^{-3}
$500 < O^* < 10\,000$ ppm	Akhtar and Teghtsoonian (1975)	6.6×10^{-3}

Eq. 4.33 with $E_o \approx \hat{T} \approx 0.1\mu b^2$, and $\mu = 3.5 \times 10^4$ Mpa. Results are displayed in Table 6.1. ΔE increases with increasing impurity content, which may correspond to an increase in size of the non-planar core structure.

6.1.2 Prismatic slip in zirconium

Prismatic slip in zirconium and titanium exhibit many common features. The CRSS measured in single crystals with different amounts of impurities decreases with increasing temperature and increasing purity level (Figure 6.8(a)). The same behaviour is observed in polycrystalline zircalloy-4 (Figure 6.9(a)). In the latter experiment, one

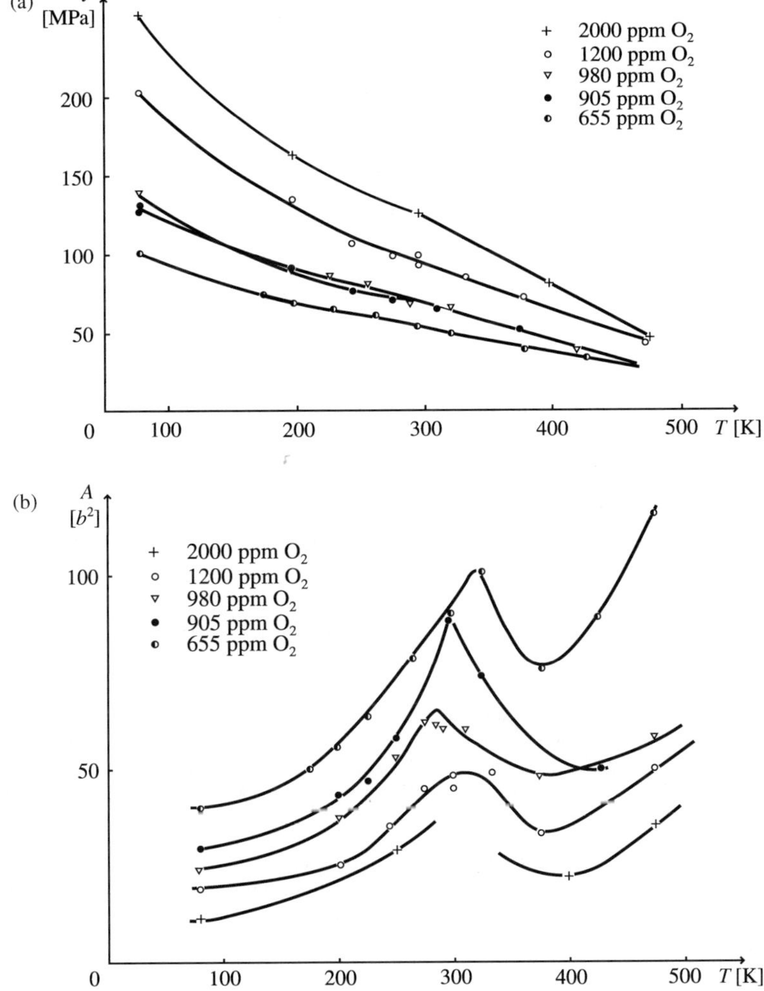

Figure 6.8. Prismatic slip in zirconium single crystals with different oxygen contents (from Soo and Higgins, 1968): (a) CRSS as a function of temperature. (b) Activation area as a function of temperature.

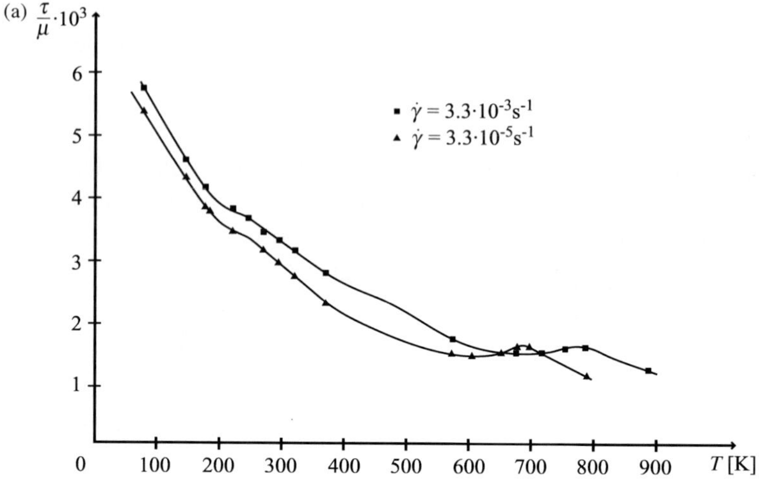

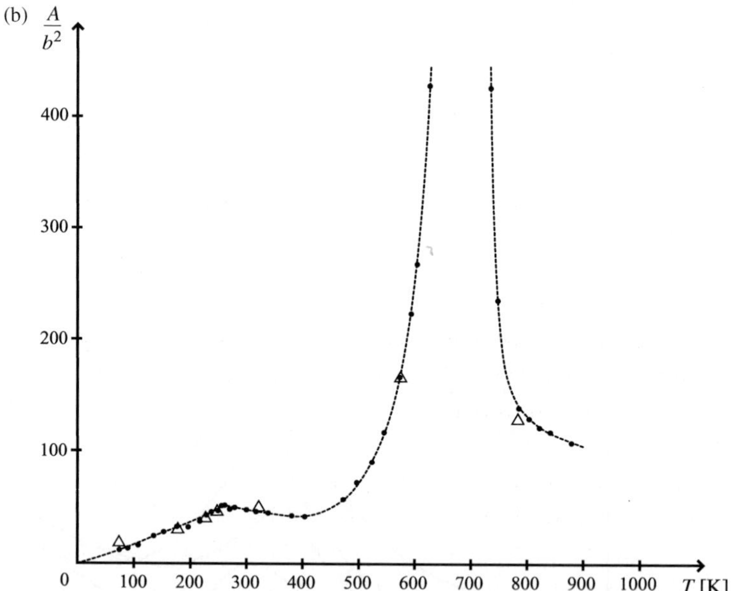

Figure 6.9. Prismatic slip in zircalloy-4 polycrystals (from Derep et al., 1980): (a) Average CRSS (scaled by the shear modulus μ) as a function of temperature. (b) Activation area as a function of temperature. (c) Stress-dependence of the activation area, A, and of $A^{-1/2}$.

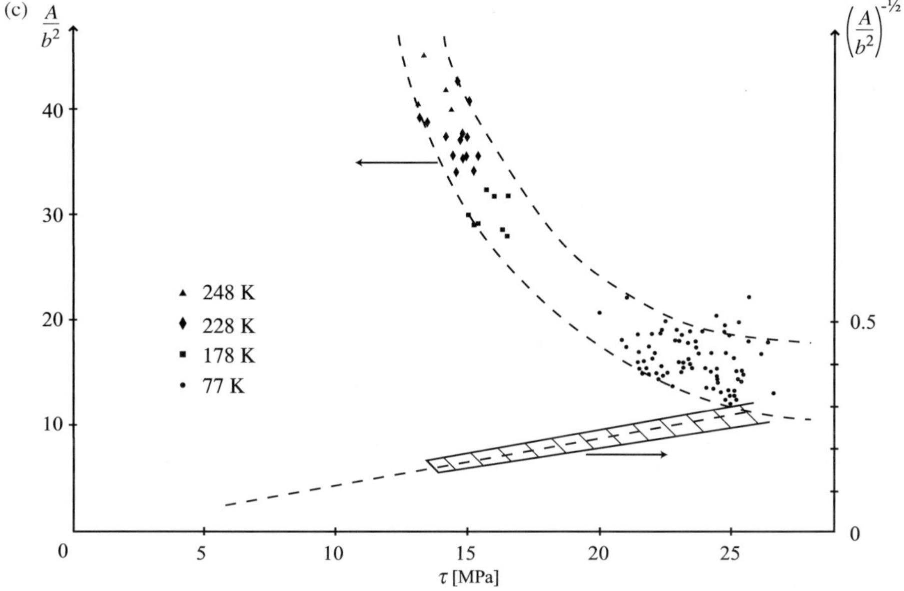

Figure 6.9. (*continued*)

can notice a small hump at about 200 K and a more pronounced one at 700–800 K. These two humps are associated with two peaks in the activation area versus temperature curves (Figure 6.9(b)). The small peak at low-temperature is also visible in Zr single crystals of different purities, as seen in Figure 6.8(b).

The high-temperature stress-hump and associated activation area peak are unambiguously due to dynamic strain ageing, according to the above authors (see also Mills and Craig, 1968; Trojanova et al., 1985).

The low-temperature stress-hump and associated activation area peak are similar to those observed in Ti. As in Ti, the activation area in the low-temperature range below 300 K varies as the inverse of the stress squared, as shown in Figure 6.9(c).

These results have predominantly been interpreted by an elastic interaction between edge dislocations and interstitial impurities (Tyson, 1967; Mills and Craig, 1968; Soo and Higgins, 1968; Derep et al., 1980). A change in the controlling mechanism is sometimes proposed above the low-temperature discontinuity: interaction with interstitial clusters according to Derep et al. (1980) and climb of jogs on screw dislocations according to Soo and Higgins (1968). The Peierls mechanism is also considered by Sastry et al. (1971) in the whole temperature range. It is not excluded by Derep et al. below 200 K.

As in Ti, only microstructural observations can help to determine the actual controlling mechanism. Akhtar and Teghtsoonian (1971) report rectilinear screw dislocations after deformation at 78 K. Figure 6.10 shows a dynamic sequence in a zirconium crystal strained at 24 K in a high voltage electron microscope. As in Ti, the less mobile dislocations are long rectilinear screws moving jerkily. These observations show that: (i) dislocation movements are controlled by Peierls-type friction forces acting on screw segments and (ii) the discontinuity observed at 200–300 K may be due to the transition between the locking–unlocking and kink-pair mechanisms as in Ti. The latter conclusion is supported by the stress-dependence of the low-temperature activation area that corresponds to the bulge mechanism (Figure 6.9(c)).

6.1.3 Prismatic slip in magnesium

Since basal slip is the easiest system in magnesium, single crystals must be strained along a direction of the basal plane in order to activate prism slip. To achieve this condition the orientation has to be very accurate. The corresponding CRSS has been measured as a function of temperature by Ward-Flynn et al. (1961), Ahmadieh et al. (1965), Akhtar and Teghtsoonian (1969) and Stohr (1972) (Figure 6.11). The stress–temperature curve for Mg–12.9 at.% Li is also shown for comparison. Both CRSSs decrease rapidly with increasing temperature, except for pure Mg between 4 and 77 K. The corresponding activation areas are shown in Figure 6.12 as a function of temperature. Measurements in pure Mg below and above 400 K have been made by different authors. They, however, clearly reveal a discontinuity that can be correlated with the small hump on the stress–temperature curve of Figure 6.11 around 400 K. A peak is also observed on the $A(T)$ curve in Mg–Li that corresponds to the small plateau on the $\tau(T)$ curve near 350 K. These characteristic features are very similar to those observed in Ti and Zr.

From the shape of $\tau(T)$ curve in pure Mg, Ward-Flynn et al. (1961) concluded that prismatic slip is controlled by the Friedel cross-slip above 450 K. Later, Ahmadieh et al. (1965), considering the shape of the $\tau(T)$ and $A(T)$ curves in Mg–Li alloys, proposed that prismatic slip might be controlled by the kink-pair mechanism below this temperature. However, these attempts do not account for the anomalous stress–temperature variation in Mg between 4 and 77 K.

The only available microscopic observations are in situ experiments by Couret and Caillard (1985a,b) and Couret et al. (1991) on Mg single crystals oriented for pure prismatic slip. Local stresses deduced from dislocation radii of curvature are close to macroscopic ones in the whole temperature range investigated, which ensures the reliability of the observations (Figure 6.13). Between 80 and 473 K the less mobile dislocations are long rectilinear screws, which indicates that their motion is controlled by a Peierls-type mechanism (Figure 6.14).

Above 300 K the movement of screws is viscous, in accordance with the kink-pair mechanism. When dislocations are pinned on extrinsic obstacles their screw parts are divided into shorter segments that still glide viscously at a lower velocity. Figure 6.15 shows that, in a restricted area and during a sufficiently short time to maintain a constant

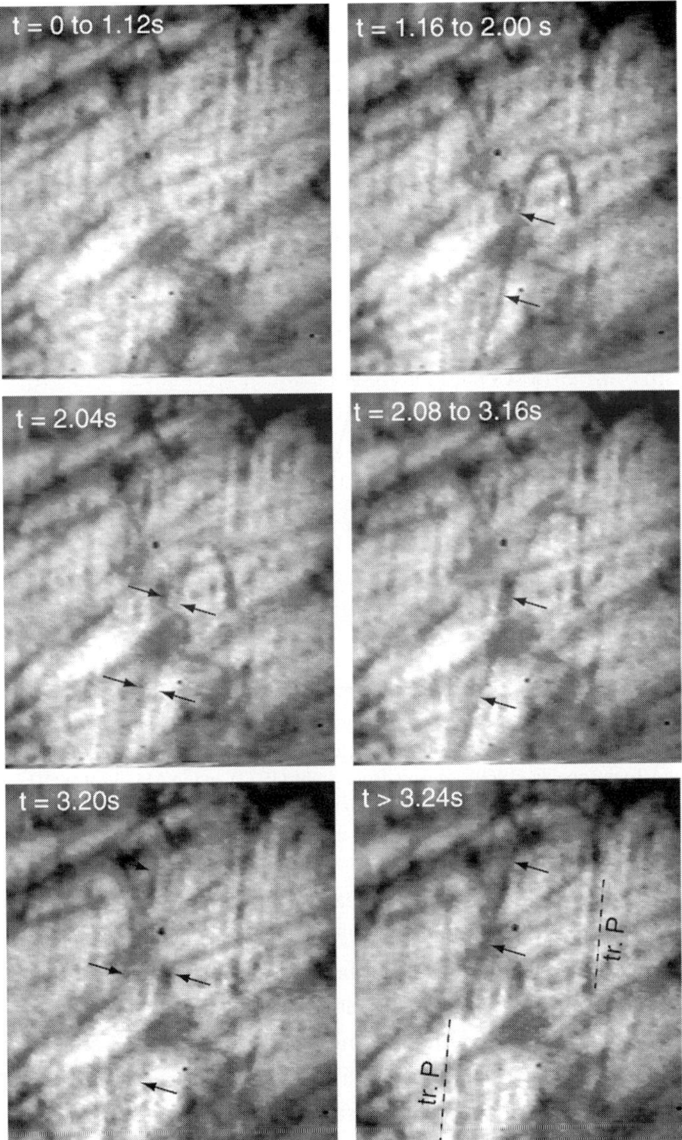

Figure 6.10. Jerky movement of a screw dislocation in a prism plane of zirconium at 24 K. In situ experiment in a TEM. Arrows indicate the successive positions of the screw part. tr. P is the trace of the prism plane. (D.Caillard, unpublished work).

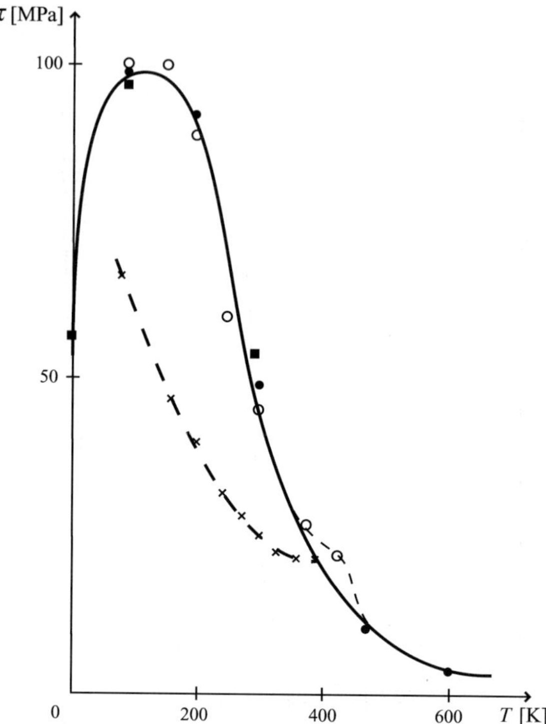

Figure 6.11. CRSS of prismatic slip in Mg single crystals (full circles from Ward-Flynn et al. (1961), open circles from Akhtar and Teghtsoonian (1969), squares from Stohr (1972) and in Mg–12.9 at.% Li single crystals (crosses from Ahmadieh et al. (1965)).

local stress, the screw velocity varies linearly with its corresponding length. L_0 is a systematic error in the determination of the dislocation length in the pure screw orientation, this result is considered to be the experimental evidence of the length effect discussed in Section 4.2.7. When the dislocation velocity is measured as a function of the local stress, and at constant length, the microscopic activation area of the kink-pair mechanism is the slope of the curve ln v as a function of τ (Figure 6.16). It is fairly small ($A = 9b^2$) in agreement with theoretical estimates (Section 4.2). The larger activation areas measured in macroscopic experiments at the same temperature ($A = 30b^2$) can be accounted for by multiplication processes that immediately increase the dislocation density (relation (2.4)) when the stress is increased (Couret and Caillard, 1985b).

The observation below 300 K of jerky movements of the rectilinear screws (Figure 6.17) suggests that, as in Ti, the peak in activation area at 350–400 K (Figure 6.12) corresponds to the transition between the locking–unlocking mechanism (Friedel cross-slip) at low temperatures and the kink-pair mechanism at higher temperatures.

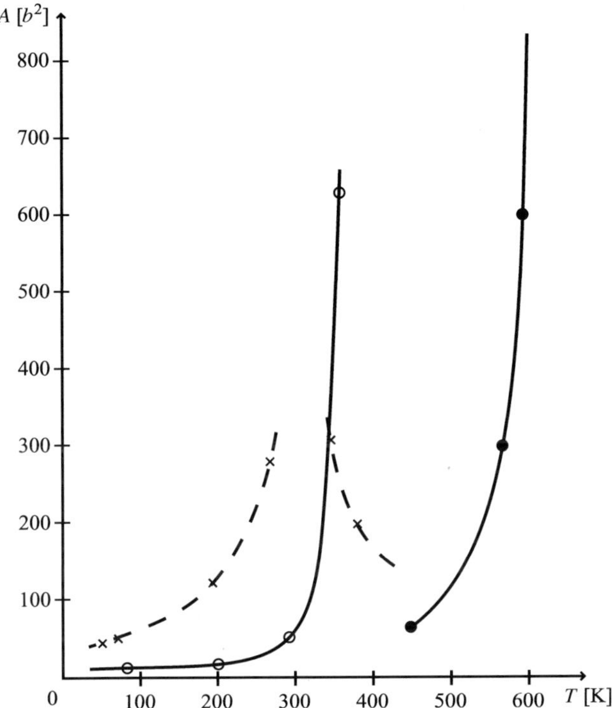

Figure 6.12. Activation areas of prismatic slip in Mg single crystals (full circles from Ahmadieh et al. (1965), open circles from Akhtar and Teghtsoonian (1969)) and in Mg–12.9 at.% Li single crystal (crosses from Ahmadieh et al. (1965)).

The movement of screw dislocations observed in situ is not planar. It consists of a net movement in the prismatic plane, under the applied stress superimposed to oscillations in the basal plane, under a small fluctuating internal stress. Dislocations are accordingly much more mobile in the basal plane than in the prismatic ones. This result is consistent with a core structure extended in the basal plane, as computed by Bacon and Martin (1981a) and Vitek and Igarashi (1991). By applying a sufficiently high-stress along the prism plane, these computations also yield a metastable core configuration, glissile in the prism plane, that may be at the origin of the low-temperature locking–unlocking mechanism.

No in situ observation has been made so far in Mg–Li alloys which prevents conclusions being drawn about the rate controlling mechanism in this compound.

6.1.4 Prismatic slip in beryllium

The temperature dependence of the CRSS of prismatic slip in beryllium has been measured in single crystals oriented so as to inhibit basal slip by Régnier and Dupouy (1970).

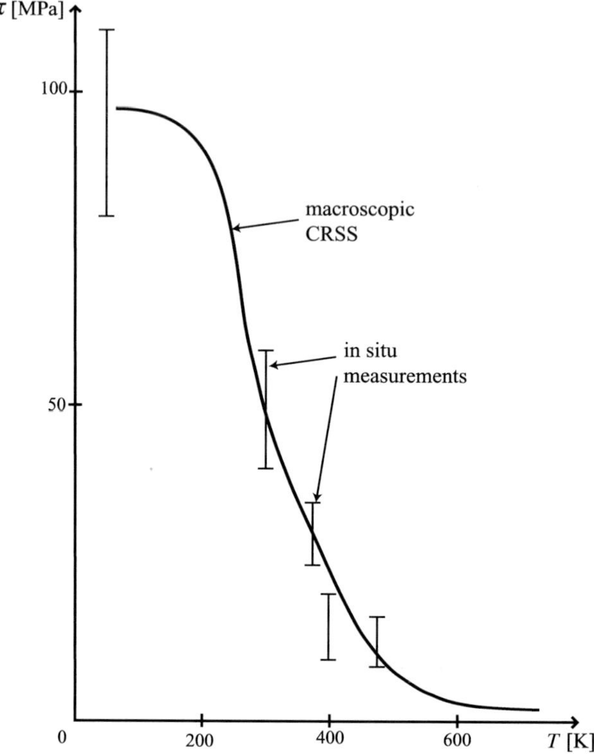

Figure 6.13. Local stress measurements on dislocations gliding in Mg single crystals (bars from in situ experiments of Couret and Caillard (1985a)). The macroscopic CRSS (Stohr, 1972) is shown for comparison.

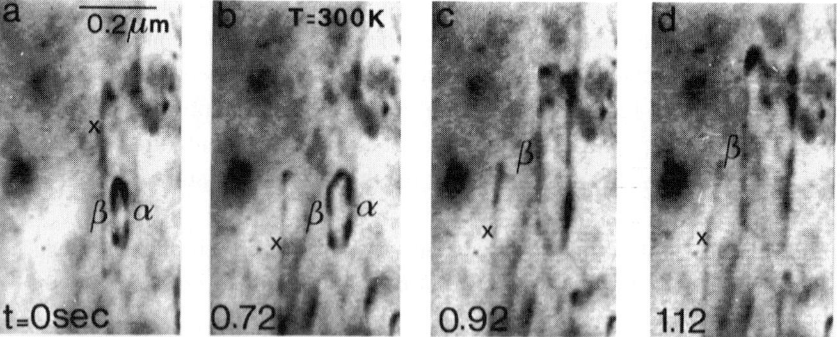

Figure 6.14. Steady movements of rectilinear screw dislocations in the prismatic planes of Mg single crystals (in situ experiment at 300 K, from Couret and Caillard (1985b)). Note the slow movement of the vertical straight screw dislocations and the fast upwards movement of one edge segment of the loop noted $\alpha\beta$ between 0.72 and 0.92 s.

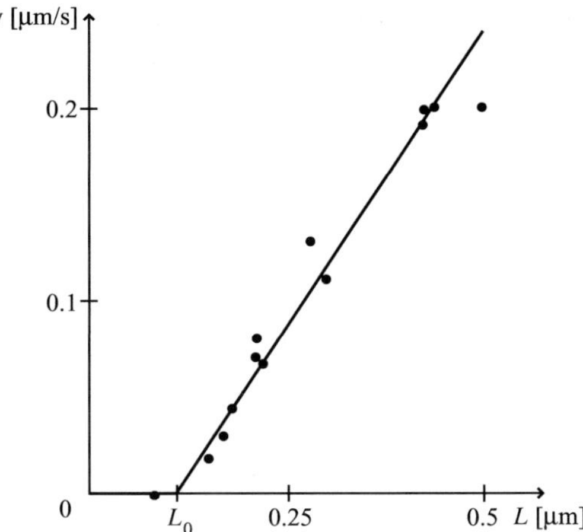

Figure 6.15. Velocities of screw dislocations in prismatic planes of Mg as a function of their length L (in situ experiment at 300 K, from Couret and Caillard (1985b)).

It exhibits a strong anomaly with an increase of stress between 150 and 300 K (Figure 6.18). The anomaly does not appear on the micro-yield stress measured at a 10^{-6} strain. A few activation area measurements indicate low values ($50b^2$) below 150 K and much higher values in the ascendant part of the flow stress curve between 150 and 300 K. Previous experiments by Treharne and Moore (1962) did not reveal the anomaly because no measurements were made between 4 and 300 K.

Observations of deformed specimens reveal straight slip lines parallel to the most activated prism plane at low-temperature and more extensive cross-slip between prism and basal planes at higher temperatures. No cross-slip can, however, be seen at micro-yield and the slip line lengths along the prism planes decrease with increasing temperature (Régnier and Dupouy, 1970). TEM observations in samples cut along the most active prism plane, after deformation at 168 K, reveal long screw dislocations connected by macrokinks lying in the prism plane (Figure 6.19, from Jonsson and Beuers, 1987).

Prismatic slip in Be has been extensively studied by TEM in situ experiments (Couret and Caillard, 1989a,b). Figure 6.20 shows that local stress estimates performed on individual gliding dislocations reproduce the anomalous temperature dependence of the CRSS fairly well. This emphasizes the reliability of the corresponding observations.

Dislocation movements are controlled by friction forces acting on the screw parts, as in the other HCP metals investigated above. Rectilinear screws move jerkily, as shown in Figure 6.21, which denotes a locking–unlocking mechanism in the whole temperature range investigated (80 to 450 K).

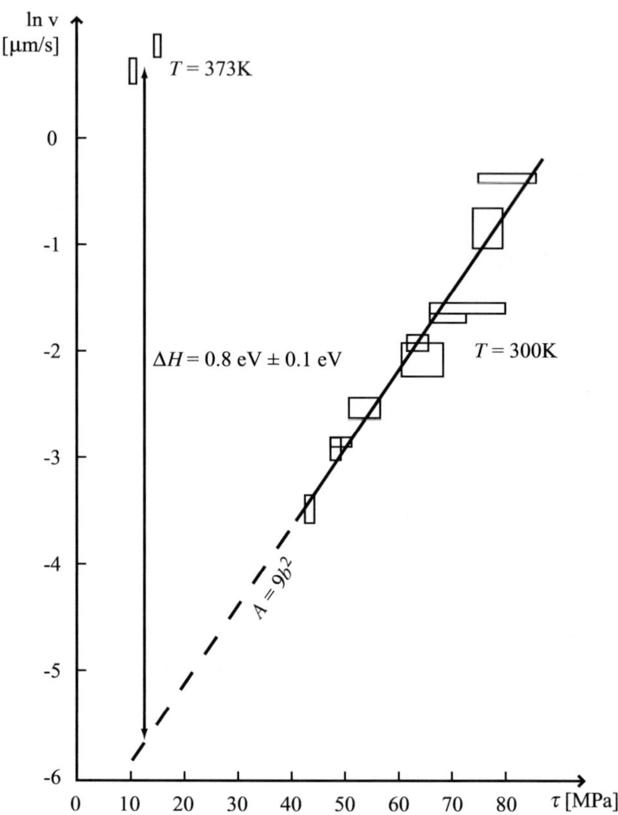

Figure 6.16. Velocities of screw dislocations in prismatic planes of Mg as a function of local stress at 300 and 373 K. The slope of the curve at 300 K yields the microscopic activation area, $A = 9b^2$, and the measurements at 300 and 373 K yield an order of magnitude of the microscopic activation enthalpy (from Couret and Caillard (1985b)).

Figure 6.22 shows that the frequency of a given value of the locking time, t_1, varies exponentially in agreement with the predictions of Section 4.2.5. The slope yields the corresponding unlocking probability per unit time, $P_{ul} = 3.5\ s^{-1}$ (Eq. (4.48)). Figure 6.23 shows that the frequency of a given jump length y_g also varies exponentially with y_g as expected from Eq. (4.47). The reciprocal of the slope yields the mean jump length $\bar{y}_g = 68$ nm. Data show that this length varies as a function of stress and temperature, which indicates that dislocation locking is not controlled by fixed obstacles. Each jump produces a pair of macrokinks (Figure 6.24) similar to those described schematically in Figure 4.28. Macrokinks seen in Figures 6.24 (in situ) and 6.19 (post mortem) exhibit similar features. Macrokinks emerging at the foil surface in Figure 6.24 produce stair-shaped traces along two directions. One direction (noted Tr.P) is parallel to the trace of

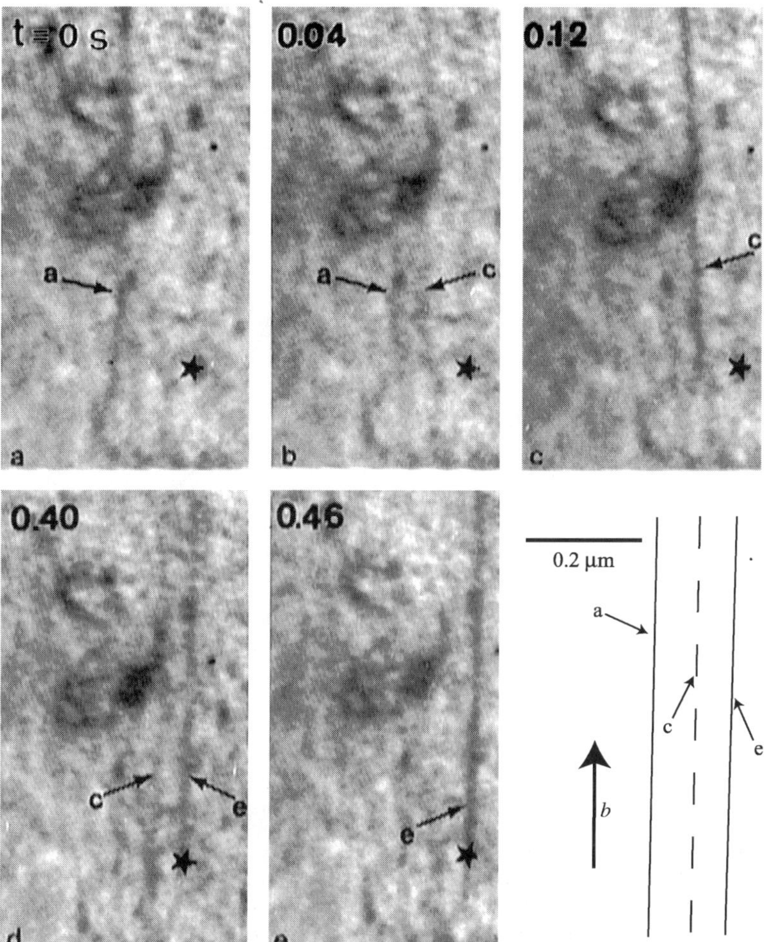

Figure 6.17. Jerky movement of a rectilinear screw dislocation in a prismatic plane of Mg at 150 K. Arrows a to e refer to successive positions of the dislocation. In situ experiment in a TEM (from Couret et al. (1991)).

the prism plane. It corresponds to a macrokink emerging at the surface which shows that jumps of the screw segments take place in the prism planes. The second direction (noted Tr.B) is parallel to the trace of the basal plane. It corresponds to motions of the screw segments in the basal plane under the internal stress as it is locked with respect to prism glide. Dislocations in the locking positions are thus dissociated in the basal plane similar to Mg. This agrees with the atomistic calculations of Bacon and Martin (1981a,b) and Vitek and Igarashi (1991).

Additional in situ experiments have been made by Beuers et al. (1987). Most microsamples were, however, cut in the basal plane, which prevents the observation of

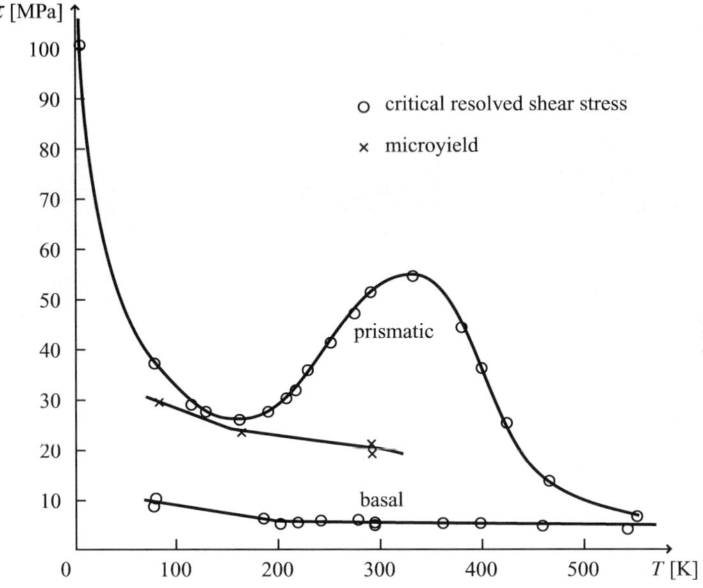

Figure 6.18. CRSS of prismatic and basal slip in Be (from Régnier and Dupouy (1970)).

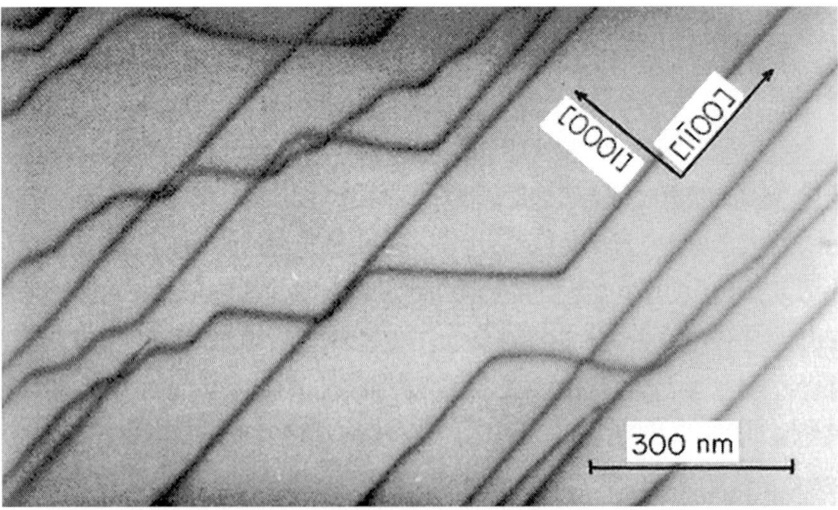

Figure 6.19. Straight screw dislocations parallel to [1$\bar{1}$00], and macrokinks, in a prismatic plane of Be strained at 168 K. (Courtesy of Jönsson and Beuers, 1987).

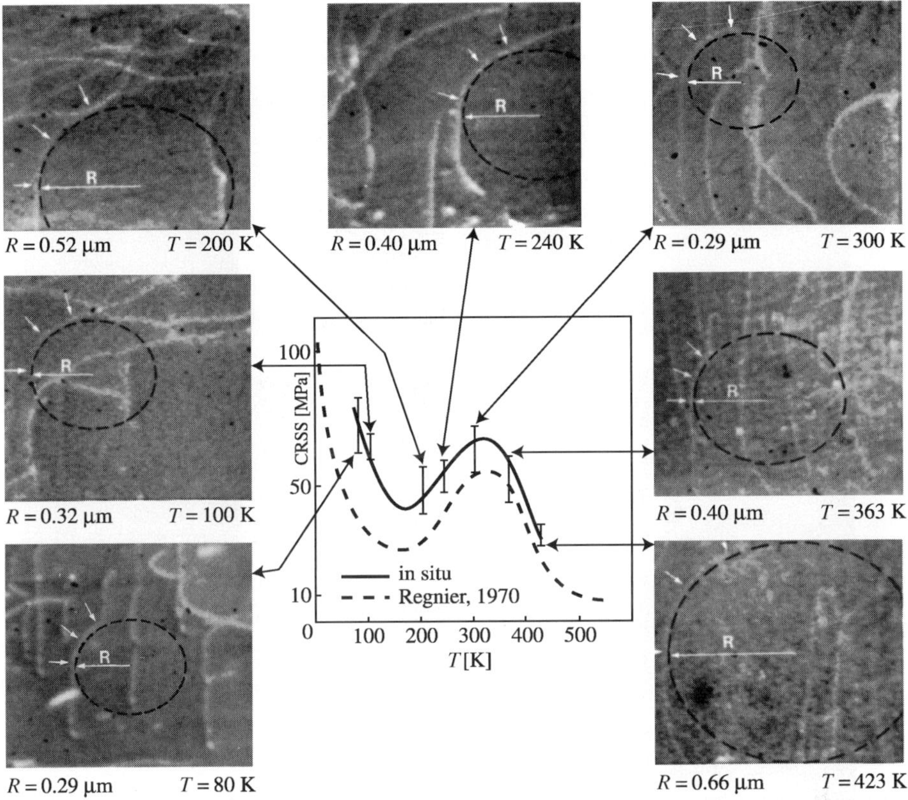

Figure 6.20. Local stress measurements at dislocations moving in prismatic planes of Be. Note that the yield stress anomaly is correctly reproduced. From Couret and Caillard (1989b).

screw dislocations. Indeed such dislocations rotate easily to the edge orientation under stress, so as to shorten their length. Under such conditions only stable and easy prismatic slip of edge segments could be observed. However, one experiment reported some cross-slip between prism and basal planes at room temperature when another foil orientation was used.

Different explanations have been proposed for the anomalous temperature dependence of the yield stress:

Régnier and Dupouy (1970) claimed that three different screw mobility mechanisms operate in three temperature domains. At low temperatures ($T < 150$ K) screw dislocations assumed to be fully glissile in the prism plane. At intermediate temperatures ($150 < T < 300$K), they should cross-slip more and more easily in the basal plane, without being able to cross-slip back to the prism plane. This would induce a thermally activated hardening that could explain the origin of the yield stress anomaly. At high

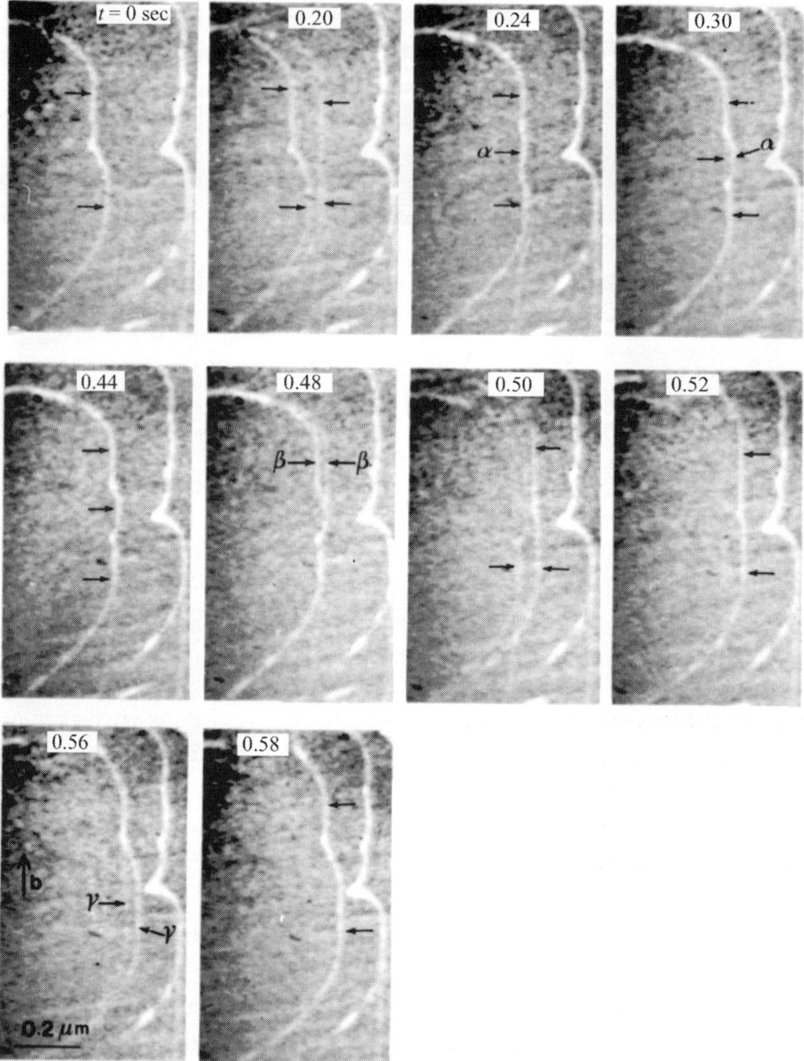

Figure 6.21. Jerky movements of a straight screw dislocation in a prismatic plane of Be at 300 K. Arrows refer to successive positions of the dislocation. Starting and final positions corresponding to one jump are both visible on the same frame, e.g. at $t = 0.2$, 0.48, 0.50 and 0.56 s. b is the Burgers vector.

temperatures ($T > 300$ K) both cross-slip processes should be possible, which corresponds in fact to the locking–unlocking mechanism.

The model proposed by Beuers et al. (1987) is also based on a thermally activated hardening by cross-slip in the intermediate temperature range.

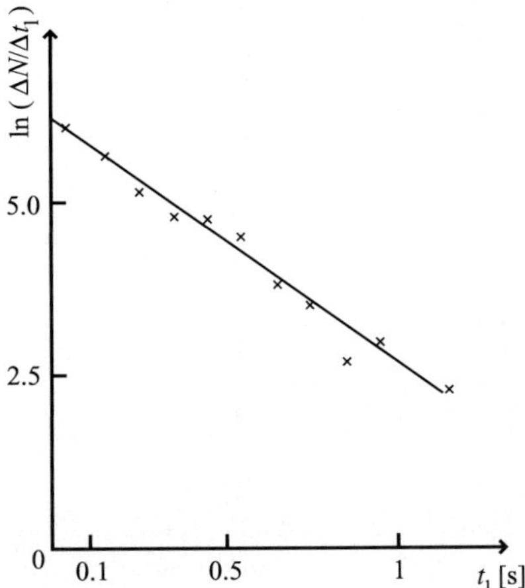

Figure 6.22. Frequency $\Delta N/\Delta t_1$ of waiting times in locked configurations during prismatic glide in Be (TEM in situ experiments at $T = 300$ K). ΔN is the number of screw dislocations locked during time t_1, within Δt_1. From Couret and Caillard (1989a).

In situ experiments show that the same locking–unlocking mechanism is operating in the whole temperature domain, under a local stress that reproduces the anomalous macroscopic behaviour (Figure 6.20). Under such conditions, the recombination energy ΔE must increase by a factor of 1.8 between 150 and 300 K in order to account for the yield stress anomaly (Couret and Caillard, 1989b). In the elastic approximation, this corresponds to a decrease with temperature of the stacking fault energy in the basal plane. Similar effects may take place in Mg at low-temperature. This conclusion is consistent with the micro-yield stress measurements shown in Figure 6.18 because only edge segments are supposed to be mobile

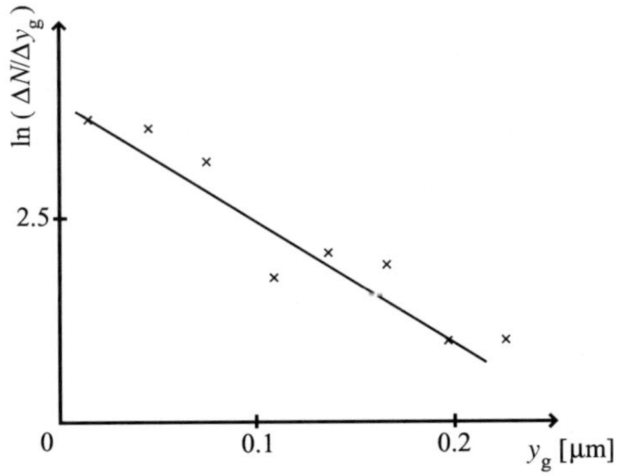

Figure 6.23. Frequency $\Delta N/\Delta y_g$ of jump lengths during prismatic slip in Be (TEM in situ experiments at $T = 300$ K). ΔN is the number of screw dislocations jumping over the length y_g, within Δy_g. From Couret and Caillard (1989a).

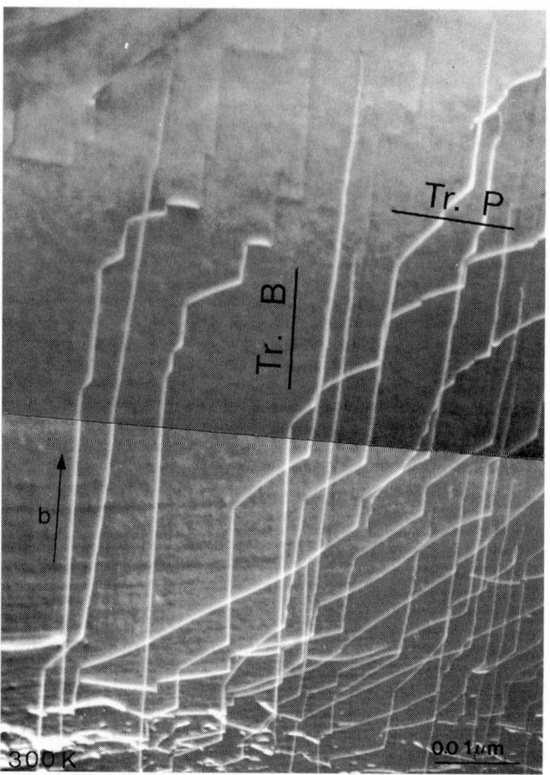

Figure 6.24. Macrokinks on screw dislocations gliding in prismatic planes of Be at 300 K (in situ experiment by Couret and Caillard (1989a)). Note the stair-shape slip traces (weak contrast on the top) along Tr. P and Tr. B and compare with Figure 6.19.

at the very beginning of deformation tests. The above interpretation cannot, however, account for the high activation areas measured by Régnier and Dupouy in the anomalous domain. Dislocation multiplication processes may explain this discrepancy (relation (2.4)).

6.1.5 Conclusions on prismatic slip in HCP metals

This study underlines the variety of the proposed dislocation mobility mechanisms. It also shows that only microstructural observations can help to determine the relevant controlling mechanisms. In particular, quantitative in situ experiments are very helpful, especially when local stress measurements are shown to be consistent with macroscopic ones. All macroscopic and microscopic data are consistent with the locking–unlocking mechanism and its transition to the kink-pair mechanism described in Chapter 4.

6.2. GLIDE ON NON-CLOSE-PACKED PLANES IN FCC METALS

After this extensive study of the macro- and microscopical aspects of prism slip in HCP metals the state of knowledge about slip on non-close-packed planes in FCC metals is exposed. This slip mode has been known to operate in aluminium for a long time, on the basis of slip trace analysis through optical observations (Schmid and Boas, 1931; Lacombe and Beaujard, 1947; Cahn, 1951; Servi et al., 1952; Johnson et al., 1956). These observations showed that slip planes such as {100}, {110}, {112}, {113} and even higher indices are activated above 200°C, while FCC metals most frequently glide along their close-packed planes. However, these results were only qualitative. In comparison to HCP metals, an experimental difficulty arises in FCC metals where the higher symmetry prevents the isolation of non-close-packed systems in a uniaxial test. Single crystal orientations are chosen which ensure a high Schmid factor on the slip system to be studied. Then various temperature conditions are imposed while the operative slip systems are determined.

6.2.1 {110} slip

A detailed study of dislocation glide on {110} planes was undertaken by Le Hazif et al. (1973) and Le Hazif and Poirier (1975). ⟨110⟩ {110} slip was first studied through compression creep experiments in ⟨001⟩ single crystals. Four ⟨110⟩ {110} slip systems can be activated (Schmid factor 0.5) and also eight ⟨110⟩ {111} systems (Schmid factor 0.45). A variety of FCC crystals were investigated (Al, Ag, Cu, Ni, Au) over a range of temperatures. Stresses were adjusted so as to obtain strain-rates between 10^{-5} and $10^{-4}\ s^{-1}$. Slip traces were identified by optical microscopy (Nomarski interferometer) and using the TEM replica technique for aluminium. For each crystal, two temperatures could be defined labelled T_1 and T_2, respectively. Below T_1 {111} slip alone was observed, above T_2 {110} slip was the only one operating while between T_1 and T_2 the crystals were gliding on both types of planes. Table 6.2 summarizes these observations.

A thorough examination of the slip traces showed the following features. Below T_1 fine {111} slip traces were observed with cross-slip evidence from {111} to {111}. Above T_2 bundles of slip traces were present corresponding to {110} glide. TEM replicas resolved

Table 6.2. Temperature domains corresponding to {111} slip exclusively ($T < T_1$) {110} slip exclusively ($T > T_2$). After Le Hazif et al. (1973). T_M is the melting temperature.

Metal	T_1 (K) ≈	T_2 (K) ≈	T_1/T_M ≈	T_2/T_M ≈
Al	350	540	0.35	0.60
Ni	700	1225	0.40	0.70
Au	650	900	0.50	0.70
Cu	820	>1270	0.60	>0.91
Ag	870	>1185	0.70	>0.96

fine slip traces inside the bundles that could never be analysed in terms of alternate slip on {111} with a net {110} slip component. This is a clear indication that dislocations do glide steadily on {110}. This effect was particularly pronounced in the aluminium crystals.

A comparison of the temperatures in Table 6.2 (normalized to T_M) with the stacking fault energies of these metals showed that the higher the stacking fault energy the lower the transition temperature. This is illustrated more quantitatively in Figure 6.25 where the parameter $\mu b/\gamma$ (proportional to the dislocation dissociation width in {111}) is plotted as a function of reduced temperature T/T_M. Stacking fault energies, γ, are from Gallagher (1970). A positive correlation is observed between the dislocation width and the onset temperature of {110} glide T_1/T_M for the five FCC metals investigated.

To get further information about the mechanism of {110} slip, the authors deformed ⟨100⟩ aluminium single crystals at constant strain-rates between 9×10^{-6} and 9×10^{-4} s^{-1}. The temperatures were between 225 and 365°C (Le Hazif and Poirier, 1975).

The stress–strain curve is schematically represented in Figure 6.26. It consists of three stages. Stage I is characterized by a strong linear hardening, with $\theta/\mu \sim 1.5 \times 10^{-2}$, which

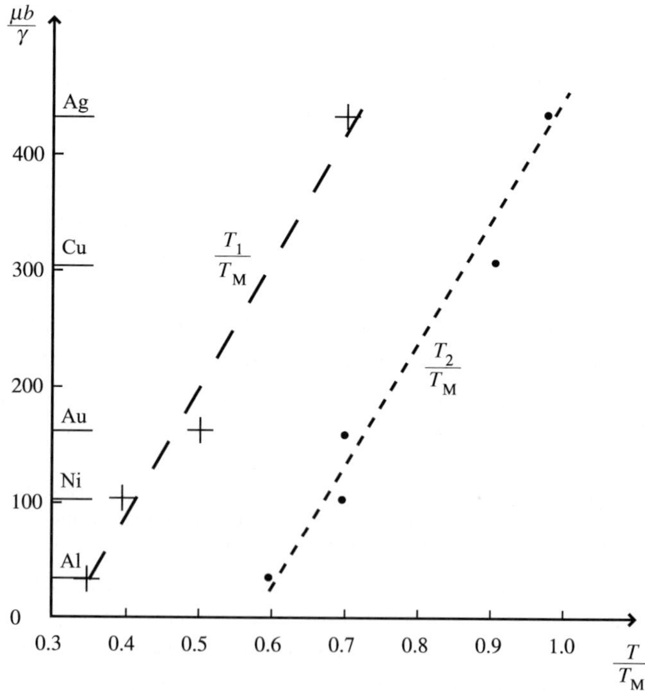

Figure 6.25. Correlation between the dissociation width $\mu b/\gamma$ on {111} as a function of reduced temperature T/T_M for five FCC metals. From Le Hazif et al. (1973).

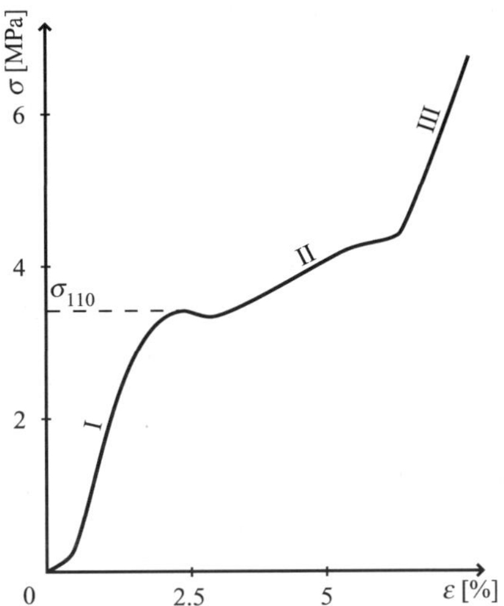

Figure 6.26. Schematics of the stress–strain curves of ⟨100⟩ Al single crystals. From Le Hazif and Poirier (1975).

does not depend too much on temperature. Fine slip on {111} is observed. At a strain of about 1 or 2×10^{-2} a stage II of lower hardening takes over. It begins with a yield drop with subsequent stress oscillations. The flow stress at the yield drop, σ_{110}, is correlated with the first {110} slip bands, visible on the lateral faces. The heterogeneity of deformation correlates well with the stress oscillations. In stage III the stress oscillations disappear and the work-hardening coefficient increases again towards the value of stage I.

Dislocation structures as observed in TEM reveal that 5–10 μm cells in stage I are still present in stage II. In stage III smaller cells (a few μm in diameter) appear inside the primary ones.

By performing tests under various conditions the variation of σ_{110} with temperature and strain-rate was determined (Figure 6.27). The trend for σ_{110} as a function of T and $\dot{\varepsilon}$ reflects a strongly thermally activated mechanism for {110} slip, which becomes athermal around 350°C for the lowest strain-rate. This mechanism will be described in Section 6.2.4.

6.2.2 {100} slip in aluminium

[112] oriented aluminium single crystals were chosen for this study (Figure 6.28). In this symmetrical orientation, the [101]($\bar{1}11$) and [011]($1\bar{1}1$) slip systems have a 0.408 Schmid factor while for [110](001) it is 0.471. For {110}, {112} and {113} planes the Schmid factors are lower. The Burger's vectors activated on the ($\bar{1}11$), ($1\bar{1}1$) and (001) glide planes are different so it is rather easy to determine which one is operating by slip trace analysis. Traces of the two octahedral systems above are visible by looking along [$1\bar{1}0$] (points B in Figure 6.28) and invisible by looking along [$\bar{1}11$] (point A). The situation is reversed for

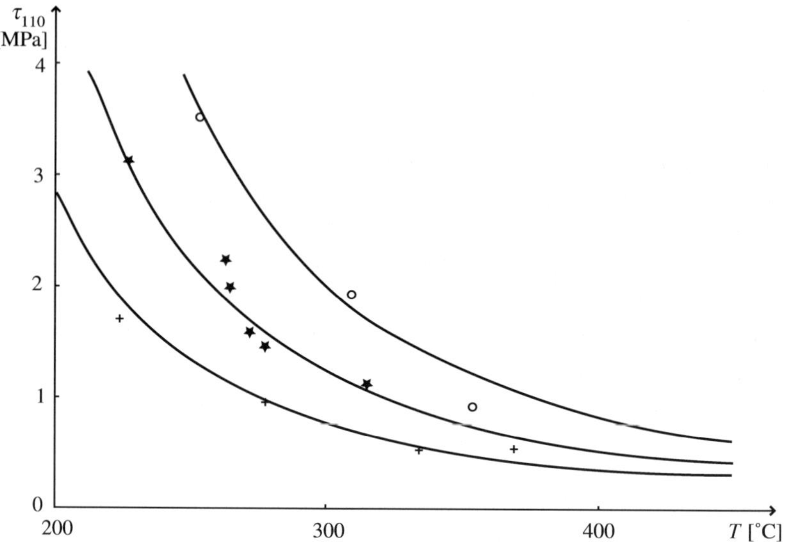

Figure 6.27. Variation of the critical stress for {110} slip in Al with temperature for various strain-rates. The lowest one is $9 \cdot 10^{-6}\ s^{-1}$ and the others are respectively 10 and 100 times higher. Data from Le Hazif and Poirier (1975).

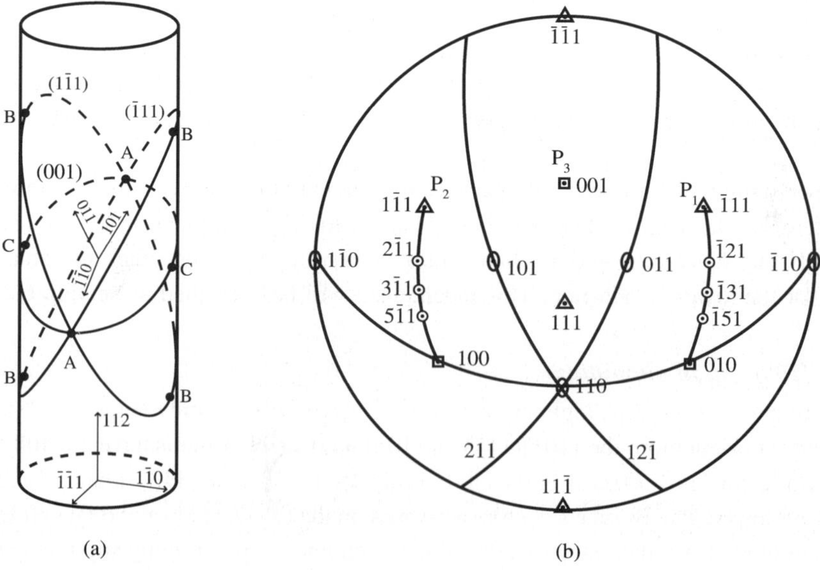

Figure 6.28. [112] single crystal geometry. (a) Slip systems with the highest Schmid factors. (b) Corresponding stereographic projection.

cube glide. Another advantage of this orientation is that the tensile axis is stable, a high degree of work-hardening being achieved by intersection between the two active octahedral systems. This is a favourable condition to activate slip on [110](001).

6.2.2.1 Creep test results. A constant stress creep machine was used at temperatures ranging from 150 to 620°C (0.45–0.96 T_M). The stress was adjusted to obtain creep rates, $\dot{\varepsilon}$, between 10^{-7} and 10^{-5} s^{-1} (Carrard and Martin, 1987).

The slip traces observed as a function of temperature are visible in Figure 6.29. At an intermediate temperature ($T = 150$°C) slip lines correspond to the two octahedral systems. Note that they become invisible when the sample surface is observed along $[\bar{1}\bar{1}1]$ as expected (see Figure 6.28). At higher temperatures (180–220°C) the cube system starts to be activated. However, the corresponding traces appear to be wavy (Figure 6.29(b)). They do not exhibit sharp angles, even at high magnification. This is an indication of multiple

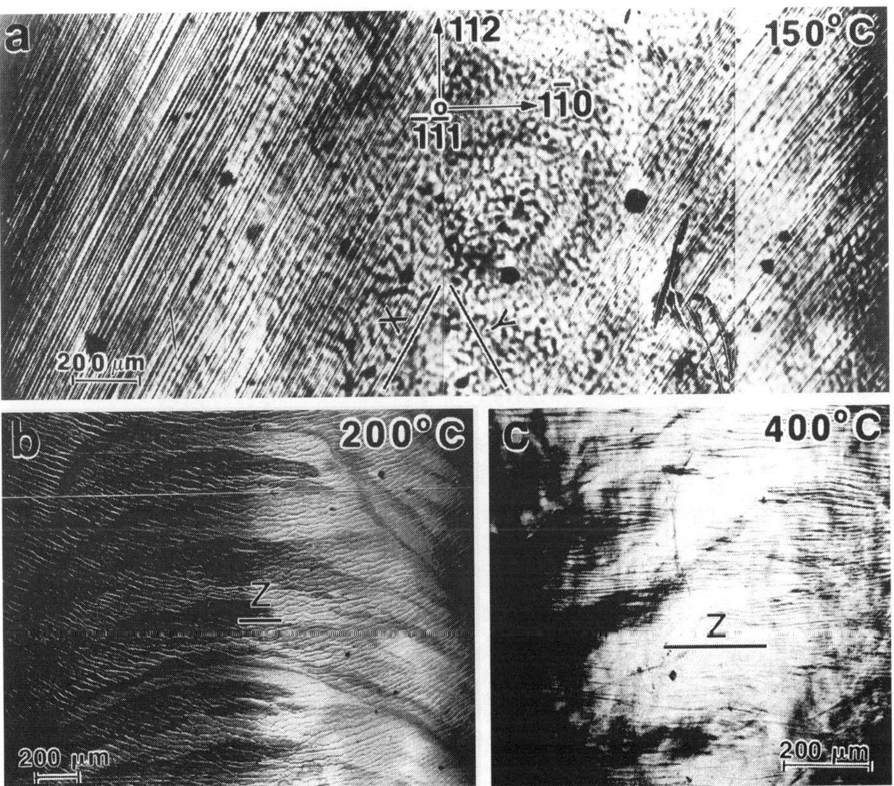

Figure 6.29. Aspect of slip traces as a function of creep temperature in ⟨112⟩ Al single crystals: (a) 150°C SEM, (b) 200°C SEM and (c) 400°C optical microscope. *X*, *Y* and *Z* are the traces of $(\bar{1}11)(1\bar{1}1)$ and (001) planes, successively. After Carrard and Martin (1987).

cross-slip between (001)($\bar{1}$11) and (1$\bar{1}$1) planes for $\frac{1}{2}$[110] dislocations. Therefore, the Schmid law is fulfilled, but slip is unstable and $\frac{1}{2}$[110] screws have a tendency to transit by cross-slip on the ($\bar{1}$11) and (1$\bar{1}$1) planes in spite of their low Schmid factors. At high-temperature ($T = 400$°C) slip lines correspond to [110](001) exclusively (Figure 6.29(c)). They are sharp and rectilinear, as compared with those at 200°C. This observation of (001) traces was also made at 620°C. Cube glide is likely to operate up to the melting point. This change of slip mode with temperature can also be evidenced by looking at the sample cross-section. From a circle, it evolves towards an elliptical shape. The major axis of the ellipse, equal to the circle diameter, is parallel to [$\bar{1}\bar{1}$1] at 150°C and to [$\bar{1}$10] at 400°C as illustrated in Figure 6.30(a) and (b), respectively. This evolution is in agreement with the different glide systems, which operate at intermediate and high temperatures.

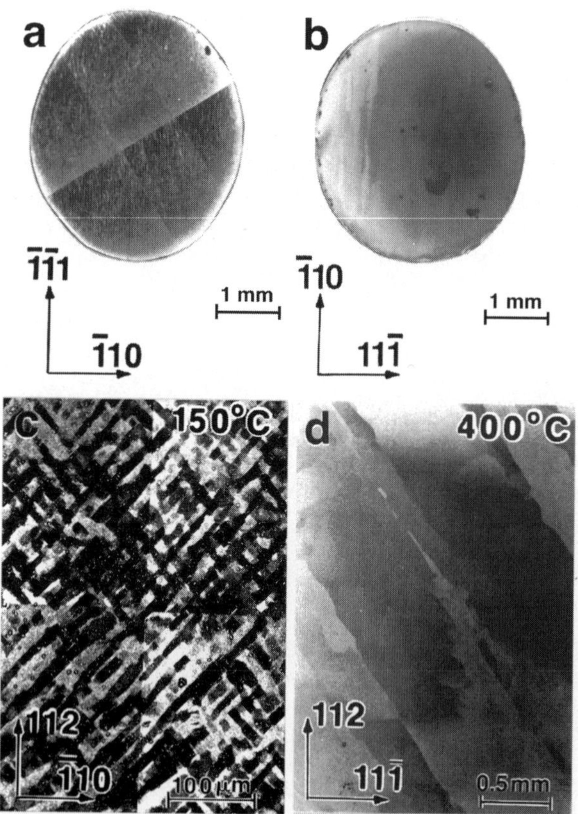

Figure 6.30. SEM observations of the creep substructure of ⟨112⟩ Al single crystals. (a) and (b) Cross-sections for 150 and 400°C, respectively. (c) and (d) Longitudinal sections at 150 and 400°C, respectively. Note that the ellipses in (a) and (b) have major axes at right angles. After Carrard and Martin (1987).

The Laue diffraction technique reveals that the tensile axis is stable at 150°C as a consequence of symmetrical double slip. On the contrary, at 400°C it is observed to rotate towards [110] as a result of slip on the single system [110](001). Consequently, the difference in Schmid factors for (001) and the {111} systems is enhanced.

The creep substructure depends drastically on temperature, i.e. on the active slip systems as illustrated in Figure 6.30.

Images of possible misorientations in the microstructure were obtained with SEM back-scattered electrons. After 150°C creep, subgrains appear (Figure 6.30(c)) as a cross-grating of two platelet families, which are symmetrical with respect to [112] ([$11\bar{1}$] longitudinal section).

The creep structure at 400°C is shown in Figure 6.30(d). Subgrains form, quite different from those at 150°C. They appear as blocks parallel to the (110) plane. They are separated by pure tilt boundaries as seen in TEM observations.

To conclude this section, for creep rates between 10^{-7} and 10^{-5} s^{-1}, the [110](001) system is not activated at 150°C. Evidence of (001) slip appears first at 180°C and operates up to at least 620°C and probably up to the melting point.

6.2.2.2 ***Results of constant strain-rate tests.*** Octahedral glide as well as cube glide can be activated in the same single crystals, depending on the conditions. It has been identified by slip-trace observation and characterization of the elliptical cross-sections for crept samples. Figure 6.31 shows a sample which is deformed by (001) slip and {111} slip in two

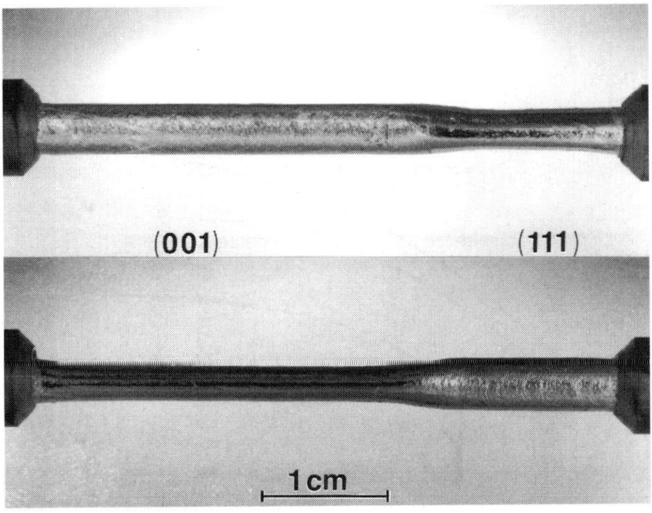

Figure 6.31. Two 90° views of a [112] Al single crystal after deformation at 314°C with $\dot{\gamma} = 12 \times 10^{-4}$ s^{-1}. (001) and {111} slip systems are activated respectively in the left and right region of the sample (see text). After Carrard and Martin (1987).

regions along its length. The major axis of the elliptical cross-section is, respectively, in the figure plane for region A and perpendicular to it in region B (top figure).

Typical stress–strain curves are shown in Figure 6.32 (Carrard and Martin, 1988). Two types of curves are observed. The first one corresponds to temperatures of 270°C in Figure 6.32(a) and to 299°C and below in Figure 6.32(b). A monotonic stress increase is observed, the curves exhibiting a more or less parabolic shape, which is typical for a crystal deforming in multiple glide. At higher temperatures, the curves look quite different. They start with a steep increase in stress and then exhibit a sudden decrease of hardening indicated by an arrow (Figure 6.32). A plateau of low hardening, typical of single glide,

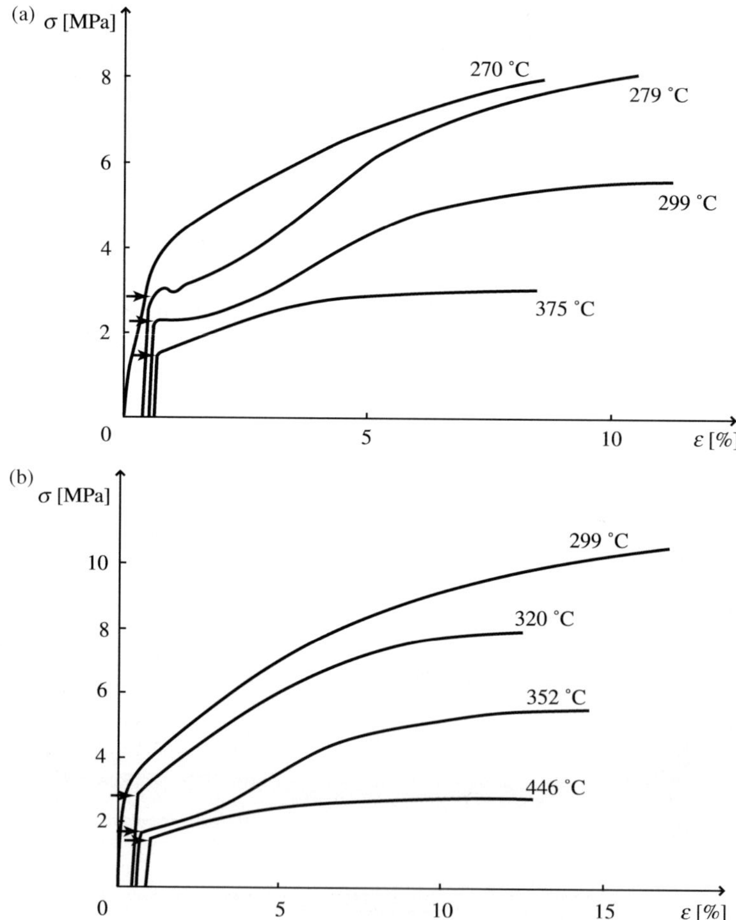

Figure 6.32. Stress–strain curves in [112] Al single crystals at different temperatures. (a) $\dot{\gamma} = 12 \times 10^{-5}\ \mathrm{s}^{-1}$. (b) $\dot{\gamma} = 12 \times 10^{-4}\ \mathrm{s}^{-1}$. When (001) glide is observed the critical stress is indicated by an arrow. After Carrard and Martin (1987).

sometimes follows up to an inflexion point where parabolic hardening starts (see, e.g. 299°C in Figure 6.32(a)). Through metallographic observations it was clearly established that the first type of curve (lower temperatures) corresponds to {111} glide only. For the second type (higher temperatures), [110](001) slip is activated when work-hardening exhibits a sudden decrease. Therefore, the arrows in Figure 6.32 correspond to the critical stress, τ_{001}, for activating cube glide for the imposed temperature and strain-rate. These results are quite similar to those reported for {110} glide (Section 6.2.1). In Figure 6.33, τ_{001} is plotted as a function of temperature and strain-rate. A regular decrease in τ_{001} as a function of temperature is observed at a given strain-rate. At a given temperature, τ_{001} increases with strain-rate. At high-temperature ($T \geq 400$ K) τ_{001} appears to be independent of the two parameters, thus reaching an athermal plateau. It corresponds to an athermal stress of about 0.6 MPa (τ_{μ} of relation (2.16)).

It is worth noting that the behaviour of the crystals is unpredictable at high temperatures. It can slip on {111} (parabolic stress–strain curve) or bifurcate on {001} without any obvious reasons. However, the curves of Figure 6.33 show that below a certain temperature no data are available for τ_{001} irrespective of the strain-rate. This suggests that the stress τ_{001} necessary to activate cube slip becomes larger than τ_{111} (critical stress for {111} slip) as temperature decreases. Under such conditions, {111} glide takes over. This transition will be discussed in Section 6.2.5.3. The boundary between the two domains of {111} and {100} slip in Figure 6.33 is approximately horizontal. This suggests that {111} slip is athermal under such conditions.

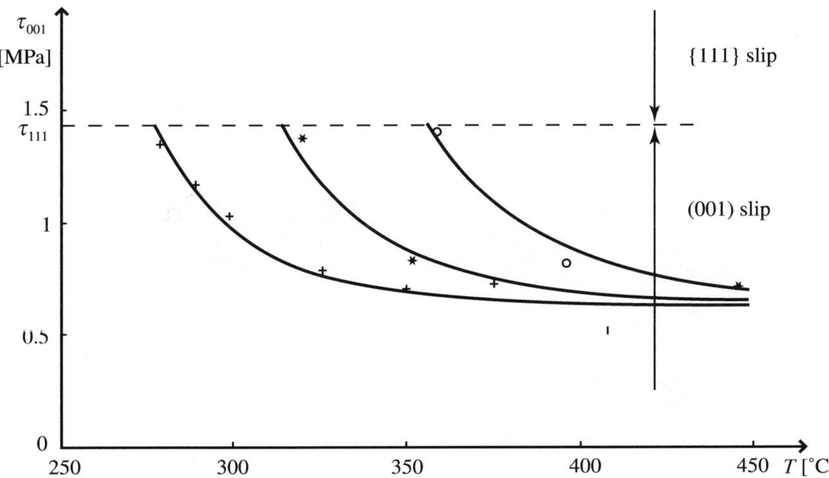

Figure 6.33. Variation with temperature of τ_{001} critical stress for {001} glide at different strain-rates $\dot{\gamma}$: (+) 12×10^{-5} s^{-1}, ($*$) 12×10^{-4} s^{-1} and (O) 12×10^{-3} s^{-1}. The domains corresponding to cube and octahedral slip, respectively are indicated. τ_{111} is the athermal stress for {111} slip (Carrard and Martin, 1988).

6.2.2.3 Features of dislocations in (001). TEM observations provide some hints about the relative mobilities of screws and edges along {001}. Since screw dislocations in aluminium can easily cross-slip from {001} onto {111} observing them along {001} requires special care. One of the methods used has been to pin them under load during creep. To achieve this, a small quantity of zinc atoms is added (Morris and Martin, 1984). The resulting Al–Zn alloy forms a solid solution at 250°C which is creep tested under a 0.8 MPa stress. Dislocations are frozen under stress during an appropriate cooling and ageing process. The creep substructure consists of subboundaries and subgrain dislocations, typical of "class M" (substructure forming) crystals. Among the dislocations, long straight screws are frequently observed as illustrated in Figure 6.34. This suggests that, under such conditions, the screw segments of gliding loops exhibit a lower mobility than the edge ones. Since the creep properties of dilute Al–Zn solid solutions are very similar to those of Al (Blum and Finkel, 1982), this feature of screw mobility can be anticipated for aluminium. This can be understood if one considers the choice of glide planes for the screws at such temperatures, which include {111}, {110}, {111}, etc.

As far as {001} edge dislocations are concerned, they can be observed in large angle tilt subboundaries as shown in Figure 6.35. This high angle asymmetric tilt boundary consists mostly of a family of $\frac{1}{2}[110](001)$ edge dislocations which are seen end on. These are Lomer dislocations. It also includes some 60° dislocations. Such subboundary geometry suggests that edges are quite mobile on cube planes. They stop by mutual interaction. Such a high mobility is related to their compact cores (Mills and Stadelmann, 1989).

The microscopic aspect of {001} slip has also been studied using TEM in situ deformation experiments. These have been performed at 150°C and between 170 and 200°C. The microsamples were extracted from the pre-crept material with the same [112] tensile axis (except for the (001) orientation presented below). This procedure ensures a sufficiently dense population of dislocations on the active slip systems of the macroscopic test.

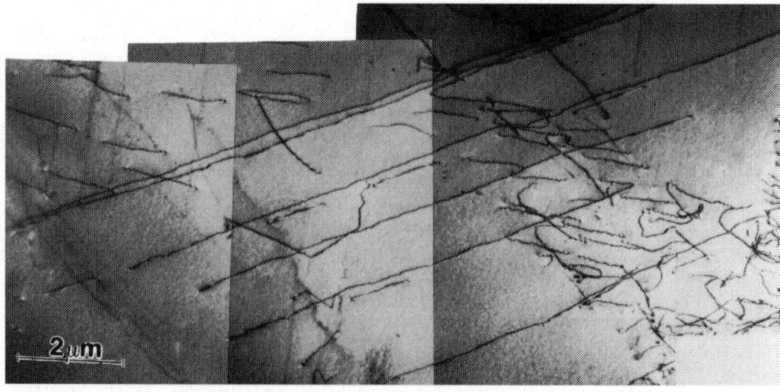

Figure 6.34. Straight screw dislocations corresponding to stage I of creep. Al–Zn solid solution. $T = 250°C$, $\sigma = 0.8$ MPa. Dislocations have been pinned under load (see text). Courtesy of F. Beltzung.

(a)

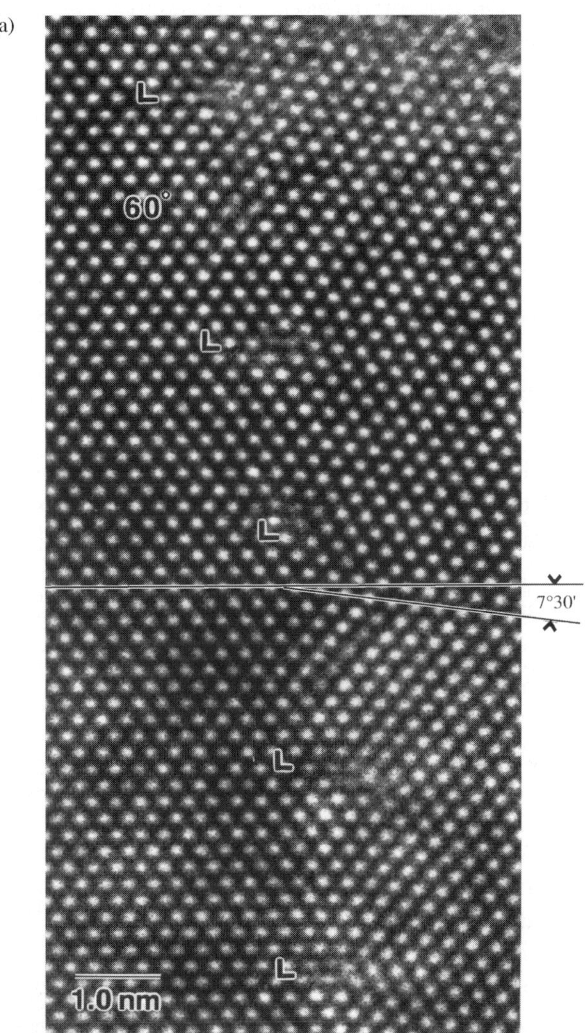

Figure 6.35. Asymmetric tilt boundary in a ⟨112⟩ aluminium crystal (creep at 400°C under 0.61 MPa). (a) High resolution electron micrograph of the boundary seen end on. Misorientation 7°30′. White dots are ⟨110⟩ atomic columns. End on Lomer and 60° dislocations are labelled L and 60°, respectively. Their cores correspond to areas of less distinct contrast (from Mills and Stadelmann, 1989). (b) Schematics illustrating the subboundary geometry. Dislocations are parallel to $[\bar{1}10]$.

In all experiments performed at 150°C, dislocations were observed to glide on {111} systems with frequent cross-slip events. Between 170 and 200°C cube slip was frequently observed. At 170°C the motion of screw segments is seen in $(11\bar{1})$ foils (Figure 6.36). Observation of slip traces on video recordings shows that dislocations originate from the

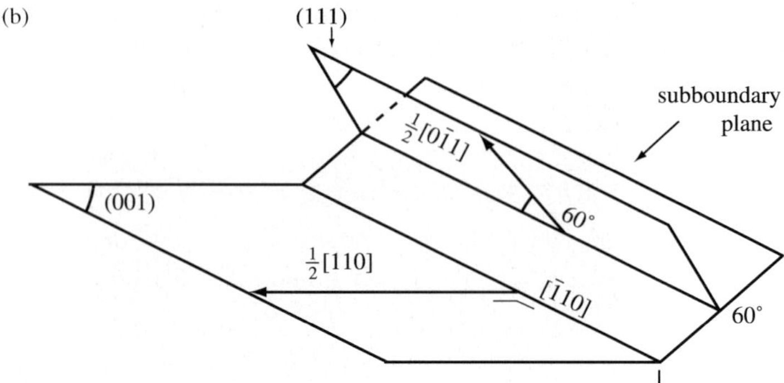

Figure 6.35. (*continued*)

thick part of the foil (bottom left). The wavy slip traces are approximately parallel to the trace of (001) indicated in Figure 6.36(a) in this area of the foil, where the stress regime is presumably identical to that in the bulk specimen (Section 2.2.2). Dislocations move by multiple cross-slip, as shown in Figure 6.36(b): detailed analysis of pairs of slip traces indicate that they are separated by a translation vector B parallel to the [110] Burgers vector. Many such sequences have been observed within this temperature range, which indicates that this deformation mode is predominant. Therefore, multiple cross-slip takes place between $(11\bar{1})$, $(\bar{1}11)$ and (001) planes as observed on the slip lines of the bulk specimen at the same temperature (Section 6.2.2.1).

Other in situ experiments in Al have been carried out with a 112 tensile axis. Under such conditions, the Schmid factor on the {111} cross-slip plane is zero, which has been confirmed by slip trace observation. The observed wavy slip traces of the primary Burgers vector are necessarily the result of some glide component on non-octahedral planes (Couret and Caillard, 1988).

6.2.3 Origin of non-octahedral glide in aluminium

The origin of non-octahedral glide is discussed with special attention to cross-slip from {111} onto the non-closed-packed plane, operation of sources on the latter plane and recombination of Lomer–Cottrell locks in the case of cube glide.

The fact that dislocations start gliding on non-octahedral planes in ⟨100⟩ crystals (Section 6.2.1) and ⟨112⟩ crystals (Section 6.2.2) probably stems for mostly geometrical reasons. In the ⟨001⟩ single crystals, four equally stressed {111} systems are active and in the ⟨112⟩ crystals, two of them are active, the other one exhibiting very low Schmid factors. The cellular dislocation structure that forms leads to a high consolidation. Primary dislocations blocked by the cell walls could eventually escape by cross-slip on non-octahedral planes which would enable deformation to proceed. This possibility was formulated by Le Hazif and Poirier (1975). In support of this view, they were able to

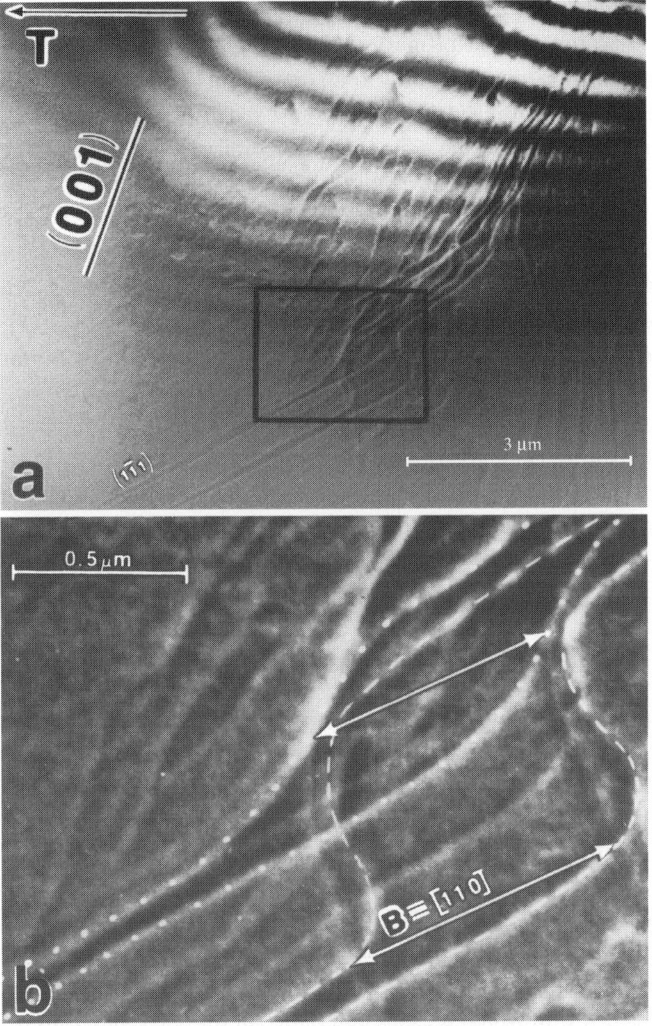

Figure 6.36. TEM micrograph at 200 kV illustrating the unstable motion of screws along (001) at 170°C. $(11\bar{1})$ Al foil. (a) General view. T is the [112] tensile axis. (b) Enlargement of an area in (a) showing two sets of wavy slip traces corresponding to multiple cross-slip. $B = [110]$ is parallel to the Burgers vector of moving dislocations. After Carrard and Martin (1987).

activate {110} slip at room temperature by shock-loading the ⟨100⟩ aluminium crystals. Very high stresses are reached under such elevated deformation rates. Presumably, these are large enough to activate {110} glide at such low temperatures.

However, the cross-slip assumption is not compatible with the behaviour of ⟨112⟩ crystals. Indeed, in the latter case, cross-slip should start from the $\frac{1}{2}[110]$ $(\bar{1}11)$ and

$\frac{1}{2}[110]$ $(1\bar{1}1)$ systems (see Figure 6.28). However, the Schmid factor for these systems is much smaller than that of the primary $\frac{1}{2}[101]$ $(\bar{1}11)$ and $\frac{1}{2}[011]$ $(1\bar{1}1)$ systems (0.272 against 0.408). Consequently, the only realistic origin for (001) slip is the activation of sources on this plane. In the $\langle 001 \rangle$ orientation used by Le Hazif and Poirier, the same Burgers vectors are active on the primary octahedral and non-octahedral slip systems at intermediate and high temperatures. Therefore, in this latter orientation, it is not conclusive that dislocation multiplication is operating exclusively on the non-close-packed planes.

In the case of {001} glide, early speculations by Friedel (1955) and Stroh (1956) postulated that it resulted from the recombination of Lomer–Cottrell barriers. Indeed they are lying in edge orientation on {001} at the intersection of two {111} planes and are expected to yield easily in view of the high stacking fault energy in aluminium. However, in the $\langle 112 \rangle$ crystals, the barriers resulting from the reaction of the two octahedral primary systems would have a Burgers vector equal to $\frac{1}{2}[1\bar{1}0]$ and would lie along [110]. On the contrary, the slip trace analysis of Section 6.2.2.1 indicates that $\frac{1}{2}[110]$ is the Burgers vector active on (001). Therefore, at least under these experimental conditions, (001) slip does not originate from the recombination of Lomer–Cottrell barriers.

All these observations suggest that dislocation multiplication operates on the cube plane in the $\langle 112 \rangle$ aluminium crystals.

6.2.4 Glide on non-close-packed planes in copper

Considering the results in aluminium (Sections 6.2.1 and 6.2.2), a study was undertaken in $\langle 112 \rangle$ Cu single crystals to check the influence of the stacking fault energy on non-octahedral glide (Anongba et al., 1993a). The temperature range investigated was between ambient and 1145 K ($0.22T_m \leq T \leq 0.84T_m$). Three different strain-rates $\dot{\gamma} \sim 2 \times 10^{-2}$, 2×10^{-3} and $2 \times 10^{-4}\ s^{-1}$ were used in tension.

6.2.4.1 Stress–strain curves. These look much more complex than the corresponding ones in aluminium (Sections 6.2.1 and 6.2.2). In particular they exhibit up to five hardening stages that depend on temperature (see, e.g. Figure 6.37). Three temperature regimes have been distinguished within which the characteristics of the stages are different (Figure 6.38).

6.2.4.2 Microstructural features. A detailed analysis of the slip traces and dislocation structures associated with each stage and temperature domain has also been performed by Anongba et al. (1993b). Figure 6.38 summarizes the observations.

As far as slip systems are concerned, the area of Figure 6.38 that corresponds to glide on non-octahedral planes corresponds to stage IV at intermediate temperatures. These operate in addition to the two primary octahedral systems observed in stage III. The new slip traces are inclined to the [110] direction on the $(1\bar{1}0)$ face of the specimen and parallel to [011] or [101] on the $(11\bar{1})$ face where they are weak. The non-octahedral glide systems are $\frac{1}{2}[011]$ $(X\bar{1}1)$ with $X = 2, 3, 5...$ and $\frac{1}{2}[011](100)$ and the symmetrical ones are associated with the

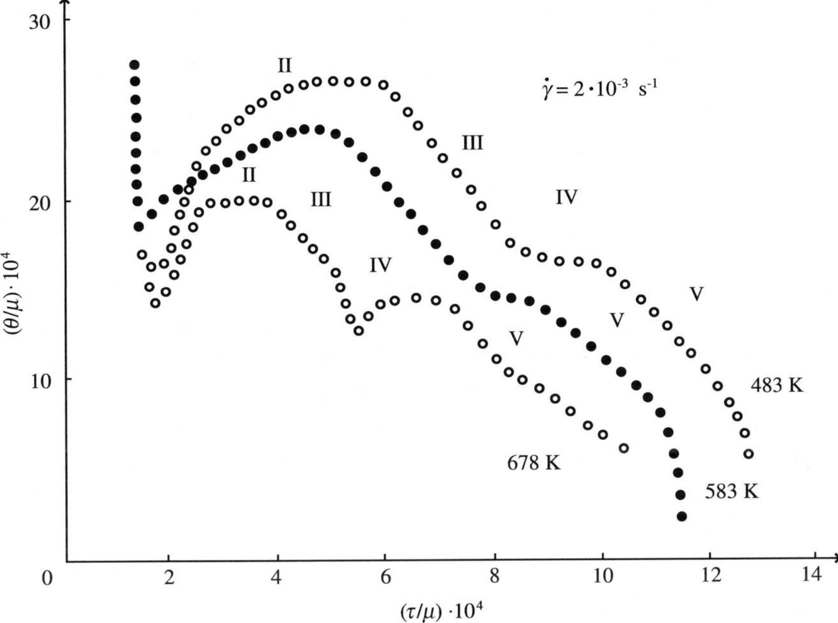

Figure 6.37. Tensile test results for [112] Cu single crystals in the lower temperature regime. $\dot{\gamma} = 2 \times 10^{-3}\ s^{-1}$. Hardening rates versus stress (θ and τ are normalized with respect to shear modulus μ). From Anongba et al. (1993a).

$\frac{1}{2}[101]$ primary Burgers vector (see Figure 6.28). These systems have a Schmid factor close to or higher than that of the primary octahedral system (0.407). Dislocation analysis also indicates non-close-packed slip systems. They lie on several of these planes with a primary Burgers vector in common. This is the signature of a pencil glide process on the corresponding planes.

6.2.4.3 Critical stress for non-octahedral glide. Since τ_{IV} corresponds to the hardening stage where glide is observed along non-close-packed systems in the intermediate temperature domain, it has been considered as the critical stress for this mechanism. Its variation with temperature and strain-rate is represented in Figure 6.39. τ_{IV} decreases as the temperature increases at a given strain-rate and as the strain-rate decreases at a given temperature. This is, again, evidence for a thermally activated mechanism. At a given strain-rate, the slope of the $\tau_{IV}(T)$ curve changes at around 770–780 K and again at around 1150 K. This corresponds to a transition between different temperature regimes where given mechanisms control τ_{IV}. The data points of interest are the intermediate temperature domain (Figure 6.39). At the upper end of this domain, τ_{IV} reaches an athermal plateau which corresponds to $\tau_{\mu} \sim 3.5$ MPa.

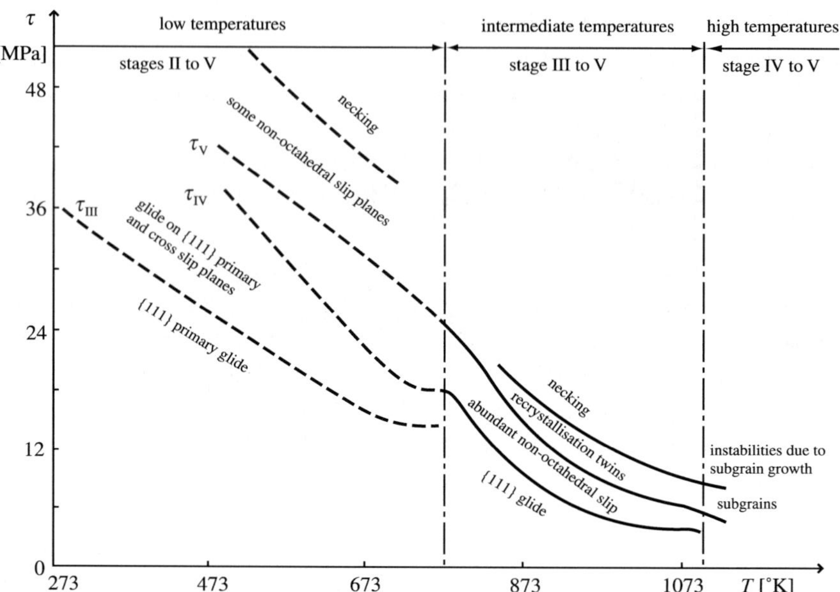

Figure 6.38. Schematic representation of the microstructural features observed in ⟨112⟩ Cu single crystals. Stress–temperature diagram. The curves illustrate the variation of τ_{III}, τ_{IV}, τ_V with temperature and the stress that corresponds to necking for $\dot{\gamma} = 2 \times 10^{-3}\ s^{-1}$. The table should be read from bottom to top. From Anongba et al. (1993b).

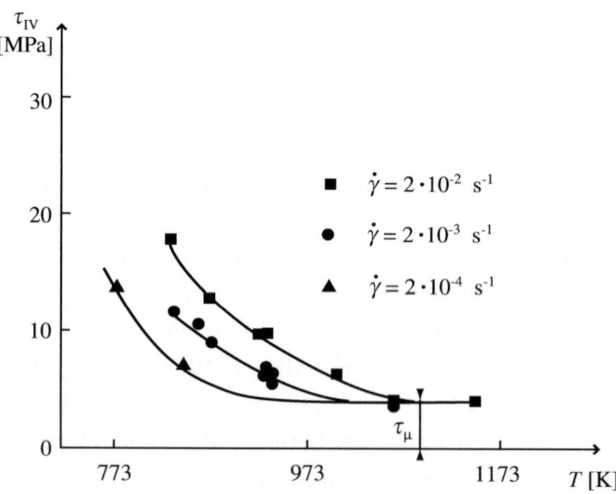

Figure 6.39. Temperature dependence of the stress τ_{IV} (critical stress for non-octahedral slip) for three strain-rates. τ_μ is the corresponding athermal stress. From Anongba et al. (1993a).

6.2.5 Modelling of non-octahedral glide in FCC metals

The microscopic mechanism which controls the motion of dislocations on non-close-packed planes, as well as the associated activation energy, is determined. Various types of mechanisms are critically examined first. Special attention is focused on the kink-pair mechanism and its derivation for HCP metals is extended to the present situation. The resulting strain-rate relation will be compared with experimental data from constant strain-rate and creep tests.

6.2.5.1 Possible mechanisms. Two different mechanisms have been proposed to explain the thermal activation of non-octahedral slip in FCC metals. We exclude the motion of recombined Lomer–Cottrell locks which is incompatible with the microstructural observations in [112] Al single crystals (Section 6.2.3).

For the first mechanism, it is possible to imagine that dislocation dissociation occurs on non-octahedral planes at high-temperature. This was proposed by Edelin (1972) for a particular stacking fault on {110} in Al, but Vitek (1975) showed that dissociation at 0 K is not favourable on this plane, by the use of pseudo-potentials appropriate for core configuration calculations. Nevertheless, it could occur at higher temperatures, owing to thermal expansion of the crystal lattice. However, this could not be simulated properly. In addition, glide on several types on non-octahedral planes has been reported, such as {001}, {110}, {112}, {113}, etc. (see Section 6.2). It seems very unlikely that a particular dislocation dissociation could exist for each type of non-conventional glide plane.

The second type of mechanism is the kink-pair mechanism, which has been suggested by Vanderschaeve and Escaig (1980) to account for {110} glide as observed by Le Hazif and Poirier (1975). This mechanism which is studied in Chapter 4 is described schematically in Figure 6.40(b). It is geometrically similar to that controlling prism slip in HCP metals (Figure 6.40(a)).

The difference between the dissociation geometry in the HCP and FCC metals is that, in the latter case, two octahedral planes are available for screw splitting, instead of one in the former case.

6.2.5.2 {001} glide in aluminium and the kink-pair mechanism. A classical method to measure the apparent activation enthalpy ΔH_a is to compare at given stress τ two strain-rates $\dot{\gamma}_1$ and $\dot{\gamma}_2$ corresponding, respectively, to two temperatures T_1 and T_2. The measurements are performed at given strain.

$$\Delta H_a = kT_1T_2 \frac{\ln(\dot{\gamma}_2/\dot{\gamma}_1)}{T_1 - T_2}\bigg|_\tau \tag{6.2}$$

For constant strain-rate tests, curves similar to those of Figure 6.33 can be considered.

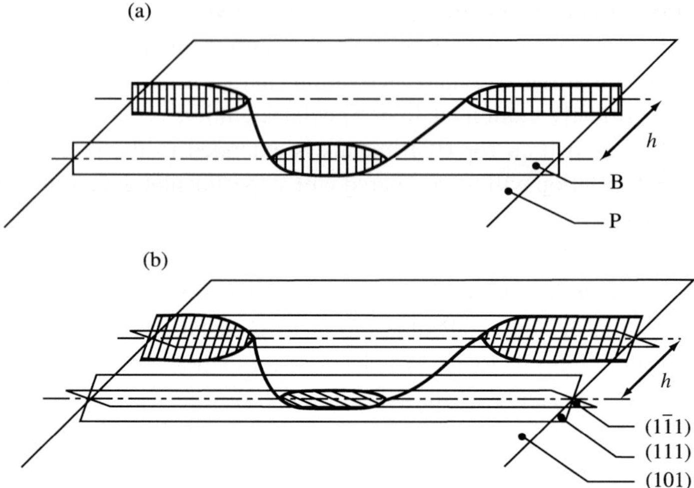

Figure 6.40. Schematic representation of the kink-pair model for a screw dislocation gliding along a non-close-packed plane in: (a) a HCP metal (P and B are, respectively, the prism and basal planes); (b) the (101) plane of a FCC metal. The distance between Peierls valleys is $h = c/2$ or c in HCP's and $h = b\sqrt{2}$ for $\{101\}$ slip.

The apparent activation enthalpy for $\{001\}$ glide in aluminium ΔH_{001} has been estimated from relation (6.2) and the curves of Figure 6.33. For a stress level of 1.4 MPa, ΔH_{001} values close to 1.58, 1.73 and 1.9 eV were found at temperatures between 280 and 320, between 280 and 360 and between 320 and 360°C, respectively.

Under creep conditions, the temperature jump method (Sherby et al., 1957) was used. In such experiments (Carrard and Martin, 1988), ΔH_a was measured between 200 and 320°C. This covers the temperature range where $\{001\}$ slip is activated but still difficult. With such a method, the strain-rate at the jump $\dot{\gamma}_2$ has to be back extrapolated from $\dot{\gamma}$ values corresponding to the new thermal equilibrium of the setting.

The results are presented in Figure 6.41. The value of the creep activation enthalpy is rather constant with temperature and stress: ΔH creep $= 1.57 \pm 0.1$ eV. Figure 6.41 shows that the present data agree within the error bar with similar measurements previously performed for polycrystalline aluminium (Sherby et al., 1957).

In order to check the validity of the kink-pair mechanism, the experimental data in Figure 6.33 were fitted with a rate relation of the form:

$$\dot{\gamma} = A(\tau_{001} - \tau_{\mu})^m \exp(-\Delta H_{001}/kT) \tag{6.3}$$

where the adjustable parameters A, τ_{μ}, m and ΔH_{001} are constant in the restricted temperature range explored. An activation enthalpy ΔH_{001} has to be considered in relation (6.3), which refers to macroscopic deformation tests.

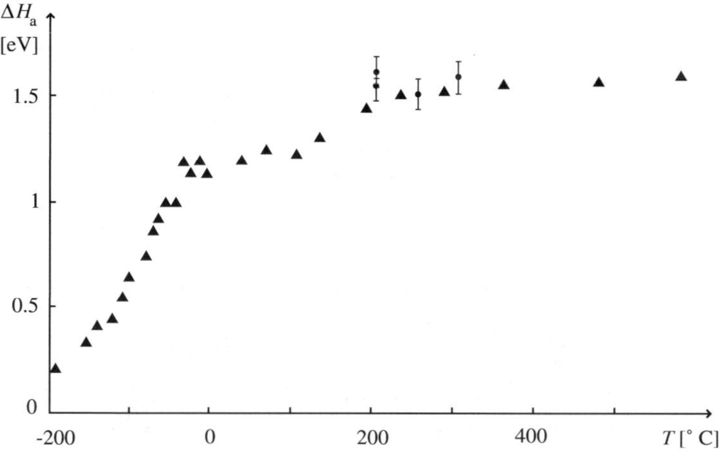

Figure 6.41. Creep activation enthalpies (•) measured in [112] Al single crystals between 200 and 320°C. For comparison, results (▲) by Sherby et al. (1957) on Al polycrystals are also reported. From Carrard and Martin (1988).

The best fit was found using a computer program of least squares with respect to stress, providing the corresponding parameters $A = 90.43 \times 10^{12}\ \mathrm{MPa}^{-m}\mathrm{s}^{-1}$, $\tau_\mu = 0.63$ MPa, $m = 1.7$ and $\Delta H_{001} = 1.74$ eV. Owing to the vicinity of the athermal plateau, the uncertainty on ΔH_{001} is rather large: $\Delta H_{001} = 1.74 \pm 0.2$ eV. This value is more accurate than the orders of magnitude above, obtained via relation (6.2).

By neglecting the terms $\tau_{001}V$ and $\tau^*_{001}V$ (error less than 0.04 eV for τ_{001} equal to 1.4 MPa), the activation energy ΔG°_{001} at zero stress for (001) glide can be estimated by using the entropy correction from Chapter 1 proposed by Schoeck (1965):

$$\Delta G^\circ_{001} = \Delta H_{001}/[1 - (1/\mu)(\mathrm{d}\mu/\mathrm{d}T)] \tag{6.4}$$

With $T = 300$°C, $\mu = 2.7 \times 10^{-4}$ MPa at 20°C and $(1/\mu)(\mathrm{d}\mu/\mathrm{d}T) = -0.43 \times 10^{-3}\ \mathrm{K}^{-1}$ (Friedel, 1964) one finds $\Delta G^\circ_{001} \sim 1.40$ eV. This value is equal to the formation energy, $2U_\mathrm{k}$, of a kink-pair (Eq. (4.36) at low stress). This yields the energy of one kink, $U_\mathrm{k} \approx 0.7$ eV. This value is close to that estimated by Caillard (1985) considering the geometry of a special type of creep subboundary and using Eq. (4.16).

Therefore, for {001} slip in aluminium, the kink-pair model predicts a strain-rate law which agrees fairly well with the experimental data and predicts reasonable values for the kink formation energy.

In the framework of this model, a description of the transition from {111} to {001} slip can be proposed for the [112] orientation as the temperature is raised. At relatively low temperatures (180°C in creep), the density of kinks in the (001) plane along a screw dislocation, proportional to $\exp(-\Delta G/kT)$ is low. The 1/2 [110] screw dislocation is not

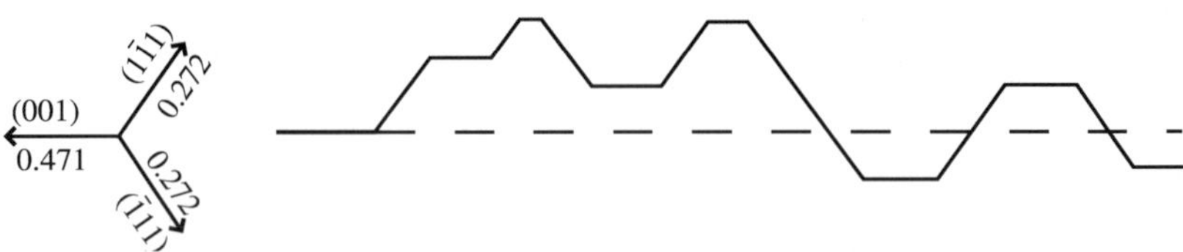

Figure 6.42. Schematic representation of a 1/2 [110] screw dislocation in (001) at the onset of (001) slip. Screws are end on. The values of the Schmid factors along the three possible slip planes involved are indicated.

stabilized on the (001) plane and cross-slips from (001) to $(\bar{1}11)$ or $(1\bar{1}1)$ although the Schmid factor is low (0.272 as against 0.471). We have verified that the stress acting on the 1/2 [110] screw and resolved on the $(\bar{1}11)$ or $(1\bar{1}1)$ planes modifies the fault width by the same amount (see Chapter 5). Consequently, the screw has equal probabilities of gliding on $(\bar{1}11)$ or $(1\bar{1}1)$, to and from the (001) plane. Figure 6.42 is a schematic illustration of the screw motion under these conditions. It moves on the average parallel to (001), but with numerous cross-slips on the two {111} planes. Since glide on the two octahedral planes is friction free, glide on (001) controls the velocity of the screw. Such a description of the onset of (001) glide agrees with various observations of Sections 6.2.2.1–6.2.2.3: wavy (001) slip traces at the onset temperature on macroscopic samples and in TEM in situ experiments and the presence of long straight screws in pinned dislocation structures.

As the temperature is raised, the kink density on (001) along the screw increases. The latter is thus stabilized on (001) as the slip traces become rectilinear. It is worth noting that the onset temperature for (001) slip is different under creep as opposed to constant strain-rate conditions (respectively, 180°C against 280°C for $\dot{\gamma} = 12 \times 10^{-5}\ \mathrm{s}^{-1}$). This difference in temperatures has also been observed for {110} glide by Le Hazif et al. (1973) and Le Hazif and Poirier (1975). This can be explained by different conditions imposed on the sample: (i) in creep, the crystal undergoes a very high deformation rate at the beginning, forcing the non-octahedral sources to operate; (ii) in creep, each sample section resists the same load so that at 400°C (001) slip traces cover the whole specimen. Conversely, at constant strain-rate, each section is able to deform independently so that {111} slip is still observed in some areas even at 450°C, the highest temperature.

6.2.5.3 Modelling {110} glide in aluminium. The data about {110} glide in Section 6.2.2 will now be analysed in terms of the kink-pair mechanism. The same fitting procedure has been used with a rate relation similar to Eq. (6.3)

$$\dot{\gamma} = A'(\tau_{110} - \tau'_{\mu})^{m'} \exp(-\Delta H_{110}/kT) \tag{6.5}$$

The following parameter values are obtained for the best agreement: $A' = 2.22 \times 10^{8} \times \mathrm{MPa}^{-m'}\mathrm{s}^{-1}$, $\tau'_{\mu} = 0.15$ MPa, $m' = 4.2$ and $\Delta H_{110} = 1.40$ eV. For the same reasons as above, the inaccuracy of ΔH_{110} is rather large and $\Delta H_{110} = 1.40 \pm 0.20$ eV. Following the same procedure yields $\Delta G^{\circ}{}_{110} \approx 1.13$ eV. The corresponding τ_{110} $(\dot{\gamma}, T)$ curves are represented in Figure 6.27 where the agreement with data is considered as satisfactory.

It is worth noting that the activation energy for {110} glide is comparable to that for {001} glide, given the scatter. This is confirmed by the datas and curves of Figures 6.27 and 6.33 which can be superimposed approximately. This can be understood considering relations 4.36 and 4.16, provided U_{constr} is the dominating term of ΔG^{0}. Consequently, the different values of the distance h between the dense rows of the slip plane defined on Figure 6.40 and in Section 6.2.5.2 ($h_{001} = b$ as against $h_{110} = b\sqrt{2}$) do not influence the ΔG^{0} value. However, the stress exponent of the strain-rate, which is proportional to the activation area, is larger for {110} glide ($m' = 4.2$ as against $m = 1.7$). This is accounted for by relation 4.37 where the activation area A_{ik} is proportional to $h^{3/2}$. This supports the kink-pair mechanism for {110} and {001} glide in aluminium.

6.2.5.4 Non-octahedral glide in copper. It has been shown in Section 6.2.4.3 that τ_{IV} is the stress at which glide on non-close-packed planes takes over. To characterize the corresponding mechanism, the activation enthalpy is computed according to relation (6.2) and the data of Figure 6.39 in the intermediate temperature range. Table 6.3 summarizes the results.

The activation enthalpy is approximately constant with respect to stress. This suggests that a rate relation of the type

$$\dot{\gamma} = A''(\tau_{\text{IV}} - \tau_{\mu})^{m''} \exp\left(-\frac{\Delta H_{\text{no}}}{kT}\right) \tag{6.6}$$

can be used. Fitting the data of Figure 6.39 and relation (6.6) yielded the best parameter values: $\Delta H_{\text{no}} = 2.65$ eV, $\tau_{\mu} = 3.5$ MPa, $m'' = 3.95$ and $A'' = 4.5 \times 10^{9}$ $\text{MPa}^{-m''}$ s^{-1} in good agreement with the values of Table 6.3. The activation energy $\Delta G^{\circ}{}_{\text{no}}$ is obtained as in Section 6.2.5.3, with $\mu = 47.4 \times 10^{3}$ MPa at room temperature and $(1/\mu)(\text{d}\mu/\text{d}T) = -4 \times 10^{-4}$ K^{-1} (Ledbetter and Naimon, 1974). This yields $\Delta G^{\circ}{}_{\text{no}} = 2.04 \pm 0.33$ eV.

Several possible mechanisms can be proposed to account for this activation energy value: (i) a diffusion controlled process, since the activation energy for self-diffusion in copper is 2 eV (Peterson, 1978); (ii) the recombination of attractive junctions, which

Table 6.3. Activation parameters for non-octahedral glide in copper (Anongba et al., 1993a).

τ_{IV} (MPa)	τ_{μ} (MPa)	ΔH (eV) ~
6.5 ± 0.6		2.6 ± 0.25
8.15 ± 0.6		2.5 ± 0.25
11.5 ± 0.6	3.5	2.56 ± 0.25
13.6 ± 0.6		2.6 ± 0.25
11.5 ± 0.6		

involves a range of activation energy values according to the problem geometry (Saada, 1960); and (iii) the kink-pair mechanism on non-close-packed planes. If processes (i) and (ii) cannot be completely discarded, they at least seem to be unlikely. Indeed, they should yield in principle to a recovery stage. On the contrary, work-hardening can be expected in the case of non-octahedral glide, since new dislocation structures are initiated, such as additional subboundaries, prismatic loops and Lomer dislocations (Anongba et al., 1991). In addition, slip trace analysis as well as TEM observations indicate clearly that numerous non-octahedral glide systems are activated in stage IV at intermediate temperatures. Therefore, the activation parameters determined for τ_{IV} are now compared with the predictions of the kink-pair model in the case of copper.

However, the situation is more complex than in aluminium single crystals since a set of different non-close-packed systems is activated simultaneously in copper. Accurate computations should take into account, for each non-octahedral plane, the corresponding recombination energy ΔE, as well as the variation of stacking fault energy with temperature. Then, instead of calculating the activation energies for each type of plane, only the orders of magnitude are estimated here. According to the non-octahedral plane, the distance h between two dissociation sites for the screw dislocation ranges between $b/2$ and $b\sqrt{2}$. The width of splitting of the screw is estimated from weak-beam observations (Cockayne et al., 1971; Stobbs and Sworn, 1971) and mechanical test data (Bonneville et al., 1988). These references yield $1.28 \leq d \leq 1.78$ nm, i.e. $d = 5 - 7b$. Using relations (4.16) and (4.36) with $\tau \sim 10$ MPa, a value of $\Delta G = 2 \pm 0.7$ eV is found for the activation energy of the kink-pair mechanism for non-octahedral glide in copper. This value is in good agreement with the experimental one of 2.04 ± 0.33 eV reported above. These results are consistent with the kink-pair mechanism to describe slip of copper along non-octahedral systems.

6.2.5.5 Comparison of FCC metals. Using the data of Le Hazif and Poirier (1975) about {011} glide in various FCC metals, it is possible to test the predictions of the kink-pair mechanism. This model allows values of $\Delta G^{\circ}{}_{110}$ to be estimated as above, using values of $\mu(T)$ given by Friedel (1964) and of γ given by Coulomb (1978). The estimated activation energies are listed in Table 6.4. In addition, the athermal temperature for the mechanism can also be estimated (T_a^{th}) using relation 1.7, written here as

$$\Delta G^{\circ}{}_{110} \sim \alpha k T_a^{th}$$

It has been proposed by Escaig (1968), with α ranging between 20 and 30. T_a^{th} can be then compared with the T_a^{exp} values, considering that the temperature T_2, at which {110} glide completely replaces {111} slip as given by Le Hazif and Poirier (1975), is the experimental athermal temperature. Values of T_a^{th} and T_a^{exp} are also given in Table 6.4. In fact, several parameters are not accurately known, such as b_o or the stacking fault energy variation with temperature. Nevertheless, the agreement between experimental and computed values of

Table 6.4. Predictions of the kink-pair mechanism for the activation energy and athermal temperature for {011} slip in various FCC metals.

Metal	$\mu b/\gamma$ at T_a	b_o	ΔG°_{110} (eV)	T_a^{th} (K)	T_a^{exp} (K) ~
Al	51	$b/3$	1.1–1.17	468–609	540
Ni	99	$b/2$	2.1–2.61	1046–1259	1225
Au	194	b	1.82–2.13	855–1095	900
Cu	193	b	2.1–2.58	1037–1262	>1270
Ag	267	b	2.14–2.76	1105–1882	>1185

the athermal temperature for {011} glide, respectively, is satisfactory. This is another point in support of the kink-pair mechanism for describing non-octahedral slip in FCC metals.

6.2.6 The relevance of slip on non-close-packed planes in close-packed metals

6.2.6.1 Optimum conditions for unconventional slip in aluminium. The conditions that dictate the slip mode at high temperatures are not clearly established. Therefore, the Schmid law has been used to check the prediction of octahedral, as opposed to non-octahedral, glide in aluminium. The deformation test results of Sections 6.2.2.1 and 6.2.2.2 show that the $\frac{1}{2}[110](001)$ system operates provided the temperature is high enough. Since it has the highest Schmid factor, the Schmid law is obeyed. The same reasoning can be used for other types of non-octahedral glide systems, such as $\frac{1}{2}\langle 110\rangle\{110\}$, {112}, {113}. It is then possible to determine the tensile axis orientations for which each type of system is expected to be activated at high temperatures. The results of this analysis are shown in Figure 6.43(a). It is interesting to note that the area of the standard triangle where non-octahedral glide is expected is larger than that for {111} glide.

For comparison, tensile axis orientations and operative slip systems deduced from the observations by Johnson et al. (1956) in high-temperature creep tests are added. The agreement is very good as a rule. The few discrepancies seen may be due to: (i) tests at temperatures that are too low; (ii) the difference in activation enthalpies according to the slip system; (iii) the action of stress on the partial dislocations according to the kink-pair model; and (iv) the ratio of the Schmid factor of the non-octahedral system to that of the most favourable {111} system. This latter point was confirmed by Johnson et al. (1956). They showed experimentally that, when this ratio increases, the onset temperature of the corresponding non-octahedral glide system decreases. The variation in this ratio with orientation has been systematically calculated and is given in Figure 6.43(b). It shows in particular that the "best" orientation for {001} slip is [111] with a Schmid factor ratio of 1.73. The best one for {110} is [001] with a ratio of 1.23. For {112} no best orientation exists since this ratio is lower than or equal to 1.1 over the corresponding area of the triangle.

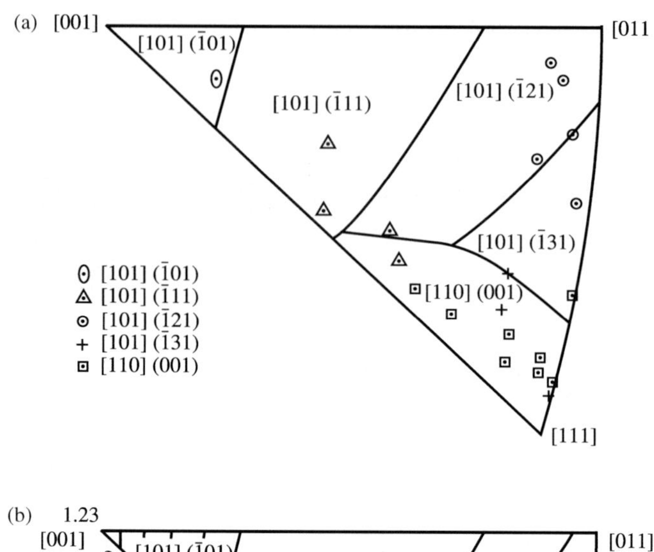

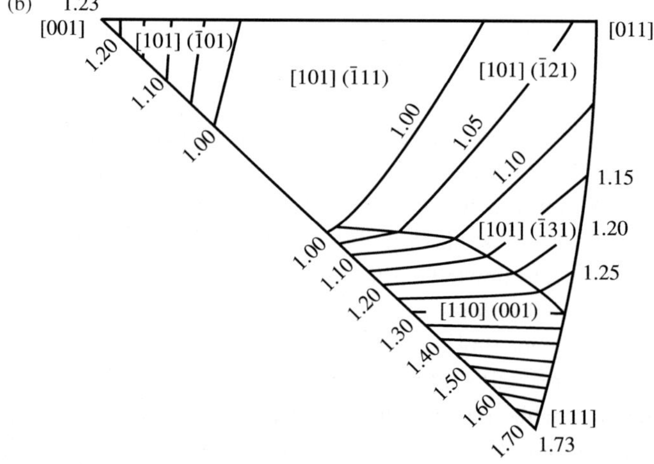

Figure 6.43. Orientation dependence of the different slip systems in aluminium. (a) Systems with maximum Schmid factors as a function of orientation. For comparison, data deduced from the results of Johnson et al. (1956) are also reported. (b) Same standard triangle with the values of the ratio of the Schmid factors of the corresponding non-octahedral systems to that of the most favourable {111} systems. After Carrard and Martin (1988).

6.2.6.2 Non-conventional glide as a rate controlling process. In the [112] aluminium single crystals, (001) glide appears to be the controlling mechanism of creep above 180°C. Indeed, fair agreement exists between the activation enthalpies ΔH_{001} and ΔH creep (1.74 ± 0.2 and 1.57 ± 0.1 eV, respectively). We think that this conclusion can also apply to other aluminium single crystals and polycrystals. Indeed, if one assumes that the activation enthalpies for {112} and {113} glide are close to those found for

{001} and {110} glide, then activation enthalpies for non-close-packed glide in aluminium are roughly between 1.4 and 1.7 eV. These values are in fair agreement with the creep activation enthalpies for single and polycrystals which range between 1.3 and 1.6 eV above 200°C (see a review by Caillard and Martin, 1983). In the case of polycrystals, the grains which are favourably oriented for non-octahedral glide are harder, deform more slowly and impose the creep rate on the crystal. This hypothesis is confirmed by the observation of long straight screw dislocations in Al–11 wt% polycrystals after creep at 250°C (Morris and Martin, 1984). This provides good agreement with the kink-pair model of non-octahedral glide. This slip process can also explain the abnormally high activation enthalpies measured in high-temperature creep of silver (Poirier, 1978) and copper (Retima and Cornet, 1986; Siethoff and Ahlborn, 1986). Values higher than the self-diffusion energy were found while no convincing interpretation was proposed.

It is worth noting that some creep experiments in Cu single crystals (773–873 K) were designed to activate non-octahedral glide (Orlova and Kucharova, 1999). Double notch creep samples were specially machined so as to impose a shear deformation parallel to the $\frac{1}{2}\langle 110\rangle\{001\}$ slip system. Slip trace analysis as well as microstructural TEM observations revealed multiple slip on {111} and restricted slip along {001}. The complicated shape of the sample, which generates complex strain fields, may be responsible for these results, which do not contradict the above.

In the case of magnesium the activation energies found in Section 6.1.3 for dislocation glide on the prism plane can be compared to creep activation energies at intermediate and high temperatures. This can provide hints about deformation conditions for which this mechanism is rate controlling. Creep activation energies determined independently by several authors are presented in Figure 6.44. A satisfactory agreement is found between 500 and 600 K while a large scatter is observed at higher temperatures. The activation energy of the dislocation velocity for prism slip is also represented. It has been estimated as follows: Figure 6.16 provides two values of v for $\tau = 12$ MPa at 300 and 373 K, respectively. This yields an activation energy of 0.8 ± 0.1 eV at a mean temperature of 336 K. It can be extrapolated towards higher temperatures using relation 1.7.

Figure 6.44 shows that the activation energy of v can be compared to the creep activation energy of Mg between 400 and 600 K. Therefore, prism slip can control the creep rate under such conditions, while self-diffusion was claimed to be the mechanism. Above 700 K no information is available so far.

The question of why non-conventional glide can control the creep rate has been investigated for aluminium at intermediate temperature (Caillard, 1985). The interpretation is based on detailed observations of the dynamic properties of the substructure performed in situ in TEM. In particular they have shown that (i) dislocations glide over distances larger than the subgrain size (i.e. they have to cut through subboundaries) and (ii) subboundary migration takes place (Caillard and Martin, 1983). The intersection of a

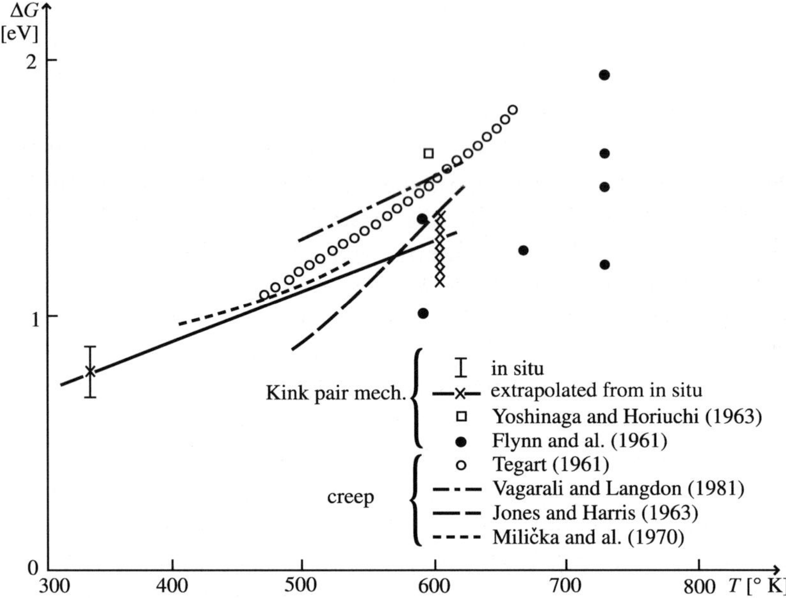

Figure 6.44. Comparison between activation energies for creep of Mg and activation energies of the dislocation velocity at various temperatures. The in situ data refer to Figure 6.16 (from Couret and Caillard, 1985a,b).

subboundary by a dislocation is a complex process (Caillard, 1984) that requires high local stresses. These are built-up during subboundary migration according to the schematics of Figure 6.45. The subboundary consists of three dislocation segments, x_1, x_2, x_3, with respective Burgers vectors, b_1, b_2, b_3, with segment x_1 in screw orientation

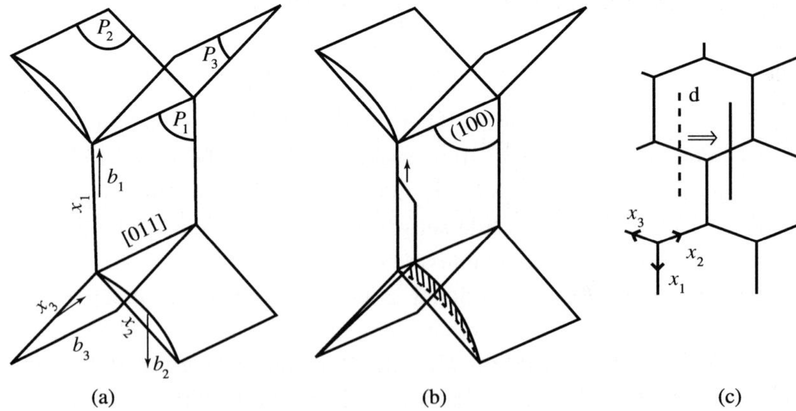

Figure 6.45. Schematics illustrating a mechanism for subboundary migration during creep (see text). From Caillard (1985).

(Caillard and Martin, 1982). Since dislocations are observed to move by glide at intermediate temperatures, the subboundary segments glide on planes P_1, P_2, P_3 (Figure 6.45(a) and (b)). P_2 and P_3 are octahedral planes intersecting along [011], while P_1 is the (100) plane. Therefore, the screw segment x_1 glides along a cube plane, according to the kink mechanism described above for {100} glide. The high local stresses necessary for dislocation emission from the subboundary are built up during migration thanks to the presence of extrinsic dislocations having met the subboundary. One of those is represented in Figure 6.45(c) (labelled d). Its glide plane is not parallel to the [011] direction as a rule so that d moves along the migrating subboundary plane. d is therefore pushed against a subboundary segment thus generating high local stresses. The activation energy of such a process, as well as its stress-dependence, have been computed (Caillard, 1985). In this case, it is related to the energy of one kink instead of a kink-pair for {001} slip in the subgrains. This mechanism could explain the lower creep activation energies at intermediate temperatures (between 100 and 200°C, Figure 6.41).

Finally, non-octahedral glide has been shown to play an important role in other deformation processes. For cold rolling it appears necessary to take into account {001} glide to explain the copper-type textures observed in FCC metals (Richards and Pugh, 1959; Haessner, 1965). More recently, it has been shown that slip and cross-slip on planes such as {001}, {110} and {112} could account for the observed rolling textures in aluminium as a function of temperature (Bacroix and Jonas, 1988). In addition, recent studies of the long term mechanical properties of the alloy 2650 (Al–Cu–Mg) has shown a degradation of the latter. During creep at 150°C, {001} glide becomes a softening mechanism from the onset of the test (Majimel et al., 2002).

It is also worth noting that cube slip is an important deformation mode in Ni_3Al (see Chapter 10).

6.3. LOW-TEMPERATURE PLASTICITY OF BCC METALS

The low-temperature mechanical properties of BCC metals were the first ones to be attributed unambiguously to a Peierls-type mechanism. Many results on Fe and Nb are available in the literature. Some of the most significant ones are summarized and discussed in this section. For more details the reader can refer to the reviews of Kubin (1982), Christian (1983), Suzuki et al. (1985), Duesberry (1989) and Taylor (1992).

6.3.1 Mechanical properties

6.3.1.1 Iron and iron alloys. Figure 6.46 shows the temperature dependence of the CRSS in pure iron single crystals, oriented for single slip with a maximum Schmid factor on a $\langle 111\rangle\{1\bar{1}0\}$ system. The CRSS has been calculated for this slip system, although the one observed may be different, as discussed in Section 6.3.2. It decreases rapidly from 400 MPa

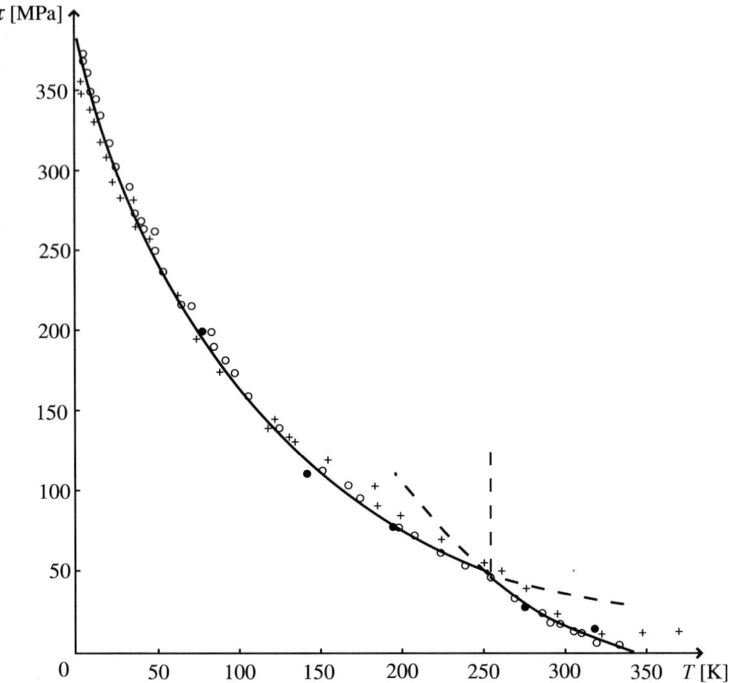

Figure 6.46. CRSS on {110} of pure Fe single crystals, as a function temperature. Note the hump at 250 K. Tensile tests by Quenel et al. (1975) (full circles), Kuramoto et al. (1979) (crosses) and Brunner and Diehl (1997) (open circles and curve).

at the lower temperatures to low values at 300–350 K. As in HCP metals, the curves exhibit a small hump corresponding to a peak in the stress and temperature dependences of the activation area (Figures 6.47(a) and (b), respectively). Similar results have been obtained in polycrystals by Tseng and Tangri (1977) and Cottu et al. (1978). An accurate determination of the activation parameters by Brunner and Diehl (1991a–d, 1997) shows that the pre-exponential factor $\dot{\varepsilon}_0/\dot{\varepsilon}$ is discontinuous at 250 K (Figure 6.47(b)).

The mechanical properties are the same in the hydrogen free "outgassed" material, according to Kuramoto et al. (1979). They are, however, very sensitive to the carbon concentration. Figure 6.48 shows that carbon addition increases the CRSS at low-temperature. It also suppresses the hump in the $\tau(T)$ curve (as well as the corresponding peak in the $A(T)$ and $A(\tau)$ curves). As a result, the CRSS is lowered between 150 and 250 K. The same tendency has been reported by Quenel et al. (1975) in single crystals and by Cottu et al. (1978) in polycrystals. If the CRSS at 300 K is considered to be the athermal component τ_μ of the flow stress, the thermally activated component τ^* is independent of the carbon concentration at low-temperature and decreases with increasing carbon content at higher temperatures. With this hypothesis, carbon softens the thermally activated

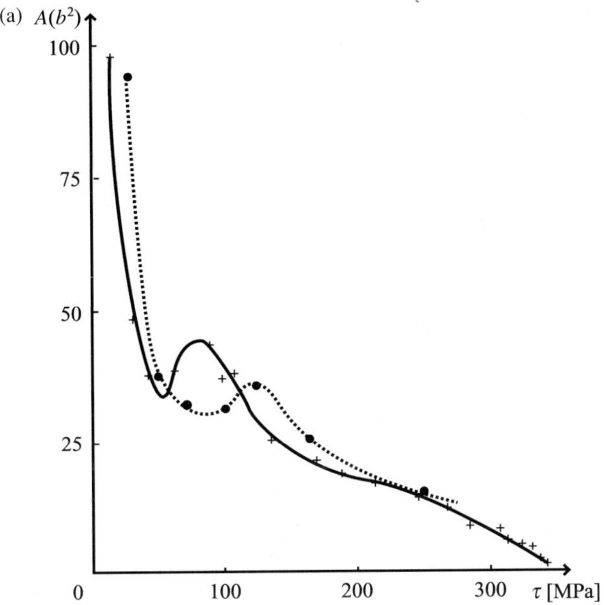

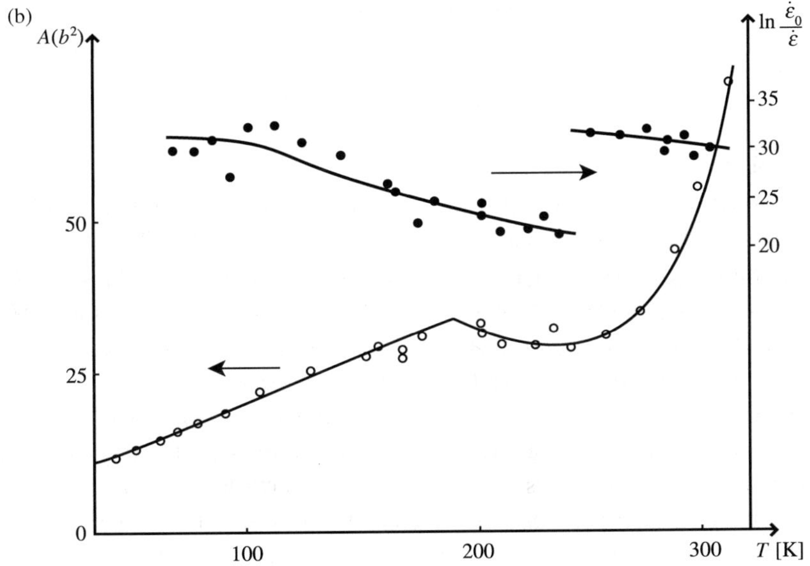

Figure 6.47. Activation areas in pure Fe single crystals: (a) versus stress, from Quenel et al. (1975) (dots) and Kuramoto et al. (1979) (crosses); (b) versus temperature, together with the corresponding pre-exponential factor, from Brunner and Diehl (1997).

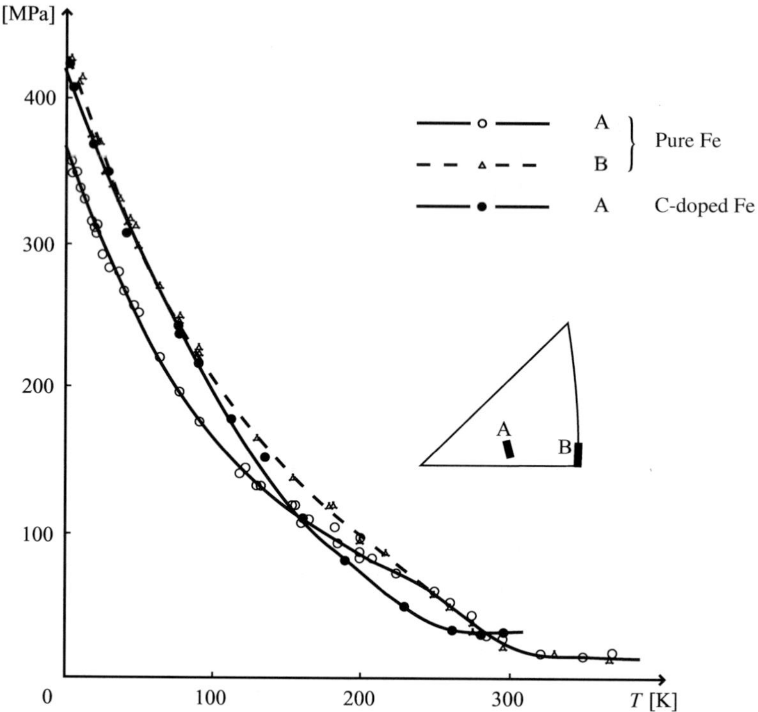

Figure 6.48. CRSS on {110} of pure Fe, with two orientations of the stress axis (A and B), and C-doped Fe, with orientation A. Tensile tests by Kuramoto et al. (1979).

component τ^* increases the athermal component τ_μ and decreases the athermal temperature (Cottu et al., 1978). This conclusion however, deserves a more careful determination of the athermal temperature.

Figure 6.48 also shows that the CRSS is different for two orientations of the straining axis. Such violations of the Schmid law, as well as tension–compression asymmetries, are very frequently observed in BCC metals. They have been discussed extensively by Christian (1983) and Duesberry (1989).

Nickel has the same effect as carbon, albeit at much larger concentrations, according to Aono et al. (1981). Indeed, Figure 6.49 shows that the peak of the $A(\tau)$ curve disappears for a 1 at.% Ni addition. A smaller concentration (0.1 at.%) only smoothes the stress-hump and the activation area peak and shifts them towards lower temperatures and higher stresses, respectively.

6.3.1.2 Niobium. The low-temperature mechanical properties of Nb are very similar to those of Fe. They have been studied in single crystals oriented for single slip with a maximum Schmid factor on a $\langle 111\rangle\{1\bar{1}0\}$ system (Statham et al., 1972; Kubin and

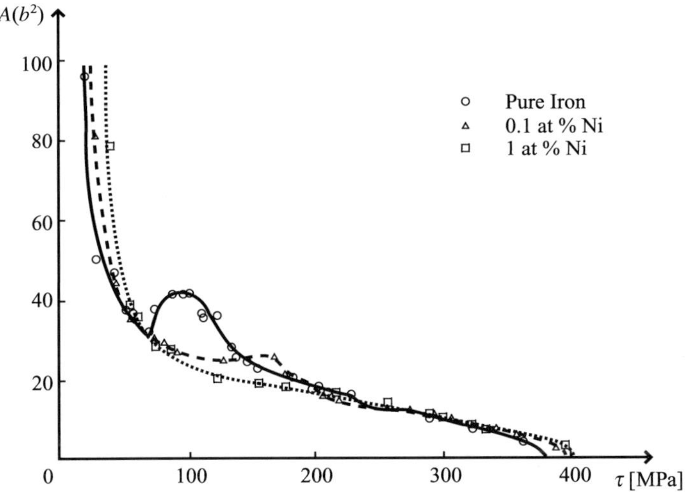

Figure 6.49. Activation area versus stress in pure Fe and Ni-containing Fe single crystals. Tensile tests by Aono et al. (1981).

Jouffrey, 1973; Bowen and Taylor, 1977). The CRSS has been calculated for this slip system, although dislocations may glide on different ones (see Section 6.3.2). Figure 6.50 shows that it exhibits a hump in the ultra high vacuum Nb used by Statham et al. (1972) not in the 50 ppm impurity Nb used by Kubin and Jouffrey (1973). This difference may be the result of different properties in tension and compression. However, when plotted as a function of the effective stress τ^* both activation volumes exhibit a peak, as seen in Figure 6.51. The differences between the values of τ^* at which the activation area peaks in pure Nb are at least partly due to different estimations of the athermal stress τ_μ.

When single crystals are oriented for single slip with a maximum Schmid factor on a $\langle 111\rangle\{11\bar{2}\}$ system, the resulting CRSS and activation areas are similar to the ones above, according to Bowen and Taylor (1977).

Alloying Nb with Mo increases the CRSS but maintains the hump at the same temperature (Figure 6.50). The corresponding activation areas exhibit very marked peaks (Figure 6.51). Addition of nitrogen up to 300 ppm increases the yield stress, according to Bowen and Taylor (1977). The activation areas remain unchanged up to 50 ppm but the peak is erased or shifted to higher stresses at 300 ppm.

6.3.1.3 Other BCC metals. The properties of Mo are very close to those of Nb, according to Aono et al. (1983). Figure 6.52 shows the temperature variation of the yield stress for three different orientations of the tensile axis. The humps are clearly seen, especially with orientation *B*. The corresponding activation area for orientation *B* also exhibits a peak.

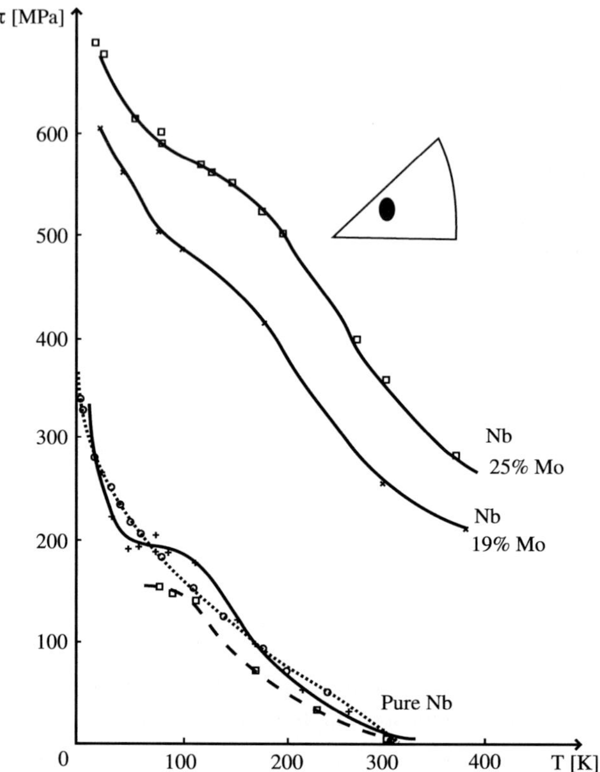

Figure 6.50. CRSS on {110} of pure Nb and Nb–Mo single crystals as a function of temperature. Tensile tests by Kubin and Jouffrey (1973) (......), compression tests by Statham et al. (1972) (full lines) and Bowen and Taylor (1977) (----).

Experiments in Li–Mg alloys with 40–60 at.% Mg yield indications of a similar behaviour (Mora-Vargas et al., 1979). Other results in Li–65% Mg (Saka and Taylor, 1981, 1982), K (Bazinski et al., 1981) and V (Carlson et al., 1979) are too incomplete to be compared with those in Fe and Nb.

6.3.2 Microstructural observations

Extensive reviews of the active slip planes have been made by Christian (1983) and Duesberry (1989). The experimental rules are rather complex, and often involve tension/compression asymmetries. In Fe–3% Si dislocations can glide either in the most stressed {110} plane or along wavy surfaces close to the non-crystallographic plane of highest Schmid factor. They can also glide along wavy surfaces corresponding to intermediate situations. Stable glide in {110} planes is, however, favoured at low temperatures or high strain-rates (Taoka et al., 1964; Sestak et al., 1967). Similar results have been obtained in Nb–5 at.% Mo and Nb–16 at.% Mo (Statham et al., 1970).

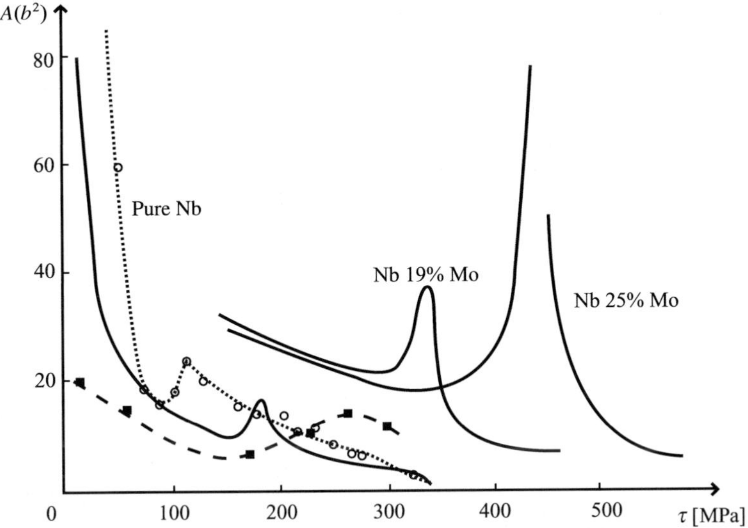

Figure 6.51. Activation areas versus stress in pure Nb and Nb–Mo single crystals corresponding to the datas of Figure 6.50.

In addition, pure Nb and Mo exhibit the so-called anomalous slip at low-temperature (50–100 K) for orientations of their tensile compression axes either in the centre of the unit triangle or close to the ⟨100⟩ direction. Anomalous slip takes place in the plane containing the primary and secondary ⟨111⟩ slip directions. It has been found in correlation with the

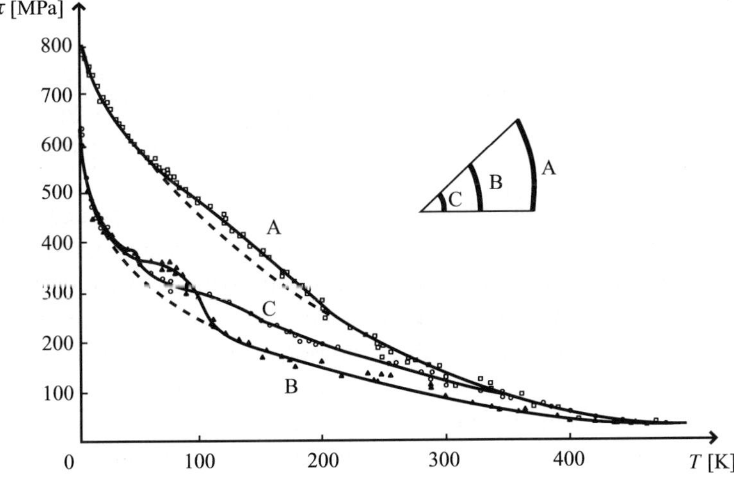

Figure 6.52. Temperature variation of the yield stress of Mo with three different orientations. Tensile tests by Aono et al. (1983).

humps in the yield-stresses versus temperature curves (Statham et al., 1972; Bowen and Taylor, 1977).

Different results have been obtained in Li–65 at.% Mg alloys. These materials exhibit wavy slip traces associated with tension–compression asymmetries, however, with a clear preference for slip on {112} planes at low-temperature (Saka and Taylor, 1981).

Post mortem TEM observations of deformed samples of Fe and Nb show long rectilinear screw dislocations with some anchoring points.

These results have been corroborated and completed by in situ TEM straining experiments[1] in Fe–3% Si at room temperature (Furubayashi, 1969) and in Nb at room temperature (Ikeno and Furabayashi, 1972) and 180 K (Ikeno and Furubayashi, 1975). These experiments show slow movements of long rectilinear screw dislocations and faster movements of short curved edges. Screw dislocations have a viscous movement in pure Fe between 150 and 180 K and a jerky one in Fe with C additions at the same temperature (Kubin and Louchet, 1979). Some jerky movements have also been reported in Nb at room temperature. Dislocations often glide on wavy surfaces, which denotes intensive cross-slip. However, they also exhibit straight slip traces which indicates that slip can be stable in {110} planes (Nb and FeSi) and possibly in {112} planes (FeSi). Intensive cross-slip generates a high density of loops and dipoles that can act as sources for dislocation multiplication (Furubayashi, 1969; Louchet and Kubin, 1979; see also Section 9.1). The transition described in Section 4.2.7 (Figures 4.19 and 4.20) between the thermally activated regimes controlled, respectively, by the Peierls mechanism on screws and by the cutting of obstacles (e.g. the dislocation forest) has been observed by Louchet et al. (1979). The corresponding temperature has been shown to depend on the density of the forest dislocations.

6.3.3 Interpretations

All experimental results indicate that deformation of pure and alloyed Fe and Nb is controlled by the motion of screw dislocations subjected to a Peierls-type friction force. This conclusion holds true for all other (less documented) BCC metals and alloys with similar properties.

Alloying or carbon/nitrogen additions may affect the athermal stress and/or the effective stress. Although the athermal stress is difficult to determine unambiguously, Figures 6.48 and 6.50 indicate that it may increase substantially with additions. This could be accounted for by a direct interaction between dislocations and clusters or solute atoms. However, this interaction should be the strongest on edge dislocations and a change would be expected in the respective mobilities of edge and screw dislocations, in contrast with observations. Effects of alloying and additions of C and N are thus at least partly due to modifications of the Peierls friction force on screw dislocations, as in Ti (Section 6.1). They have been interpreted either as softening or hardening effects, depending on the choice of the athermal stress. The effect of carbon in iron discussed in Section 6.3.1.1 illustrates this indetermination. If C is assumed to increase the athermal stress, it also decreases the effective stress according to Cottu et al. (1978). This softening effect can then be interpreted by an easier kink-pair nucleation near carbon atoms according to Kubin et al. (1979). On the contrary, if the athermal stress is assumed to be negligible, the same effect can be interpreted by the combination of an increase of the Peierls force and the vanishing or the shift of the hump on the stress–temperature curve (Figure 6.48).

The non-planar core structures at the origin of the friction force have been computed using various methods and potentials. The main results have been reviewed by Duesberry (1989), Vitek (1992) and Duesberry and Vitek (1998). In brief, degenerate and non-degenerate cores may exist for screw dislocations in BCC structures (Figure 6.53). The degenerate one has a three-fold symmetry with extensions in the three {110} planes intersecting along the ⟨111⟩ direction of the screw dislocation. It has two different but equivalent variants that can be deduced from each other by a 180° rotation. The non-degenerate core is more compact and has only one variant. According to the most recent review of Duesberry and Vitek (1998), dislocation cores are non-degenerate in VB group metals (V, Nb, Ta) and degenerate in VIB group metals (Cr, Mo, W). This difference is mainly due to the higher shear modulus of VIB metals along the ⟨111⟩ directions.

When a stress is applied, degenerate cores may glide by elementary steps on different {110} planes with periodic shape reversions. The resulting average motion should be on {112} planes, as shown in Figure 6.54(a) (Duesberry, 1989; Ngan and Wen, 2001). The alternative motion described in Figure 6.54(b) has been proposed by Brunner and

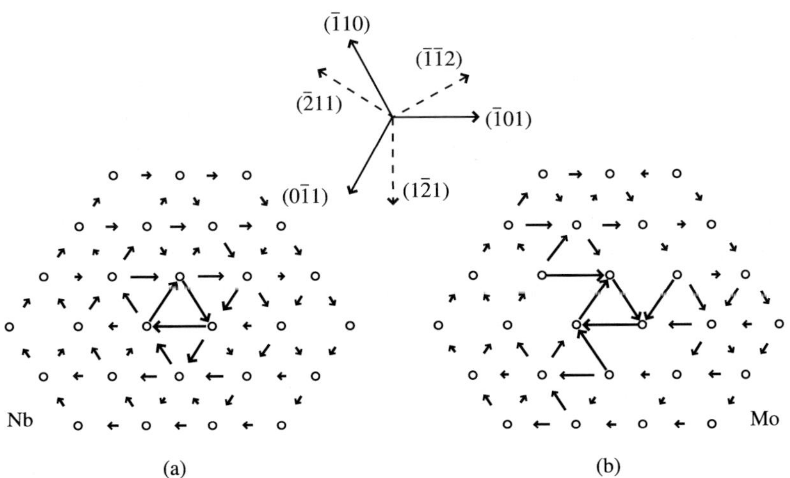

Figure 6.53. Computed core structures of screw dislocations: (a) non-degenerate in Nb; and (b) degenerate in Mo. From Duesberry and Vitek (1998).

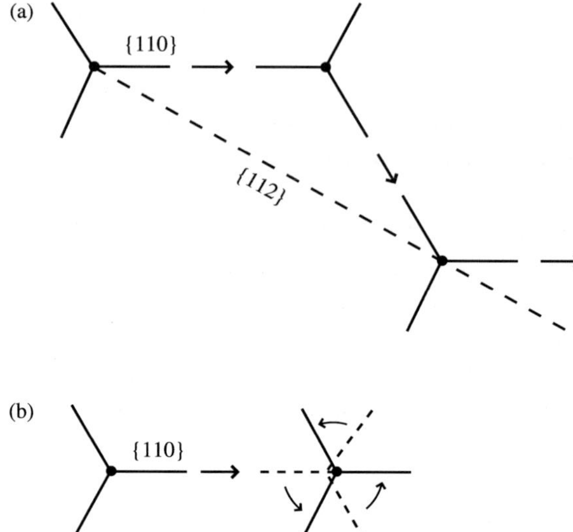

Figure 6.54. Two possible glide processes of screw dislocations with degenerate cores: (a) in an average {112} plane; and (b) in a {110} plane, with a static core reversion.

Diehl (1991c,d). It will be discussed later in this section. Non-degenerate cores may glide in a single {110} plane, possibly via a metastable intermediate configuration (Takeuchi, 1979). More recent calculations (Duesberry and Vitek, 1998), however, show that non-degenerate cores (Nb and Ta) may glide in {112}, whereas degenerate ones may glide either in a single {110} plane (Mo, W) or in an average {112} plane (Cr). Experimental observations of glide in {110} planes in Nb (Section 6.3.2) are at variance from these core simulations.

Violations of the Schmid law and tension–compression asymmetries can be accounted for by these core calculations. Two types of effects are expected: intrinsic ones, resulting from the particular symmetry of the BCC structure, and extrinsic ones, due to components of the stress tensor different from the glide shear stress. Both effects are smaller for non-degenerate cores that have a higher degree of symmetry and smaller edge components, according to Duesberry and Vitek (1998) and Bassani et al. (2001).

The discontinuities on the $\tau(T)$ and $A(\tau)$ curves do not, however, result straightforwardly from core calculations. Several explanations have been proposed and a new one is suggested. They are described in what follows.

– *Anomalous slip*:

Since anomalous slip is often reported in the same temperature range as humps, these two phenomena may be related to each other (Statham et al., 1972; Bowen and Taylor,

[1] In situ experiments for which cross-slip was allowed.

1977; Aono et al., 1983). However, this possible explanation is restricted to pure Nb and Mo with specific orientations of the deformation axis. Indeed, it cannot account for discontinuities on the $\tau(T)$ and $A(\tau)$ curves in materials that do not exhibit anomalous slip (e.g. Fe). Unless these discontinuities have different origins in different situations there must be another more general explanation.

– *Camel-hump potential*:

Takeuchi (1979) suggested that dislocations have non-degenerate cores in materials exhibiting the above discontinuities. Such dislocations may indeed glide in {110} planes via a metastable intermediate configuration in such a way that the corresponding Peierls potential may have the shape of a camel-hump (see Figure 4.5). Guyot and Dorn (1967, see Section 4.2.6), Aono et al. (1981) and Koizumi et al. (1993a,b) showed that discontinuities then appear on the $\tau(T)$ and $A(\tau)$ curves, as seen in Figure 4.18.

– *Transition between the low-stress and high-stress regimes of the kink-pair mechanism*:

This explanation has been proposed by Kubin and Jouffrey (1973), Statham et al. (1972) and Brunner et al. (1984). As discussed in Section 4.2.6, the corresponding discontinuities are, however, expected to be rather small, or even absent. Kubin and Jouffrey (1973) found an activation enthalpy varying as $((\tau_p - \tau)/\mu)^2$ in the high-stress regime. They suggested that it may correspond to the constriction energy that is the most significant part of the activation energy at high-stress in the Escaig approximation (cf. Eq. (4.30)). This is not, however, an experimental confirmation of the Escaig model because (i) the theoretical proportionality coefficient in Eq. (4.30b) is found to be too small and (ii) the experimental activation energy is also consistent with the critical bulge energy $U_b^{(c)}$ calculated with the antiparabolic potential (Eq. (4.24)).

–*Change of the kink-pair plane*:

On the basis of further results in Fe, in particular the temperature variation of the pre-exponential term plotted in Figure 6.47(b), Brunner and Diehl (1991b–d) changed their first interpretation, which was based on the transition between the low-stress and high-stress regimes of the kink-pair mechanism (see the preceding paragraph). They pointed out that the discontinuity in the pre-exponential factor at 250 K denotes a change between two clearly different mechanisms.

In the low-stress regime, above 250 K, the activation parameters are well described by the elastically interacting kinks approximation (Brunner and Diehl, 1991b, cf. Section 4.2.5). They estimated $2U_k$ by extrapolating the experimental activation enthalpy to zero stress and concluded that it corresponds to kink-nucleations in {112} planes.

In the high-stress regime, the same authors identify two subregimes, respectively, below and above 100 K (Brunner and Diehl, 1991c,d). These two subregimes are, however, not easily seen on the curves shown in Figures 6.46 and 6.47(b). Without going into too much detail, glide is proposed by these authors to take place in the high-stress regime by kink-pair nucleations in {110} planes. The activation energy extrapolated to zero stress does indeed correspond to the energy of two kinks in {110} planes (although the extrapolation from the high-stress range is questionable). In order to account for glide in {110}, kink-pair nucleations and propagations are assumed to be followed by a core reversion, in such a way that the final core remains identical to the initial one (Figure 6.54(b)).

This interpretation, based on very precise experimental results and analyses, does not, however, give a clear explanation for the discontinuity of the pre-exponential term. It is also inconsistent with the lack of extensive {112} slip at high temperatures and low stresses.

– *Transition between kink-pair and locking–unlocking mechanisms*:

Discontinuities could be accounted for by the same transition as in HCP metals, although there is no direct proof of that. Vitek (1966) was the first one to suggest that, if screw dislocations take a glissile configuration during the thermally activated unlocking process, the reverse glissile–sessile transition may also be thermally activated, as a result dislocations may glide over several interatomic distances before being locked in another Peierls valley. More details on this locking–unlocking process can be found in Section 4.3. The expected $\tau(T)$, $A(T)$ and $A(\tau)$ curves are shown in Figure 4.27.

The main advantage of a transition between the kink-pair and the locking–unlocking mechanisms, described in Section 4.3.4, is that it may readily explain the decrease in the pre-exponential term at the onset of the low-temperature regime (see Section 4.3.5). It may also account for the jerky movements of screw dislocations sometimes observed in situ (see Section 6.3.2).

6.3.4 *Conclusions on the low-temperature plasticity of BCC metals*

Many experimental data are available on the activation parameters of glide in Fe and Nb. They show that low-temperature plasticity is controlled by Peierls forces acting on screw dislocations in pure metals and in less-pure metals and alloys. There is, however, a lack of information on the microscopic aspects, e.g. the shape of dislocation cores (degenerate or not), the planes of kink-pair nucleations, the planes of stable slip at a microscopic scale, the microscopic mechanisms responsible for the discontinuities in the $\tau(T)$ and $A(\tau)$ curves. The effect of C or N additions remains, accordingly, poorly understood. Additional in situ experiments should be made on Nb, Fe and other BCC metals, e.g. Mo and Ta, to elucidate the above questions.

6.4. THE IMPORTANCE OF FRICTION FORCES IN METALS AND ALLOYS

This chapter emphasizes the relevance of Peierls friction forces on gliding dislocations in several types of metal lattices. Other examples will be provided in Chapter 10, such as cube slip in Ni_3Al.

The models for such forces derived in Chapter 4 are well adapted to describe the corresponding velocity data. In particular, there is a fair agreement between the description of Dorn and that of Hirsch and Escaig. This is confirmed by recent atomistic core structure calculations under stress. They show in particular that the core structure is actually an important component of the mobility. However, the evolution of this structure under stress suggests that (i) the description in terms of partial dislocations interacting elastically is oversimplified and (ii) the corresponding Peierls potential is smoother than expected from the dissociation model.

The above data provide strong analogies between Peierls forces in different structures. With respect to the core structure, two groups can be defined: (i) magnesium, beryllium and FCC metals gliding on non-close-packed planes exhibit a planar core along a single intersecting close-packed plane; (ii) in titanium, zirconium and BCC metals, the core is extended along several intersecting dense planes. In spite of these different core structures, all the HCP and BCC metals investigated exhibit some common features in their mechanical behaviour. They are characterized by comparable discontinuities on the temperature dependence of their yield stress and activation area. A common interpretation has been proposed on the basis of a transition from the locking–unlocking mechanism at low-temperature to the kink-pair mechanism at higher temperatures. However, this has to be confirmed by additional experiments in BCC metals.

REFERENCES

Ahmadieh, A., Mitchell, J. & Dorn, J.E. (1965) *Trans. Met. Soc. AIME*, **233**, 1130.

Akhtar, A. & Teghtsoonian, E. (1969) *Acta Met.*, **17**, 1351.

Akhtar, A. & Teghtsoonian, A. (1971) *Acta Met.*, **19**, 655.

Akhtar, A. & Teghtsoonian, E. (1975) *Met. Trans.*, **6**, 2201.

Anongba, P.N.B. & Martin, J.L. (1991) *in Proc. 9th Int. Conf. On Strength of Metals and Alloys* vol. **1**, Eds. Brandon, D.G., Chaim, R. & Rosen A., Freund, London, p. 203.

Anongba, P.N.B., Bonneville, J., Douin, J., Martin, J.L. (1991). in *Proc. 9th Int. Conf. On Strength of Metals and Alloys, Ibidem*, p. 303.

Anongba, P.N.B., Bonneville, J. & Martin, J.L. (1993a) *Acta Met. Mater.*, **41**, 2897.

Anongba, P.N.B., Bonneville, J. & Martin, J.L. (1993b) *Acta Met. Mater.*, **41**, 2907.

Aono, Y., Kitajima, K. & Kuramoto, E. (1981) *Scripta Met.*, **15**, 275.

Aono, Y., Kuramoto, E. & Kitajima, K. (1983), in *Proc. 6th Int. Conf. on the Strength of Metals and Alloys*, Ed. Gifkins, R.C., Pergamon Press, New York, p. 9.

Bacon, D.J. & Martin, J.W. (1981a) *Phil. Mag. A*, **43**, 883.

Bacon, D.J. & Martin, J.W. (1981b) *Phil. Mag. A*, **43**, 901.

Bacroix, B. & Jonas, J.J. (1988) *Textures Microstruct.*, **8–9**, 267.

Bassani, J.L., Ito, K. & Vitek, V. (2001) *Mater. Sci. Eng. A*, **319–321**, 97.

Basinski, Z.S., Duesberry, M.S. & Murty, G.S. (1981) *Acta Met.*, **29**, 801.

Beuers, J., Jönsson, S. & Petzow, G. (1987) *Acta Met.*, **35**, 2277.

Biget, M.P. & Saada, G. (1989) *Phil. Mag.*, **59**, 747.

Blum, W. & Finkel, A. (1982) *Acta Met.*, **30**, 1705.

Bonneville, J., Escaig, B. & Martin, J.L. (1988) *Acta Met.*, **36**, 1989.

Bowen, D.K. & Taylor, G. (1977) *Acta Met.*, **25**, 417.

Brion, H.G., Siethoff, H. & Schröter, W. (1981) *Phil. Mag.*, **43**, 1505.

Brunner, D. & Diehl, J. (1991a) *Phys Stat. Sol (a)*, **125**, 203.

Brunner, D. & Diehl, J. (1991b) *Phys. Stat. Sol (a)*, **124**, 155.

Brunner, D. & Diehl, J. (1991c) *Phys. Stat. Sol (a)*, **124**, 455.

Brunner, D. & Diehl, J. (1997) *Phys. Stat. Sol (a)*, **160**, 355.

Brunner, D., Diehl, J. & Seeger, A. (1984) in *Structure and Properties of Crystal Defects*, Eds. Paidar, V. & Lejcek L., Elsevier Publ. Co, Amsterdam, New York, p. 175.

Cahn, R.W. (1951) *J. Inst. Met.*, **79**, 129.

Caillard, D. (1984) *Acta Met.*, **32**, 1483.

Caillard, D. (1985) *Phil. Mag.*, **51**, 157.

Caillard, D. & Martin, J.L. (1982) *Acta Met.*, **30**, 791.

Caillard, D. & Martin, J.L. (1983) *Acta Met.*, **31**, 813.

Carlson, O.N., Rehbein, D.K. & Bogacik, K.E. (1979) in *Strength of Metals and Alloys*, Eds. Haasen, P., Gerold, V. & Kostorz G., Pergamon, Toronto, p. 995.

Carrard, M. & Martin, J.L. (1987) *Phil. Mag. A*, **56**, 391.

Carrard, M. & Martin, J.L. (1988) *Phil. Mag. A*, **58**, 491.

Christian, J.W. (1983) *Met. Trans.*, **14A**, 1237.

Cockayne, D.J.H., Jenkins, M.L. & Ray, I.L.F. (1971) *Phil. Mag.*, **24**, 1383.

Cottu, J.P., Peyrade, J.P., Chomel, P. & Groh, P. (1978) *Acta Met.*, **26**, 1179.

Coulomb, P. (1978) *J. Micr. Spectr. Electron.*, **3**, 295.

Couret, A. & Caillard, D. (1985a) *Acta Met.*, **33**, 1447.

Couret, A. & Caillard, D. (1985b) *Acta Met.*, **33**, 1455.

Couret, A. & Caillard, D. (1988) *Acta Met.*, **36**, 2515.

Couret, A. & Caillard, D. (1989a) *Phil. Mag. A*, **59**, 783.

Couret, A. & Caillard, D. (1989b) *Phil. Mag. A*, **59**, 801.

Couret, A., Caillard, D., Pushl, W. & Schoeck, G. (1991) *Phil. Mag. A*, **63**, 1045.

Derep, J.L., Ibrahim, S., Rouby, R. & Fantozzi, G. (1980) *Acta Met.*, **28**, 607.

Duesberry, M.S. (1989) in *Dislocations in Solids*, vol. **8**, Ed. Nabarro, F.R.N., North Holland, Amsterdam, p. 67.

Duesberry, M.S. & Vitek, V. (1998) *Acta Met.*, **46**, 1481.

Edelin, G. (1972) *Scripta Met.*, **6**, 1185.

Escaig, B. (1968) *Phys. Stat. Sol.*, **28**, 463.

Farenc, S., Caillard, D. & Couret, A. (1993) *Acta Met. Mater.*, **41**, 2701.

Farenc, S., Caillard, D. & Couret, A. (1995) *Acta Met. Mater.*, **43**, 3669.

Foreman, A.J.E. & Lomer, W.M. (1955) *Phil. Mag.*, **46**, 73.

Friedel, J. (1955) *Phil. Mag.*, **46**, 1169.

Friedel, J. (1959) in *Internal Stresses and Fatigue*, Eds. Rasseiler, G.M. & Grube W.L., Elsevier, London, p. 220.

Friedel, J. (1964) *Dislocations*, Pergamon, Paris, p. 454.

Furubayashi, E. (1969) *J. Phys. Soc. Jpn*, **27**, 130.

Gallagher, P.C.J. (1970) *Met. Trans.*, **1**, 2429.

Gil Sevillano, J., Van Houtte, P. & Aernoudt, E. (1980) *Progr. Mater. Sci.*, **25**, 69.

Guyot, P. & Dorn, J.E. (1967) *Can. J. Phys.*, **45**, 983.

Haessner, F. (1965) *Recrystallisation, Grain Growth and Textures*, American Soc. For Metals, Metals Park, Ohio, p. 386.

Ikeno, S. & Furabayashi, E. (1972) *Phys. Stat. Sol (a)*, **12**, 611.

Ikeno, S. & Furubayashi, E. (1975) *Phys. Stat. Sol (a)*, **27**, 581.

Jaffe, N. & Dorn, J.E. (1962) *Trans. Met. Soc. AIME*, **224**, 1167.

Johnson, R.D., Young, A.P. & Schwope, A.D. (1956) *Proceedings of a Symp. on the Creep and Fracture of Metals at High Temperatures*, Her Majesty's Stationary Office, London, p. 25.
Jonsson, S. & Beuers, J. (1987) *Mater. Sci. Eng.*, **91**, 111.
Koizumi, H., Kirschner, H.O.K. & Suzuki, T. (1993a) *Acta Met. Mater.*, **41**, 3483.
Koizumi, H., Kirschner, H.O.K. & Suzuki, T. (1993b) *Acta Met. Mater.*, **41**, 3483.
Kubin, L. (1982) in *Reviews on the Deformation Behaviour of Materials*, vol. **IV**, no. 3, Ed. Feltham, P., Scientific Pub. Div. Freund Pub. House Ltd, Tel Aviv, Israël, p. 181.
Kubin, L.P. & Jouffrey, B. (1973) *Phil. Mag.*, **27**, 1369.
Kubin, L.P. & Louchet, F. (1979) *Acta Met.*, **27**, 337.
Kubin, L.P., Louchet, F., Peyrade, J.P., Groh, P. & Cottu, J.P. (1979) *Acta Met.*, **27**, 343.
Kuramoto, E., Aono, Y. & Kitajima, K. (1979) *Scripta Met.*, **13**, 1039.
Lacombe, P. & Beaujard, L. (1947) *J. Inst. Met.*, **74**, 1.
Le Hazif, R. & Poirier, J.P. (1975) *Acta Met.*, **23**, 865.
Le Hazif, R., Dorizzi, P. & Poirier, J.P. (1973) *Acta Met.*, **21**, 903.
Ledbetter, H.M. & Naimon, E.R. (1974) *J. Phys. Chem. Ref. Data*, **3**, 897.
Legrand, B. (1984) *Phil. Mag. B*, **49**, 171.
Legrand, B. (1985) *Phil. Mag. A*, **52**, 83.
Levine, E.D. (1966) *Trans. Met. Soc. AIME*, **236**, 1558.
Louchet, F. & Kubin, L.P. (1979) *Phys. Stat. Sol (a)*, **56**, 169.
Louchet, F., Kubin, L.P. & Vesely, D. (1979) *Phil. Mag. A*, **39**, 433.
Majimel, J., Casanove, M.J., Lapasset, G. & Molenat, G. (2002) in *Proceed. ICAA8 396–402*, Eds. Gregson, P.J. & Harris S.J., Materials Science Forum, p. 1383.
Mills, D. & Craig, G.B. (1968) *Trans. Met. Soc. AIME*, **242**, 1881.
Mills, M.J. & Stadelman, P. (1989) *Phil. Mag. A*, **60**, 355.
Mora-Vargas, M.I., Saka, H. & Taylor, G. (1979) in *Strength of Metals and Alloys*, Eds. Haasen, P., Gerold, V. & Kostorz G., Pergamon, New York, p. 1007.
Morris, M.A. & Martin, J.L. (1984) *Acta Met.*, **8707**, 5505.
Naka, S., Lasalmonie, A., Costa, P. & Kubin, L.P. (1988) *Phil. Mag. A*, **57**, 717.
Naka, S., Kubin, L.P. & Perrier, C. (1991) *Phil. Mag.*, **63**, 1035.
Ngan, A.H. & Wen, M. (2001) *Phys. Rev. Lett.*, **87**, 675505.
Orlova, A. & Kucharova, K. (1999) *Phil. Mag.*, **A79**, 339.
Peterson, N.L. (1978) *J. Nucl. Mater.*, **69–70**, 3.
Poirier, J.P. (1972) *Phil. Mag.*, **26**, 701.
Poirier, J.P. (1978) *Acta Met.*, **26**, 629.
Quenel, D.J., Sato, A. & Meshii, M. (1975) *Mater. Sci. Eng.*, **18**, 199.
Régnier, P. & Dupouy, J.M. (1970) *Phys. Stat. Sol.*, **39**, 79.
Retima, M. & Cornet, M. (1986) *Acta Met.*, **34**, 573.
Richards, T.L. & Pugh, S.F. (1959) *J. Inst. Met.*, **88**, 399.
Saada, G. (1960) *Acta Met.*, **8**, 841
Saka, H. & Taylor, G. (1981) *Phil. Mag.*, **43**, 1377.
Saka, H. & Taylor, G. (1982) *Phil. Mag.*, **45**, 973.
Sakai, T. & Fine, M.E. (1974) *Acta Met.*, **22**, 1359.
Sastry, D.H. & Vasu, K.I. (1972) *Acta Met.*, **20**, 399.
Schmid, E. & Boas, W. (1931) *Z. Phys.*, **71**, 703.
Schoeck, G. (1965) *Phys. Stat. Sol.*, **8**, 499.
Sastry, D.H., Prasad, Y.V.R.K. & Vasu, K.I. (1971) *Journal of Materials Science*, **6**, 332.
Servi, I.S., Norton, J.T. & Grant, N.J. (1952) *Trans. Am. Inst. Min. Met. Engrs*, **194**, 965.

Sestak, B., Zarubova, N. & Sladek, V. (1967) *Can. J. Phys.*, **45**, 1031.
Sherby, O.D., Lytton, J.L. & Dorn, J.E. (1957) *Acta Met.*, **5**, 219.
Siethoff, H. & Ahlborn, K. (1986) *Scripta Met.*, **20**, 144.
Soo, P. & Higgins, T. (1968) *Acta Met.*, **16**, 177.
Statham, C.D., Vesely, D. & Christian, J.W. (1970) *Acta Met.*, **18**, 1243.
Statham, C.D., Koss, D.A. & Christian, J.W. (1972) *Phil. Mag.*, **26**, 1089.
Stobbs, W.M. & Sworn, C.H. (1971) *Phil. Mag.*, **24**, 1365.
Stohr, J.F (1972). Thèse d' Etat no 968, Orsay University.
Stroh, A.N. (1956) *Phil. Mag.*, **1**, 491.
Suzuki, T., Takeuchi, S. & Yoshinaga, H. (1985) *Dislocation Dynamics and Plasticity*, vol. **12**, Springer Series in Materials Science, Springer-Verlag, Berlin.
Takeuchi, S. (1979) *Phil. Mag. A*, **39**, 661.
Takeuchi, S. & Argon, A.S. (1976) *J. Mater. Sci.*, **11**, 1542.
Tanaka, T. & Conrad, H. (1972) *Acta Met.*, **20**, 1019.
Taoka, T., Takeuchi, S. & Furubayashi, E. (1964) *J. Phys. Soc. Jpn*, **19**, 701.
Taylor, G. (1992) *Progr. Mater. Sci.*, **36**, 29.
Treharne, P.I. & Moore, A. (1962) *J. Less Common Met.*, **4**, 275.
Trojanova, Z., Luckac, P., Dlouhy, A. & Heinitz, H.J. (1985) *Czech. J. Phys. B*, **35**, 298.
Trojanova, Z., Luckac, P. & Hammersky, M. (1991) *Acta Universitatis Carolinae—Mathematica et Physica*, **32**, 69.
Tseng, D. & Tangri, K. (1977) *Scripta Met.*, **11**, 719.
Tung, P.P. & Sommer, A.W. (1970) *Met. Trans.*, **1**, 947.
Tyson, W.R. (1967) *Can. Met. Quarty*, **6**, 301.
Vanderschaeve, G. & Escaig, B. (1980) in *Dislocations et Déformation Plastique*, Eds. Groh, P., Kubin, L. & Martin J.L., Les Editions de Physique, Orsay, p. 141.
Vitek, V. (1966) *Phys. Stat. Sol.*, **18**, 687.
Vitek, V. (1975) *Scripta Met.*, **9**, 611.
Vitek, V. (1992) *Progr. Mater. Sci.*, **36**, 1.
Vitek, V. & Igarashi, M. (1991) *Phil. Mag.*, **A63**, 1059.
Ward-Flynn, P., Mote, J. & Dorn, J.E. (1961) *Trans. Met. Soc. AIME*, **221**, 1148.
Yoshinaga, H. & Horiuchi, R. (1963) *Trans. Jpn Inst. Met.*, **5**, 14.

Chapter 7

The Peierls–Nabarro Mechanism in Covalent Crystals

Chapter 7
The Peierls–Nabarro Mechanism in Covalent Crystals

Dislocations in covalent materials are subjected to Peierls friction stresses as in metals and alloys, although their physical origin is markedly different. In contrast with the situation described in Section 4, dislocation cores are compact or extended in their slip-plane. The friction force is due to the breaking of strongly directional covalent bonds, followed by their restoration at adjacent first neighbours, each time dislocations move over one interatomic distance.

This friction is essentially different from that in metals and alloys because it only depends on the density of bonds to be cut, namely on the slip-plane and Burgers vector. In that sense, it is more a property of the perfect crystal alone than in metals and alloys where it is also a property of the (more or less extended) core structure.

In the following, the modelling of dislocation mobility is presented. The Hirth and Lothe theory of kink nucleation and diffusion along the dislocation line is revisited and extended to dissociated dislocations. It is then compared with available experimental data in semi-conducting materials. The theory also applies to other covalent materials and various types of ceramics.

7.1. DISLOCATION CORE STRUCTURES AND PEIERLS–NABARRO FRICTION FORCES

Peierls valleys, where dislocations have a markedly lower energy, correspond to directions and positions where covalent bonds are not disturbed much. Straight dislocations in deep Peierls valleys are difficult to move as a whole because this would involve the co-operative breaking of a large number of bonds. Like in metals and alloys it is thus easier to nucleate and propagate kink pairs. Kinks between adjacent Peierls valleys are expected to be more mobile than straight dislocations because their motion involves, due to their short length, the simultaneous breaking of fewer bonds. Kinks annihilate at the corners defined by the intersections between the valleys. Parts of gliding dislocations that are not parallel to Peierls valleys are formed by the collection of kinks provided by the less mobile segments. Their motion is accordingly controlled by the lateral glide of these kinks (Figure 7.1).

In covalent crystals, there is no possibility of moving any dislocation, with any core structure or character, without breaking bonds. Dislocations are thus expected never to be fully glissile. Two consequences that will be discussed in the following are the existence

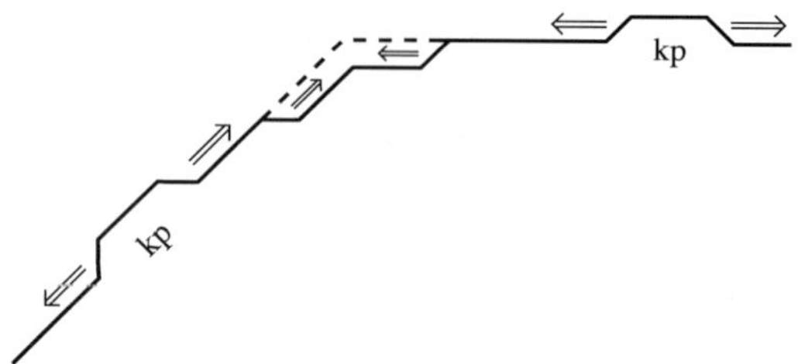

Figure 7.1. Dislocation motion by nucleation and propagation of kink-pairs (kp).

of friction forces on kinks and the absence of any transition towards a locking–unlocking mechanism, unlike in metals and alloys.

Most experimental studies have been conducted in semiconductors and (to a lesser extent) in some ceramic materials. The crystal structure of elemental semiconductors (ESC) is a diamond cubic structure, which can be described as an FCC lattice with a two-atom basis. In compound semiconductors (CSC), the two atoms of the basis are different, which, in the case of the cubic symmetry, gives the so-called sphalerite (or zinc-blende) structure (Figure 7.2). Some other, less extensively studied, CSC and ceramics have a HCP symmetry with a two different atom basis, called the wurtzite structure. Covalent bonds in CSC have a partial ionic character.

Dislocations in ESC and CSC structures have the same Burgers vectors and slip planes as in FCC and HCP close-packed metals, respectively. In particular, the behaviour of $\frac{1}{2}\langle 01\bar{1}\rangle$ dislocations in $\{111\}$ planes of diamond cubic and sphalerite structures should be similar to that of $\frac{1}{3}\langle 1\bar{1}00\rangle$ dislocations in the basal plane of the wurtzite structure. As in close-packed structures, a low energy stacking fault can be defined between the narrowly spaced planes parallel to $\{111\}$ (e.g. II and 1 in Figure 7.2). Dislocations in these planes can, accordingly, lower their energy by dissociating into two Shockley partials. These are called "glide dislocations". Their motion, however, appears difficult because it involves the cutting of a high density of covalent bonds.

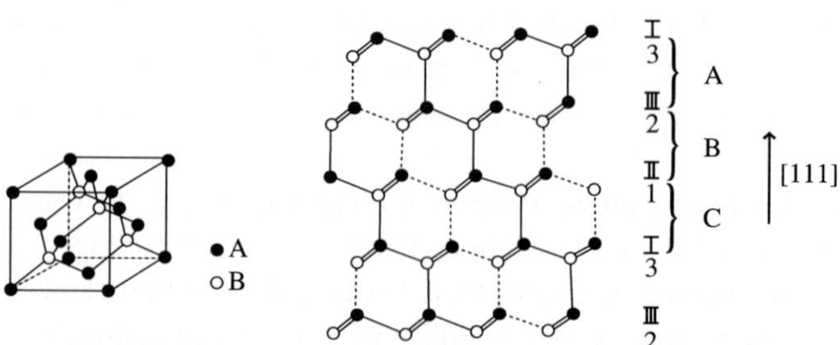

Figure 7.2. The sphalerite structure. I-1, II-2 and III-3 layers correspond to the ABC stacking of the FCC structure. The two atoms are identical in the diamond structure.

Glide can also take place between the widely spaced planes (e.g. I and 1 in Figure 7.2), where dissociation is not allowed but the motion involves the cutting of three times less covalent bonds and so should be easier. The corresponding dislocations are called "shuffle".

Whether dislocations glide in the shuffle set or in the glide set has been extensively discussed (Hirth and Lothe, 1982; George and Rabier, 1987; Louchet and Thibault-Desseaux, 1987; Koizumi et al., 2000). Electron microscopy observations reveal that gliding dislocations are dissociated, at least at high temperature. This shows that dislocations are made of two Shockley partials separated by a stacking fault that is necessarily in the glide set plane. Shockley partials initially in the glide set plane may, however, move by climb over one interatomic distance to the shuffle set plane, leaving the stacking fault unchanged. Such a complex structure could then move the stacking fault in the glide set plane by glide and the Shockley partials in the shuffle set via short-range atomic displacements (George and Rabier, 1987). Note that this process is different from climb because it does not involve diffusion over long distances. Under such conditions, the higher mobility of kinks in the shuffle set due to the lower density of cut bonds, would be at least partly cancelled by the additional friction due to the short-range atomic displacements. Some authors believe that Shockley partials may be partly glide, partly shuffle (Louchet and Thibault-Desseaux, 1987).

Perfect and Shockley dislocations of the glide and shuffle sets can decrease their core energies by reconstructing their dangling bonds (Figure 7.3). This increases the depth of Peierls valleys and the corresponding friction forces. Different types of reconstruction have been discussed by George and Rabier (1987), Louchet and Thibault-Desseaux (1987) and Bulatov et al. (1995). Reconstructions are expected to be more difficult in CSC because they should take place between atoms of the same type. Friction forces are accordingly expected to be weaker.

7.2. DISLOCATION VELOCITIES

Different models have been proposed for the motion of dislocations parallel to Peierls valleys. They are all based on the nucleation and propagation of kink-pairs, but they rely on different hypotheses for the mobility of kinks. They are presented in order of decreasing kink mobilities.

7.2.1 High kink mobility (metal-like model of Suzuki et al., 1995)

If the kinks are assumed to be as mobile as in metals and alloys, the models described in Section 4 apply. The treatment proposed by Suzuki et al. (1995) is based on a "wall-like" potential with flat minima and very steep barriers at $y = \pm h/2$ (compare with Figure 4.5). This potential has a very high Peierls stress τ_p close to the shear modulus μ, that induces a very pronounced stress increase at decreasing temperature. The critical bulge shape in the

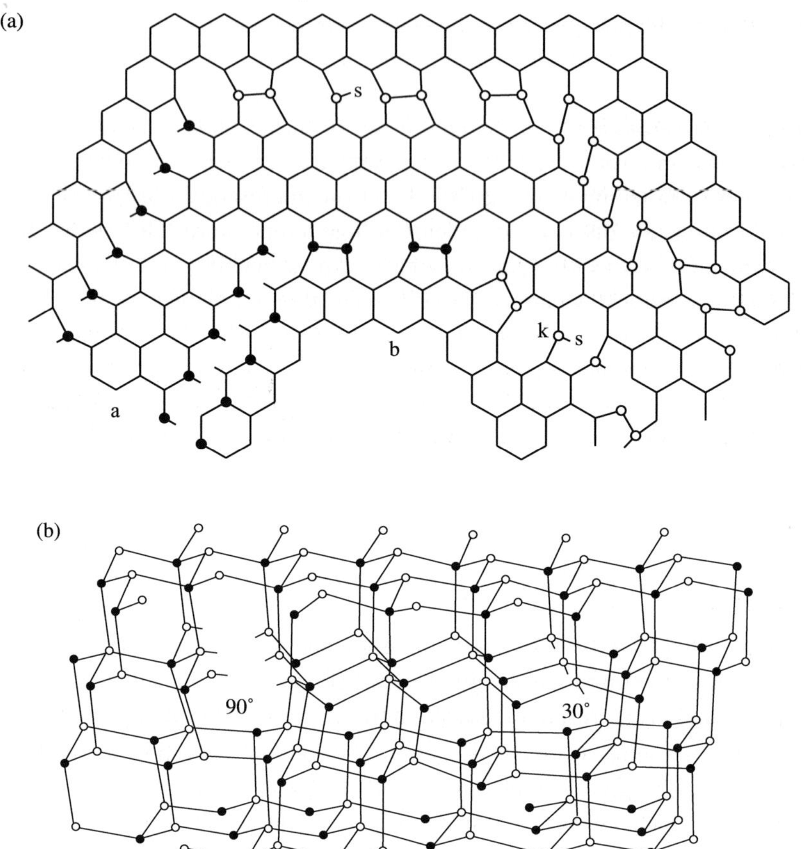

Figure 7.3. Dislocation half-loop in the glide set of the sphalerite structure (a) in plane view: unreconstructed core in A, reconstructed cores in B, with kinks k and antiphase reconstruction defects s (the screw segment is horizontal). (b) In three dimensions. From George and Rabier (1987).

high stress range is assumed to be trapezoidal with a constant slope. This model accounts for the large stress increase measured at low temperature under hydrostatic confining pressure (see Suzuki et al., 1994). It must be noted, however, that any potential with an abrupt variation (e.g. with a locally high slope) would yield similar results at high stresses. In addition, this model does not account for experimental results at higher temperatures and lower stresses (see Section 7.3.2). Its interest is thus very limited.

7.2.2 Low kink mobility: case of undissociated dislocations

The mobility of kinks is assumed to be limited. It is then necessary to introduce the mean free path of kink pairs, X, before they annihilate with kinks of opposite signs or before they pile up or disappear at dislocation extremities.

Then, the dislocation velocity can be expressed as:

$$\mathrm{v} = P_{\mathrm{kp}} X h \tag{7.1}$$

where P_{kp} is the probability of nucleating kink pairs per unit dislocation length and unit time and h is the distance between Peierls valleys.

Two regimes can be defined (Figure 7.4):

(1) In the kink-collision regime, the dislocation length, L, is larger than the kink pair mean-free-path, X. This means that the time between two kink pair nucleations, $1/P_{\mathrm{kp}}L$, is shorter than the time for propagating kinks to dislocation extremities $L/2\mathrm{v}_{\mathrm{k}}$, where v_{k} is the kink velocity. In other words, kinks annihilate with kinks of opposite sign coming from other kink-pair nucleations along the same dislocation line. The time between two successive nucleations on a straight segment of average length $X/2$, is $2/P_{\mathrm{kp}}X$. It must be equal to the time $X/2\mathrm{v}_{\mathrm{k}}$ for propagating kinks until they annihilate, (Figure 7.4). This condition yields:

$$X^2 = 4\frac{\mathrm{v}_{\mathrm{k}}}{P_{\mathrm{kp}}} \tag{7.2}$$

which can be inserted in Eq. (7.1), whence:

$$\mathrm{v} = 2h\sqrt{\mathrm{v}_{\mathrm{k}} P_{\mathrm{kp}}} \tag{7.3a}$$

(2) In the length-effect regime, the dislocation length L is smaller than X calculated above (Eq. (7.2)). In other words, there is only one kink-pair moving along the dislocation segment at the same time, in such a way that the effective value of X to be inserted in Eq. (7.1) is L. This yields a dislocation velocity proportional to length:

$$\mathrm{v} = P_{\mathrm{pk}} L h \tag{7.3b}$$

The kink-collision (length-independent) regime is never obtained when the velocity of kinks is high (metal-like behaviour). Accordingly, the existence of a length-independent regime is the only way to prove that the velocity of kinks is actually low. This point will be discussed in Section 7.3.4.

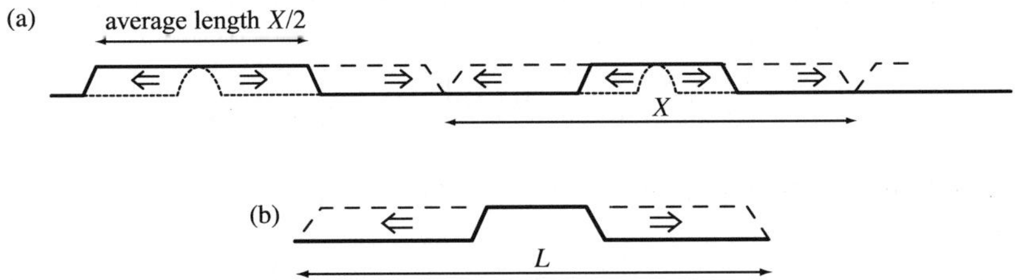

Figure 7.4. Nucleation and annihilation of kinks (a) in the collision regime and (b) in the length-effect regime.

7.2.2.1 ***Point-obstacle model of Celli et al. (1963).*** Point obstacles are assumed to be present along dislocation lines. Their role is to slow down the movement of kinks—not to stop them—as they can be crossed with the help of thermal activation. This model has essentially an historical importance because it is not supported by the most recent experimental results (Section 7.3). The average distance between point obstacles is x_{d} and the activation energy of the crossing process is U_{d}. The corresponding average kink velocity is

$$\mathrm{v}_{\mathrm{k}} = \nu_{\mathrm{k}} x_{\mathrm{d}} \exp\left(-\frac{U_{\mathrm{d}}}{kT}\right) \tag{7.4}$$

where ν_{k} is the kink vibration frequency ($\nu_{\mathrm{k}} \approx \nu_{\mathrm{D}}$).

The energy diagram for nucleating a kink pair is shown in Figure 7.5. A kink pair is successfully formed when (i) the maximum energy of the kink pair is attained for the critical kink separation x_{c} and (ii) the first obstacle of height U_{d} is crossed for the kink separation x_{d}. The rate of nucleation per unit length P_{kp} can be obtained by multiplying Eq. (4.1) by a factor $f(\tau)$ that takes into account the effect of the point obstacle:

$$P_{\mathrm{kp}} = \nu_{\mathrm{D}} \frac{b}{x_{\mathrm{c}}^2} f(\tau) \exp\left(-\frac{U_{\mathrm{kp}}^{(\mathrm{c})}(\tau)}{kT}\right) \tag{7.5}$$

with $f(\tau) \sim 1$ for $\tau bh(x_{\mathrm{d}} - x_{\mathrm{c}}) \gg U_{\mathrm{d}}$ (no backward movement during the waiting time to cross the obstacle, Figure 7.5(a)) and $f(\tau) \to 0$ for $\tau bh(x_{\mathrm{d}} - x_{\mathrm{c}}) \ll E_{\mathrm{d}}$ (easy backward movement from x_{d} to 0, Figure 7.5(b)).

For the whole stress range, Celli proposes

$$f(\tau) = \left(1 + \frac{U_{\mathrm{d}}}{\tau bax_{\mathrm{d}}} + \frac{x_{\mathrm{c}}}{x_{\mathrm{d}}}\right) \exp\left[-\left(\frac{U_{\mathrm{d}}}{\tau bax_{\mathrm{d}}} + \frac{x_{\mathrm{c}}}{x_{\mathrm{d}}}\right)\right] \tag{7.6}$$

which is consistent with the above two asymptotic conditions.

In the kink-collision regime, the kink pair mean free path is (using Eq. (7.2))

$$X = 2\left(\frac{x_{\mathrm{d}}}{bf(\tau)}\right)^{1/2} x_{\mathrm{c}} \exp\left(\frac{U_{\mathrm{kp}}^{(\mathrm{c})}(\tau) - U_{\mathrm{d}}}{2kT}\right)$$

Then, for $L > X$, the dislocation velocity is (combining Eqs. (7.3a), (7.4) and (7.5))

$$\mathrm{v} = 2h \frac{\sqrt{bx_{\mathrm{d}} f(\tau)}}{x_{\mathrm{c}}} \nu_{\mathrm{D}} \exp\left(-\frac{U_{\mathrm{kp}}^{(\mathrm{c})}(\tau) + U_{\mathrm{d}}}{2kT}\right) \tag{7.7a}$$

In the length-effect regime, i.e. for $L < X$, it is (combining Eqs. (7.3b) and (7.5)):

$$\mathrm{v} = L \frac{hb}{x_{\mathrm{c}}^2} f(\tau) \nu_{\mathrm{D}} \exp\left(-\frac{U_{\mathrm{kp}}^{(\mathrm{c})}(\tau)}{kT}\right) \tag{7.7b}$$

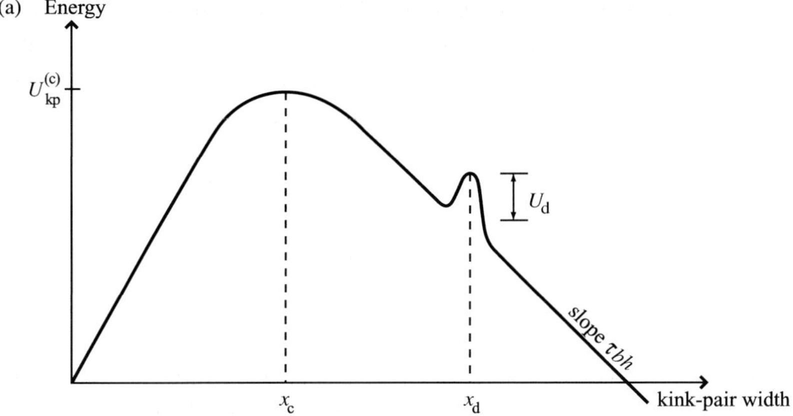

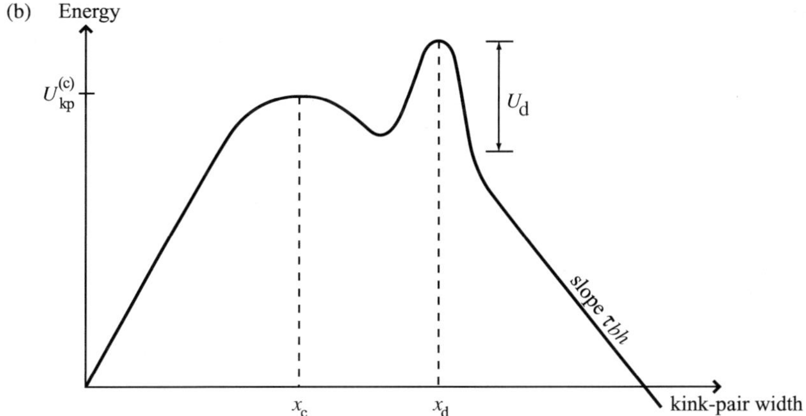

Figure 7.5. Energy profile for kink-pair nucleation according to Celli et al. (1963): (a) high stress and (b) low stress.

The stress dependence of the dislocation velocity lies both in the pre-exponential term $f(\tau)$ and in the exponential term.

7.2.2.2 Kink diffusion model of Hirth and Lothe (1982).

(a) *Kink mobility*

The movement of kinks along straight dislocations in deep Peierls valleys is assumed to be controlled by the breaking of covalent bonds at every interatomic distance. Only a few covalent bonds are involved in each jump over the distance b along the dislocation line, in such a way that the process can be thermally activated, with an activation energy U_m. U_m may also include the short-range diffusion of atoms, according to Section 7.1.

U_m is expected to be higher when the kinks are reconstructed, e.g. in strongly covalent ESC such as Si.

When there is no applied force, the probability of a jump in both directions is:

$$\nu = \nu_D \exp\left(-\frac{U_m}{kT}\right) \tag{7.8}$$

When there is an applied stress τ, the probabilities of forward and backward jumps become different, because the work done by the stress over the area bh (with $h = b(\sqrt{3}/2)$) must be taken into account:

$$\nu^+ = \frac{1}{2}\nu_D \exp\left(-\frac{U_m - \tau b^2 h}{kT}\right)$$

$$\nu^- = \frac{1}{2}\nu_D \exp\left(-\frac{U_m + \tau b^2 h}{kT}\right)$$

and the net jump frequency is:

$$\nu = \nu_D sh\frac{\tau b^2 h}{kT}\exp\left(-\frac{U_m}{kT}\right) \tag{7.9a}$$

Since the quantity $\tau b^2 h$ is usually much smaller than kT, it reduces to:

$$\nu = \nu_D\frac{\tau b^2 h}{kT}\exp\left(-\frac{U_m}{kT}\right) \tag{7.9b}$$

Since the jump distance of kinks is b, their average velocity is accordingly:

$$\mathrm{v}_k = \nu_D\frac{\tau b^3 h}{kT}\exp\left(-\frac{U_m}{kT}\right) \tag{7.9c}$$

Introducing the diffusion coefficient of kinks,

$$D_k = \nu_D b^2 \exp\left(-\frac{U_m}{kT}\right) \tag{7.10}$$

the kink velocity becomes:

$$\mathrm{v}_k = D_k\frac{\tau bh}{kT} \tag{7.11a}$$

(Einstein mobility relation). The corresponding transport flux of kinks along the dislocation line is:

$$\Phi_t = \frac{c_k \mathrm{v}_k}{b} = D_k c_k \frac{\tau h}{kT} \tag{7.11b}$$

where c_k is the dimensionless kink density per atomic site of length b. The motion of kinks can thus be treated like a diffusion process. In particular, when there is a gradient of concentrations of kinks $\partial c_k/\partial x$ along a dislocation line, there is also a net total drift flux that can be computed as the difference between the number of jumps of a concentration c_k in one direction and the number of jumps of a concentration $(c_k - (\partial c_k/\partial x)b)$ in the opposite direction. This second, diffusional, flux is accordingly:

$$\Phi_d = -\frac{D_k}{b}\frac{\partial c_k}{\partial x} \tag{7.12}$$

(Fick's first law).

(b) *Rate of kink-pair nucleations*

The nucleation of a kink-pair of critical size that can expand under applied stress is a complex problem that involves the elastic interaction between the two kinks, calculated in Section 4 (Figure 4.4), and the diffusion-controlled kink mobility.

The corresponding energy profile as a function of the kink separation Δx is described schematically in Figure 7.6. It is the sum of the energy profile $U_{kp}(\tau, \Delta x)$ discussed in Section 4 (Figure 4.4(c)) and the periodic potential of height U_m controlling the kink motion. As in Section 4, we consider the movement of kink-pairs across this energy profile, the new variable describing the movement being Δx. The kinks here are considered to have a small width of the order of b (Figure 7.4). Then, in contrast to the case of metals, where the width of kinks can be much larger than b, the energy profile is controlled by the elastic interaction of kinks up to high stresses due to the small critical kink separation x_c. Accordingly, the energy profile is $U_{kp} = U_{ik}$, the maximum of which is $U_{ik}^{(c)}(\tau)$, given by Eq. (4.36),

$$U_{ik}^{(c)}(\tau) = 2U_k - hb\left(\frac{hb}{2\pi}\mu\tau\right)^{1/2}$$

for the critical separation distance x_c given by Eq. (4.35). This expression has also been deduced from atomistic calculations by Marklund (1984).

The kink energy, U_k, is, however, different from those calculated in Section 4.2.3 in the case of metals. Another estimation of U_k can be made, starting from the relations derived

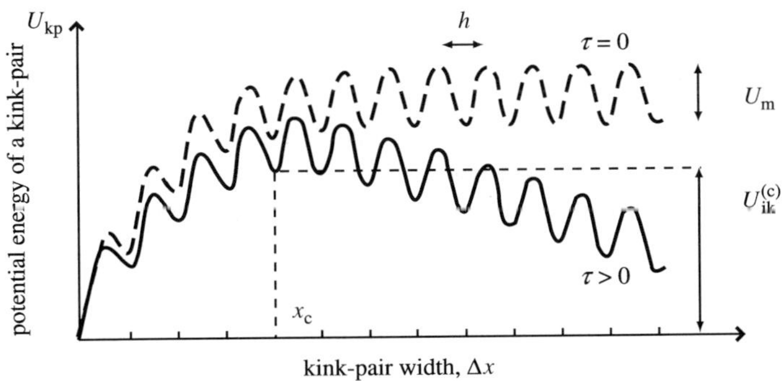

Figure 7.6. Schematic representation of the energy profile for the nucleation of a kink-pair, with and without stress (Hirth and Lothe model).

for an abrupt potential, with the following differences:

- The term ΔE (core energy) is no longer a recombination energy. It is connected to the change in the energy of distorted or cut covalent bonds, due to the local change in line direction.
- The angle θ_c is large. It can be considered as constant and equal to 60°.
- The change in the dislocation line energy, E_0, due to the change in the dislocation character, θ, can now be estimated.

Let E_0 and E_0' be the dislocation line energies in the direction of the dislocation and in the direction of the kink, respectively. According to Section 4.2.3, both energies are in the range of

$$\frac{\mu b^2}{4\pi} \ln \frac{x_c}{b}$$

The kink energy is:

$$U_k = (E_0' + \Delta E)\frac{h}{\sin\,\theta_c} - E_0 \frac{h}{tg\theta_c}$$

Its elastic component (with $\Delta E = 0$) is, with $\theta_c = 60°$:

$$U_k^{(el)} = \frac{hE_0}{\sqrt{3}} + \frac{2h}{\sqrt{3}}(E_0' - E_0) \tag{7.13}$$

The second term can be either positive or negative, depending on the orientation of the gliding dislocation. Marklund (1984) has shown by atomistic calculations that U_k and $U_{ik}(\tau)$ are about two times larger for 30° dislocations than for 90° ones.

The energy barrier described in Figure 7.6 is not crossed by kink-pairs in a single thermally activated event that would require the simultaneous breaking of all bonds over the critical length x_c, with the high activation energy

$$U_{ik}^{(c)}(\tau) + \frac{x_c}{b} U_m$$

It is on the contrary crossed by a series of thermally activated kink jumps over interatomic distances, namely by the diffusion-controlled glide process of kinks, over the energy profile $U_{ik}(\tau, \Delta x)$. We estimate the flux of kink pairs (concentration times velocity) across the energy profile during the different attempts that lead to a successful kink pair expansion. Kink pair concentrations, c_{kp}, as a function of the separation distance, Δx, are average values over a sufficiently long time including many unsuccessful attempts. They must be understood as probabilities for successively emitted kink pairs to reach a given separation distance, Δx, because only one kink pair moves at the same time along the energy profile at a given nucleation site. Many kink pairs are nucleated with a separation distance $\Delta x = b$. Only a few of them are able to move apart, namely to pass over the increasing part of the energy profile, and to expand over large distances.

The following calculations are slightly different from those proposed by Hirth and Lothe (1982) although they follow the same basic principles.

The elastic energy profile is given by Figure 7.7. It is approximated by three straight lines:

– a linear increase from the origin to $U_{\mathrm{ik}}^{(\mathrm{c})}$ for $\Delta x = x_{\mathrm{c}}$
– a plateau between x_{c} and $x_{\mathrm{c}} + x'$, where it intersects the asymptote of the decreasing part
– a linear decrease along the asymptote for $\Delta x > x_{\mathrm{c}} + x'$.

Figure 7.7 shows that $x' = x_{\mathrm{c}}$ because of the hyperbolic variation of $U_{\mathrm{ik}}^{(0)}(\Delta x)$ (given by Eq. (4.34), with $\tau = 0$). The influence of the exact shape of the energy profile, including the value of x' and x_{c}, will be discussed at the end of the calculation.

To estimate the flux of kink-pairs across the energy profile, it is first calculated along the decreasing part. Then, its continuity will be expressed along the other portions of the profile.

The average velocity of a kink pair, v_{kp}, is defined as $\partial \Delta x/\partial t$. It is twice the velocity of an individual kink, $\partial x/\partial t$. According to Eq. (7.11a), it is:

$$\mathrm{v}_{\mathrm{kp}} = 2D_{\mathrm{k}} \frac{\tau b h}{kT} \tag{7.14}$$

and the corresponding transport flux is:

$$\Phi_{\mathrm{t}} = 2D_{\mathrm{k}} c_{\mathrm{kp}} \frac{\tau h}{kT}$$

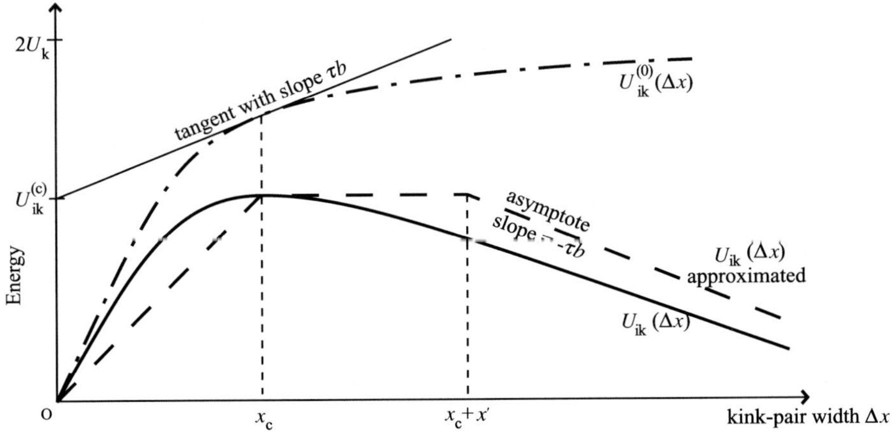

Figure 7.7. Exact and approximated shapes of the energy profile of $U_{\mathrm{ik}}(\tau, \Delta x)$. $U_{\mathrm{ik}}^{(0)}$ is defined in Section 4.2.1.

In the same way, the diffusional flux of kink-pairs is (from Eq. (7.12)):

$$\Phi_{\mathrm{d}} = -\frac{2D_{\mathrm{k}}}{b}\frac{\partial c_{\mathrm{kp}}}{\partial \Delta x}$$

All kink pairs in this part of the energy profile have been emitted successfully. There is accordingly no backward movement and the total flux of kink-pairs, $\Phi_{\mathrm{t}} + \Phi_{\mathrm{d}}$, must be independent of Δx. This implies that $\partial c_{\mathrm{kp}}/\partial \Delta x$ is constant, namely $\partial c_{\mathrm{kp}}/\partial \Delta x = 0$, i.e. c_{kp} is a constant equal to $c_{\mathrm{kp}}^{(\mathrm{s})}$. The total flux is accordingly:

$$\Phi_{\mathrm{t}} = 2D_{\mathrm{k}}c_{\mathrm{kp}}^{(\mathrm{s})}\frac{\tau h}{kT} \tag{7.15}$$

There is no driving mechanical stress between x_{c} and $x + x'$. Therefore, the flux in this part is necessarily due to a gradient of concentration, assumed to be constant as a function of Δx. It is:

$$\Phi_{\mathrm{d}} = -2\frac{D_{\mathrm{k}}}{b}\frac{\partial c_{\mathrm{kp}}}{\partial \Delta x} \approx 2\frac{D_{\mathrm{k}}}{b}\frac{c_{\mathrm{kp}}(x_{\mathrm{c}}) - c_{\mathrm{kp}}(x_{\mathrm{c}} + x')}{x'} \tag{7.16}$$

with $c_{\mathrm{kp}}(x_{\mathrm{c}} + x') = c_{\mathrm{kp}}^{(\mathrm{s})}$.

At steady state, the same flux of kink pairs moves across the whole energy profile. Eqs. (7.15) and (7.16) can thus be combined ($\Phi_{\mathrm{t}} = \Phi_{\mathrm{d}}$) which yields:

$$c_{\mathrm{kp}}^{(\mathrm{s})} = \frac{c_{\mathrm{kp}}(x_{\mathrm{c}})}{1 + \dfrac{\tau b h x'}{kT}}$$

and, using Eq. (7.15):

$$\Phi = 2D_{\mathrm{k}}c_{\mathrm{kp}}(x_{\mathrm{c}})\frac{\tau h}{kT}\frac{1}{1 + \dfrac{\tau b h x'}{kT}} \tag{7.17a}$$

Using the analytical calculations in the elastic kink interaction regime (Eq. (4.35)) we have:

$$x' = x_{\mathrm{c}} = \left(\frac{hb\mu}{8\pi\tau}\right)^{1/2}$$

Order of magnitude calculations yield (with $\mu b^3 = 20$ eV and $T \sim 800$ K)

$$\frac{\tau b h x'}{kT} = 1.5 \text{ for } \tau = \frac{\mu}{1000} \text{ and } \frac{\tau b h x'}{kT} = 0.45 \text{ for } \tau = \frac{\mu}{10\,000}$$

For the sake of simplification, $\tau b h x'/kT$ is assumed to be close to 1 and:

$$\Phi = D_{\mathrm{k}}c_{\mathrm{kp}}(x_{\mathrm{c}})\frac{\tau h}{kT} \tag{7.17b}$$

The last step is to estimate $c_{kp}(x_c)$. This can be done by calculating the same flux Φ in the increasing part of the energy profile. In this region, the flux results from a gradient of concentration superimposed on a local driving stress. Note that the latter opposes to the effect of the gradient and tends to inhibit the emission process. This flux is:

$$\Phi = \Phi_d + \Phi_t = \frac{D_k}{b}\left[-2\frac{\partial c_{kp}}{\partial \Delta x} - 2c_{kp}\frac{1}{kT}\frac{\partial U_{ik}}{\partial \Delta x}\right] \tag{7.18}$$

Then, according to Figure 7.7

$$\frac{\partial U_{ik}}{\partial \Delta x} = \frac{U_{ik}^{(c)}}{x_c}$$

and combining Eqs. (7.17b) and (7.18) yields

$$-\frac{\partial c_{kp}}{\partial \Delta x} = \frac{c_{kp}}{kT}\frac{U_{ik}^{(c)}}{x_c} + c_{kp}(x_c)\frac{\tau hb}{2kT}$$

After integration:

$$c_{kp} = \left(\exp\left(-\frac{U_{ik}^{(c)}}{kT}\frac{\Delta x}{x_c}\right)\right) - \frac{\tau hbx_c}{2U_{ik}^{(c)}}c_{kp}(x_c)$$

For $\Delta x = x_c$, c_{kp} is:

$$c_{kp}(x_c) = \frac{1}{1 + \dfrac{\tau hbx_c}{2U_{ik}^{(c)}}}\exp\left(-\frac{U_{ik}^{(c)}}{kT}\right) \tag{7.19a}$$

This result is close to that given by Hirth and Lothe (1982)

$$c_{kp}(x_c) = \frac{1}{2}\exp\left(-\frac{U_{kp}^{(c)}}{kT}\right)$$

For a strongly asymmetric profile ($U_{ik}^{(c)} \gg \tau hbx_c$), it reduces to:

$$c_{kp}(x_c) = \exp\left(-\frac{U_{ik}^{(c)}}{kT}\right) \tag{7.19b}$$

The flux of successful kink pairs is thus on the average (Eqs. (7.17b) and (7.19b)):

$$\Phi = D_k\frac{\tau h}{kT}\exp\left(-\frac{U_{ik}^{(c)}}{kT}\right) \tag{7.20}$$

This is the probability per unit time to nucleate an expanding kink-pair at a site of size b. Using Eq. (7.10), the corresponding net kink-pair nucleation rate per unit dislocation length is:

$$P_{\mathrm{kp}} = \frac{\Phi}{b} = \frac{D_{\mathrm{k}}\tau h}{bkT}\exp\left(-\frac{U_{\mathrm{ik}}^{(\mathrm{c})}}{kT}\right) = \nu_{\mathrm{D}}\frac{\tau bh}{kT}\exp\left(-\frac{U_{\mathrm{ik}}^{(\mathrm{c})}+U_{\mathrm{m}}}{kT}\right) \tag{7.21}$$

This expression is similar to that proposed by Hirth and Lothe. The kink-distances, x_{c} and x', are no longer present is this expression. They are indeed not important as long as $\tau bhx' \approx kT$ (Eqs. (7.17a) and (7.17b)) and $U_{\mathrm{ik}}^{(\mathrm{c})} \gg \tau bhx_{\mathrm{c}}$ (Eqs. (7.19a) and (7.19b)). The only important parameter is the height of the energy profile, $U_{\mathrm{ik}}^{(\mathrm{c})}$.

(c) *Dislocation velocity*

From Eqs. (7.2), (7.11a) and (7.21), the mean free path of kink pairs in the collision regime is:

$$X = 2b\exp\left(\frac{U_{\mathrm{ik}}^{(\mathrm{c})}}{2kT}\right)$$

In the length-effect regime ($L < X$) the dislocation velocity is, according to Eqs. (7.3b), and (7.21):

$$\mathrm{v} = P_{\mathrm{kp}}Lh = n_{\mathrm{D}}\frac{\tau h^2 bL}{kT}\exp\left(-\frac{U_{\mathrm{ik}}^{(\mathrm{c})}+U_{\mathrm{m}}}{kT}\right) \tag{7.22a}$$

In the kink-collision regime (dislocation length $L > X$), it is (Eqs. (7.3a) and (7.21)):

$$\mathrm{v} = 2h\sqrt{\mathrm{v}_{\mathrm{k}}P_{\mathrm{kp}}} = 2\nu_{\mathrm{D}}\frac{\tau h^2 b^2}{kT}\exp\left(-\frac{\frac{1}{2}U_{\mathrm{ik}}^{(\mathrm{c})}+U_{\mathrm{m}}}{kT}\right) \tag{7.22b}$$

In the length-effect regime, the higher activation energy is compensated by a larger pre-exponential factor. Both expressions yield the same result for $L = X$ as expected.

(d) *Stress dependence of the dislocation velocity*

In the kink diffusion model of Hirth and Lothe, the stress dependence of the dislocation velocity is shared between the pre-exponential term and the elastic component of the activation energy, $U_{\mathrm{ik}}^{(\mathrm{c})}$. As discussed above, $U_{\mathrm{ik}}^{(\mathrm{c})}$ is given by Eq. (4.36):

$$U_{\mathrm{ik}}^{(\mathrm{c})}(\tau) = 2U_{\mathrm{k}} - hb\left(\frac{hb}{2\pi}\mu\tau\right)^{1/2}$$

In the length-effect regime, the corresponding activation area is:

$$A = -\frac{1}{b}\frac{\partial U_{\mathrm{ik}}^{(\mathrm{c})}(\tau)}{\partial\tau} = h\left(\frac{hb\mu}{8\pi\tau}\right)^{1/2}$$

This is the area swept by the kinks during the kink-pair nucleation.

The apparent stress exponent, $m = \partial(\ln \mathrm{v})/\partial(\ln \tau)$, is:

$$m = 1 + \tau\frac{bA}{kT} = 1 + \frac{(\tau\mu)^{1/2}(hb)^{3/2}}{\sqrt{8\pi}kT} \tag{7.23a}$$

This expression is the same as that included in the denominator of Eq. (7.17a). Considering the orders of magnitude estimated below Eq. (7.17a), m appears to be larger than 2 for stresses larger than $\mu/2000$ (with $\mu b^3 = 20$ eV and $T \sim 800$ K).

In the kink-collision regime, the area swept by the kinks during the kink-pair nucleation is the same as in the above regime but the activation area equal to Eq. (7.22b):

$$A = -\frac{1}{b}\frac{\partial \frac{1}{2}U_{\mathrm{ik}}^{(\mathrm{c})}}{\partial\tau}$$

is two times smaller. The apparent stress exponent then becomes:

$$m = 1 + \frac{(\tau\mu)^{1/2}(hb)^{3/2}}{2\sqrt{8\pi}kT} \tag{7.23b}$$

7.2.3 *Low kink mobility: case of dissociated dislocations*

Gliding dislocations are usually dissociated into two Shockley partials, with Burgers vectors of magnitude b_{p}, separated by a stacking fault of surface energy γ.

Each partial is assumed to move independently (uncorrelated kink-pair nucleations). Accordingly, the velocities of leading and trailing partials respectively, v_{pl} and v_{pt}, are determined by the total effective stresses acting on them (Figure 7.8). For the leading and trailing partials they are, per unit length:

$$\tau_{\mathrm{eff}}^{(\mathrm{l})} = \tau_{\mathrm{l}} - \frac{\gamma}{b_{\mathrm{p}}} + \tau_{\mathrm{i}} \tag{7.24a}$$

and

$$\tau_{\mathrm{eff}}^{(\mathrm{t})} = \tau_{\mathrm{t}} + \frac{\gamma}{b_{\mathrm{p}}} - \tau_{\mathrm{i}} \tag{7.24b}$$

where τ_{l} and τ_{t} are the components of the stress tensor parallel to b_{pl} and b_{pt}, respectively, and τ_{i} is the elastic interaction stress. Taking into account that $b_{\mathrm{pl}}b_{\mathrm{pt}} = \frac{1}{2}b_{\mathrm{p}}^2$, we have $\tau_{\mathrm{i}} \approx \mu b_{\mathrm{p}}/4\pi d$, where d is the dynamic dissociation width (Figure 7.8). The interaction stress can also be written

$$\tau_{\mathrm{i}} = \frac{\gamma}{b_{\mathrm{p}}}\frac{d_0}{d} \tag{7.25}$$

where d_0 is the static equilibrium dissociation width ($d_0 = \mu b_{\mathrm{p}}^2/4\pi\gamma$).

Note that $\mathrm{v}_{\mathrm{p}}^{(\mathrm{l})}$ and $\mathrm{v}_{\mathrm{p}}^{(\mathrm{t})}$ do not necessarily involve the same activation energies because the corresponding dislocation lines can be either pure edge or with a 30° character. When the two partials have different characters, d and τ_{i} adjust in order to increase the total stress

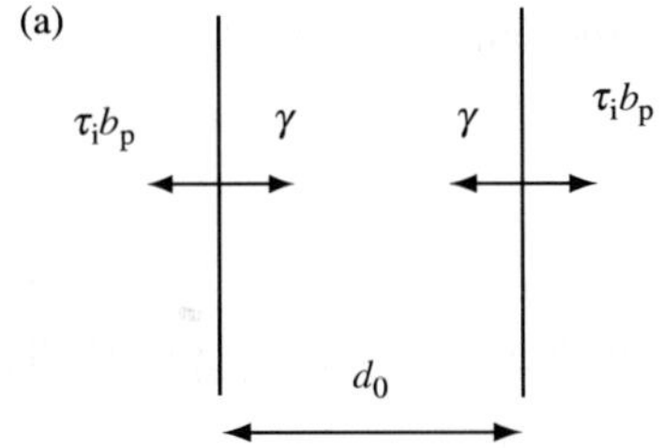

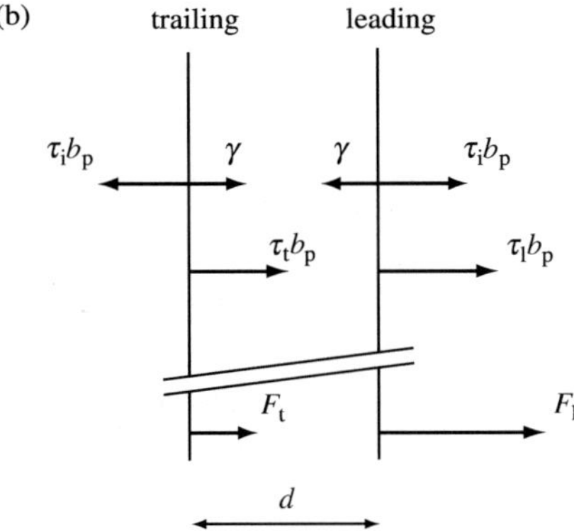

Figure 7.8. Forces on the Shockley partials of a dissociated dislocation (a) without stress and (b) with applied stress. The leading partial is here assumed to be the less mobile one (constricted stacking fault ribbon, $d < d_0$). See text.

τ_{eff} on the less mobile partial and to decrease it on the most mobile one. $\tau_i > \gamma/b_p$ if the leading partial is the less mobile and $\tau_i < \gamma/b_p$ in the opposite case. τ_i can be obtained by writing that the velocities of the two partials are equal (see below).

The main relations that are necessary to determine v_{pl} and v_{pt} can be deduced from those established for perfect dislocations, taking into account that the Burgers vector is now $b_p = b/\sqrt{3}$ (the length of an activation site remains, however, equal to b). In addition, the distance between Peierls valleys, h, has also been replaced by its actual value $b\sqrt{3}/2$. The following relations are derived assuming that each partial obeys the mobility laws of the Hirth and Lothe theory revisited above.

From Eq. (7.9c) (where b^3 has been replaced by $b^2 b_p$), the velocity of a kink on a Shockley partial is:

$$v_k^{(p)} = \frac{1}{2} \nu_D \tau_{eff} b^4 \exp\left(-\frac{U_m^{(p)}}{kT}\right)$$

where τ_{eff} is given by Eqs. (7.24a) and (7.24b).

From Eq. (7.10) (keeping the same pre-exponential term), the diffusion coefficient of kinks is

$$D_{\mathrm{k}}^{(\mathrm{p})} = \nu_{\mathrm{D}} b^2 \exp\left(-\frac{U_{\mathrm{m}}^{(\mathrm{p})}}{kT}\right)$$

From Eq. (7.21) (keeping the same pre-exponential term), the net kink-pair nucleation rate per unit length on Shockley partials is:

$$P_{\mathrm{kp}}^{(\mathrm{p})} = \nu_{\mathrm{D}} \frac{\sqrt{3}}{2} \frac{\tau_{\mathrm{eff}} b^2}{kT} \exp\left(-\frac{U_{\mathrm{ik}}^{(\mathrm{c,p})} + U_{\mathrm{m}}^{(\mathrm{p})}}{kT}\right)$$

with

$$U_{\mathrm{ik}}^{(\mathrm{c,p})} = 2U_{\mathrm{k}}^{(\mathrm{p})} - \frac{1}{4} b^3 \left(\frac{\mu \tau_{\mathrm{eff}}}{\pi}\right)^{1/2}$$

The mean free path of kinks in the collision regime is then:

$$X^{(\mathrm{p})} = \frac{2}{3^{1/4}} b \exp \frac{U_{\mathrm{ik}}^{(\mathrm{c,p})}}{2kT}$$

In the length-effect regime ($L < X^{(\mathrm{p})}$), from Eq. (7.22a), the dislocation velocity becomes:

$$\mathrm{v}_{\mathrm{p}} = \frac{3}{4} \nu_{\mathrm{D}} \frac{\tau_{\mathrm{eff}} b^3 L}{kT} \exp\left(-\frac{U_{\mathrm{ik}}^{(\mathrm{c,p})} + U_{\mathrm{m}}^{(\mathrm{p})}}{kT}\right) \tag{7.26a}$$

Considering the values of $U_{\mathrm{ik}}^{(\mathrm{c,p})}$ given above, the activation area, here equal to the area swept during the thermally activated kink-pair nucleation, is:

$$A_{\mathrm{p}} = -\frac{1}{b_{\mathrm{p}}} \frac{\partial U_{\mathrm{ik}}^{(\mathrm{c,p})}}{\partial \tau_{\mathrm{eff}}} = \frac{b^2}{4\sqrt{\pi}} \left(\frac{\mu}{\tau_{\mathrm{eff}}}\right)^{1/2} \tag{7.26b}$$

In the kink-collision regime ($L > X^{(\mathrm{p})}$), from Eq. (7.22b), the dislocation velocity becomes:

$$\mathrm{v}_{\mathrm{p}} = \frac{3^{3/4}}{2} \nu_{\mathrm{D}} \frac{\tau_{\mathrm{eff}} b^4}{kT} \exp\left(-\frac{\frac{1}{2} U_{\mathrm{ik}}^{(\mathrm{c,p})} + U_{\mathrm{m}}^{(\mathrm{p})}}{kT}\right) \tag{7.26c}$$

The area swept by the kinks during kink-pair nucleation is the same as above, but the activation area, defined by

$$A_{\mathrm{p}} = -\frac{1}{b_{\mathrm{p}}} \frac{\partial U_{\mathrm{ik}}^{(\mathrm{c,p})}}{\partial \tau_{\mathrm{eff}}}$$

is two times smaller.

It is necessary to determine τ_i in order to derive τ_{eff} and the velocity of the dissociated dislocation. This value can be obtained by writing that the velocities of the two partials are equal, $v_p^{(l)} = v_p^{(t)}$.

If the two partials have equal characters and if the corresponding components of the shear stress are also identical ($\tau_l = \tau_t = \tau_{lt}$), then:

$$\tau_{eff} = \tau_{lt} \tag{7.27a}$$

Conversely, if one partial is much more mobile than the other, the work done by both stresses on both partials is spent on moving the less mobile one and the velocity of the dissociated dislocation is that of the less mobile partial with

$$\tau_{eff} = \tau_l + \tau_t \tag{7.27b}$$

The corresponding apparent stress exponents are, respectively, the following:
In the length-effect regime:

$$m = 1 + \frac{1}{8\sqrt{\pi}} \frac{\tau_{eff}^{1/2} \mu^{1/2} b^3}{kT} \tag{7.28a}$$

where τ_{eff} is given by Eq. (7.27a) or (7.27b).
It yields $m > 2$ for $\tau_{eff} > \mu/400$, if $\mu b^3 \approx 20$ eV and $T \sim 800$ K.
In the kink-collision regime:

$$m = 1 + \frac{1}{16\sqrt{\pi}} \frac{\tau_{eff}^{1/2} \mu^{1/2} b^3}{kT} \tag{7.28b}$$

where τ_{eff} is given by Eq. (7.27a) or (7.27b).
It yields $m > 2$ for $\tau_{eff} > \mu/100$.

Note that the second term of m is $\sqrt{8}$ times smaller than for non-dissociated dislocations (Eqs. (7.23a) and (7.23b)).

The velocity of dissociated dislocations is easier to compute if one assumes that it is proportional to stress ($m = 1$) (Wessel and Alexander, 1977) which corresponds to the low stress approximation. Under such conditions, dislocation velocities can be written as:

$$v_p^{(l)} = M_l b_p \tau_{eff}^{(l)} \tag{7.29a}$$

$$v_p^{(t)} = M_t b_p \tau_{eff}^{(t)} \tag{7.29b}$$

where M_l and M_t are the mobilities of the two partials.

Since these two velocities are necessarily equal, we obtain, using Eqs. (7.24a) and (7.24b):

$$\tau_i - \frac{\gamma}{b_p} = \frac{M_t \tau_t - M_l \tau_l}{M_t + M_l}$$

and the dislocation velocity is:

$$\mathrm{v} = \left(\frac{1}{M_\mathrm{l}} + \frac{1}{M_\mathrm{t}}\right)^{-1} (\tau_\mathrm{t} + \tau_\mathrm{l}) b_\mathrm{p}$$

Since $\tau b = (\tau_\mathrm{t} + \tau_\mathrm{l}) b_\mathrm{p}$ (total force acting on the dislocation), this yields:

$$\mathrm{v} = \left(\frac{1}{M_\mathrm{l}} + \frac{1}{M_\mathrm{t}}\right)^{-1} \tau b \tag{7.30}$$

If one partial (e.g. the leading one) is much less mobile than the other, the total dislocation has the mobility of the less mobile partial ($\mathrm{v} \approx M_\mathrm{l}\tau$). Conversely, if the two partials are equally mobile, the total dislocation velocity is $\mathrm{v} = \frac{1}{2} M_\mathrm{l}\tau$. This result is the same as in the general case treated above (Eqs. (7.27a) and (7.27b)). At this stage, the dynamic dissociation width, d, can be derived. This will be useful for the discussion of the experimental results. In the general case $m \neq 1$, the ratio of the dynamic and static dissociation widths is (Eqs. (7.24a,b) and (7.25)):

$$\frac{d}{d_0} = \left[1 + \frac{b_\mathrm{p}}{2\gamma}\left(\tau_\mathrm{t} - \tau_\mathrm{l} + \tau_\mathrm{eff}^{(\mathrm{l})} - \tau_\mathrm{eff}^{(\mathrm{t})}\right)\right]^{-1}$$

In the particular case $m = 1$, using Eqs. (7.29a,b) and (7.30), it becomes: (Wessel and Alexander, 1977)

$$\frac{d}{d_0} = \left[1 + \frac{\tau b}{2\gamma}\left(f + \frac{M_\mathrm{t} - M_\mathrm{l}}{M_\mathrm{t} + M_\mathrm{l}}\right)\right]^{-1} \tag{7.31}$$

where f is such that $(\tau_\mathrm{t} - \tau_\mathrm{l}) b_\mathrm{p} = f \tau b$. ($f$ can be positive or negative, depending on the orientation of the straining axis, but $|f| < 1$).

The leading partial can also move alone and trail a long stacking fault ($d = \infty$) for a stress larger than:

$$\tau_{d\infty} = \frac{2\gamma}{b}\left[\frac{M_\mathrm{l} - M_\mathrm{t}}{M_\mathrm{l} + M_\mathrm{t}} - f\right]^{-1} \tag{7.32}$$

This takes place only when M_l is sufficiently larger than M_t.

Moller (1978) pointed out that additional effects may arise from the discretization of the possible dissociation widths. If the theoretical dissociation width, d, is a multiple of h, the first kink-pair nucleation introduces a reaction stress that tends to decrease the corresponding driving stress, τ_eff. This reaction stress varies from zero, for the starting (equilibrium) dissociation width d, to $\gamma(1 - d/(d + h))$, for the final (out of equilibrium) dissociation width $d \pm h$. Its average value is accordingly $\approx \frac{1}{2}\gamma h/d$. The driving stress for the second kink-pair nucleation on the second partial is increased by the same quantity, but since the velocity of the overall dissociated dislocation is controlled by the most difficult step, this second process appears to be unimportant.

Under such conditions, the overall dislocation velocity is still given by Eqs. (7.26a) or (7.26c) (or by Eqs. (7.29a) or (7.29b) in case of a linear dependence on stress) but τ_{eff} must be replaced by $\tau_{\mathrm{eff}} - \frac{1}{2}\gamma h/d$. It should accordingly decrease to zero for $\tau_{\mathrm{eff}} \approx \frac{1}{2}\gamma h/d$, namely, assuming that the two partials are equally mobile and subjected to the same applied resolved shear stress τ_{tl} for $\tau_{\mathrm{tl}} \approx \frac{1}{2}\gamma h/d$. Then, dislocations should be able to move only by the more difficult process of correlated kink-pair nucleation, under the effective stress τ_{eff}, but with a larger activation energy equal to the sum of the activation energies for the two partials. This slowing down effect does not, however, appear when the dissociation distance is a half-integral multiple of h because partials jump between two equivalent out-of-equilibrium positions and as a result the average driving stress is unchanged.

The same results have been obtained by Cai et al. (2000) by kinetic Monte-Carlo simulations. These authors also showed that, taking into account the variation of the dynamic dissociation width with stress (the so-called Escaig effect, cf. Section 5.1.3.3), the velocity should oscillate between maximum values for d half-integral multiple of h and minimum values for d integral multiple of h (Figure 7.9). Such oscillations have never been measured (see Section 7.3 below). It is plausible that moving dislocations always contain some parts where the dissociation is favourable, so that the Moller effect cannot be observed.

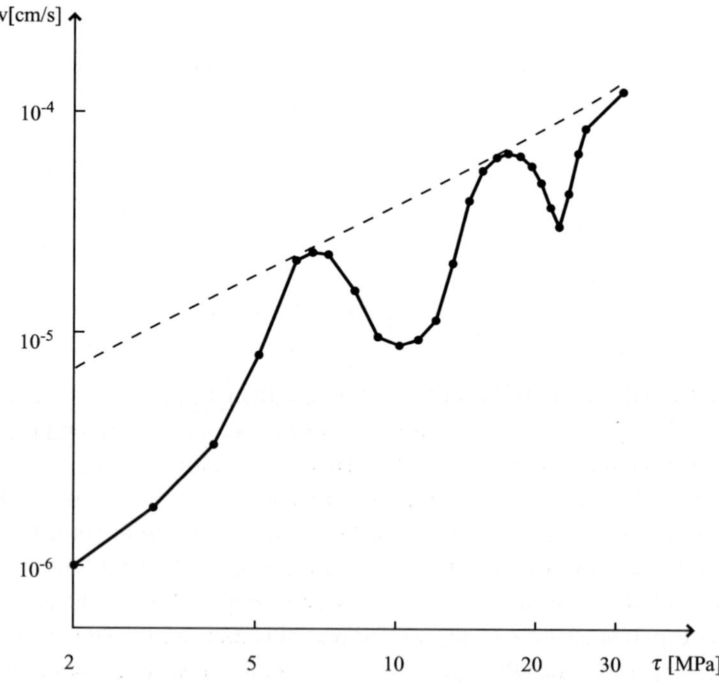

Figure 7.9. Simulation of the Moller effect on the dislocation velocity as a function of stress (from Cai et al., 2000).

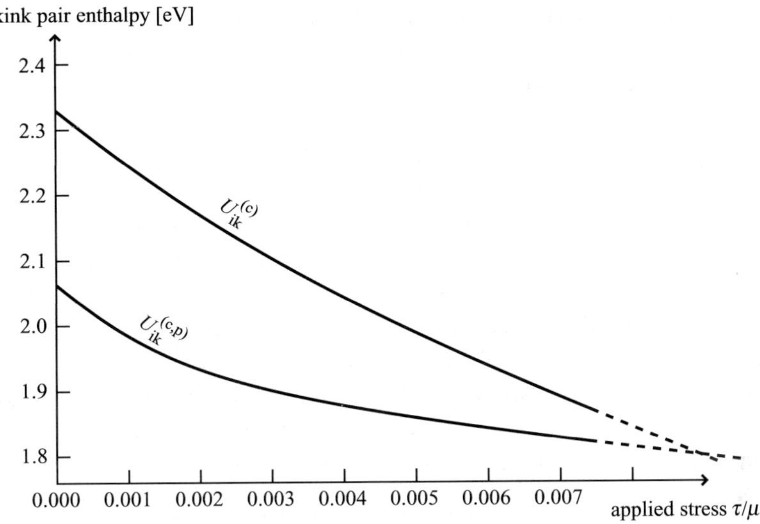

Figure 7.10. Elastic components of the activation energies of glide for dissociated and recombined dislocations. The dissociated glide set is favoured except at very high stresses where the two curves may cross each other (from Duesberry and Joos, 1996).

Lastly, we can discuss the respective mobilities of dislocations in the glide and shuffle sets. Dissociated dislocations of the glide set are expected to be less mobile than non-dissociated ones of the shuffle set, considering the number of covalent bonds to cut (Section 7.1). This corresponds to the condition $U_{\rm m}^{(\rm p)} > U_{\rm m}$. Conversely, dissociated dislocations are expected to be more mobile considering the elastic part of the activation energy. Indeed, $U_{\rm ik}^{(\rm c,p)}$ is expected to be smaller than $U_{\rm ik}^{(\rm c)}$ because they depend on the Burgers vectors squared, respectively, $b_{\rm p}^2$ and b^2. Then, the respective velocities of the two possible configurations will depend on the relative importance of the elastic and covalent parts of the total activation energies. Assuming that the elastic part dominates, this provides a good explanation for the experimental observations of dissociated mobile dislocations that likely belong to the glide set (see Section 7.3). Calculations by Duesberry and Joos (1996) confirm that $U_{\rm ik}^{(\rm c)} > U_{\rm ik}^{(\rm c,p)}$ over a large stress range (Figure 7.10). However, the two curves may intersect each other when extrapolated to $\tau > \mu/100$. This indicates that glide in the shuffle set may take over at very high stresses.

7.3. EXPERIMENTAL RESULTS ON DISLOCATION VELOCITIES

In this section we have collected experimental results that can be directly compared with the mobility relations demonstrated above. Quantitative data about velocities of dissociated dislocations are given in Sections 7.3.1–7.3.5. Then, dislocation behaviour at low temperatures and high stresses, under a confining pressure, is discussed in Section 7.3.6. More detailed results can be found in several reviews (Alexander, 1986; George and Rabier, 1987; Maeda and Takeuchi, 1996a,b; Yonenaga, 1997).

7.3.1 Mobility as a function of character

7.3.1.1 Elemental semiconductors (Si). Figure 7.11 shows an expanding dislocation loop in silicon. The two arms, respectively, pinned in A and B, rotate in two opposite directions, in two parallel {111} planes. Dislocation lines exhibit straight segments parallel to ⟨110⟩ directions, namely with screw and 60° orientations. They are dissociated into Shockley partials, of 30° character for screws, and of edge and 30° character, respectively, for 60° orientations (Figure 7.12). This figure shows that a dislocation loop contains two types of 60° segments: the $90°_l/30°_t$ segments have the leading partial in the edge orientation and the trailing one in the 30° orientation and $30°_l/90°_t$ segments have the reversed partial positions.

Figure 7.13 shows that screws are less mobile than 60° dislocations. Some authors report different behaviours for $30°_l/90°_t$ and $90°_l/30°_t$ dislocations. Among them, George et al. (1972) observed in "in situ" Lang topography (XRT) that one set of moving 60° dislocations takes a zigzag shape whereas the second set remains straight along the ⟨110⟩

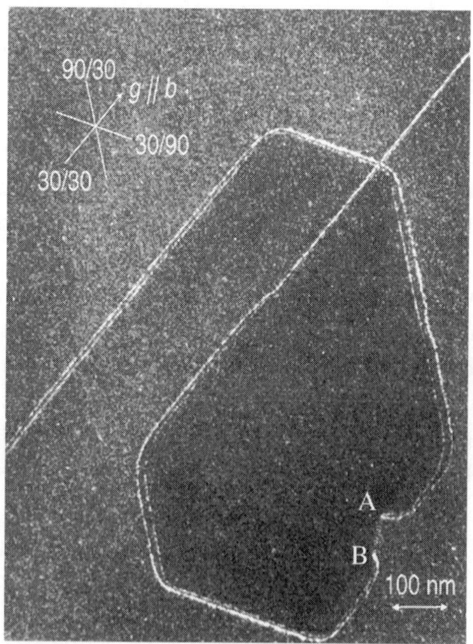

Figure 7.11. Dissociated dislocation loop in Si deformed at 400°C under a shear stress $\tau = 260$ MPa. Weak beam TEM. From Wessel and Alexander (1977).

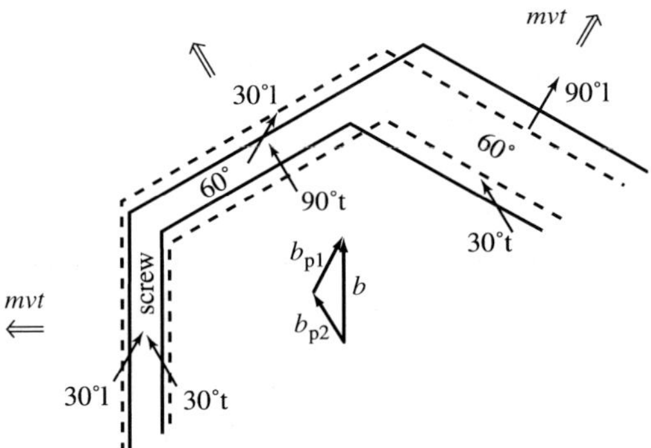

Figure 7.12. Schematics of a dissociated dislocation loop in an ESC under stress (full line) and without stress (dotted line). Black arrows indicate the Burgers vectors of the partial dislocations.

direction. Alexander et al. (1983) reported that the velocities of opposite 60° dislocations are in the ratio 4/3. These effects are not understood clearly but they may be due to interactions with impurities. Indeed these interactions are very important at low stresses, as discussed in Section 7.3.2 below. The most reliable results, for which dislocation movements are controlled by the pure Peierls mechanism, should accordingly be obtained either at high stresses or in high-purity materials.

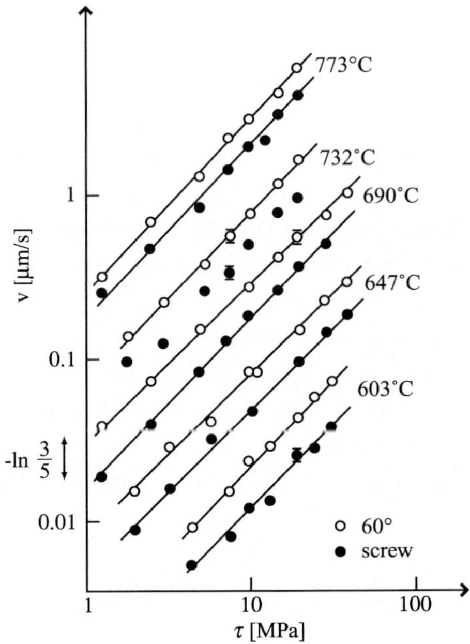

Figure 7.13. Dislocation mobilities in high-purity silicon (XRT, Imai and Sumino, 1983).

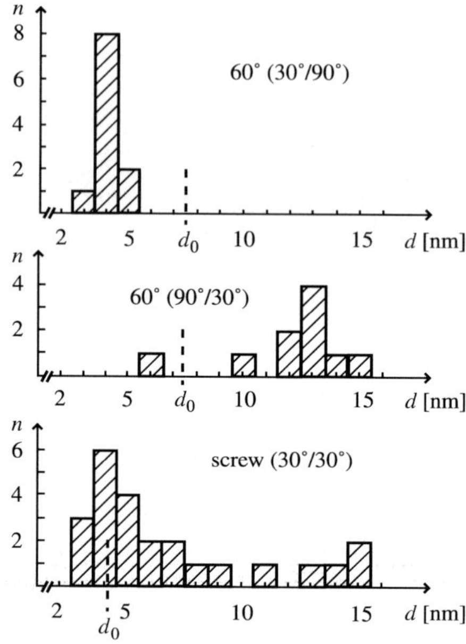

Figure 7.14. Dissociation widths under high stress in silicon ($\tau = 260$ MPa, $T = 420°C$). $d_0 = 6.4$ and 4.1 nm are, respectively, the 60°-orientation and screw dissociation widths, without stress. TEM. From Wessel and Alexander (1977).

Important information on the relative mobilities of Shockley partials can be deduced from the experiments of Wessel and Alexander (1977). Figure 7.14 shows dissociation widths measured on the screws and on the two sets of 60° dislocations in Si deformed at a fairly high stress (260 MPa) and then cooled rapidly under load. These data can be analysed using Eq. (7.31). The much smaller dissociation widths of $30°_l/90°_t$ segments as compared to $90°_l/30°_t$ segments indicate that 30° partials have a lower mobility than 90° ones (Figure 7.12). The wide distribution of the dissociation widths of screws exhibits an average value of 5.8 nm, larger than the stress-free value ($d_0 = 4.1$ nm, according to Ray and Cockayne, 1971). This was interpreted by Wessel and Alexander as evidence of different mobilities for leading and trailing 30° partials. In reality, a wide distribution is necessarily asymmetrical for the following reasons, detailed by Paidar and Caillard (1994) and Paidar et al. (1994). Figure 7.15 shows the back force acting on both partials when their distance deviates from the equilibrium value d_0. This force is equal to

$$\frac{A}{d} - \gamma$$

where A is defined below Eq. (5.1) for screw segments. It is asymmetrical as shown in Figure 7.15. The experimentally observed distribution of the dissociation distances depends on the mobility of the more mobile partial.

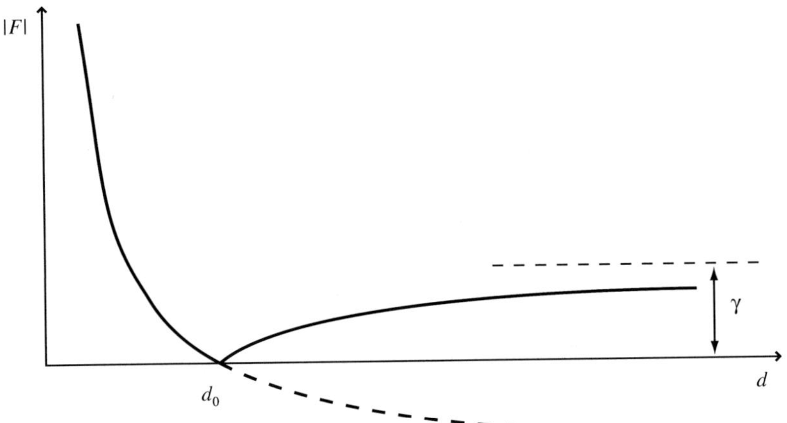

Figure 7.15. Back force on partials as a function of separation d. d_0 is the equilibrium dissociation distance. Note the asymmetrical variation of $|F|$.

A low mobility allows a more pronounced deviation from equilibrium. This distribution, however, reflects the asymmetry of the curve shown in Figure 7.15, i.e. more metastable dislocations can be found with $d > d_0$ than with $d < d_0$. Under such conditions, the equilibrium dissociation does not correspond to the average value but to the most frequent one. This applies to the present situation, especially to screw dislocations that are made of two poorly mobile 30° Shockley partials. Then, all results can be explained in a consistent way without the assumption of different mobilities for leading and trailing 30° partials. Using Eq. (7.31), with $f = 1/6$ and $\tau = 260$ MPa (from Wessel and Alexander), $\gamma/b = 140$ MPa, $d_0 = 4.1$ nm for screws and $d_0 = 6.4$ nm for 60° orientations (from Ray and Cockayne, 1971), the correct dynamic dissociations are obtained for

$$\frac{M_{90^\circ} - M_{30^\circ}}{M_{90^\circ} + M_{30^\circ}} = 0.7$$

(Table 7.1). This shows that 90° partials are about five times more mobile than 30° ones (Caillard and Vanderschaeve, 2003). The same results are obtained in pSi and nSi (Alexander et al., 1980). This analysis shows that the mobility of Shockley partials

Table 7.1. Experimental and calculated dynamic dissociation widths in silicon deformed under the stress $\tau = 260$ MPa.

	screw	30°/90°	90°/30°
experimental (max on Fig. 7.13)	4 nm	4 nm	13 nm
calculated	3.5 nm	3.5 nm	12.5 nm

does not depend on their leading/trailing position, at least within the accuracy of measurements. In particular, the influence of climb forces (put forward in order to account for the apparent different mobilities of leading and trailing 30° partials) is not a priori necessary.

According to Eqs. (7.27a) and (7.27b), the velocity of dissociated dislocations in Si equals the velocity of 30° Shockley partials, with $\tau_{eff} = \tau_{lt}$ for screws and $\tau_{eff} = \tau_l + \tau_t$ for 60° orientations.

Using Eq. (7.30), we find that the mobility of 60° dissociated dislocations must be 5/6 of that of 30° Shockley partials. Similarly, the mobility of screw dissociated dislocations must be half that of 30° Shockley partials, namely of 3/5 that of 60° dissociated dislocations. This is verified approximately by in situ X-ray topography experiments on high-purity Si (Figure 7.13).

The apparent stress exponent (Eq. (7.28a) or (7.28b)) must also be calculated using the above τ_{eff} values.

In germanium, the results are much more contradictory, as discussed by Patel and Freeland (1971) and Schaumburg (1972). Important pinning effects are expected below 20 MPa, especially for 60° dislocations (Section 7.3.2). They may account for the different velocities measured by these two groups. Measurements at high stresses, however, show that, as in *silicon*, 90° partials are about 5.5 times more mobile than 30° ones (Caillard and Vanderschaeve, 2003).

7.3.1.2 ***Compound semiconductors.*** Dislocations are also dissociated into two Shockley partials. One difference with ESC is that the cores of non-screw dislocations (namely the last row of the extra-half planes) are different for two opposite-sign dislocations. In the shuffle set, the α-dislocations have their extra-half plane ending on a row of group III or group II atoms (e.g. Ga in GaAs, Zn in ZnS), respectively. The situation is reversed for the glide set. For the β-dislocations of the shuffle set, the extra-half plane ends on a row of group V or group VI atoms (e.g. As in GaAs, S in ZnS). Perfect dislocations are made of screw, 60°α and 60°β rectilinear segments (Figure 7.16).

Dislocation mobilities in III–V compounds have been investigated by in situ transmission electron microscopy and by double etch pits experiments. Figure 7.17 shows that dislocation velocities measured by double etch pits in several III–V compounds (except InP) are in the sequence: $V_s \leq V_{60°\beta} \ll V_{60°\alpha}$ (Yonenaga, 1997). Figure 7.18 shows a dislocation source working in GaAs in a transmission electron microscope. The dissociation is too small to be seen on this dynamic sequence. Short 60°α segments move very rapidly to the right and trail long screw segments noted S. Then, 60°β and screws move much more slowly. Similar in situ observations in InSb, GaAs and InP (Fnaiech et al., 1987; Caillard et al., 1987; Zafrany et al., 1992; Louchet et al., 1993) showed that the velocities of 60°β and screw dislocations of both signs are identical and proportional to their respective lengths (cf. Section 7.3.4). In particular, observations of long screw

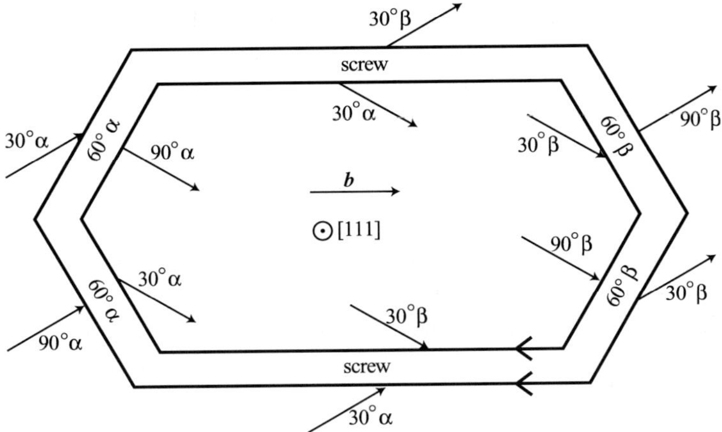

Figure 7.16. Schematic description of an expanding dissociated dislocation loop in a III–V compound. Arrows indicate the Burgers vectors.

segments such as those in Figure 7.18 does not imply that screws are less mobile than both 60°α and 60°β ones.

These results can be used to estimate the mobilities of the different partials in dissociated dislocations. We assume that Eq. (7.30) is valid, i.e. that the velocity is proportional to stress. Even if this assumption is not valid, the orders of magnitude determined below will be significant.

Eq. (7.30) shows that the mobility of a dissociated dislocation is close to that of its less mobile partial. Accordingly, the less mobile partial is that involved in the slowest—screw and 60°β—dissociated segments, i.e. it is the 30°β one. The following considerations show that the 30°β partial is in fact much less mobile than all the others.

Since 60°α dissociated dislocations are much more mobile than 60°β ones, we have, according to Eq. (7.30):

$$\left(\frac{1}{M_{30^\circ\alpha}}+\frac{1}{M_{90^\circ\alpha}}\right)^{-1} \gg \left(\frac{1}{M_{30^\circ\beta}}+\frac{1}{M_{90^\circ\beta}}\right)^{-1}$$

$$\text{i.e.}\qquad \frac{1}{M_{30^\circ\alpha}}+\frac{1}{M_{90^\circ\alpha}} \ll \frac{1}{M_{30^\circ\beta}}+\frac{1}{M_{90^\circ\beta}}$$

Since $M_{90^\circ\beta} \gg M_{30^\circ\beta}$, this implies,

$$\frac{1}{M_{30^\circ\alpha}}+\frac{1}{M_{90^\circ\alpha}} \ll \frac{1}{M_{30^\circ\beta}}$$

whence

$$M_{30^\circ\alpha} \text{ and } M_{90^\circ\alpha} \gg M_{30^\circ\beta} \tag{7.33}$$

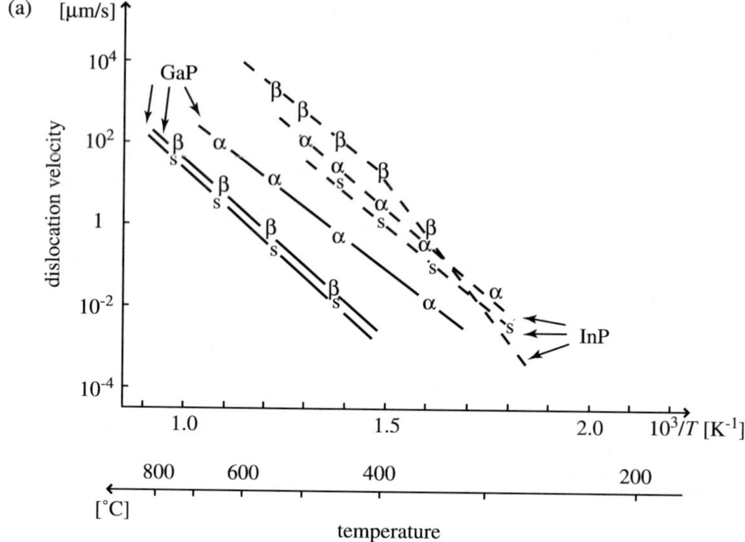

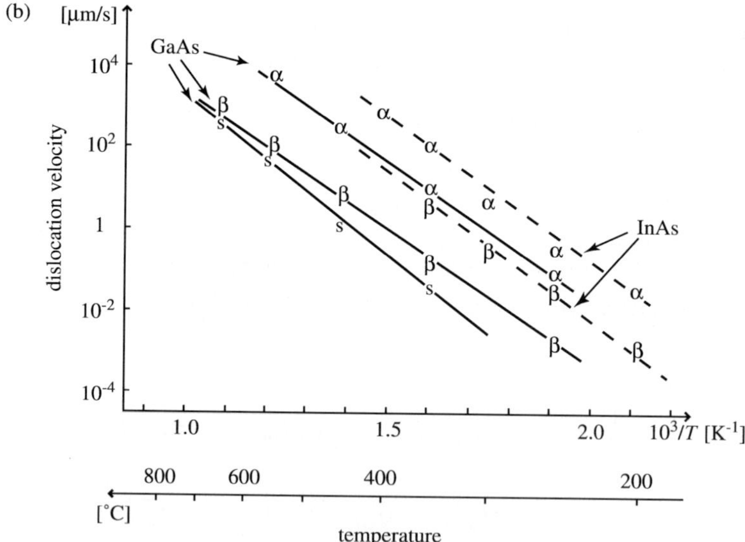

Figure 7.17. Dislocation velocities in several III–V compounds (double etch pits). From Yonenaga (1997).

This inequality implies that the mobility of screws is very close to $M_{30\beta}$. Then, the mobility of 60°β segments is necessarily identical or lower than that of screw segments. This shows that double etch pit experiments, that yield the opposite result, are probably not reliable. According to in situ experiments in transmission electron microscopy, screw and 60°β

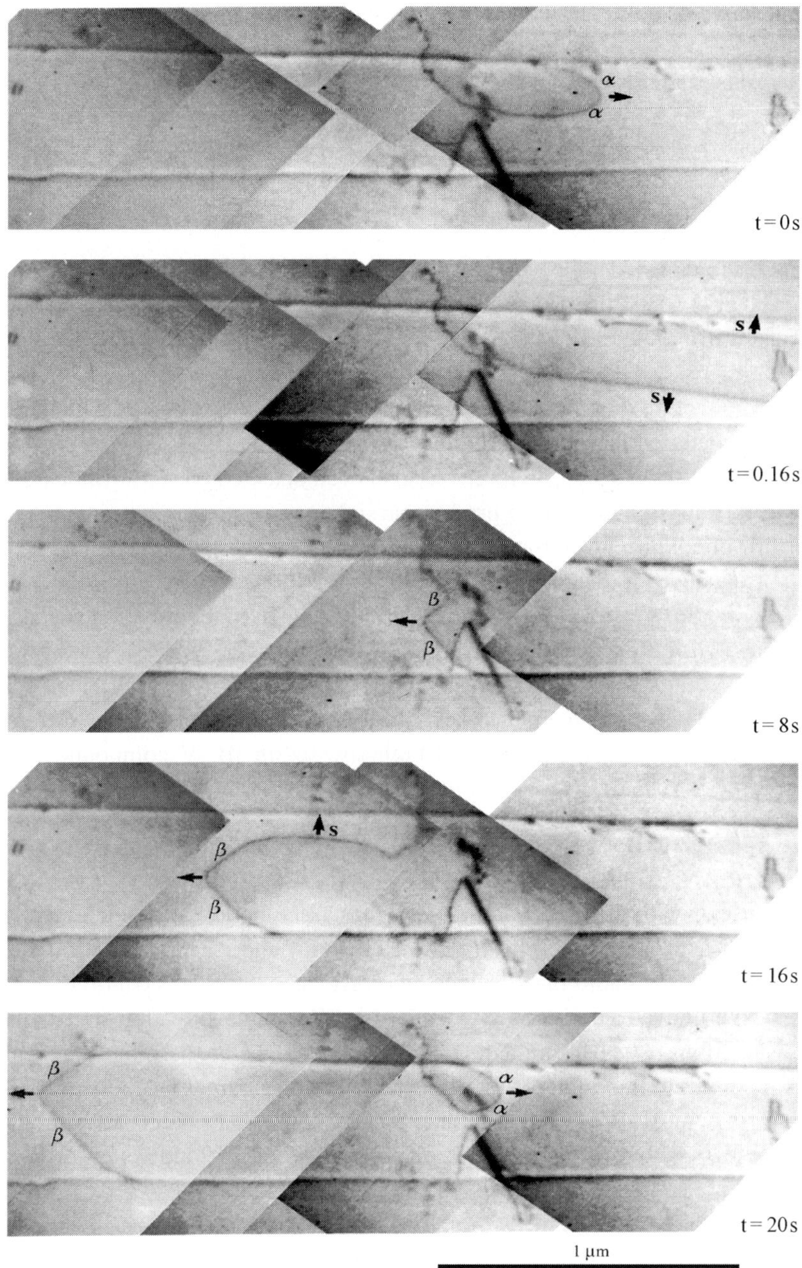

Figure 7.18. Dislocation source in GaAs. TEM in situ experiment at 350°C, 200 kV. (See text) From Vanderschaeve et al. (2001).

orientations have actually similar mobilities. This yields:

$$\left(\frac{1}{M_{30^\circ\beta}} + \frac{1}{M_{90^\circ\beta}}\right)^{-1} \approx \left(\frac{1}{M_{30^\circ\beta}} + \frac{1}{M_{90^\circ\alpha}}\right)^{-1}$$

whence

$$M_{90^\circ\alpha},\ M_{90^\circ\beta} \text{ and } M_{30^\circ\alpha} \gg M_{30^\circ\beta} \tag{7.34}$$

Relation (7.34) is more detailed than Eq. (7.33).

Androussi et al. (1987) showed that the relative velocities of partial dislocations can be also estimated from their respective densities in micro-twins. They conclude that

$$M_{90^\circ\alpha},\ M_{90^\circ\beta} \gg M_{30^\circ\alpha} \gg M_{30^\circ\beta} \tag{7.35}$$

which is more detailed than Eqs. (7.33) and (7.34).

By the same method, a friction force has also been detected on screw Shockley partials. It is, however, not large enough to produce straight segments on dissociated dislocations along the 30° orientation (Lefebvre and Vanderschaeve, 1989).

One consequence of the different mobilities of 30°α and 30°β partials is that, in agreement with Eq. (7.30), screws with 30°β leading partials are constricted, whereas opposite ones, with 30°α leading partials, are extended (Eq. (7.31)). Cross-slip is thus easier for the former, as observed in in situ TEM experiments in *InSb* (Vanderschaeve and Caillard, 1994). This phenomenon can also partly be the origin of the extensive cross-slip that is reported at low temperatures and high stresses in III–V compounds, but not in silicon (see Section 7.3.6).

7.3.2 *Velocity as a function of stress*

Many measurements of the velocity of dislocations as a function of character and stress are now available in ESC and CSC. Those obtained by XRT and scanning electron microscopy (SEM) are the most reliable ones because the displacement of dislocations or emerging points can be followed continuously. XRT also allows the direction and the shape of moving dislocations to be determined. Double etch pits (DEP) experiments are on the other hand less precise and they must be considered with some care.

In silicon, the most recent and complete measurements have been carried out by Imai and Sumino (1983), by means of XRT. Their results for 60° and screw dislocations in high-purity Si are shown in Figure 7.12. Dislocation velocities are exactly proportional to stress, in excellent agreement with the Hirth and Lothe theory in the low-stress regime (Eqs. (7.28a) and (7.28b)). Previous measurements showed a more pronounced decrease of the dislocation velocity with decreasing stress, that was sometimes related to a transition between uncorrelated and correlated kink-pair nucleations on Shockley partials (see end of Section 7.2.3). Figure 7.19 shows that it is, in reality, due to the presence of impurities:

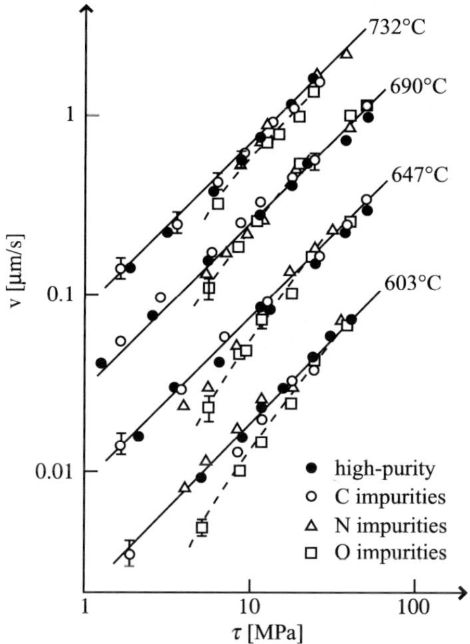

Figure 7.19. Comparison between dislocation mobilities in high-purity and impurity-containing silicon (XRT). From Imai and Sumino (1983).

straight lines refer to high-purity Si grown by the floating-zone technique, whereas dotted lines refer to different crystals grown by the Czochralski technique and containing dopants or impurities such as oxygen. It is clear that the non-linear dependence at low stress of the less pure silicon crystals is due to interactions with foreign atoms that take over Peierls forces. This is confirmed by the corresponding X-ray images that exhibit wavy dislocations in impure Si. The locking of dislocations by oxygen atoms in Czochralski silicon has been studied in detail by Senkader et al. (2001). Only experimental results obtained in high-purity materials or at high stresses should accordingly be compared with the relations derived in Section 7.2.

According to the experimental results of Alexander et al. (1987), m is smaller for screw dislocations than for 60° ones at high stresses. Figure 7.20 shows velocities of 60° dislocations as a function of stress in high-purity silicon or at high stress, i.e. when no interaction with foreign atoms is expected. The slope m tends to increase from 1 at low stresses to $m = 1.5-2$ at about 100 MPa. All these data are consistent with theoretical estimates (Eqs. (7.28a) and (7.28b)). Since τ_{eff} is larger for 60° dislocations ($\tau_{\text{eff}} \approx \tau_{\text{l}} + \tau_{\text{t}}$ because the 90° partial is more mobile than the 30° one) than for screw dislocations ($\tau_{\text{eff}} \approx \tau_{\text{lt}}$ because the two 30° partials are equally mobile), the corresponding m must be larger. For 60° segments, an applied stress $\tau = 100$ MPa yields $\tau_{\text{eff}} \approx 200$ MPa, hence, for

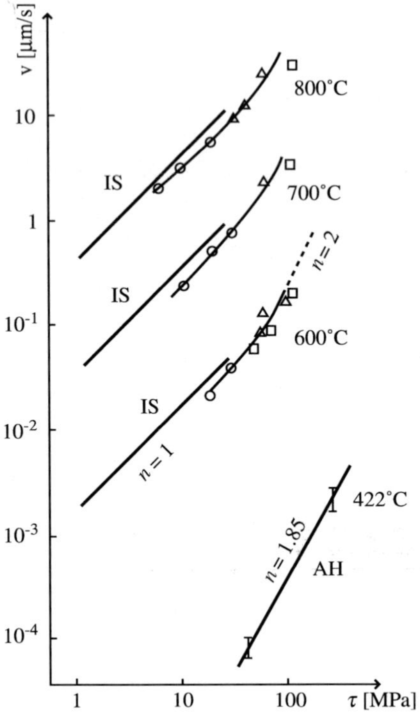

Figure 7.20. Dislocation velocity measurements in silicon for which impurity effects are not expected. AH: from Alexander et al. (1987, DEP). IS: from Imai and Sumino (1983, XRT). ○: from George and Champier (1979, XRT). △: from Chaudhuri et al. (1962, DEP). □: from Kabler (1963, DEP).

$\mu b^3 = 20$ eV and $T = 800$ K, $m = 2$ in the length-effect regime (Eq. (7.28a)) and $m = 1.5$ in the kink-collision regime (Eq. (7.28b)). These values are close to those deduced from Figure 7.20 for the same stress ($m = 1.5-2$, see above). The accuracy on m is, however, not sufficient to determine the exact glide regime.

In germanium, the stress exponent varies between 1 and 2.1 for $20 < \tau < 40$ MPa (Patel and Freeland, 1971; Schaumburg, 1972). It is clear, however, that pinning by impurities becomes important below 20 MPa, especially for 60° dislocations, because m increases to much higher values (the effect of impurities is confirmed by activation energy measurements—see Section 7.3.3). It is thus not possible to make a reliable comparison with theory.

In GaAs, dislocation velocities exhibit the same pronounced decrease at low decreasing stresses as in impurity-containing silicon (Figure 7.21(a) and (b)), with an average slope m ranging between 1.4 and 1.8 (Table 7.2). In InSb, the average values of m range between 1 and 2.6, with a large uncertainty (Table 7.2). In both compounds, m values are definitely higher than those expected from Eq. (7.28) ($m = 1-1.3$).

This discrepancy is likely to be due to interactions between dislocations and impurities, as in impurity-containing Si.

In Figure 7.21(c), two different sets of etch pit data are compared with results of in situ experiments on dislocations with different lengths at 350°C (see Section 7.3.4). This comparison indicates that velocity measurements by the double etch pit technique are probably performed on short dislocations gliding in the length-effect regime.

7.3.3 *Velocity as a function of temperature*

Arrhenius plots in high-purity silicon yield constant values of the activation energy of glide, independent of stress and temperature, within the accuracy of measurements (Figure 7.22, Table 7.3). The activation energy for 60° dislocations is slightly lower than for screw dislocations. This difference is accounted for by the different activation energies of 30° and 90° Shockley partials. As a matter of fact, the activation energy of 60° dislocations is a mixture of the activation parameters of 30° partials (lower mobility, higher activation energy) and 90° partials (higher mobility, lower activation energy), whereas the activation energy of screw dislocations is equal to that of 30° partials.

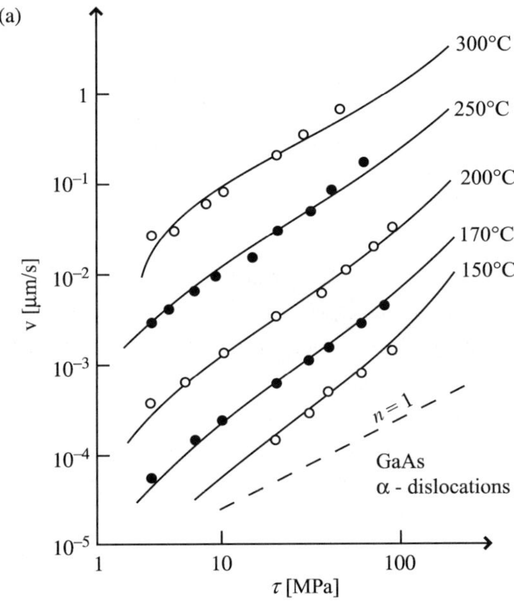

Figure 7.21. Dislocation velocities in GaAs. (a) α dislocations, (b) β dislocations (DEP, from Choi et al., 1987) and (c) at 350°C, comparison between DEP measurements by Erofeeva and Osspyan (1973, EO), Choi et al. (1987, CMN) and Yonenaga and Sumino (1987, YS) and in situ TEM measurements for different dislocation lengths by Caillard et al. (1987, horizontal bars).

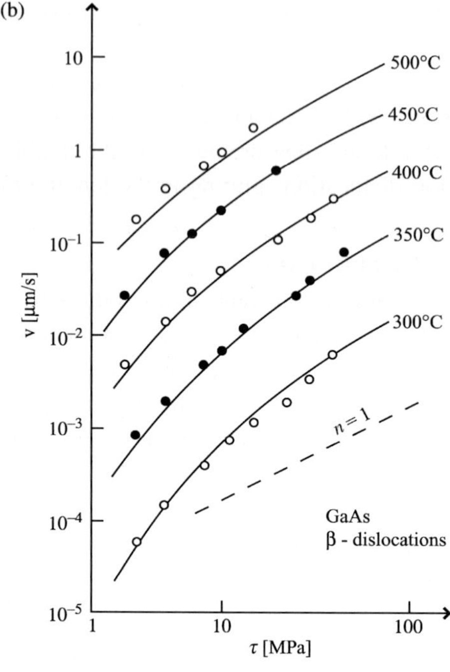

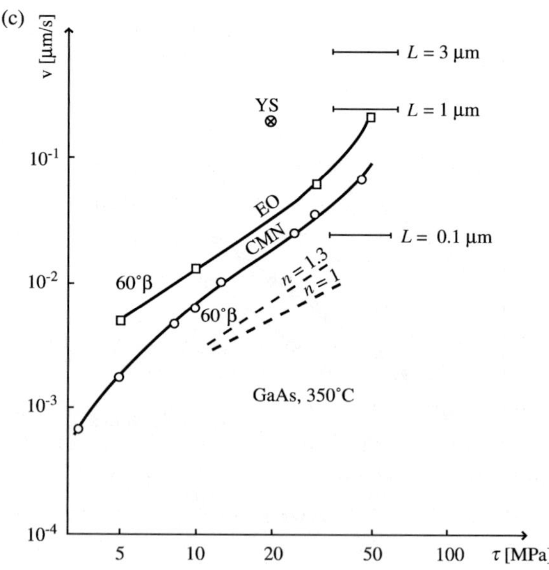

Figure 7.21. (*continued*)

Table 7.2. Experimental values of the stress exponent, m, and activation energy, Q, in GaAs and InSb (DEP and TEM in situ experiments).

Material	Dislocations	m	Q (eV)	T (°C)	Reference
GaAs	60°α	1.4	1–1.1	300	Choi et al. (1987)
	60°α	1.7	1.3	550	Yonenaga and Sumino (1989)
	60°α	1.7	1–1.3	200–400	Erofeeva and Osspyan (1973)
	60°β	1.6	1.3–1.6	300–500	Choi et al. (1987)
	60°β	1.6	1.2	550	Yonenaga and Sumino (1989)
	60°β	1.6	1.3–1.5	250–500	Erofeeva and Osspyan (1973)
	60°β		1.6[a]	350	Vanderschaeve et al. (2001)
	Screws	1.8	1.4	550	Yonenaga and Sumino (1989)
			1.6[a]	350	Vanderschaeve et al. (2001)
InSb	60°α	2	0.75–0.9	50–165	Mihara and Ninomiya (1975)
	60°α	2	0.75–1	50–250	Erofeeva and Osspyan (1973)
	60°α	1	0.8	20–330	Steinhardt and Schafer (1971)
	60°β	2.6	1.15	130–190	Mihara and Ninomiya (1975)
	60°β	1.9	0.8–1	80–300	Erofeeva and Osspyan (1973)
	60°β		1.1	270–330	Steinhardt and Schafer (1971)
	60°β		1.2[a]	250	Vanderschaeve et al. (2001)
	Screws	1.3	1–1.1	130–190	Mihara and Ninomiya (1975)
			1.2[a]	250	Vanderschaeve et al. (2001)

[a] From in situ TEM experiments, using the Hirth and Lothe theory.

Previous activation energy measurements in less pure silicon exhibit a pronounced increase as the stress decreases to low values (Figure 7.23). This corresponds to the non-linear decrease of the corresponding dislocation velocity in the same low stress range (Figure 7.19). Activation energies of 60° dislocations at high stresses are, however, similar to those measured in high-purity silicon and of the order of 2.2 eV.

In pure silicon, an activation energy independent of stress indicates that the work done by the stress is small during the thermally activated kink-pair nucleations. According to Eq. (7.23a), this work is $W = (m - 1)kT$, with $kT \approx 6.7 \times 10^{-2}$ eV at 800 K. It is thus negligible as long as $m \approx 1$ (see Figure 7.23).

Germanium exhibits the same increase of activation energy Q with decreasing stress as impure silicon (Figure 7.23). This confirms that impurity effects are also important. The most reliable values obtained at high stresses are in the range 1.5–1.6 eV.

Some results on III–V compounds are listed in Table 7.2. The experimental scatter is rather large but there is a tendency for $Q_{60°\alpha} < Q_{60°\beta} \leq Q_s$. However, it is shown in Section 7.3.1 that v_s is probably very close to $v_{60°\beta}$, so that Q_s and $Q_{60°\beta}$ are expected to be equal. The results in Table 7.2 should accordingly be better interpreted as: $Q_{60°\alpha} < Q_{60°\beta} \approx Q_s$. More precise measurements are needed, excluding the too inaccurate double etch pit experiments.

Another interesting point to discuss is the relative contributions of U_{ik} and U_m to the total activation energy. There is an elastic contribution that is a part of U_{ik} and a

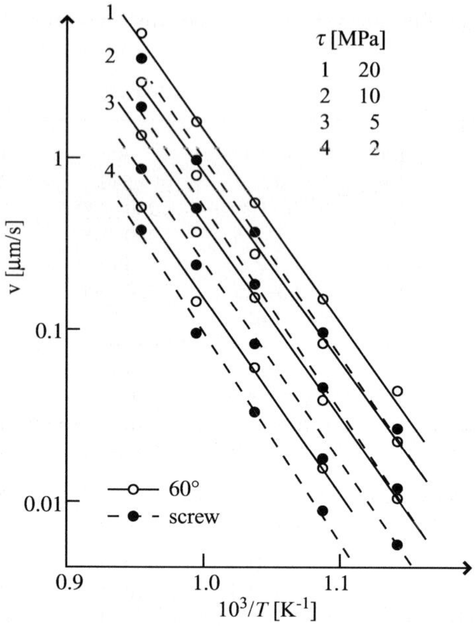

Figure 7.22. Arrhenius plot of dislocation velocities versus temperature in high-purity silicon (from Imai and Sumino, 1983).

covalent contribution that is shared between the core energy involved in U_{ik} and the kink-migration energy U_m. Several results indicate that the elastic contribution is probably larger:

- Measured activation energies in different tetrahedrally coordinated crystals are more or less proportional to $\mu b_p^2 h$, with a slope ~ 0.25 (Figure 7.24) (Maeda and Takeuchi, 1996a,b). Assuming $U_{ik}^{(c,p)} \approx 2U_k^{(p)}$, these values can be compared with twice

Table 7.3. Experimental activation energies in high-purity silicon (XRT, from Imai and Sumino, 1983).

Stress (MPa)	Activation energies Q (eV)	
	60°	Screws
2	2.25	2.45
5	2.24	2.30
10	2.19	2.38
20	2.24	2.31

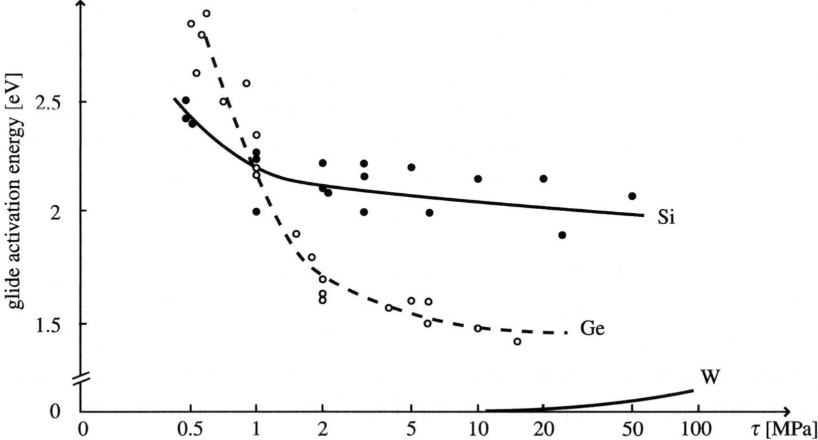

Figure 7.23. Activation energies as a function of stress in impurity-containing Si and Ge (60° dislocations). From Alexander (1986). *W* is the computed work done by the stress, in Si.

the elastic component of the kink energy, $U_k^{(el)}$, given by Eq. (7.13). Taking

$$E_0' \approx E_0 \approx \frac{\mu b_p^2}{4\pi} \ln \frac{x_c}{b} \approx 0.2\mu b_p^2$$

we obtain: $2U_k^{(p,el)} \approx 0.23\mu b_p^2$, which is close to the experimental result of Figure 7.24.

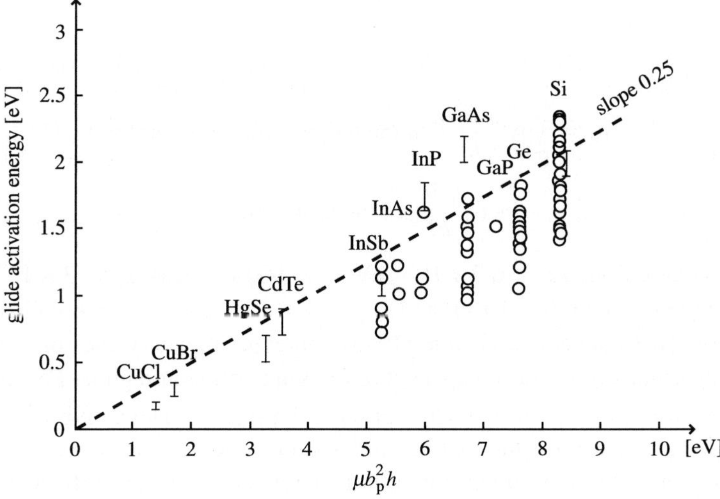

Figure 7.24. Glide activation energies as a function of the elastic term $\mu b_p^2 h$ in several tetrahedrally coordinated crystals. From Maeda and Takeuchi (1996a,b).

- The preponderance of the dissociated glide set over the non-dissociated shuffle set can be explained easily if the elastic component of $U_{\mathrm{ik}}^{(\mathrm{c,p})}$ is large (cf. Section 7.2.3).
- Relation (7.13) shows that the elastic component of $2U_{\mathrm{k}}^{(\mathrm{p})}$ is expected to be higher for 30° partials (one 30° kink, one edge kink) than for edge partials (two 30° kinks) because the dislocation line energy is higher along the edge direction. This is in agreement with the lower mobility of 30° partials, provided the elastic component of $2U_{\mathrm{k}}^{(\mathrm{p})}$ is a substantial part of the total activation energy.

The importance of the covalent contribution to the activation energy of glide can be estimated from the difference in the mobilities of α and β-type 60° dislocations in III–V compounds. The difference in the corresponding activation energies is indeed a minimum value for this contribution. From Table 7.2, we can estimate that the contribution of the covalency to the total activation energy is at least 25%.

At this stage, we still ignore if dislocations move in the length-effect or in the kink-collision regime, i.e. if the measured activation energies correspond to $U_{\mathrm{ik}}^{(\mathrm{c,p})} + U_{\mathrm{m}}$ or to $\frac{1}{2}U_{\mathrm{ik}}^{(\mathrm{c,p})} + U_{\mathrm{m}}$. Hints to the answer are given below (Section 7.3.4), which aims at further improving the understanding of activation energies.

7.3.4 Regimes of dislocation movements

The length-effect regime is characterized by a dislocation velocity proportional to its length. In silicon, an in situ experiment has shown that the velocity of dislocations was length-dependent below 0.4 μm and length-independent above (Louchet, 1981). Similar experiments are required at different temperatures in order to confirm this result.

The two regimes can also be distinguished on the basis of the different pre-exponential terms of the corresponding velocities.

Referring to Eqs. (7.26a) and (7.26b) this term can be expressed as $\mathrm{v}_0(\tau_{\mathrm{eff}}/\mu)$, with:

$$\mathrm{v}_0 = \frac{3}{4}\nu_{\mathrm{D}}\frac{\mu b^3 L}{kT} \quad \text{in the length-effect regime, and}$$

$$\mathrm{v}_0 = \frac{3^{3/4}}{2}\nu_{\mathrm{D}}\frac{\mu b^4}{kT} \quad \text{in the kink-collision regime.}$$

Taking $\mu b^3 = 20$ eV and $kT = 6.7 \times 10^{-2}$ eV ($T = 800$ K) yields $\mathrm{v}_0 \approx 3 \times 10^9$ m/s in the length-effect regime (with $L = 1$ μm) and $\mathrm{v}_0 \approx 1.2 \times 10^6$ m/s in the kink-collision regime. As pointed out by Louchet and George (1983), the experimental values of v_0 are several orders of magnitude above those expected in the kink-collision regime. For instance, the most reliable measurements of Imai and Sumino (1983) performed by XRT in high-purity silicon yields $\mathrm{v}_0 = 6.8 \times 10^8$ m/s for 60° dislocations and $\mathrm{v}_0 = 2.4 \times 10^9$ m/s for screw dislocations. Unless there is an entropy term as large as $S = 8k$ (Marklund, 1985), these measurements indicate that dislocations most often glide in the length-effect regime, although their length (observed by XRT) can be as large as 1 mm.

There is, thus, a very large discrepancy between the results of Louchet (no length effect above 0.4 μm) and those of Imai and Sumino (possible length effect at 1 mm). A confirmation of both results, and what may be the key of this apparent discrepancy, has been given by Maeda and Yamashita (1993) and Maeda and Takeuchi (1996a,b). These authors have measured the velocity of dislocations in $Si_{0.9}$–$Ge_{0.1}$ hetero-epitaxial thin films of different thicknesses by SEM in the cathodo-luminescence mode, under the stress due to the lattice misfit. They have also measured the corresponding activation energies. The results are shown in Figure 7.25. The velocity of dislocations increases with length and tends to saturate above 1 μm, in agreement with the in situ results of Louchet. However, the corresponding activation energy remains constant, close to 2.2 eV in contrast with the two different predicted activation energies in Eqs. (7.26a) and (7.26c). The authors concluded that the dislocations move in the length-effect regime in all cases but that the mean-free-path of kinks, X, is restricted to 1 μm by interactions with impurities. The kink-collision regime, for which the mean-free-path X would be larger than 1 μm, cannot thus exist. With such an interpretation, the transition observed by Louchet would also be due to interactions with impurities. The good correspondence between experimental and theoretical pre-exponential factors, for $L = 1$ μm, is in excellent agreement with this statement.

A length effect has also been observed at one temperature in germanium, by TEM in situ experiments with an indication of a transition towards a length-independent

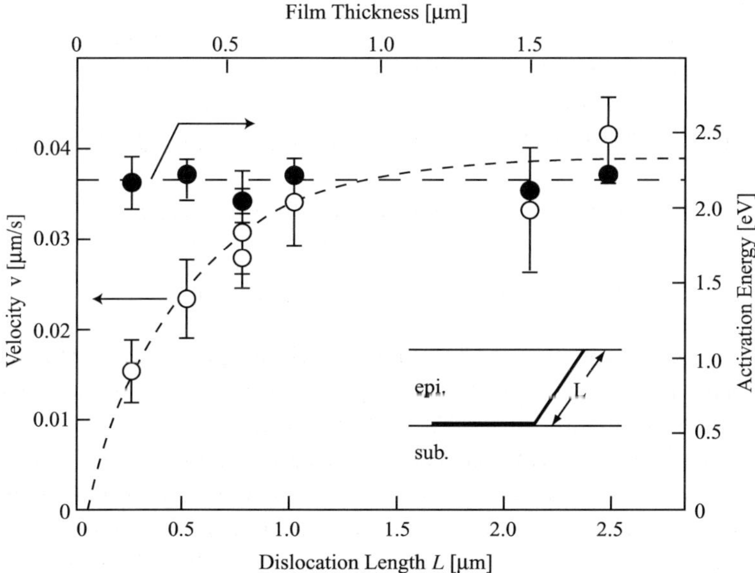

Figure 7.25. Dislocation velocity and corresponding activation energy as a function of length in $Si_{0.9}Ge_{0.1}$ epitaxial layers. From Maeda and Takeuchi (1996a,b).

velocity above 1 μm, especially for 60° segments (Louchet et al., 1988). This behaviour has been interpreted as a transition between length-effect and kink-collision regimes but interactions with impurities like in Si cannot be ruled out.

A length effect has been clearly shown in several III–V compounds (Figure 7.26) by means of in situ TEM experiments (Caillard et al., 1987, 1989; Fnaiech et al., 1987; Zafrany et al., 1992). The results show that there is no significant difference between the velocities of screw and 60°β dislocations. Experiments in a high voltage electron microscope, which allow for the observation of long dislocations, have shown that this velocity remains proportional to length up to 3 μm in GaAs (Figure 7.26(b)) and up to 5 μm in InSb (Gauffier 1992), without any indication of a transition towards a maximum constant value.

A length effect has also been observed in the II–VI compound ZnS (Figure 7.27) when dislocation velocities are enhanced by electron irradiation (see Section 7.3.5 below).

These data allow better estimations of $U_{\mathrm{ik}}^{(\mathrm{c,p})}$ and $U_{\mathrm{m}}^{(\mathrm{p})}$, assuming that the relations of the Hirth and Lothe theory (Section 7.2) are valid.

In the length-effect regime, the velocity of dislocations is approximately given by Eq. (7.26a), with $\tau_{\mathrm{eff}} \approx \tau$, whence

$$U_{\mathrm{ik}}^{(\mathrm{c,p})} + U_{\mathrm{m}}^{(\mathrm{p})} = kT \ln\left(\frac{3}{4}\nu_{\mathrm{D}}\frac{\tau b^3}{kT}\frac{L}{\mathrm{v}}\right)$$

where v/L can be measured on the v(L) curve.

The velocity in the kink-collision regime is given by Eq. (7.26c), with $\tau_{\mathrm{eff}} \approx \tau$. It is larger than the maximum velocity v_{M} measured on the velocity versus length curves, whence

$$\frac{1}{2}U_{\mathrm{ik}}^{(\mathrm{c,p})} + U_{\mathrm{m}}^{(\mathrm{p})} < kT \ln\left(\frac{3^{3/4}}{2}\nu_{\mathrm{D}}\frac{\tau b^4}{kT\mathrm{v}_{\mathrm{M}}}\right)$$

This allows maximum values for $U_{\mathrm{m}}^{(\mathrm{p})}$ and limiting values for $U_{\mathrm{ik}}^{(\mathrm{p})}$ and $U_{\mathrm{k}}^{(\mathrm{p})}$to be proposed. In all III–V compounds investigated, the kink migration energy, $U_{\mathrm{m}}^{(\mathrm{p})}$, is much smaller than the energy of a critical kink-pair, $U_{\mathrm{ik}}^{(\mathrm{c,p})}$.

On the other hand, the Hirth and Lothe theory is based on the assumption that kink-migration is a difficult thermally activated process. If it was not the case, the velocity of dislocations (in situations where it is not disturbed by impurities) would not be proportional to stress. This implies that $U_{\mathrm{m}}^{(\mathrm{p})} > kT$, namely $U_{\mathrm{m}}^{(\mathrm{p})} > 0.05$ eV or $U_{\mathrm{m}}^{(\mathrm{p})} \geq$ 0.1 eV.

All these values are gathered in Table 7.4 for different III–V compounds.

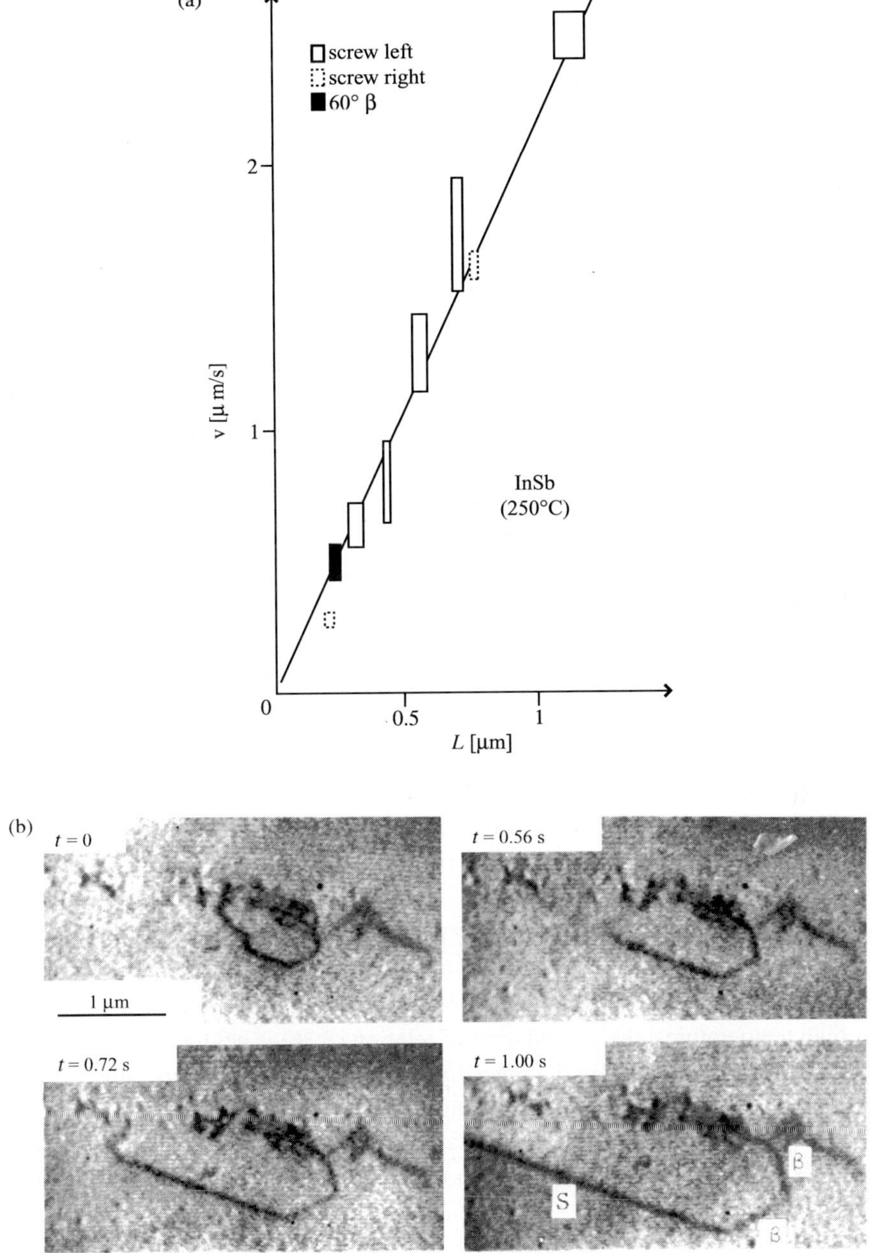

Figure 7.26. Length effect, from in situ experiments in (a) InSb at 250°C in a 200 kV electron microscope (Fnaiech et al., 1987) and (b) and (c) GaAS at 350°C, screw dislocations, in a 1 MeV electron microscope (Caillard et al., 1989).

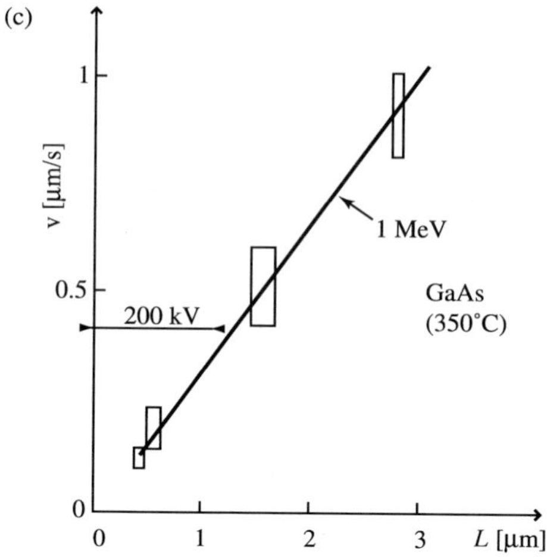

Figure 7.26. (*continued*)

7.3.5 Velocity enhancement under irradiation

Dislocation velocities at low temperatures are often strongly increased under irradiation by an electron beam. This radiation-enhanced dislocation glide (REDG) effect has been studied in detail in the review of Maeda and Takeuchi (1996a,b).

Experiments in a scanning electron microscope (cathodo-luminescence mode) have shown that the velocity of dislocations can be described by the empirical relation:

$$\mathrm{v} = \mathrm{v}_{0\mathrm{D}}\left(\frac{\tau}{\tau_0}\right)^m \exp\left(-\frac{U_\mathrm{D}}{kT}\right) + \mathrm{v}_{0\mathrm{E}}\left(\frac{\tau}{\tau_0}\right)^m \frac{I}{I_0} \exp\left(-\frac{U_\mathrm{dark} - \Delta U}{kT}\right) \tag{7.36}$$

where I is the electron beam intensity, U_dark is the activation energy "in the dark" ($U_\mathrm{dark} = U_\mathrm{ik}^{(c,p)} + U_\mathrm{m}^{(p)}$ or $U_\mathrm{dark} = \frac{1}{2}U_\mathrm{ik}^{(c,p)} + U_\mathrm{m}^{(p)}$), $U_\mathrm{dark} - \Delta U$ is the reduced activation energy under irradiation and the parameters $\mathrm{v}_{0\mathrm{D}}$, $\mathrm{v}_{0\mathrm{E}}$, τ_0 and I_0 are constant (Figure 7.28).

At high temperatures, the second term of relation (7.36) is negligible. This corresponds to the situations investigated in the preceding paragraph. At low temperatures, the second term becomes much higher than the first one and the dislocation velocity is proportional to the beam intensity, with the reduced activation energy $U_\mathrm{dark} - \Delta U$. Figure 7.29(a) shows that in ZnS the screw velocity depends on the beam intensity in the length-effect regime. Other experiments do not allow determination of the regime in which the REDG effect takes place. Figure 7.29(b) shows that the velocity tends to saturate for beam intensities larger than 5×10^2 A/m^2 in ZnS. This indicates that Eq. (7.36) is only valid for low and moderate beam intensities corresponding to experiments in a scanning electron microscope.

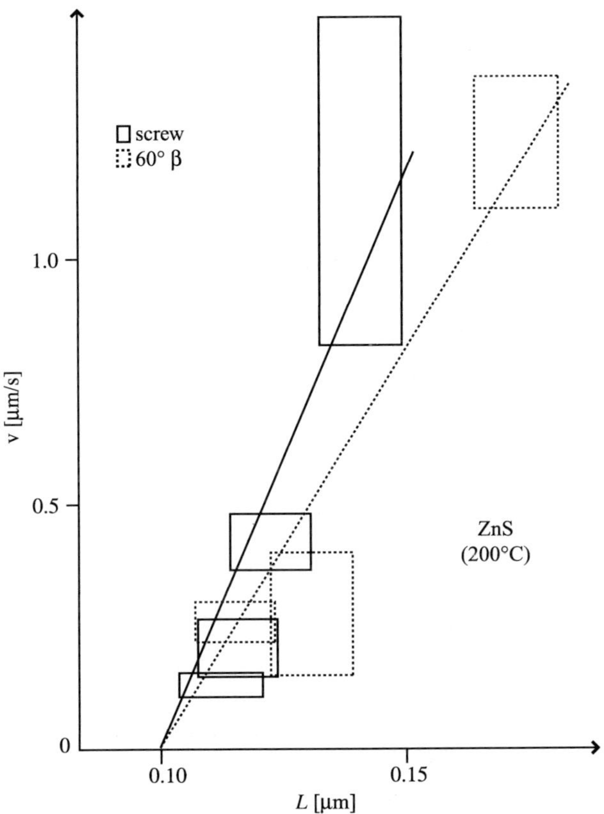

Figure 7.27. Length effect in ZnS at 200°C. In situ TEM experiment at 200 kV under a beam intensity $I = 1700$ A/m^2. The lines do not go through the origin probably because of a systematic error in the length measurements. From Levade et al. (1994).

Table 7.4. Activation parameters for glide of screw and 60°β dislocations in several III–V compounds, as determined from in situ TEM experiments and using the Hirth and Lothe theory.

Material	Temperature (°C)	Stress (MPa)	X (μm)	$U_{dark} = U_{ik}^{(p)} + U_m^{(p)}$ (eV)	$U_{ik}^{(p)}$ (τ) (eV)	$U_m^{(p)}$ (eV)	$U_k^{(p)}$ (eV)	$\Delta U/U_{dark}$ (eV)
InSb	250	~50	>5	1.2	0.8–1.1	0.1–0.4	0.5–0.65	0
InP	350	~50	>0.7	1.5	0.8–1.4	0.1–0.7		60%
GaAs	350	~50	>3	1.6	1.1–1.5	0.1–0.5	0.65–0.9	64%

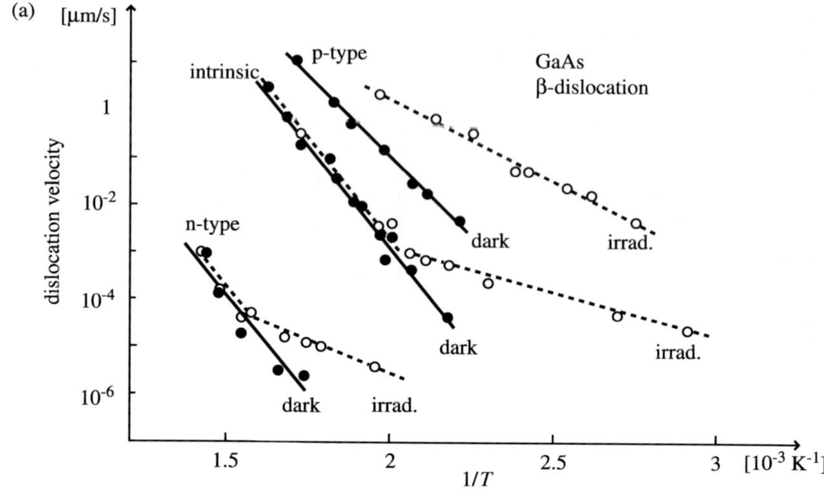

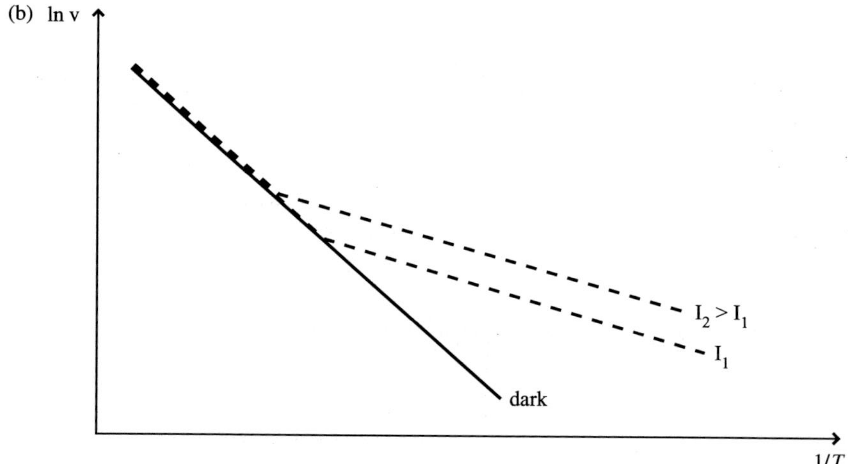

Figure 7.28. REDG effect: (a) experimental results in GaAs and (b) schematic variation of the dislocation velocity versus stress and beam intensity. From Maeda and Takeuchi (1996a,b).

Table 7.4 shows that there is no REDG effect in InSb. In GaAs and InP, however, the reduction in the total activation energy, ΔU, is fairly large (64 and 60%, respectively), larger than the migration energy $U_{\mathrm{m}}^{(\mathrm{p})}$. Thus, the REDG effect probably corresponds to a decrease in the critical kink-pair energy $U_{\mathrm{ik}}^{(\mathrm{c,p})}$ due to the recombination of electrons and holes. The possible underlying mechanisms have been described in detail by Maeda and Takeuchi (1996a,b).

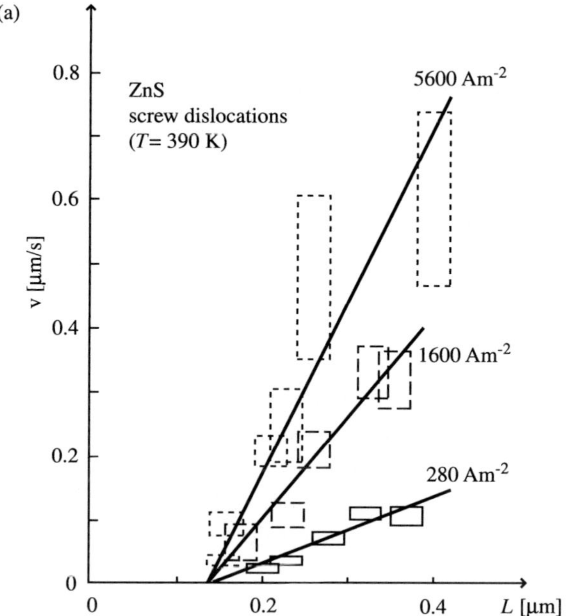

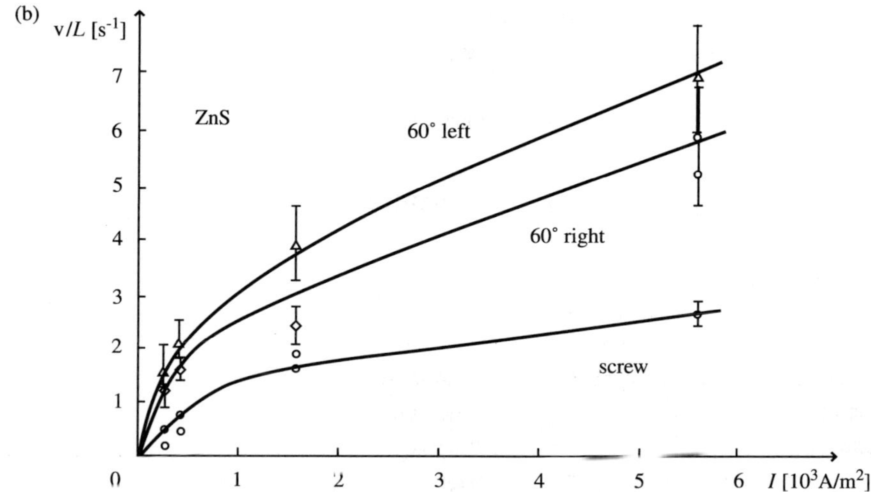

Figure 7.29. REDG effect in ZnS. Results from in situ experiment in TEM. (a) Same length effect as in Figure 7.27. The lines do not go through the origin probably because of a systematic error in the length measurements. (b) Saturation of the REDG effect at high beam intensities.
From Vanderschaeve et al. (1991).

7.3.6 Experiments at very high stresses

Very high stresses and low deformation temperatures cannot be explored in conventional tests because of brittleness. This requires a dedicated equipment that allows a monotonic compression test under a hydrostatic pressure. Figure 7.30 shows the corresponding critical resolved shear stresses for single slip in several III–V compounds as a function of temperature (Edagawa et al., 2000). The curves exhibit a sharp transition at approximately room temperature that indicates that another easier mechanism takes over from the usual high-temperature one. The same data has been reported in Figure 7.31 (from Rabier and Demenet, 2000) together with similar results in 4H–SiC (from Pirouz et al., 2000). The results in Si have been completed with more recent ones (Rabier et al., 2001) showing that some plastic deformation can take place at 473 and 300 K under a deviatoric compression stress that is not measured but is of the order of the confining stress (5 GPa).

The corresponding microscopic observations are the following:

- In SiC, dislocations are dissociated above the transition temperature and isolated Shockley partials trail stacking faults below (Pirouz et al., 2000).

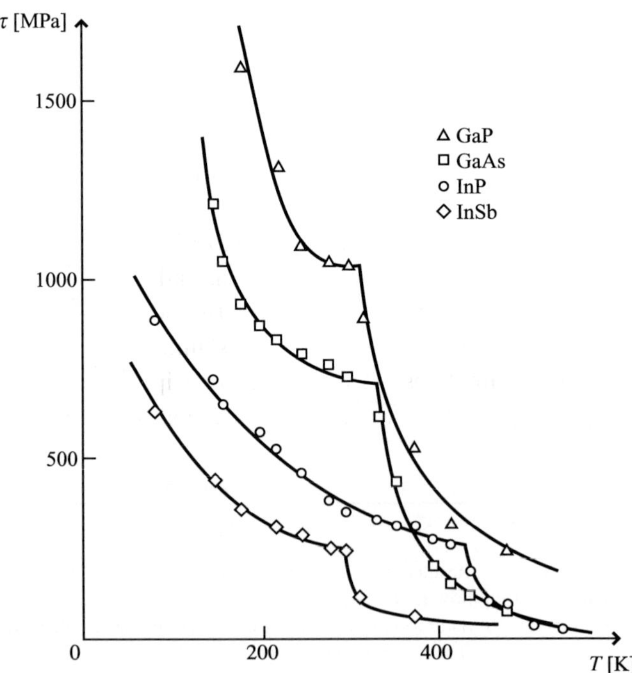

Figure 7.30. Temperature dependence of the critical resolved shear stress in several III–V compounds, including the high-stress regime. From Edagawa et al. (2000).

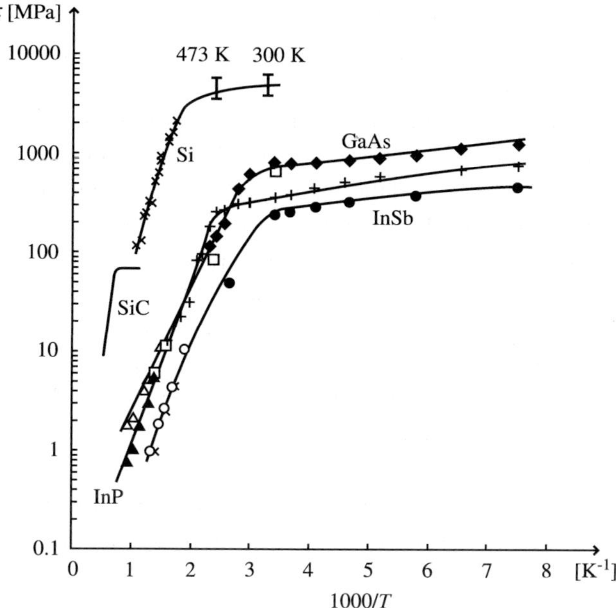

Figure 7.31. Same data as in Figure 7.30, with additional results on Si (Demenet, 1987; Rabier and Demenet, 2000; Rabier et al., 2001) and SiC (Pirouz et al., 2000).

- In Si, dislocations are dissociated just above the transition temperature but the stacking fault ribbon can be either narrowed or widened (depending on the direction of the compression axis). Widening can lead to movements of isolated Shockley partials trailing stacking faults (Castaing et al., 1981; Rabier and Demenet, 2000). Below the transition temperature, i.e. at 300 and 473 K, dislocations no longer appear dissociated at the scale of weak beam observations and they tend to be straight along the screw and the more unusual $\langle 112 \rangle$ directions (Rabier and Demenet, 2000). These dislocations may, thus, have non-dissociated shuffle cores, expected to be less energetic along these directions (Hornstra, 1958). Slip line analyses reveal that no extensive cross-slip takes place, although dislocations appear to be undissociated (Rabier et al., 2001).
- In III–V compounds many different mechanisms are observed at the same time: movements of isolated Shockley partials, twinning, locally undissociated dissociations and cross-slip (Androussi et al., 1987; Boivin et al., 1990; Suzuki et al., 1999a).

The mechanical properties at high stresses can a priori be explained in two ways:

- If the leading Shockley partial has a higher mobility than the trailing one it can escape, move alone and trail a stacking fault of surface energy γ. Its velocity is given by Eq. (7.26a) or (7.26c), or Eq. (7.24a) with (7.29a), with $\tau_{eff} = \tau_l - \gamma/b_p$, $\tau_i = 0$). The critical stress for the uncoupling of the two partials, $\tau_{d\infty}$, is given by Eq. (7.32). This process can account for the lower deformation stress at low temperature, if the leading partial is much more mobile than the trailing one ($M_l \gg M_t$). Then, the transition stress is $\tau_{d\infty} \approx 2\gamma/b \approx \gamma/b_p$. This is sketched in Figure 7.32(a).

– If the activation energies of glide have large elastic components, as assumed by Duesberry and Joos (1996), the non-dissociated shuffle set may be more mobile at higher stresses because of a lower activation energy (see Figure 7.10). The corresponding transition is sketched in Figure 7.32(b).

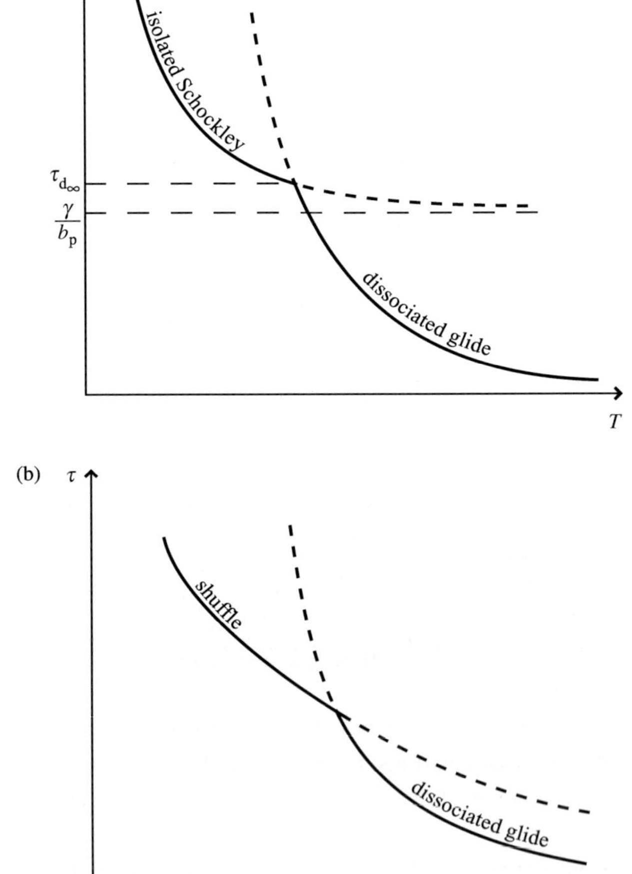

Figure 7.32. Schematic description of the transition towards the high-stress regime, with two different assumptions (see text).

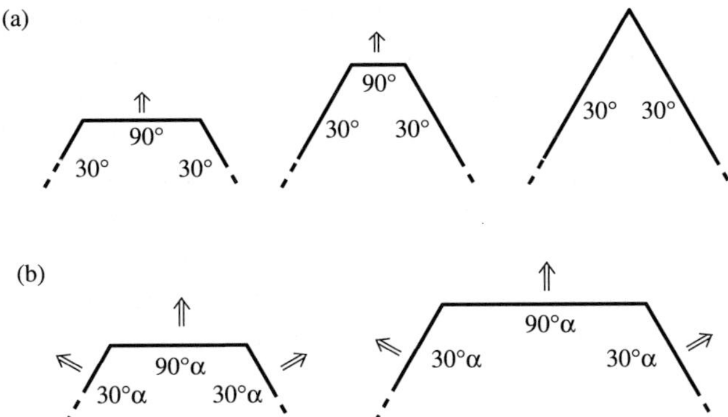

Figure 7.33. (a) Short-distance (ESC) and (b) long-distance (III–V compound) movements of isolated Shockley partials, in the high-stress regime. From Rabier and Demenet (2000).

In silicon, only 90° Shockley partials have a higher mobility than dissociated dislocations. However, the leading 90° Shockley segments are connected with two 30° ones that exhibit as low a mobility as dissociated dislocations (Section 7.3.1.1). Then, only short-range movements of 90° partials can take place, as described schematically in Figure 7.33(a). Therefore, the first explanation does not hold. On the contrary, a glide-shuffle transition based on experimental observations of non-dissociated dislocations with unusual directions is a plausible explanation (Rabier et al., 2001).

In III–V compounds the situation is less clear. The easy movement of α-type leading partials with three different orientations could account for large amounts of plasticity at high stresses (Figure 7.33(b)). On the other hand, although the existence of shuffle dislocations has not been proven unambiguously, the observation of cross-slip traces indicates that some dislocations at least are perfect. The motion of a different type of dislocation (such as shuffles) could explain why glide occurs under stresses lower than those extrapolated from the high-temperature regime. Observations indicate that both mechanisms could be involved.

– In 4H–SiC, the transition may be explained by isolated Shockley partial movements, although their mobilities are not known.

7.4. CONCLUSIONS

The reformulated Hirth and Lothe theory considers a simplified energy barrier profile for kink-pair nucleation. The exact shape of the profile is subsequently shown to be unimportant. The difficult step based on the "Zeldovich treatment" (Zeldovich, 1943) in the original formulation of Hirth and Lothe is thus avoided. The theory is then extended to the real case of split dislocations. This leads to significantly different expressions for the activation parameters of the dislocation velocities.

Abundant experimental datas are available in silicon as compared to other covalent materials. For high purity silicon, a fairly good agreement is found with the revisited theory provided the dissociation is taken into account. The mobilities of perfect and Shockley dislocations can then be understood consistently without introducing ad hoc assumptions.

For less pure silicon, germanium and CSC, the interaction of dislocations with impurities makes the comparison between experiment and theory less straightforward.

Several problems are still not completely solved, concerning the extension of the length-effect regime, the effect of electron irradiation and the plastic properties at very high stresses.

REFERENCES

Alexander, H. (1986) in *Dislocations in Solids*, vol. **7**, Ed. Nabarro, F.R.N., Elsevier, Amsterdam, p. 113.

Alexander, H., Eppenstein, H., Gottschalk, H. & Wendler, S. (1980) *J. Microsc.*, **118**, 13.

Alexander, H., Kisielovski-Kemmerich, C. & Weber, E.R. (1983) *Physica*, **116B**, 583.

Alexander, H., Kisielowski-Kemmerich, C. & Swalski, A.T. (1987) *Phys. Stat. Sol (a)*, **104**, 183.

Androussi, Y., Vanderschaeve, G. & Lefebvre, A. (1987) *Inst. Phys. Conf. Ser.*, **87**(sect. 4), 291.

Boivin, P., Rabier, J. & Garem, H. (1990) *Phil. Mag. A*, **61**, 647.

Bulatov, V.V., Yip, S. & Argon, A.S. (1995) *Phil. Mag. A*, **72**, 453.

Cai, W., Bulatov, V.V., Justo, J.F., Argon, A.S. & Yip, S. (2000) *Phys. Rev. Lett.*, **84**, 3346.

Caillard, D. & Vanderschaeve, G. (2003) to be published.

Caillard, D., Clément, N., Couret, A., Androussi, Y., Lefebvre, A. & Vanderschaeve, G. (1987) *Inst. Phys. Conf. Ser.*, **87**(sect. 5), 361.

Caillard, D., Clément, N., Couret, A., Androussi, Y., Lefebvre, A. & Vanderschaeve, G. (1989) *Inst. Phys. Conf. Ser.*, **100**, 403.

Castaing, J., Veyssière, P., Kubin, L.P. & Rabier, J. (1981) *Phil. Mag. A*, **44**, 1407.

Celli, V., Kabler, M., Ninomiya, T. & Thomson, R. (1963) *Phys. Rev.*, **131**, 58.

Chaudhuri, A.R., Patel, J.R. & Rubin, L.G. (1962) *J. Appl. Phys.*, **33**, 2736.

Choi, S.K., Mihara, M. & Ninomiya, T. (1987) *Jpn. J. Appl. Phys.*, **16**, 737.

Demenet, J.L. (1987) *Thèse*, Université de Poitiers, France.

Duesberry, M.S. & Joos, B. (1996) *Phil. Mag. Lett.*, **74**, 253.

Edagawa, K., Koizumi, H., Kamimura, Y. & Suzuki, T. (2000) *Phil. Mag. A*, **80**, 2591.

Erofeeva, S.A. & Osspyan, Y.A. (1973) *Sov. Phys. Sol. State*, **15**, 538.

Fnaiech, M., Reynaud, F., Couret, A. & Caillard, D. (1987) *Phil. Mag. A*, **55**, 405.

Gauffier, J.L. (1992) *Thèse*, INSA Toulouse, France.

George, A. & Champier, G. (1979) *Phys. Stat. Sol (a)*, **53**, 529.

George, A. & Rabier, J. (1987) *Rev. Phys. Appl.*, **22**, 941, see also p. 1327.

George, A., Escaravage, C., Champier, G. & Schröter, W. (1972) *Phys. Stat. Sol (b)*, **53**, 483.

Hirth, J.P. & Lothe, J. (1982) *Theory of Dislocations*, 2nd edition, Krieger Pub. Comp., Florida.

Hornstra, J. (1958) *J. Phys. Chem. Solids*, **5**, 129.

Imai, M. & Sumino, K. (1983) *Phil. Mag. A*, **47**, 599.
Kabler, M.N. (1963) *Phys. Rev.*, **131**, 54.
Koizumi, H., Kaminura, Y. & Suzuki, T. (2000) *Phil. Mag.*, **80**, 609.
Lefebvre, A. & Vanderschaeve, G. (1989) *Inst. Phys. Conf. Ser.*, **100**(sect. 5), 397.
Levade, C., Faress, A. & Vanderschaeve, G. (1994) *Phil. Mag. A*, **69**, 855.
Louchet, F. (1981) *Phil. Mag. A*, **43**, 1289.
Louchet, F. & George, A. (1983) *J. Phys. Colloque*, **44**, C4-51.
Louchet, F. & Thibault-Desseaux, J. (1987) *Rev. Phys. Appl.*, **22**, 207.
Louchet, F., Cochet-Muchy, D., Bréchet, Y. & Pélissier, J. (1988) *Phil. Mag. A*, **57**, 327.
Louchet, F., Pélissier, J., Caillard, D., Peyrade, J.P., Levade, C. & Vanderschaeve, G. (1993) *Microsc. Microanal. Microstruct.*, **4**, 199.
Maeda, K. & Takeuchi, S. (1996a) in *Dislocations in Solids*, vol. **10**, Eds. Nabarro, F.R.N. & Duesberry M.S., Elsevier, Amsterdam, p. 443.
Maeda, K. & Takeuchi, S. (1996b) in *Dislocations in Solids*, vol. **10**, Eds. Nabarro, F.R.N. & Duesberry M.S., Elsevier, Amsterdam, ch. 54, p. 43.
Maeda, K. & Yamashita, Y. (1993) *Phys. Stat. Sol (a)*, **138**, 523.
Marklund, S. (1984) *Solid State Commun.*, **50**, 185.
Marklund, S. (1985) *Solid State Commun.*, **54**, 555.
Mihara, M. & Ninomiya, T. (1975) *Phys. Stat. Sol (a)*, **32**, 43.
Moller, H.J. (1978) *Acta Met.*, **26**, 963.
Paidar, V. & Caillard, D. (1994) *Phil. Mag. Lett.*, **69**, 305.
Paidar, V., Legros, M., Couret, A. & Caillard, D. (1994) *Key Eng. Mat.*, **97–98**, 201.
Patel, J.R. & Freeland, P.E. (1971) *J. Appl. Phys.*, **42**, 3298.
Pirouz, P., Kubin, L.P., Demenet, J.L., Hong, M.H. & Samant, A.V. (2000) *MRS Symp. Proc.*, **578**, 205.
Rabier, J. & Demenet, J.L. (2000) *Phys. Stat. Sol (b)*, **222**, 63.
Rabier, J., Cordier, P., Demenet, J.L. & Garem, H. (2001) *Mat. Sci. Eng.*, **309–310**, 74.
Ray, I.L.F. & Cockayne, D.J.H. (1971) *Proc. Roy. Soc. A*, **325**, 543.
Schaumburg, H. (1972) *Phil. Mag.*, **25**, 1429.
Senkader, S., Jurkschat, K., Gambaro, D., Falster, R.J. & Wilshaw, P.R. (2001) *Phil. Mag. A*, **81**, 759.
Steinhardt, H. & Schafer, S. (1971) *Acta Met.*, **19**, 65.
Suzuki, T., Koizumi, H. & Kirchner, H.O.K. (1994) in *Strength of Materials*, Eds. Oikowa, H., Maruyama, K., Tatenchi, S. & Yomoguchi M., The Japan Inst. of Metals, p. 89.
Suzuki, T., Koizumi, H. & Kirchner, H.O.K. (1995) *Phil. Mag. A*, **71**, 389.
Suzuki, T., Yasutomi, T., Takuoka, T. & Yonenaga, I. (1999a) *Phil. Mag. A*, **79**, 2637.
Suzuki, T., Yasutomi, T., Tokuoka, T. & Yonenaga, I. (1999b) *Phys. Stat. Sol (a)*, **171**, 47.
Vanderschaeve, G. & Caillard, D. (1994) in *Electron Microscopys*, Vol. **2**, Eds. Jouffrey, B., Colliex, C., Chevalier, J.P., Glas, F. & Hawkes P.W., Les Editions de Physique, p 65.
Vanderschaeve, G., Levade, C., Faress, A., Couderc, J.J. & Caillard, D. (1991) *J. Phys. IV (Paris)*, **1**, C6.305.
Vanderschaeve, G., Levade, C. & Caillard, D. (2001) *J. Microsc.*, **203**, 72.
Wessel, K. & Alexander, H. (1977) *Phil. Mag.*, **35**, 1523.
Yonenaga, I. (1997) *J. Phys. III France*, **7**, 1435.
Yonenaga, I. & Sumino, K. (1987) *J. Appl. Phys.*, **62**, 1212.
Yonenaga, I. & Sumino, K. (1989) *J. Appl. Phys.*, **65**, 85.
Zafrany, M., Voillot, F., Peyrade, J.P., Caillard, D., Couret, A. & Cocquillet, R. (1992) *Phil. Mag. A*, **65**, 195.
Zeldovich, J.B. (1943) *Acta Physicochem. URSS*, **18**, 1.

Chapter 8
Dislocation Climb

Chapter 8

[illegible]

Chapter 8

Dislocation Climb

Dislocation climb becomes the dominant deformation mechanism in high temperature plasticity. It also controls the kinetics of prismatic loop evolution under high point defect supersaturations. This chapter rationalizes various attempts at modelling the dislocation climb velocity in the frame of several approximations. These correspond to various stress, temperature ranges and point defect concentrations. A few reliable experiments are also reviewed, in which climb has been properly isolated. They are compared with the theoretical estimates of dislocation velocities.

The material in this chapter is organized as follows. The elementary mechanisms of point defect diffusion and their interactions with jogs are first recalled and the concept of "chemical force" is introduced. Then the process of climb is described for dislocations containing, respectively, high and low jog densities. In the latter case, the dislocation velocity is computed when only bulk diffusion operates and then in the presence of pipe diffusion. Then climb through vacancy or interstitial absorption is described. This corresponds to prismatic loops of various shapes that grow or shrink in the presence of point defect supersaturations. The experimental studies of dislocation climb which are then described are related to high temperature creep data in HCP metals, intermetallics and quasicrystals. They also include loop-annealing experiments under various circumstances.

8.1. INTRODUCTION: BASIC MECHANISMS

This section introduces the definition and the description of the elementary mechanisms involved in the climb processes. They will be used in Sections 8.2 and 8.3 to obtain the climb rate equations.

8.1.1 Definition of climb

Dislocations climb when they move in planes that do not contain their Burgers vector. Under such conditions, it is necessary to add or remove atoms at the dislocation core, by a diffusion process that usually takes place only at high temperature. We will consider pure climb along the plane perpendicular to the Burgers vector. Other situations corresponding to a mixture of glide and climb will be considered in Section 8.2.3.

Figure 8.1 illustrates the two processes that yield the same climb motion in the direction arrowed: emission of a vacancy in (a) and absorption of an interstitial in (b). The opposite

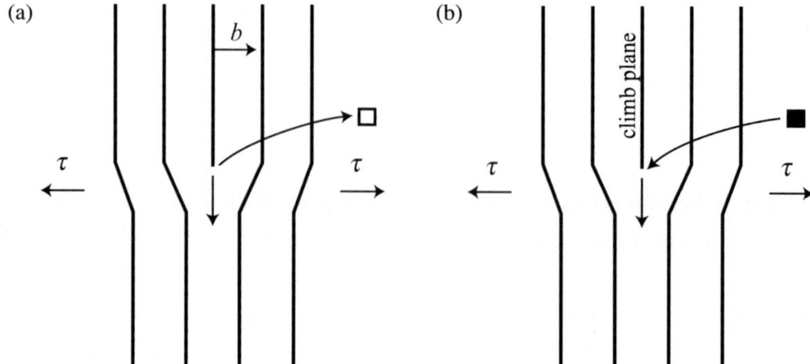

Figure 8.1. Climb of an edge dislocation by (a) emission of a vacancy and (b) absorption of an interstitial atom.

motion would be achieved either by the emission of an interstitial or by the absorption of a vacancy. Both mechanisms will be considered in what follows.

8.1.2 *Mechanical forces for pure climb*

The Peach and Koehler formula (1950) indicates that under a normal stress τ, perpendicular to the climb plane, the "mechanical" force per unit dislocation length is

$$F = \tau b \tag{8.1a}$$

where b is the Burgers vector.

This force is oriented along the climb plane and perpendicular to the dislocation.

It is important to note that climb is not sensitive to the same components of the stress tensor as glide. In the case of curved dislocations, the driving force includes the line tension effects and the eventual surface energy of a fault, γ. Consequently, the effective normal stress is, in the most general case:

$$\tau = \tau^{(\text{appl.})} + \frac{\hat{T}}{bR} + \frac{\gamma}{b} \tag{8.1b}$$

where $\tau^{(\text{appl.})}$ is the applied normal stress.

8.1.3 *Diffusion of point defects*

The equilibrium atomic vacancy (interstitial) concentration far from the dislocation is:

$$c_{\text{v}}^{(0)} = \exp\left(-\frac{U_{\text{f}}^{(\text{v})}}{kT}\right) \tag{8.2}$$

where $U_{\text{f}}^{(\text{v})}$ is the energy to form a vacancy. The same equation holds for interstitials with subscript or superscript i.

The frequency ν of vacancy jumps from one site to a neighbouring one is:

$$\nu = \nu_D \exp\left(-\frac{U_d^{(v)}}{kT}\right) \tag{8.3}$$

where ν_D is the Debye frequency and $U_d^{(v)}$ the vacancy diffusion energy. This frequency allows one to define the diffusion coefficient $D_v = a^2\nu$ (Philibert, 1985):

$$D_v = a^2 \nu_D \exp\left(-\frac{U_d^{(v)}}{kT}\right) \tag{8.4}$$

where a is an average interatomic distance.

Under these conditions, the self-diffusion coefficient for vacancies is:

$$D_v^{(sd)} = c_v^{(0)} D_v = a^2 \nu_D \exp\left(-\frac{U_{sd}^{(v)}}{kT}\right) \tag{8.5}$$

where $U_{sd}^{(v)} = U_f^{(v)} + U_d^{(v)}$ is the self-diffusion activation energy for vacancies.

Non-equilibrium point defect concentrations can result from preliminary treatments such as quenching or irradiation. They also arise from the climb mechanism itself, because the point defects emitted by the dislocations do not immediately dissolve by diffusion away from the core. In the latter case, a gradient in the concentration of point defects builds up near dislocations. This gradient induces an average vacancy drift velocity v_v, which obeys Einstein's formula (Balluffi and Granato, 1979):

$$\mathrm{v}_v = -D_v\left[\frac{\operatorname{grad} c_v}{c_v} + \frac{\Delta\Omega}{KT}\operatorname{grad} p\right] \tag{8.6a}$$

where $\Delta\Omega$ is the volume change due to the point defect and p the hydrostatic component of the stress. The second term is generally negligible, whence

$$\mathrm{v}_v = -D_v \frac{\operatorname{grad} c_v}{c_v} \tag{8.6b}$$

The same laws are valid for interstitials where subscript or superscript v is replaced by i.

8.1.4 Jog–point defect interactions

The emission or the absorption of point defects always takes place at jogs (see Figure 8.2). We consider the vacancy–jog interactions first and then the interstitial–jog interactions.

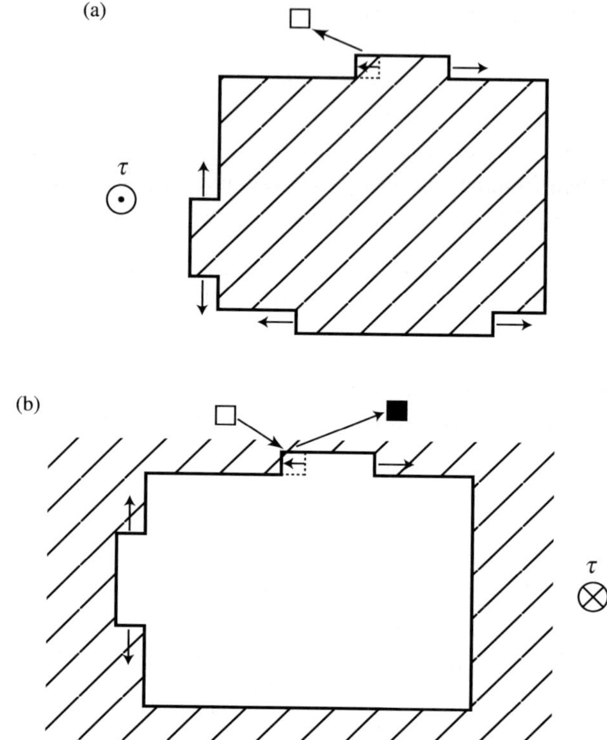

Figure 8.2. (a) Growth of an interstitial-type loop assisted by a tensile stress normal to the picture. The supplementary plane is hatched. The vacancy loop in (b) would shrink under the same conditions. (b) Growth of a vacancy-type loop assisted by a compressive stress normal to the picture. The loop in (a) would shrink under the same conditions.

8.1.4.1 Jog–vacancy interactions. In the absence of any driving stress, jog-jumps through vacancy emission require the formation of a vacancy and its diffusion to one of the n neighbouring sites. The corresponding frequency is:

$$\nu_{\mathrm{j}} = n\nu_{\mathrm{D}} \exp\left(-\frac{U_{\mathrm{f}}^{(\mathrm{v})} + U_{\mathrm{d}}^{(\mathrm{v})}}{kT} \right) = n\frac{D_{\mathrm{v}}^{(\mathrm{sd})}}{a^2}$$

where n is the number of first nearest neighbors in the lattice ($n = 12$ in FCC lattices). This emission process can also be decomposed into several steps (Schoeck, 1980, see p. 120): this does not change the result, except that n becomes the number of next nearest neighbors.

Jog-jumps through vacancy absorption require the diffusion of one of the neighbouring vacancies, in concentration $c_{\mathrm{v}}^{(0)}$, to the jog. The corresponding frequency is $n\nu_{\mathrm{D}}c_{\mathrm{v}}^{(0)} \times \exp(-U_{\mathrm{d}}^{(\mathrm{v})}/kT)$.

Since $c_{\mathrm{v}}^{(0)}$ is given by Eq. (8.2), these two frequencies are equal, in agreement with a zero average jog velocity, under thermal fluctuations.

Under a mechanical normal stress, the activation energy for jog motion in the direction imposed by the stress is reduced by $\tau\Omega$, where Ω is the atomic volume. *This mechanical normal stress is taken as positive when it tends to favour vacancy emission at the jogs (Figure 8.2(a))*. Under such conditions, the frequency of jog motion in the direction favoured by the stress is:

$$\nu_{\mathrm{j}}^{+} = n\nu_{\mathrm{D}} \exp\left(-\frac{U_{\mathrm{f}}^{(\mathrm{v})} + U_{\mathrm{d}}^{(\mathrm{v})} - \tau\Omega}{kT}\right) = \frac{nD_{\mathrm{v}}^{(\mathrm{sd})}}{a^2}\exp\left(\frac{\tau\Omega}{kT}\right) \tag{8.7}$$

The emitted vacancies, with local concentration $c_{\mathrm{v}}^{(\mathrm{j})}$ in the n neighboring sites around the jog, tend to jump back. This process takes place at a frequency of:

$$\nu_{\mathrm{j}}^{-} = n\nu_{\mathrm{D}}c_{\mathrm{v}}^{(\mathrm{j})} \exp\left(-\frac{U_{\mathrm{d}}^{(\mathrm{v})}}{kT}\right) = \frac{nD_{\mathrm{v}}^{(\mathrm{sd})}}{a^2}\frac{c_{\mathrm{v}}^{(\mathrm{j})}}{c_{\mathrm{v}}^{(0)}} \tag{8.8a}$$

The frequency ν_{j}^{-} resembles that of the stress free situation (ν_{j}), except that $c_{\mathrm{v}}^{(0)}$ has been replaced by $c_{\mathrm{v}}^{(\mathrm{j})}$.

One should note that the activation energy required in backward jumps does not involve the work done by the applied stress, τ, because it is assumed that the vacancies do not feel the applied stress in their threshold positions. In the alternative case, the activation energy would be $U_{\mathrm{d}}^{(\mathrm{v})} + \tau\Omega$ and a factor of 2 would be introduced in the final result (Schoeck, 1980, see p. 120).

By analogy with Eq. (8.7), ν_{j}^{-} can be written:

$$\nu_{\mathrm{j}}^{-} = \frac{nD_{\mathrm{v}}^{(\mathrm{sd})}}{a^2}\exp\left(\frac{F_{\mathrm{c}}^{(\mathrm{j})}\Omega}{bkT}\right) \tag{8.8b}$$

where

$$F_{\mathrm{c}}^{(\mathrm{j})} = \frac{kTb}{\Omega}\ln\frac{c_{\mathrm{v}}^{(\mathrm{j})}}{c_{\mathrm{v}}^{(0)}} \tag{8.9a}$$

$F_{\mathrm{c}}^{(\mathrm{j})}$ thus defined has the dimension of a force per unit dislocation length. It is called a "chemical force" because it arises from a spatial variation of the vacancy concentration.

This chemical force opposes τb at the jog.

For $c_{\mathrm{v}}^{(\mathrm{j})}/c_{\mathrm{v}}^{(0)}$ close to 1, it can be developed to the first order:

$$F_{\mathrm{c}}^{(\mathrm{j})} \approx \frac{kTb}{\Omega}\left(\frac{c_{\mathrm{v}}^{(\mathrm{j})}}{c_{\mathrm{v}}^{(0)}} - 1\right) \tag{8.9b}$$

Combining Eqs. (8.7) and (8.8), the net jump frequency in the direction of the mechanical stress is:

$$\nu_j^+ - \nu_j^- = \frac{nD_v^{(sd)}}{a^2}\left[\exp\left(\frac{\tau\Omega}{kT}\right) - \frac{c_v^{(j)}}{c_v^{(0)}}\right]$$

$$= \frac{nD_v^{(sd)}}{a^2}\left[\exp\left(\frac{\tau\Omega}{kT}\right) - \exp\left(\frac{F_c^{(j)}\Omega/b}{bkT}\right)\right] \quad (8.10a)$$

The parameter $\nu_j^+ - \nu_j^-$ is a significant one because it is proportional to the *climb velocity* of dislocations.

For $\tau\Omega \ll kT$ and $F_c^{(j)}/b \ll kT$, Eq. (8.10a) reduces to:

$$\nu_j^+ - \nu_j^- \approx \frac{nD_v^{(sd)}}{a^2}\left[1 + \frac{\tau\Omega}{kT} - \frac{c_v^{(j)}}{c_v^{(0)}}\right] \approx \frac{n\Omega D_v^{(sd)}}{a^2 kT}\left[\tau - \frac{F_c^{(j)}}{b}\right] \quad (8.10b)$$

A variety of kinds of behaviour can take place according to the relative values of the mechanical stress τ and the chemical force on the jog, $F_c^{(j)}$. $\tau > 0$ and $F_c^{(j)}/b < 0$ tend to promote the emission of vacancies, whereas $\tau < 0$ and $F_c^{(j)}/b > 0$ tend to promote their absorption.

The three main situations of interest are going to be examined with reference to Figure 8.3. In this figure, the stress and the vacancy concentration at jogs have arbitrary values, independent of each other. In reality, these two parameters are related, as will be shown in Section 8.2. The first two situations correspond to plasticity problems and the third one to the annealing of quenched specimens.

(i) *Emission of vacancies assisted by a positive mechanical stress* (Figure 8.2(a)):

The net jump frequency (Eq. (8.10a)) is positive, which is equivalent to $\exp(\tau\Omega/kT) > c_v^{(j)}/c_v^{(0)}$. Since the mechanical stress that favours the emission of vacancies is positive, we also have $\exp(\tau\Omega/kT) > 1$. For the present conditions, there is a local excess of vacancies near the jog, i.e. $c_v^{(j)}/c_v^{(0)} > 1$. Putting all inequalities together yields: $1 < (c_v^{(j)}/c_v^{(0)}) < \exp(\tau\Omega/kT)$.

The local high vacancy concentration due to the accumulation of emitted vacancies slows down the motion.

A high stress can induce rapid climb because Eq. (8.10a) yields $\nu_j^+ - \nu_j^- \gg (nD_v^{(sd)}/a^2)$.

(ii) *Absorption of vacancies assisted by a negative mechanical stress (Figure 8.2(b))*:

The net jump frequency (Eq. (8.10a)) is negative, which is equivalent to $\exp(\tau\Omega/kT) < c_v^{(j)}/c_v^{(0)}$. Since the mechanical stress that favours the absorption of vacancies is negative, we also have $\exp(\tau\Omega/kT) < 1$. There is a local depletion of vacancies near the jogs where they are trapped, so that $c_v^{(j)}/c_v^{(0)} < 1$. Putting all inequalities together we obtain: $0 < \exp(\tau\Omega/kT) < c_v^{(j)}/c_v^{(0)} < 1$.

Climb velocities are expected to be similar to those of case (i) at low stresses. However, it is important to note that, in contrast with case (i), a high (negative) stress cannot induce rapid climb because Eq. (8.10a) yields $\nu_j^- - \nu_j^+ < \dfrac{nD_v^{(sd)}}{a^2}$ in all cases.

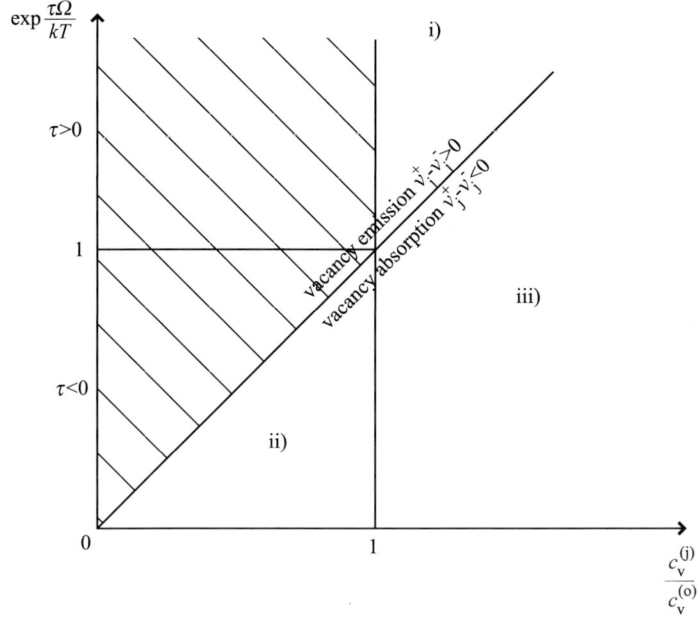

Figure 8.3. Diagram illustrating the three domains (i) to (iii) of interest for dislocation climb, by either emission or absorption of vacancies at jogs. At given temperature, the vertical axis refers to the influence of stress τ and the horizontal one to that of the vacancy concentration at jogs.

(iii) *Absorption of vacancies assisted by a chemical force:*

The net jump frequency is negative, which is equivalent to $\exp(\tau\Omega/kT) < c_v^{(j)}/c_v^{(0)}$. The dominant chemical force must be consistent with a negative net jump frequency $\nu_j^+ - \nu_j^-$, i.e. $c_v^{(j)}/c_v^{(0)} > 1$ (local excess of vacancies near the jogs). A high supersaturation $(c_v^{(j)}/c_v^{(0)} \gg 1)$ can induce rapid climb, because the dominant term ν_j^- can be high ($\nu_j^- \gg nD_v^{(sd)}/a^2$), according to Eq. (8.8a).

8.1.4.2 ***Jog–interstitial interactions.*** To be consistent with the preceding section, *the mechanical stress is taken as negative when it tends to favour the emission of interstitials by the jogs* (compression stress across the loop plane, cf. Figure 8.2(b)). Under such conditions, by analogy with Eq. (8.10), the frequency of jog motion in the direction favoured by the stress is:

$$\nu_j^+ - \nu_j^- = \frac{nD_i^{(sd)}}{a^2}\left[\exp\left(\frac{-\tau\Omega}{kT}\right) - \frac{c_i^{(j)}}{c_i^{(0)}}\right]$$

$$= \frac{nD_i^{(sd)}}{a^2}\left[\exp\left(\frac{-\tau\Omega}{kT}\right) - \exp\left(\frac{F_c^{(j)}\Omega}{bkT}\right)\right] \qquad (8.11)$$

where $F_c^{(j)} = kTb/\Omega \ln c_i^{(j)}/c_i^{(0)}$ is the chemical force for interstitials.

Since it is well known that $D_i^{(sd)} \ll D_v^{(sd)}$, substantial dislocation velocities can be achieved only when one of the two available driving forces (mechanical or chemical) is high. This leads to the two following situations:

(i) *Emission of interstitials assisted by a negative mechanical stress:*

The net jump frequency (Eq. (8.11)) is positive, which is equivalent to $\exp(-\tau\Omega/kT) > c_i^{(j)}/c_i^{(0)}$. For the present conditions, there is a local excess of interstitials near the jogs, i.e. $c_i^{(j)}/c_i^{(0)} > 1$. Putting all inequalities together yields:

$$1 < \frac{c_i^{(j)}}{c_i^{(0)}} < \exp\left(\frac{-\tau\Omega}{kT}\right)$$

The local interstitial concentration tends to slow down the jog motion. In other words, the chemical force opposes the mechanical one. High stresses can induce moderate climb velocities, in spite of a low $D_i^{(sd)}$. For a given dislocation, under the conditions of Section 8.1.4.1 (ii), this mechanism is an alternative to vacancy absorption assisted by the mechanical stress, because the large value of $\exp(-\tau\Omega/kT)$ can compensate for the small $D_i^{(sd)}$.

(ii) *Absorption of interstitials assisted by a chemical force:*

The net jump frequency is negative, i.e. $\exp(-\tau\Omega/kT) < c_i^{(j)}/c_i^{(0)}$. The dominant chemical force must be consistent with a negative net jump frequency, i.e. $c_i^{(j)}/c_i^{(0)} > 1$ (local excess of interstitials near the jogs).

A high supersaturation ($c_i^{(j)}/c_i^{(0)} \gg 1$) can induce rapid climb because $\nu_j^- \gg nD_i^{(sd)}/a^2$. Climb is, however, more difficult than for the absorption of vacancies because the self-diffusion coefficient $D_i^{(sd)}$ is smaller.

8.1.4.3 Summary. The mechanisms of jog–point defect interactions are not all equivalent and they operate in different situations.

In plasticity mechanisms, the main driving force is the applied stress. From the above, the jog mechanism at a dislocation changes when the stress direction is reversed. If it consists of vacancy emission it becomes either vacancy absorption or interstitial-emission (Figure 8.2). At high stresses, climb is less rapid after stress reversal because the two available mechanisms are slower. Consequently, *tension–compression asymmetries are expected under such conditions when plasticity is controlled by climb.*

During irradiation or after quenching the main driving force is a high concentration of vacancies or interstitials.

8.2. VACANCY EMISSION CLIMB MECHANISM

The calculations derived in this section are valid for mechanisms where the main driving force for climb is the applied stress, including the line tension forces on curved dislocations. In particular, we assume that there is no supersaturation of point defects (this situation is studied in Section 8.3). All equations of this section are also valid for interstitial-emission and vacancy-absorption climb, following the remarks of Sections 8.1.4.1 (ii) and 8.1.4.2 (i) (see conclusions of §8.2.3).

To obtain reliable values of the dislocation climb velocity, point defect migration away from the jogs, as well as their emission, is now considered.

To estimate the concentration of vacancies near jogs, $c_v^{(j)}$, we use the method proposed by Edelin (1971). The flux Φ_e of vacancies emitted by the jogs, computed by using Eq. (8.10a), is set equal to the flux Φ_d diffusing away from them. This condition allows the vacancy concentration, c_v, as a function of the distance from the jogs, to be determined and the dislocation velocity to be estimated without any assumption on the intensity of the chemical force F_c.

Different treatments are used depending on the density of jogs. At the beginning of the climb process, the latter may be determined by thermal equilibrium or by the initial dislocation shape. In the first case:

$$c_j = \exp\left(-\frac{U_j}{kT}\right) \tag{8.12a}$$

where U_j is the jog energy. The corresponding jog separation distance, x, is

$$x = \frac{a}{c_j} = a\exp\left(\frac{U_j}{kT}\right) \tag{8.12b}$$

At low stresses and high temperatures, the climb velocity is so low that the density of jogs remains close to thermal equilibrium, as in Figure 8.4(a). This is studied in Section 8.2.1. However, at higher stresses, jogs climb rapidly towards dislocation extremities. The jog concentration is no longer close to equilibrium and jog-pair nucleations must be considered (Figure 8.3(b) and (c)). This case will be dealt with in Section 8.2.2.

8.2.1 High jog density

This section considers dislocations climbing under high temperatures and low stresses. It includes dislocations with a small average curvature (Section 8.2.1.1) as well as small rounded prismatic loops (Section 8.2.1.2).

8.2.1.1 Climbing dislocations with a small average curvature. The small dislocation curvature is compatible with a large jog density of either sign, as illustrated in Figure 8.4(a). This density is assumed to be so high that the vacancy concentration along the whole dislocation core is uniform and equal to $c_v^{(j)}$. The diffusion process of point defects, from the dislocation to the crystal, has a cylindrical symmetry described schematically in Figure 8.5.

The velocity of diffusing vacancies is deduced from Eq. (8.6b):

$$\mathrm{v}_v = -D_v\frac{1}{c_v(r)}\frac{\partial c_v}{\partial r}$$

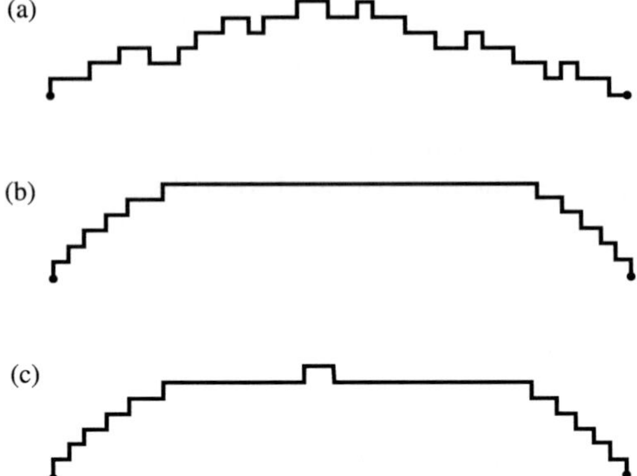

Figure 8.4. Jogged dislocations under stress. (a) Low stress, high temperature: high jog density at thermal equilibrium. (b) and (c) Low jog density at high stress and jog-pair nucleation.

The corresponding flux across the lateral surface of a cylinder of radius r and length L is:

$$\Phi_{\mathrm{d}} = \frac{c_{\mathrm{v}}(r)}{\Omega} 2\pi L r \mathrm{v}_{\mathrm{v}} = -2\pi \frac{D_{\mathrm{v}} L}{\Omega} \frac{\partial c_{\mathrm{v}}}{\partial r} r$$

The steady state condition $\partial \Phi_{\mathrm{d}}/\partial r = 0$ yields $r(\partial^2 c_{\mathrm{v}}/\partial r^2) + (\partial c_{\mathrm{v}}/\partial r) = 0$, which can be integrated as:

$$c_{\mathrm{v}}(r) = c_{\mathrm{v}}^{(0)} + (c_{\mathrm{v}}^{(\mathrm{j})} - c_{\mathrm{v}}^{(0)}) \frac{\ln R/r}{\ln R/a},$$

assuming that $c_{\mathrm{v}} = c_{\mathrm{v}}^{(\mathrm{j})}$ for $r = a$ and $c_{\mathrm{v}} = c_{\mathrm{v}}^{(0)}$ at large distances ($r = R$).

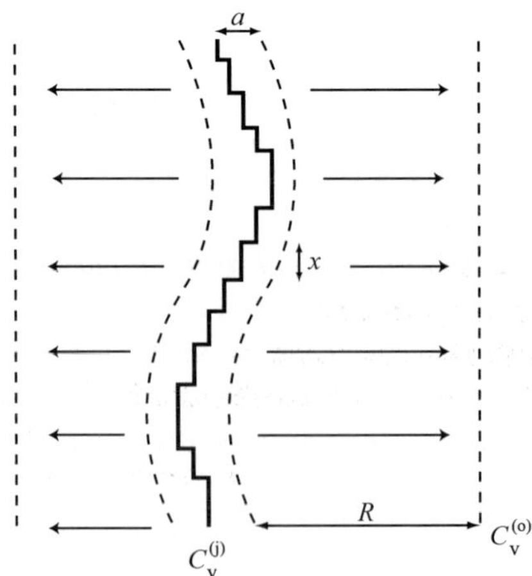

Figure 8.5. Cylindrical diffusion of vacancies emitted by a dislocation containing a high density of jogs.

Hence:

$$\Phi_{\mathrm{d}} = \frac{2\pi}{\ln(R/a)} \frac{D_{\mathrm{v}}L}{\Omega}(c_{\mathrm{v}}^{(\mathrm{j})} - c_{\mathrm{v}}^{(0)}) \tag{8.13a}$$

or

$$\Phi_{\mathrm{d}} \approx \frac{L}{\Omega} D_{\mathrm{v}}^{(\mathrm{sd})} \left(\frac{c_{\mathrm{v}}^{(\mathrm{j})}}{c_{\mathrm{v}}^{(0)}} - 1 \right) \tag{8.13b}$$

assuming $2\pi/(\ln(R/a)) \approx 1$ and considering Eq. (8.5).

The flux Φ_{e} emitted at the dislocation is:

$\Phi_{\mathrm{e}} = (\nu_{\mathrm{j}}^{+} - \nu_{\mathrm{j}}^{-})L/x$, where $\nu_{\mathrm{j}}^{+} - \nu_{\mathrm{j}}^{-}$ is given by Eq. (8.10a), whence:

$$\Phi_{\mathrm{e}} = \frac{nD_{\mathrm{v}}^{(\mathrm{sd})}L}{a^2 x} \left(\exp\left(\frac{\tau\Omega}{kT} \right) - \frac{c_{\mathrm{v}}^{(\mathrm{j})}}{c_{\mathrm{v}}^{(0)}} \right) \tag{8.14}$$

Then, setting $\Phi_{\mathrm{d}} = \Phi_{\mathrm{e}}$ yields the vacancy concentration at jogs, $c_{\mathrm{v}}^{(\mathrm{j})}$:

$$\frac{c_{\mathrm{v}}^{(\mathrm{j})}}{c_{\mathrm{v}}^{(0)}} - 1 = \frac{1}{1 + \dfrac{a^2 x}{n\Omega}} \left(\exp\left(\frac{\tau\Omega}{kT} \right) - 1 \right) \tag{8.15}$$

Inserting (8.15) into (8.13b) yields:

$$\Phi_{\mathrm{d}} = \frac{L}{\Omega} \frac{D_{\mathrm{v}}^{(\mathrm{sd})}}{1 + \dfrac{a^2 x}{n\Omega}} \left(\exp\left(\frac{\tau\Omega}{kT} \right) - 1 \right)$$

The diffusing flux is thus expressed as a function of stress and temperature only. The dislocation velocity can now be computed, considering that for each emitted vacancy the dislocation sweeps the area Ω/b. The total area swept per unit time is thus $\Phi_{\mathrm{d}}\Omega/b$ and the corresponding dislocation velocity is:

$$\mathrm{v} = \frac{\Phi_{\mathrm{d}}\Omega}{Lb} = \frac{D_{\mathrm{v}}^{(\mathrm{sd})}}{b} \frac{1}{1 + \dfrac{a^2 x}{n\Omega}} \left(\exp\left(\frac{\tau\Omega}{kT} \right) - 1 \right) \tag{8.16a}$$

For $\tau\Omega \ll kT$, it becomes:

$$\mathrm{v} = \frac{D_{\mathrm{v}}^{(\mathrm{sd})}}{b} \frac{1}{1 + \dfrac{a^2 x}{n\Omega}} \frac{\tau\Omega}{kT} \tag{8.16b}$$

Assuming $\Omega \approx a^3$ and $b \approx a$, it reduces to

$$\mathrm{v} \approx \frac{D_{\mathrm{v}}^{(\mathrm{sd})}}{a + \dfrac{x}{n}} \frac{\tau a^3}{kT} \tag{8.16c}$$

Under these conditions, and using relations (8.9b) and (8.15), the chemical force is

$$F_c^{(j)} \approx \frac{kTb}{\Omega}\left(\frac{c_V^{(j)}}{c_V^{(0)}} - 1\right) = \frac{\tau b}{1 + \dfrac{a^2 x}{n\Omega}} \tag{8.17}$$

When the density of jogs is very high ($x \approx a \Rightarrow x/n \ll a$), the chemical force $F_c^{(j)}$ approaches the mechanical force τb and opposes it. The dislocation is said to be completely saturated with vacancies. The total driving force on jogs is zero and the dislocation velocity is entirely controlled by the lattice diffusion process, with

$$\mathrm{v} = \frac{D_V^{(sd)}}{b}\frac{\tau\Omega}{kT} \tag{8.18}$$

where relation (8.16b) has been used.

In the latter case, v is characterized by an activation energy $U_{sd}^{(v)}$ and a linear dependence on stress. The condition $x/n \ll a$ is, however, difficult to satisfy, because it corresponds approximately to one jog per atomic distance on the dislocation.

For a lower jog density, such that $x/n \gg a$, the assumption of a cylindrical diffusion is no longer valid and a different treatment must be used (see Section 8.2.2).

8.2.1.2 Growth or shrinking of small prismatic dislocation loops. A small loop contains a high jog density. It is assumed to be a torus of radius r and circular cross-section of diameter a, containing a uniform density of vacancies $c_V^{(j)}$.

The flux diffusing out of the torus has been computed by Seidman and Balluffi (1966) and is:

$$\Phi_d = \frac{4\pi^2\sqrt{3}}{\ln\dfrac{8r}{a}}\, r\, \frac{D_V^{(sd)}}{\Omega}\left(\frac{c_V^{(j)}}{c_V^{(0)}} - 1\right)$$

Taking $2\pi\sqrt{3}/\ln(8r/a) \approx 1$), we obtain

$$\Phi_d = 2\pi r\frac{D_V^{(sd)}}{\Omega}\left(\frac{c_V^{(j)}}{c_V^{(0)}} - 1\right)$$

(same expression as (8.13b) with $L = 2\pi r$)

The flux emitted by the jogs is estimated as in the above case and is:

$$\Phi_e = \frac{nD_V^{(sd)} 2\pi r}{a^2 x}\left(\exp\left(\frac{\tau\Omega}{kT}\right) - \frac{c_V^{(j)}}{c_V^{(0)}}\right)$$

(This expression is similar to (8.14)),

Then, the condition $\Phi_d = \Phi_e$ yields:

$$v = \frac{\Phi_d \Omega}{2\pi r b} = \frac{D_v^{(sd)}}{b} \frac{1}{1 + \dfrac{a^2 x}{n\Omega}} \left(\exp\left(\frac{\tau\Omega}{kT} \right) - 1 \right)$$

where x is given by Eq. (8.12b).

The result is the same as for straighter dislocations (Eq. (8.16a)). v is also given by Eqs. (8.16b) and (8.16c) under the corresponding hypotheses and by Eq. (8.18) for $x/n \ll a$. Due to the small radius of curvature, the mechanical stress contains a high line tension component.

8.2.2 Low jog density

This situation is met at lower temperatures and higher stresses because high stresses tend to sweep jogs away and push them towards the dislocation ends (Figure 8.4(b)). Dislocation segments are straight and parallel to dense crystallographic rows. The diffusion process is assumed to take place independently at each jog. Pipe diffusion is ignored in the first step, for the sake of simplification, and will be included later on (Section 8.2.2.2).

8.2.2.1 No pipe diffusion. The diffusion process now has a spherical symmetry, centred around each jog as described schematically in Figure 8.6. The distance R at which $c_v = c_v^{(0)}$ is assumed to be smaller than the jog distance x.

Following the same procedure as above, and using Eq. (8.3b), the diffusion flux across the surface of a sphere of radius r is:

$$\Phi_d = \frac{c_v(r)}{\Omega} 4\pi r^2 v_v = -\frac{4\pi D_v}{\Omega} r^2 \frac{\partial c_v}{\partial r}$$

The steady state condition yields $r(\partial^2 c_v/\partial r^2) + (2\partial c_v/\partial r) = 0$, which can be integrated and yields:

$$c_v(r) \approx c_v^{(0)} + \frac{a(c_v^{(j)} - c_v^{(0)})}{r},$$

assuming that $c_v = c_v^{(j)}$ for $r = a$ and $c_v \approx c_v^{(0)}$ for large values of $r (r = R)$.

This yields:

$$\Phi_d = 4\pi \frac{D_v a}{\Omega} (c_v^{(j)} - c_v^{(0)}) = 4\pi \frac{D_v^{(sd)} a}{\Omega} \left(\frac{c_v^{(j)}}{c_v^{(0)}} - 1 \right) \tag{8.19}$$

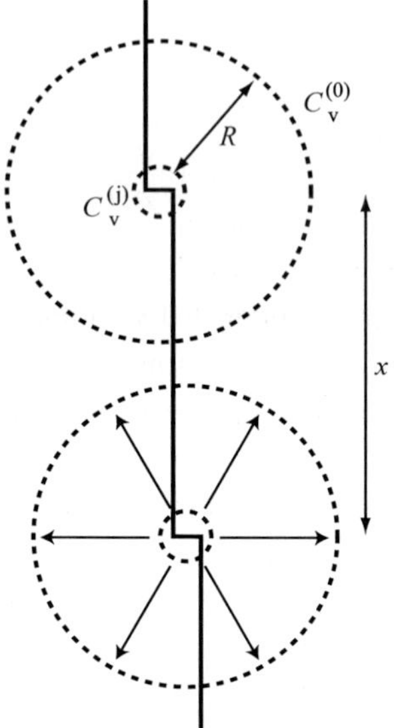

Figure 8.6. Spherical diffusion of vacancies emitted by individual jogs separated by large distances.

This flux is equal to the flux emitted by one jog, given by Eq. (8.10a), whence:

$$\frac{c_v^{(j)}}{c_v^{(0)}} - 1 = \frac{n}{n + 4\pi}\left(\exp\left(\frac{\tau\Omega}{kT}\right) - 1\right) \tag{8.20}$$

Inserting (8.20) into (8.19) allows the dislocation climb velocity to be determined:

$$v = \Phi_d \frac{\Omega}{bx} = \frac{4\pi n}{n + 4\pi} \frac{a}{bx} D_v^{(sd)}\left(\exp\left(\frac{\tau\Omega}{kT}\right) - 1\right) \tag{8.21a}$$

For $\tau\Omega \ll kT$, it becomes:

$$v = \frac{4\pi n}{n + 4\pi} \frac{a}{bx} D_v^{(sd)} \frac{\tau\Omega}{kT} \tag{8.21b}$$

Assuming $\Omega \approx a^3$, $b = a$ and $n \approx 4\pi$, it reduces to

$$v \approx 2\pi \frac{D_v^{(sd)}}{x} \frac{\tau a^3}{kT} \tag{8.21c}$$

This velocity is two times smaller than that given by Eq. (8.16c) in the limit $x/n \gg a$. For large x values, the present approximation is, however, better justified.

The chemical force can be deduced from Eqs. (8.9b) and (8.20). At low stresses and high temperatures ($\Omega\tau \ll kT$), $c_v^{(j)}/c_v^{(0)}$ is necessarily close to unity and the chemical force is:

$$F_c^{(j)} = \frac{n}{n + 4\pi}\tau b \approx \frac{\tau b}{2} \tag{8.22}$$

$F_c^{(j)}$ is half the value corresponding to saturation. The jogs are thus not saturated with vacancies. This arises from the fact that a spherical diffusion ensures a fairly rapid decrease of the local excess of point defects.

These results are modified by the presence of pipe diffusion which is examined now.

8.2.2.2 The role of pipe diffusion. In experimental studies of creep, pipe diffusion is often considered whenever the measured activation energy is lower than the self-diffusion energy. Therefore, the corresponding mechanism is now critically studied. Pipe diffusion has been treated by Balluffi (1969) and Hirth and Lothe (1982, p. 571). Diffusion of vacancies is expected to be faster, with a smaller activation energy $U_d^{(v)} - \Delta U_d^{(v,p)}$, along the core of dislocations where the crystal is heavily distorted. The pipe diffusion coefficient $D_v^{(p)}$ is accordingly higher than the bulk one:

$$D_v^{(p)} = D_v \exp\left(\frac{\Delta U_d^{(v,p)}}{kT}\right) \tag{8.23}$$

The energy to create a vacancy in the core is $U_f^{(v)} - \Delta U_f^{(v)}$ and thus the equilibrium concentration of vacancies along the core, $c_v^{(0,p)}$, is higher than in the bulk material:

$$c_v^{(0,p)} = c_v^{(0)} \exp\left(\frac{\Delta U_f^{(v,p)}}{kT}\right) \tag{8.24}$$

When a vacancy is emitted at a jog, it diffuses rapidly along the core over an average distance Δx before being transferred into the surrounding perfect lattice (Figure 8.7(a)). The mean free path Δx can be estimated as follows:

The average drift velocity of vacancies along the core is given by Eq. (8.6b):

$$\mathrm{v}_v = D_v^{(p)} \frac{1}{c_v^{(p)}} \frac{\partial c_v^{(p)}}{\partial x} = \frac{D_v^{(p)}}{\mathrm{v}_v} \frac{1}{c_v^{(p)}} \frac{\partial c_v^{(p)}}{\partial t} \tag{8.25}$$

To extract a vacancy from the core, it is necessary to complement its formation energy by $\Delta U_f^{(v,p)}$ and to move it away in the perfect crystal. The corresponding activation energy is $U_d^{(v)} + \Delta U_f^{(v,p)}$. Accordingly, the average life-time of a vacancy in the core is $\Delta t = 1/\nu_D \exp(U_d^{(v)} + \Delta U_f^{(v,p)})/kT)$ and the concentration of vacancies emitted at $t = 0$ at the jog decreases exponentially as they drift away according to:

$$c_v^{(p)} = c_v^{(j,p)} \exp\left(-\frac{\Delta t}{t}\right) \tag{8.26}$$

where $c_v^{(j,p)}$ is the vacancy concentration at the jog, when pipe diffusion is considered.

(a)

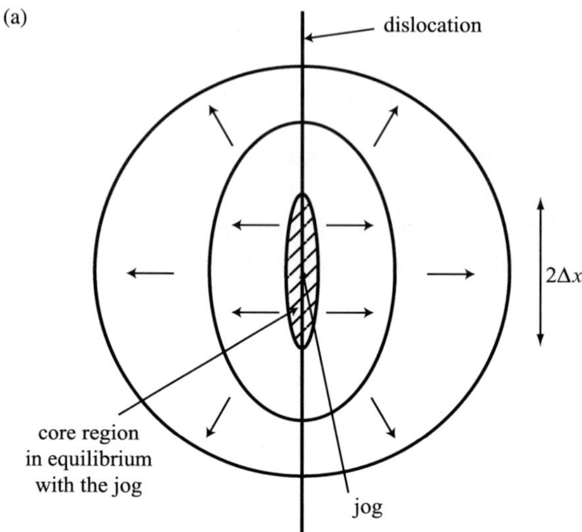

(b)

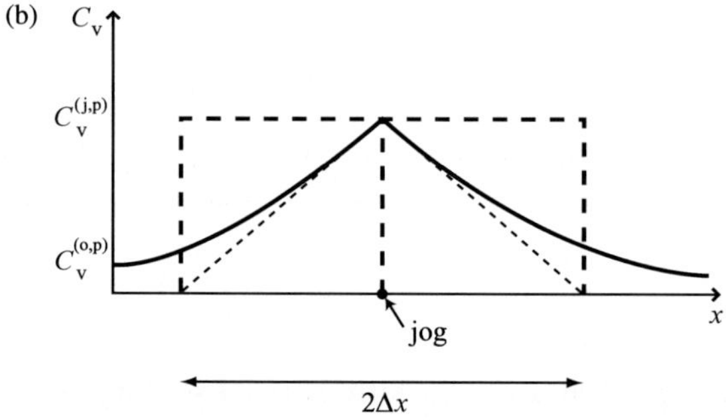

Figure 8.7. Jog climb assisted by pipe diffusion. (a) Diffusion of vacancies emitted by the jog. The hatched core region contains the vacancy concentration $c_v^{(j,p)}$. Arrows indicate volume diffusion: (b) Real (full line) and approximate (thick dotted line) vacancy concentration profiles along the core.

Inserting this relation into (8.25) yields a constant drift velocity in the core of $v_v = (D_v^{(p)}/\Delta t)^{1/2}$.

Then, using the variable $x = v_v t$ in Eq. (8.26) yields

$$c_v^{(p)} = c_v^{(j,p)} \exp\left(-\frac{x}{\Delta x}\right), \text{ with } \Delta x = (D_v^{(p)}\Delta t)^{1/2} \tag{8.27}$$

Using Eqs. (8.5), (8.23) and the above expression of Δt, the mean-free-path becomes:

$$\Delta x = a \exp\left(\frac{\Delta U_d^{(v,p)} + \Delta U_f^{(v,p)}}{2kT}\right), \quad \text{or} \quad \Delta x = a \exp\left(\frac{\Delta U_{sd}^{(v,p)}}{2kT}\right),$$
$$\text{with} \quad \Delta U_{sd}^{(v,p)} = \Delta U_f^{(v,p)} + \Delta U_d^{(v,p)} = U_{sd}^{(v)} - U_{sd}^{(v,p)} \tag{8.28}$$

In order to estimate the vacancy concentration near the jog, $c_{\rm v}^{(\rm j,p)}$, we write, as previously, that the flux of vacancies emitted by the jog in the dislocation core, $\Phi_{\rm e}^{(\rm p)}$, is equal to the flux diffusing along the core, $\Phi_{\rm d}^{(\rm p)}$. Using Eq. (8.25), and taking into account that vacancies diffuse away from the jog in two opposite directions along the dislocation line, the diffusing flux is:

$$\Phi_{\rm d}^{(\rm p)} = 2D_{\rm v}^{(\rm p)}\frac{1}{a}\frac{\partial c_{\rm v}^{(\rm p)}}{\partial x} = 2\frac{D_{\rm v}^{(\rm p)}}{a\,\Delta x}\left(c_{\rm v}^{(\rm j,p)} - c_{\rm v}^{(0,\rm p)}\right) = 2\frac{D_{\rm v}^{(\rm sd,p)}}{a\,\Delta x}\left(\frac{c_{\rm v}^{(\rm j,p)}}{c_{\rm v}^{(0,\rm p)}} - 1\right) \tag{8.29}$$

$$\text{with } D_{\rm v}^{(\rm sd,p)} = c_{\rm v}^{(0,\rm p)}D_{\rm v}^{(\rm p)} = a^2\nu_{\rm D}\exp\left(-\frac{U_{\rm sd}^{(\rm v,p)}}{kT}\right)$$

From Eq. (8.10a), the emitted flux is (with $n = 2$ along the dislocation line):

$$\Phi_{\rm e}^{(\rm p)} = 2\frac{D_{\rm v}^{(\rm sd,p)}}{a^2}\left(\exp\left(\frac{\tau\Omega}{kT}\right) - \frac{c_{\rm v}^{(\rm j,p)}}{c_{\rm v}^{(0,\rm p)}}\right)$$

Then, taking into account that $\Delta x \gg a/2$, the condition $\Phi_{\rm d}^{(\rm p)} = \Phi_{\rm e}^{(\rm p)}$ yields:

$$c_{\rm v}^{(\rm j,p)} = c_{\rm v}^{(0,\rm p)}\exp\left(\frac{\tau\Omega}{kT}\right) \tag{8.30}$$

According to Eqs. (8.9a) and (8.30), the chemical force on the jogs is $F_{\rm c}^{(\rm j)} = \tau b$. The jogs are thus always saturated with vacancies, although they are separated by large distances unlike in Section 8.2.1.1. This happens because vacancy core-diffusion is a one-dimensional process, whereas it is a two dimensional process in the case of cylindrical diffusion and a three dimensional process in the case of spherical diffusion.

In conclusion, the equilibrium vacancy concentration along the core decreases from $c_{\rm v}^{(\rm j,p)} = c_{\rm v}^{(0,\rm p)}\exp(\tau\Omega/kT)$ at the jog to $c_{\rm v}^{(0,\rm p)}$ at distance Δx, on both sides of the jog. We approximate it by a uniform density $c_{\rm v}^{(\rm j,p)}$ over the length $2\,\Delta x$ (Figure 8.6(b)).

The last step is to compute the diffusion of this one-dimensional distribution of vacancies into the three-dimensional surrounding crystal. Just after escaping the core, the vacancy energy increases by $\Delta U_{\rm f}$. Accordingly, its concentration falls to:

$$c_{\rm v}^{(\rm j,p)}\exp\left(-\frac{\Delta U_{\rm f}}{kT}\right) = c_{\rm v}^{(\rm j,p)}\frac{c_{\rm v}^{(0)}}{c_{\rm v}^{(0,\rm p)}} = c_{\rm v}^{(0)}\exp\left(\frac{\tau\Omega}{kT}\right).$$

If r is the distance to the jog, the diffusing flux $\Phi_{\rm d}$ has a cylindrical symmetry close to the jog ($r < 2\,\Delta x$) and a spherical symmetry far from the jog ($r > 2\,\Delta x$) (see Figure 8.7(a)).

For $r < 2\,\Delta x$ (cylindrical diffusion), the same calculations as in Section 8.2.1.1 yield the concentration profile

$$c_{\rm v}(r) = c_{\rm v}^* + \left(c_{\rm v}^{(0)}\exp\left(\frac{\tau\Omega}{kT}\right) - c_{\rm v}^*\right)\frac{\ln 2\,\Delta x/r}{\ln 2\,\Delta x/a}$$

where c_v^* is the concentration for $r = 2\,\Delta x$ and the corresponding flux is

$$\Phi_d = \frac{2\pi}{\ln\frac{2\,\Delta x}{a}} \frac{D_v 2\,\Delta x}{\Omega}\left(c_v^{(0)} \exp\frac{\tau\Omega}{kT} - c_v^*\right) \tag{8.31}$$

(expression similar to (8.13a)).

For $r > 2\,\Delta x$ (spherical diffusion), the same calculations as in Section 8.2.2.1 yield the concentration profile $c_v(r) = c_v^{(0)} + (2\,\Delta x(c_v^* - c_v^{(0)}))/r$ and the corresponding flux is

$$\Phi_d = 4\pi\frac{D_v 2\,\Delta x}{\Omega}(c_v^* - c_v^{(0)}) \tag{8.32}$$

(expression similar to (8.19)).

Then, eliminating c_v^* between Eqs. (8.31) and (8.32) yields

$$\Phi_d = \frac{4\pi}{\ln\left(\frac{2\,\Delta x}{a}\right) + \frac{1}{2}} D_v \frac{\Delta x}{\Omega} c_v^{(0)}\left(\exp\left(\frac{\Omega\tau}{kT}\right) - 1\right) \tag{8.33}$$

When $\Delta x \gg 0$, the corresponding dislocation velocity is close to:

$$\mathrm{v} = \Phi_d\frac{\Omega}{bx} = \frac{4\pi}{\ln\frac{\Delta x}{a}} \frac{\Delta x}{ax} D_v^{(sd)}\left[\exp\left(\frac{\tau\Omega}{kT}\right) - 1\right]$$

$$= \frac{4\pi}{\ln\frac{\Delta x}{a}} \frac{D_v^{(sd)}}{x}\left[\exp\left(\frac{\tau\Omega}{kT}\right) - 1\right]\exp\left(\frac{\Delta U_{sd}^{(v,p)}}{2kT}\right) \tag{8.34a}$$

For small stresses ($\tau\Omega/kT \ll 1$) it reduces to

$$\mathrm{v} = \frac{4\pi}{\ln\frac{\Delta x}{a}} \frac{D_v^{(sd)}}{x} \frac{\tau\Omega}{kT} \exp\left(\frac{\Delta U_{sd}^{(v,p)}}{2kT}\right) \tag{8.34b}$$

This expression is valid as long as the distance x between the jogs is larger than Δx, i.e. when the diffusion segments of length Δx do not overlap. It replaces relation (8.21c) (no pipe diffusion) when $\Delta x > a$ but it is equivalent to this equation when Δx amounts to a few times a (negligible pipe diffusion). When $\Delta x > x$ the effect of pipe diffusion vanishes and Eq. (8.16) (cylindrical diffusion) must be used.

8.2.2.3 Jog-pair nucleations. The jog density a/x must now be estimated. When the climb velocity is low, and the stress moderate, the jog concentration is close to thermal equilibrium. Combining Eqs. (8.12b) and (8.34b), the climb velocity is:

$$\mathrm{v} = \frac{4\pi}{\ln\dfrac{\Delta x}{a}} \nu_{\mathrm{D}} a \frac{\tau\Omega}{kT} \exp\left(-\frac{U_{\mathrm{sd}}^{(\mathrm{v})} + U_{\mathrm{j}} - (\Delta U_{\mathrm{sd}}^{(\mathrm{v,p})}/2)}{kT}\right) \tag{8.35}$$

When the climb velocity is high, namely when the stress is high (but still fulfils the condition $\tau\Omega \ll kT$), nucleations of jog-pairs on straight dislocations must be considered (Figure 8.4(b) and (c)). This process is similar to the nucleation of kink-pairs on gliding dislocations in covalent materials (Chapter 7).

The elastic energy of a jog-pair is $U_{\mathrm{jp}}(\tau) = U_{\mathrm{ik}}(\tau)$ given by Eq. (4.34). Its maximum value for the critical jog separation x_{c} is $U_{\mathrm{jp}}^{(\mathrm{c})}(\tau)$, given by Eq. (4.36). As in covalent materials, jog-pairs move across this energy profile by a diffusion process.

Since the stress is high, and the jog density low, the jog velocity is expressed in the frame of the pipe diffusion model.

From Eq. (8.34), it is:

$$\mathrm{v}_{\mathrm{j}} = \mathrm{v}\frac{X}{a} = \frac{D_{\mathrm{j}}}{a}\frac{\Omega\tau}{kT} \tag{8.36}$$

where D_{j} is the diffusion coefficient of jogs along the dislocation line:

$$D_{\mathrm{j}} = \frac{4\pi}{\ln\dfrac{\Delta x}{a}} D_{\mathrm{v}}^{(\mathrm{sd})} \exp\left(\frac{\Delta U_{\mathrm{sd}}^{(\mathrm{v,p})}}{2kT}\right)$$

It is worth noting that replacing D_{k} by D_{j} in Eq. (7.21) results in two similar regimes for the dislocation velocity, as a function of its length, L. Similarly, the mean free path of a jog issued from a pair is:

$$X = a\exp\left(\frac{U_{\mathrm{jp}}^{(\mathrm{c})}(\tau)}{2kT}\right) \tag{8.37}$$

X here plays the role of the average jog distance along the line.

For $L > X$, the dislocation climb velocity is:

$$\mathrm{v} = \frac{4\pi}{\ln\dfrac{\Delta x}{a}} \nu_{\mathrm{D}} a \frac{\tau\Omega}{kT} \exp\left(-\frac{U_{\mathrm{sd}}^{(\mathrm{v,p})} + \frac{1}{2}U_{\mathrm{jp}}^{(\mathrm{c})}(\tau) - \frac{1}{2}\Delta U_{\mathrm{sd}}^{(\mathrm{v,p})}}{kT}\right),$$

$$\text{or} \quad \mathrm{v} = \frac{4\pi}{\ln\dfrac{\Delta x}{a}} \frac{D_{\mathrm{v}}^{(\mathrm{sd})}}{X} \frac{\tau\Omega}{kT} \exp\left(-\frac{\Delta U_{\mathrm{sd}}^{(\mathrm{v,p})}}{2kT}\right) \tag{8.38a}$$

This expression is similar to Eq. (8.34b) for low stresses because in the latter case $U_{\mathrm{jp}}^{(\mathrm{c})} \approx 2U_{\mathrm{j}}$, whence $X \approx x$.

For $L < X$, the dislocation climb velocity is:

$$\mathrm{v} = \frac{4\pi}{\ln\dfrac{\Delta x}{a}} \nu_{\mathrm{D}} L \frac{\tau\Omega}{kT} \exp\left(-\frac{U_{\mathrm{sd}}^{(\mathrm{v})} + U_{\mathrm{jp}}^{(\mathrm{c})}(\tau) - (\Delta U_{\mathrm{sd}}^{(\mathrm{v,p})}/2)}{kT}\right),$$

$$\text{or} \quad \mathrm{v} = \frac{4\pi}{\ln\dfrac{\Delta x}{a}} \frac{D_{\mathrm{v}}^{(\mathrm{sd,p})} L}{X^2} \frac{\tau\Omega}{kT} \exp\left(-\frac{\Delta U_{\mathrm{sd}}^{(\mathrm{v,p})}}{2kT}\right) \tag{8.38b}$$

This expression differs from Eq. (8.38a) by its higher activation energy compensated by its higher pre-exponential term. *A length effect similar to that occurring for the kink-pair mechanism is expected in this case.*

It is important to note that although passing over the critical jog-pair configuration of size x_c requires the emission of several vacancies, the energy $U_{\mathrm{sd}}^{(\mathrm{v})}$ is involved only once in the total activation energy. This is because the vacancies are not emitted in a single thermally activated event but in several sequential steps. The same remark holds true for covalent materials, in which several bonds must be cut to create a kink-pair of critical size, but the total activation energy involves only once the activation energy U_{m} corresponding to one elementary kink movement. Since the pipe diffusion mechanism requires the condition $X > \Delta x$, this implies that $(1/2)U_{\mathrm{jp}}^{(\mathrm{c})} \approx U_{\mathrm{j}} > (1/2)\Delta U_{\mathrm{sd}}^{(\mathrm{v,p})}$ (Eqs. (8.28) and (8.37)), the total activation energies in Eq. (8.38a and b) are larger than $U_{\mathrm{sd}}^{(\mathrm{v})}$ although pipe diffusion takes place.

8.2.2.4 ***Stress dependence of the climb velocity.*** The total stress τ, including the applied stress and the line tension stress, appears in Eqs. (8.38a) and (8.38b) both in the pre-exponential term and in the activation energy.

As in the kink-pair mechanism (Eq. (4.37)), we have:

$$\frac{\partial U_{\mathrm{jp}}^{(\mathrm{c})}}{\partial \tau} = -(hb)^{3/2}\sqrt{\frac{\mu}{8\pi\tau}},$$

where h is the jog height.

Then, the apparent stress exponent of the dislocation velocity, $m = \partial(\ln \mathrm{v})/\partial(\ln \tau)$, is:

$$\text{For } L > X: \quad m = 1 + \frac{\tau}{2kT}\frac{\partial U_{\mathrm{jp}}^{(\mathrm{c})}}{\partial \tau} \text{ or } m = 1 + \frac{1}{4\sqrt{2\pi}}\frac{\sqrt{\mu\tau}(hb)^{3/2}}{kT} \tag{8.39a}$$

and

$$\text{For } L < X: \ m = 1 + \frac{\tau}{kT}\frac{\partial U_{\mathrm{jp}}^{(\mathrm{c})}}{\partial \tau} \text{ or } m = 1 + \frac{1}{2\sqrt{2\pi}}\frac{\sqrt{\mu\tau}(hb)^{3/2}}{kT} \tag{8.39b}$$

The second term in Eq. (8.39) may be important for large stresses so that m can be significantly larger than 1. For example, taking $\mu(hb)^{3/2} \approx \mu a^3 \approx 10$ eV and $kT \approx 0.1$ eV, we obtain that $m > 2$ for $\tau > \mu/400$ (namely $\Omega\tau > kT/4$) in the case $L < x$. The assumption $\Omega\tau \ll kT$ is not verified in this case, but this order-of-magnitude estimate remains significant for $\Omega\tau < kT/2$ (see Section 8.2.3 below). The stress exponent m tends to 1 at small stresses, in agreement with Eq. (8.35).

8.2.3 Conclusion on the vacancy-emission climb mechanism

These conclusions are also valid for insterstitial-emission and vacancy-absorption climb mechanisms, according to Sections 8.1.4.1 (ii) and 8.1.4.2 (i). Equations for vacancy-absorption climb can also be deduced from Section 8.3, with $c^{(s)} = c^{(0)}$.

The validity domains of the various mechanisms described above are now discussed as a function of stress. At high stresses, the average jog distance is $X = a \exp(U_{jp}^{(c)}(\tau)/2kT)$ (Eq. (8.37)) while at low stresses, it is $x \approx a \exp(U_j/kT)$ (Eq. (8.12b)). When pipe diffusion is effective, the diffusion length along dislocation cores is $\Delta x = a \exp(\Delta U_{sd}^{(v,p)}/2kT)$ (Eq. (8.28)).

In the following we assume $\tau\Omega \ll kT$. All results are summarized in Table 8.1.

In the extreme case of very high stresses ($\tau\Omega \sim kT$), the two columns on the right of Table 8.1 are still valid, provided $\tau\Omega/kT$ is replaced by $\exp(\tau\Omega/kT) - 1$ in the expression of the dislocation velocity. Similarly, in the expression of m, the term 1 is replaced by

$$\frac{\tau\Omega}{kT} \frac{\exp(\tau\Omega/kT)}{\exp(\tau\Omega/kT) - 1} > 1.$$

For mixed dislocations, the mechanical stress, τ, equals the applied stress times the corresponding Schmid factor (equal, in this case, to $\cos\alpha \cos\beta$ where α and β are the angles between the applied stress and, respectively, the Burgers vector and the normal to the plane of motion). The atomic concentration of point defects created along this plane of motion is lower than for pure climb. Accordingly, the area swept by jogs, for each vacancy (or interstitial) created or absorbed, is multiplied by $\cos\Psi$, where Ψ is the angle between the Burgers vector and the plane of motion. Therefore, the work term $\Omega\tau/kT$ in Eq. (8.10), is $(\cos\Psi)^{-1}$ times higher. For geometrical reasons, the dislocation velocity is multiplied by $(\cos\Psi)^{-1}$ a second time (Eq. (8.16)), so that v is multiplied by $(\cos\Psi)^{-2}$.

The important predictions of Table 8.1 are:

- *The climb activation energy is predicted to be equal to, or larger than, the bulk self-diffusion energy. This conclusion is also valid in the presence of pipe diffusion.*
- *At large stresses, the velocity-stress exponent* $m = \partial(\ln v)/\partial(\ln \tau)$ *is definitely larger than 1.*

Table 8.1. Vacancy emission climb mechanisms without supersaturation effects: dislocation velocity and activation parameters.

Conditions	Low stress high jog density, no effect of pipe diffusion ($x < \Delta x \Leftrightarrow U_j < (1/2)\Delta U_{sd}^{(v,p)}$)		High stress (but still $\Omega\tau \ll kT$) low jog density, pipe diffusion ($X > \Delta x \Leftrightarrow U_{jp}^{(c)} > \Delta U_{sd}^{(v,p)}$, $X > na$)	
	Saturation of the whole dislocation ($x < na$)	Partial saturation ($na < x < \Delta x$)	No length effect ($L > X$)	Length effect ($L < X$)
Dislocation velocity	$v = \frac{D_v^{(sd)}}{b}\frac{\tau\Omega}{kT}$ (Eq. (8.18))	$v = \frac{D_v^{(sd)}}{b}\frac{1}{1+\frac{a^2x}{n\Omega}}\frac{\tau\Omega}{kT}$ (Eq. (8.16b))	$v = \frac{4\pi}{\ln\frac{\Delta x}{a}}\frac{D_v^{(sd,p)}}{X}\frac{\tau\Omega}{kT}$ (Eq. (8.38a))	$v = \frac{4\pi}{\ln\frac{\Delta x}{a}}\frac{D_v^{(sd,p)}L}{X^2}\frac{\tau\Omega}{kT}$ (Eq. (8.38b))
Activation energy	$U_{sd}^{(v)}$	Between $U_{sd}^{(v)}$ and $U_{sd}^{(v)} + U_j$	$U_{sd}^{(v)} + \frac{U_{jp}^{(c)}}{2} - (\Delta U_{sd}^{(v,p)}/2) > U_{sd}^{(v)}$	$U_{sd}^{(v)} + U_{jp}^{(v)} - (\Delta U_{sd}^{(v,p)}/2) > U_{sd}^{(v)}$
Stress exponent	$m = 1$	$m = 1$	$m \approx 1 + \frac{a^3}{4\sqrt{2\pi}}\frac{\sqrt{\mu\tau}}{kT}$	$m \approx 1 + \frac{a^3}{2\sqrt{2\pi}}\frac{\sqrt{\mu\tau}}{kT}$

– For low jog densities and small dislocation lengths (such that $L < x$), the dislocation climb velocity is proportional to L and the pre-exponential term is several orders of magnitude larger than in the other cases.

All the above results are also valid for interstitial-emission climb mechanisms.

8.3. VACANCY OR INTERSTITIAL-ABSORPTION CLIMB MECHANISM

This situation corresponds to climb mechanisms where the main driving force is a supersaturation of point defects at large distances from the dislocations. Large supersaturations (concentration $c^{(s)}$) are usually due to quenching (vacancies) or irradiation (interstitials or vacancies). Contrary to the vacancy (or interstitial) emission climb mechanism, the flux towards the dislocation is taken as positive. The mechanical stress superimposed on the chemical stress may be an applied stress (e.g. creep under irradiation) or the line tension of curved dislocation loops (annealing conditions). It is considered as positive when acting in the same direction as the supersaturation effect. Equations for climb with no supersaturation are obtained by setting $c^{(s)} = c^{(0)}$.

The relations are the same as in Section 8.2, but the concentration at a large distance from the jogs that is involved in the diffusion flux, Φ_d, is $c_v^{(s)}$ instead of $c_v^{(0)}$ (case of vacancies).

8.3.1 High jog density (e.g. curved dislocations)

The following relations are valid for either interstitials or vacancies.

The flux that diffuses towards the dislocation of length L is:

$$\Phi_{\mathrm{d}} = \frac{L}{\Omega} D^{(\mathrm{sd})} \frac{c^{(\mathrm{s})} - c^{(\mathrm{j})}}{c^{(0)}} \tag{8.40}$$

equivalent to Eq. (8.13b).

It is equal to the flux Φ_{abs} absorbed by the dislocation:

$\Phi_{\mathrm{abs}} = (\nu_{\mathrm{j}}^{-} - \nu_{\mathrm{j}}^{+})L/x$, where $\nu_{\mathrm{j}}^{-} - \nu_{\mathrm{j}}^{+}$ is given by Eq. (8.10a), whence:

$$\Phi_{\mathrm{abs}} = \frac{nD^{(\mathrm{sd})}}{a^2} \frac{L}{x} \left[\frac{c^{(\mathrm{j})}}{c^{(0)}} - \exp\left(\frac{-\tau\Omega}{kT} \right) \right] \tag{8.41}$$

equivalent to Eq. (8.14).

The equilibrium between these two fluxes yields:

$$\mathrm{v} = \frac{D^{(\mathrm{sd})}}{b} \frac{1}{1 + \dfrac{a^2 x}{\Omega n}} \left[\frac{c^{(\mathrm{s})}}{c^{(0)}} - \exp\left(-\frac{\tau\Omega}{kT} \right) \right] \tag{8.42a}$$

equivalent to Eq. (8.16a).

For $\Omega\tau \ll kT$, it reduces to:

$$\mathrm{v} = \frac{D^{(\mathrm{sd})}}{b} \frac{1}{1 + \dfrac{a^2 x}{\Omega n}} \left[\frac{\tau\Omega}{kT} + \frac{c^{(\mathrm{s})}}{c^{(0)}} - 1 \right] \tag{8.42b}$$

equivalent to Eq. (8.16b).

Another kind of chemical force is due to the supersaturation. It is defined as:

$$F_{\mathrm{c}}^{(\mathrm{s})} = \frac{kTb}{\Omega} \ln \frac{c^{(\mathrm{s})}}{c^{(0)}}$$

The dislocation velocity for small supersaturation levels ($c^{(\mathrm{s})}/c^{(0)}$ close to unity) is:

$$\mathrm{v} = \frac{D^{(\mathrm{sd})}}{b} \frac{1}{1 + \dfrac{a^2 x}{\Omega n}} \frac{\Omega}{kT} \left[\tau + \frac{F_{\mathrm{c}}^{(\mathrm{s})}}{b} \right] \tag{8.42c}$$

Assuming $\Omega \approx a^3$ and $b \approx a$, it reduces to:

$$\mathrm{v} \approx D^{(\mathrm{sd})} \frac{a}{1 + \dfrac{x}{na}} \left[\frac{\tau a + F_{\mathrm{c}}^{(\mathrm{s})}}{kT} \right] \tag{8.42d}$$

This relation is equivalent to Eq. (8.16c). The supersaturation chemical force adds to the mechanical force.

For a very high jog density ($x/n \ll a$), the dislocation velocity becomes (Eq. (8.42c)):

$$\mathrm{v} \approx \frac{D^{(\mathrm{sd})}}{b} \frac{\Omega}{kT}(\tau + F_{\mathrm{c}}^{(\mathrm{s})}/b) \approx \frac{D^{(\mathrm{sd})}}{b}\left[\frac{\Omega\tau}{kT} + \frac{c^{(\mathrm{s})}}{c^{(0)}} - 1\right] \tag{8.43}$$

equivalent to Eq. (8.18).

The v-component which is proportional to $(\Omega\tau/kT) - 1$ has the activation energy U_{sd}. If the supersaturation $c^{(\mathrm{s})}$ is fixed by external conditions, the other component, which is proportional to $c^{(\mathrm{s})}/c^{(0)}$, has the activation energy U_{d} (Eq. (8.2)). The total apparent activation energy thus depends on the relative magnitudes of these two terms. It is U_{d} for very high supersaturations.

8.3.2 Low jog density (e.g. polygonal dislocations)

The calculations carried out in Section 8.2.2 can be transposed easily. It is just necessary to replace $\exp(\tau\Omega/kT) - 1$, in Eqs. (8.21a) or (8.34a), by $(c^{(\mathrm{s})}/c^{(0)}) - \exp(-\tau\Omega/kT)$. Consequently, when pipe diffusion takes place, the climb velocity, in the kink-collision regime ($L > X$) is:

$$\mathrm{v} = \frac{4\pi}{\ln\dfrac{\Delta x}{a}} \frac{D^{(\mathrm{sd})}}{X}\left[\frac{c^{(\mathrm{s})}}{c^{(0)}} - \exp\left(-\frac{\tau\Omega}{kT}\right)\right]\exp\left(\frac{\Delta U_{\mathrm{sd}}^{(\mathrm{p})}}{2kT}\right) \tag{8.44a}$$

equivalent to Eq. (8.34a).

For small stresses and supersaturation levels (cf Eq. (8.38a)):

$$\mathrm{v} = \frac{4\pi}{\ln\dfrac{\Delta x}{a}} \frac{D^{(\mathrm{sd})}}{X}\left(\frac{\Omega\tau}{kT} + \frac{c^{(\mathrm{s})}}{c^{(0)}} - 1\right)\exp\left(-\frac{\Delta U_{\mathrm{sd}}^{(\mathrm{p})}}{2kT}\right)$$

$$= \frac{4\pi}{\ln\dfrac{\Delta x}{a}} \frac{D^{(\mathrm{sd})}}{X} \frac{\Omega}{kT}\left(\tau + \frac{F_{\mathrm{c}}^{(\mathrm{s})}}{b}\right)\exp\left(-\frac{\Delta U_{\mathrm{sd}}^{(\mathrm{p})}}{2kT}\right) \tag{8.44b}$$

Here again, the apparent activation energy is complex. The v-component which is proportional to $(\Omega\tau/kT) - 1$ has the activation energy $U_{\mathrm{sd}} - (\Delta U_{\mathrm{sd}}^{(\mathrm{p})}/2) + (1/2)U_{\mathrm{jp}}^{(\mathrm{c})}$ (Eq. (8.37)). However, if the supersaturation $c^{(\mathrm{s})}$ is fixed by external conditions, the other component, proportional to $c^{(\mathrm{s})}/c^{(0)}$, has a different activation energy equal to $U_{\mathrm{d}} - (\Delta U_{\mathrm{sd}}^{(\mathrm{p})}/2) + (1/2)U_{\mathrm{jp}}^{(\mathrm{c})}$.

In the length effect regime, $1/X$ must be replaced by L/X^2 (cf. Eq. (8.38b)) and the activation energies are higher ($U_{\mathrm{jp}}^{(\mathrm{c})}$ instead of $(1/2)U_{\mathrm{jp}}^{(\mathrm{c})}$—see Table 8.1).

The energy of the critical jog-pair, $U_{\mathrm{jp}}^{(\mathrm{c})}(\tau)$, and the stress exponent, m, are the same as in the absence of point defect supersaturation (Eq. (8.39) and (4.36), respectively. These parameters are indeed independent of the chemical force $F_{\mathrm{c}}^{(\mathrm{s})}$.

8.3.3 Growth and shrinking of prismatic loops during annealing

For the sake of simplicity, the case of no point defect supersaturation is considered first. This situation is met in thin foils where the surfaces provide perfect sinks.

For rounded loops containing a very high jog density, the shrinking velocity is given by Eq. (8.18), where the mechanical stress τ is the sum of two terms only (cf Eq. (8.1b)): the line tension stress $\hat{T}/Rb$, where R is the loop radius, and the surface tension stress of the inner fault, γ/b. This yields:

$$\frac{\mathrm{d}R}{\mathrm{d}t} = -\frac{D^{(\mathrm{sd})}}{b^2}\frac{\Omega}{kT}\left(\frac{\hat{T}}{R} + \gamma\right) \tag{8.45}$$

For perfect loops ($\gamma = 0$), this expression can be integrated, which yields:

$$R = R_0\left(1 - \frac{t}{t_0}\right)^{1/2}, \text{ with } t_0 = 2R_0^2\frac{b^2kT}{D_{\mathrm{sd}}\hat{T}\Omega} \tag{8.46}$$

For polygonal loops, the shrinking velocity can be deduced in the same way as from Eq. (8.38a) or (8.38b). However, the line tension stress is a poor approximation in that case and an estimation of the elastic interaction between segments would be more appropriate.

The case of a point defect supersaturation is considered now. This corresponds to the annealing of bulk materials. Since dislocation loops provide the only available sources and sinks for point defects, the total number of point defects in the crystal volume remains approximately constant. Then a dynamic equilibrium sets in between shrinking and growing loops, the shrinking loops feeding the growing ones with point defects. Calculations modelling this process have been proposed by Kirschner (1973) and Burton and Speight (1986). They are summarized in the following, in the approximation of rounded loops.

Climb velocities are given by Eq. (8.42a), where the mechanical stress is $\tau = (1/b)\times((\hat{T}/R) + \gamma)$, i.e. (assuming $a^2x \ll n\Omega$):

$$\frac{\mathrm{d}R}{\mathrm{d}t} = -\frac{D^{(\mathrm{sd})}}{b}\frac{\Omega}{kT}\left[\frac{c^{(\mathrm{s})}}{c^{(0)}} - \exp\left(\frac{\Omega}{bkT}\left(\frac{\hat{T}}{R} + \gamma\right)\right)\right] \tag{8.47}$$

Note that the mechanical stress is negative because it acts in the shrinking direction, whereas the supersaturation acts in the growing direction.

Consequently, small loops shrink because

$$\exp\left(\frac{\Omega}{bkT}\left(\frac{\hat{T}}{R} + \gamma\right)\right) > \frac{c^{(\mathrm{s})}}{c^{(0)}},$$

and large ones grow for the opposite reason.

Since the supersaturation concentration $c^{(\mathrm{s})}$ is approximately constant for a short time, the net point defect quantity emitted and absorbed is zero, which can be written as

$\sum_\ell \mathrm{v}_\ell R_\ell = 0$, or:

$$\frac{c^{(\mathrm{s})}}{c^{(0)}} \sum_\ell R_\ell = \sum_\ell R_\ell \exp \frac{\Omega}{bkT}\left(\frac{\hat{T}}{R_\ell} + \gamma\right) \tag{8.48a}$$

If $(\hat{T}/R_\ell) + \gamma < (bkT/\Omega)$, the exponential can be developed, whence:

$$\left(\frac{c^{(\mathrm{s})}}{c^{(0)}} - 1 - \frac{\Omega\gamma}{bkT}\right) \sum_\ell R_\ell = \frac{N\hat{T}\Omega}{bkT},$$

where N is the number of loops, and:

$$\frac{c^{(\mathrm{s})}}{c^{(0)}} = 1 + \frac{\Omega\gamma}{bkT} + \frac{\hat{T}}{\bar{R}b}\frac{\Omega}{kT} \tag{8.48b}$$

Eq. (8.48b) connects the supersaturation concentration $c^{(\mathrm{s})}$ to the mean radius of loops, $\bar{R}$. Combining Eqs. (8.47) and (8.48b) yields a climb velocity for the loop number, ℓ, proportional to

$$\frac{\hat{T}\Omega}{bkT}\left[\frac{1}{\bar{R}} - \frac{1}{R_\ell}\right]$$

Loops with radius $\bar{R}$ are metastable as long as the supersaturation $c^{(\mathrm{s})}$ remains constant, whereas loops with $R_\ell < \bar{R}$ shrink and loops with $R_\ell > \bar{R}$ grow. The supersaturation, however, progressively tends to $c^{(0)}$ as $\bar{R}$ increases and previously metastable loops start to shrink.

The evolution of the mean radius $\bar{R}$ with time has been computed by Kirschner (1973) and Burton and Speight (1986). A simplified calculation is proposed in what follows.

If the distribution of the loop sizes is assumed to vary homothetically with time (steady state distribution), the ratio of the maximum radius R_M to the average radius $\bar{R}$ is constant, $R_\mathrm{M}/\bar{R} = K$. Then, using the climb velocity of the largest loops in the form

$$\frac{\mathrm{d}R_\mathrm{M}}{\mathrm{d}t} = \alpha\left(\frac{1}{\bar{R}} - \frac{1}{R_\mathrm{M}}\right)$$

yields:

$$\frac{\mathrm{d}\bar{R}}{\mathrm{d}t} = \alpha\frac{K-1}{K^2}\frac{1}{\bar{R}},$$

which can be integrated into:

$$\bar{R}^2 = \bar{R}_0^2 + \frac{2\alpha(K-1)}{K^2}t \tag{8.49}$$

Relation (8.49) provides the average growth rate, where $\bar{R}_0$ is the initial mean loop radius. With additional hypotheses, Kirschner (1973) and Burton and Speight (1986) obtain $K = 2$.

If loops are polygonal, the competitive growth process is more difficult to describe. It should be based on Eq. (8.38a) or (8.38b), preferably not using the line tension approximation. The main results should, however, be close to those for rounded loops.

8.4. EXPERIMENTAL STUDIES OF CLIMB PROCESSES

Experimental quantitative studies of dislocation climb are surprisingly very scarce and not recent, although this process is obviously of fundamental importance for explaining mechanical properties at high temperatures. The results on climb-plasticity and annealing of prismatic loops are described below. They are completed by fragmentary results on irradiation and creep under irradiation. The corresponding data allow the validity of the dislocation climb velocity laws of Sections 8.2 and 8.3 to be checked.

8.4.1 Pure climb-plasticity

A few experiments are available in which pure climb has been isolated and studied quantitatively.

8.4.1.1 Climb in HCP magnesium and beryllium. The most comprehensive study of climb in deformed materials has been carried out by Le Hazif et al. (1968) in Be, and by Edelin and Poirier (1973a,b) in Mg. When these two HCP metals are deformed by compression creep along the c-axis, all slip systems involving a-dislocations are inhibited and pyramidal slip involving $c + a$-dislocations is too hard to be activated. Then, the authors have shown that deformation takes place by the pure climb of dislocations with c Burgers vectors in the basal plane. The creep rate increases with increasing time, as well as the density of c-dislocations, as shown in Figure 8.8. In compression, c-dislocations climb by absorption of vacancies. a-dislocations, which are either curved in their climb planes (Be) or in subboundaries (Mg) are thought to provide vacancies for the c-dislocations. The main results are as follows.

In Be (Le Hazif et al., 1968), the creep activation energy obtained by temperature jumps is close to the self-diffusion energy and the stress dependence of the strain-rate, deduced from experiments at different strain-rates, is $\dot{\epsilon} \propto \sigma^{3.5}$.

In Mg (Edelin and Poirier, 1973a,b), the density of c-dislocations has been measured as a function of strain, for various stresses and temperatures. The dislocation velocity deduced from the Orowan relation is independent of strain. This allows the activation parameters of the climb velocity of individual dislocations to be determined (Figure 8.9). The activation energy is 1.80 eV, larger than the self-diffusion energy ($U_{sd} = 1.43$ eV), and the stress dependence is $v \propto \sigma^{2.8}$. These activation parameters are the same as for the creep rate, which ensures that they are not influenced by dislocation multiplication.

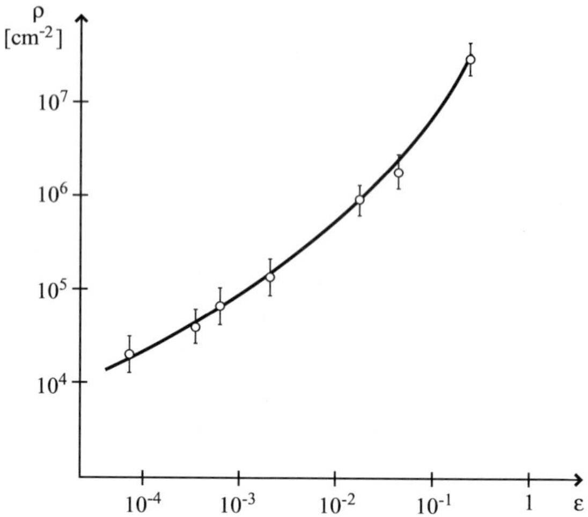

Figure 8.8. c-dislocation density as a function of strain, in Be. Compression test along the c-axis at 415°C, and 500 MPa (from Le Hazif et al., 1968).

Activation parameters of the creep rate in Be are thus probably also characteristic of the dislocation climb velocity.

These results have been discussed by Edelin and Poirier, on the basis of a theoretical approach (Edelin, 1971) that has been included in Section 8.2. The dislocation velocity expected by the authors corresponds to Eq. (8.16c) (high jog density approximation). If the jog-density is close to unity, Eq. (8.16c) reduces to Eq. (8.18). The corresponding activation energy is predicted to be equal to the self-diffusion energy, which is consistent with the experimental results in Be, not in Mg. In Mg, the authors assume that $x/n \gg a$, in such a way that the total activation energy, given by Eqs. (8.16) and (8.12b), is $U_{sd}^{(v)} + U_j$.

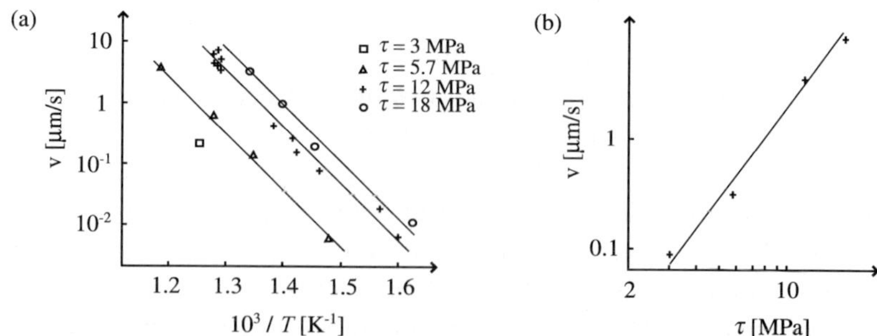

Figure 8.9. Climb velocities in Mg (a) as a function of temperature, and (b) as a function of stress. From Edelin and Poirier (1973a,b).

This can explain the experimental activation energy, provided $U_j = 0.37$ eV. However, in both Mg and Be, the experimental pre-exponential factors are several orders of magnitude too large, as compared to relation (8.16), and the stress-exponent of the velocity is definitely larger than unity. No alternative solution was proposed. Below, a tentative interpretation of these results is given, in the frame of the low jog density approximation (Section 8.2.2).

In Be, the creep stress is fairly high ($\tau \sim 500$ MPa) and the temperature is moderate ($T \sim 700$ K), so that climb is definitely in the high stress regime ($\Omega\tau \approx (kT/2)$). Dislocation loops exhibit slightly polygonal shapes, which suggests a difficult jog-pair nucleation. Then, Eq. (8.38) can be used. The corresponding activation energy is, according to Table 8.1, of the order of $U_{sd}^{(v)}$ in the kink-collision regime or larger than $U_{sd}^{(v)}$ in the length effect regime. An experimental activation energy close to $U_{sd}^{(v)}$ may correspond to the kink-collision regime. Then, using Eq. (8.39a) with $\mu = 14.7 \times 10^4 \times$ MPa, $b = 0.37$ nm and $h \sim 0.2$ nm, the predicted stress exponent is $m = 2.82$ (instead of 3.5). The discrepancy is not too high. The predicted stress exponent would be higher in the length effect regime ($m = 4.64$) but the experimental activation energy would be too large as compared to the predicted one.

In Mg, climb takes place in the low-stress regime ($\tau \sim 10$ MPa, $T \sim 720$ K, $\Omega\tau \approx kT/40$). Eqs. (8.38) and (8.39) can nevertheless be tentatively used. Climb in the length effect regime could explain the experimental pre-exponential term being higher than expected by Edelin and Poirier and the experimental activation energy being larger than U_{sd}. With $\mu = 1.75 \times 10^4$ MPa, $b = 0.55$ nm and $h \sim 0.28$ nm, the predicted stress exponent $m = 1.52$ is, however, too low as compared to the experimental value, $m = 2.8$.

In conclusion, the agreement between the experimental results and the relations derived in the high stress–low jog density approximation is satisfactory in beryllium. It is less convincing in magnesium, where problems similar to those arising in low-purity semiconductors (e.g. Ge, see Chapter 7) could modify the apparent activation parameters of the climb process. In all cases, activation energies larger than $U_{sd}^{(v)}$, and stress exponents larger than unity, should not be considered as incompatible with climb.

8.4.1.2 Climb in intermetallic alloys. Dislocations are expected to move by climb at a sufficiently high temperature. This process has, however, never been studied extensively. Fragmentary observations show that prismatic dislocation loops with $\langle 100 \rangle$ Burgers vectors parallel to the strain axis develop during the high-temperature deformation of NiAl (Fraser et al., 1973a,b; Srinivasan et al., 1997) and of the γ' phase of superalloys (Louchet and Ignat, 1986; Eggeler and Dlouhy, 1997). These microstructural features are similar to those encountered in HCP metals, which indicates that pure climb may be an important plasticity mechanism in intermetallics at high temperatures. Additional experiments would be welcome to elucidate dislocation climb mechanisms in such compounds.

8.4.1.3 Climb in quasicrystals. Recent observations have shown that AlPdMn icosahedral quasicrystals deform by dislocation movements like in crystals (see reviews by Feuerbacher et al. (1997) and Caillard (2000)). Without going into the details of the particular crystallography of quasicrystals—which must be described in a six-dimensional space—dislocations can a priori glide or climb, depending on the direction of the component of their Burgers vector in the physical space, $b_{\|}$, with respect to the plane of movement. As in crystals, dislocations glide when $b_{\|}$ is contained in the plane of movement and climb in all other situations. Dislocations were first considered to move by glide, although this was never verified by electron microscope observations. Several models were aimed at describing the kinetics of glide and at interpreting the corresponding mechanical properties. However, more recent investigations show that in reality dislocations move by climb, and that glide seems impossible at all temperatures (Figure 8.10) (Caillard et al., 2000). Figure 8.11 shows a penrose two-dimensional lattice (equivalent to

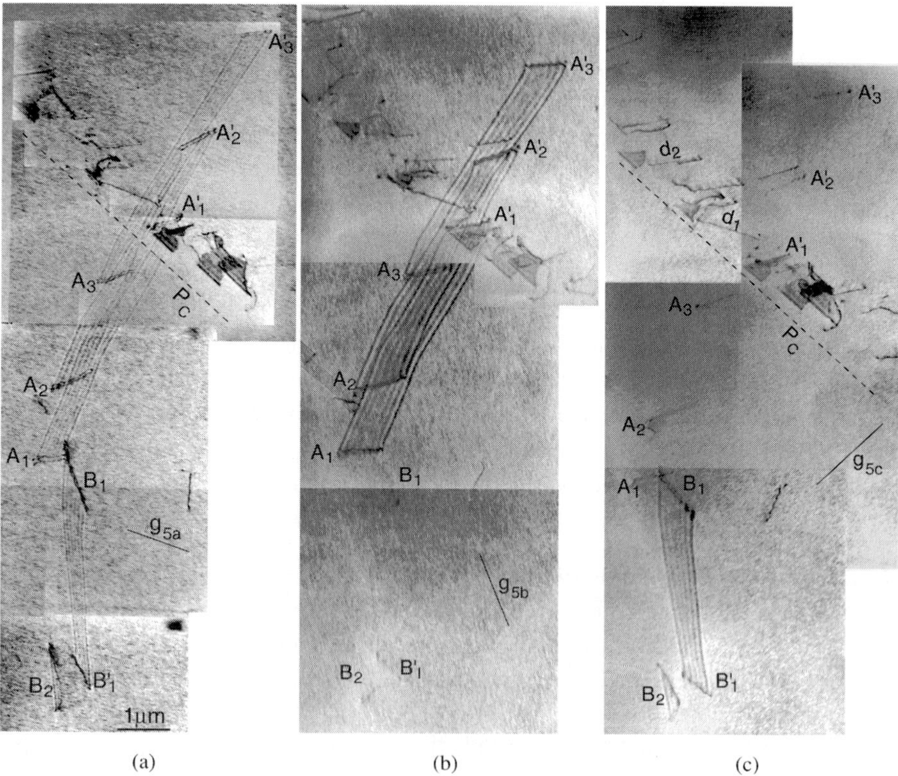

Figure 8.10. Climbing dislocations in $Al_{70.6}Pd_{21.1}Mn_{8.3}$. (a) All dislocations noted A and B are in contrast. (b) B dislocations are invisible. (c) A dislocations exhibit a residual contrast. Note that dislocations trail phason faults (fringe contrast). From Caillard et al. (2000).

Figure 8.11. Schematics of a two-dimensional Penrose lattice deformed by pure shear (horizontal arrows) and by pure climb (vertical arrows).

a cut of the icosahedral structure). Glide along a corrugated plane (dark areas) by pure shear along the horizontal arrows results in highly energetic topological defects (Mikulla et al., 1998). However, if the corrugated dense plane is considered as an extra-half plane of an edge dislocation, climb proceeding by the annihilation or the extension of this plane appears topologically easier. The corresponding displacement is shown by vertical arrows. Quasicrystals are thus materials in which pure climb can be studied over a large temperature range.

Moving dislocations have been observed during in situ experiments by Messerschmidt et al. (1999) and Mompiou et al. (2003). Movements take place in planes perpendicular to either 2-, 3- or 5-fold directions. Dislocations exhibit polygonal shapes with sides parallel to 2-fold directions and they move in a viscous way. This indicates that climb is partly controlled by the nucleation of jog-pairs, like in growing loops (see Section 8.4.2.3).

It is then possible to compare experimental results in AlPdMn with the theoretical climb velocity in the low jog density approximation (Section 8.2.2). Several experimental values of the stress exponent m are plotted in Figure 8.12 as a function of the square root of the deformation stress. Taking $\mu \sim 66.5$ GPa, $b_{\|} = 0.29$ or 0.47 nm (two actual values), m is seen to vary according to Eq. (8.39b) (length effect regime).

The corresponding activation energies are difficult to identify because of the large uncertainties on the self diffusion energies.

In conclusion, activation parameters in icosahedral AlPdMn appear to be consistent with a climb process in the low jog density and high stress approximation. Further work is needed to analyse other quasicrystalline alloys.

8.4.2 Growth and shrinking of loops during annealing

Many experiments have been carried out in quenched metals (essentially Al, but also Cu, Au, Mg, Zn, etc.) in which thermal vacancies coalesce to form faulted and perfect

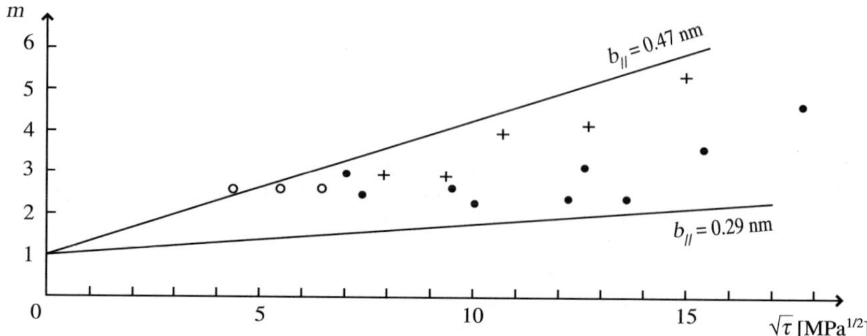

Figure 8.12. Experimental stress exponent of the creep rate of AlPdMn quasicrystals as a function of square root of stress. Crosses from Feuerbacher et al. (1995), closed circles from Brunner et al. (1997), open circles courtesy of L. Bresson. The two straight lines correspond to Eq. (8.39b) with $h = b$.

prismatic loops. The main results have been reviewed by Washburn (1972) and Smallman and Westmacott (1972). More recently, similar experiments have been conducted in ion-implanted silicon, in which interstitial atoms coalesce to form the same types of loops. The kinetics of shrinking or growth can be compared with the theoretical relations of climb involving supersaturation of point defects.

8.4.2.1 Shrinking of vacancy loops in thin foils.

When heated in the electron microscope, vacancy loops formed by quenching in the bulk state tend to shrink because the surfaces of thin foils are very efficient sinks for vacancies. Before heating in the electron microscope, loops have polygonal shapes, which corresponds to a growing process partly controlled by the nucleation of jog-pairs (low jog density approximation). As soon as thin foils are heated, the supersaturation disappears (i.e. $c_{\mathrm{V}}^{(\mathrm{s})} \rightarrow c_{\mathrm{V}}^{(0)}$) because vacancies are pumped by the surfaces. Then loops are observed to take a rounded shape and to shrink.

The climb velocity of large faulted Frank loops with $1/3\langle 111\rangle$ Burgers vectors is constant because, $\hat{T}/bR$ being negligible in Eq. (8.45), the driving force is constant and equal to γ/b, where γ is the stacking fault energy (Figure 8.13). When the radius of curvature decreases below 30 nm, the climb velocity increases because the line tension force becomes important. Tartour and Washburn (1968) have shown that the experimental variation of $\bar{R}$ obeys Eq. (8.45).

The climb velocity data of perfect prismatic loops with $1/2\langle 110\rangle$ Burgers vectors obey Eq. (8.46) (Silcox and Whelan, 1960, Tartour and Washburn, 1968, see Figure 8.14).

These results show that the climb velocity is proportional to stress, in agreement with the hypotheses that yield Eqs. (8.45) and (8.46). Such a linear dependence is consistent with the rounded shape of shrinking loops, which indicates that jog nucleation is easy.

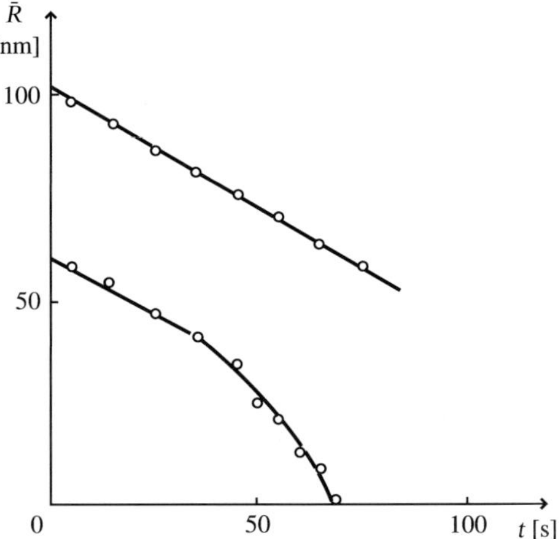

Figure 8.13. Average radius $\bar{R}$ as a function of time of quenched faulted loops in Al. The two curves refer to two different loop sizes. Annealing experiment in TEM (Dobson et al., 1967).

8.4.2.2 Competitive loop growth in bulk materials. This behavior has been described theoretically in Section 8.3.3. Experimental results are available in Al (vacancy loops, see Silcox and Whelan, 1960) and in Si (interstitial loops, see Bonafos et al., 1998). It is interesting to note that shrinking loops (i.e. the smallest ones) have generally rounded

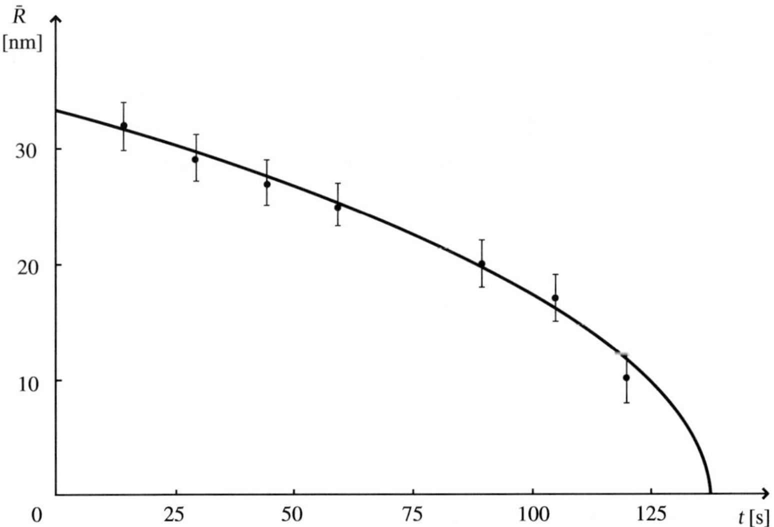

Figure 8.14. $\bar{R}$ as a function of time of quenched perfect prismatic loops in Al. Annealing experiment in TEM (Silcox and Whelan, 1960).

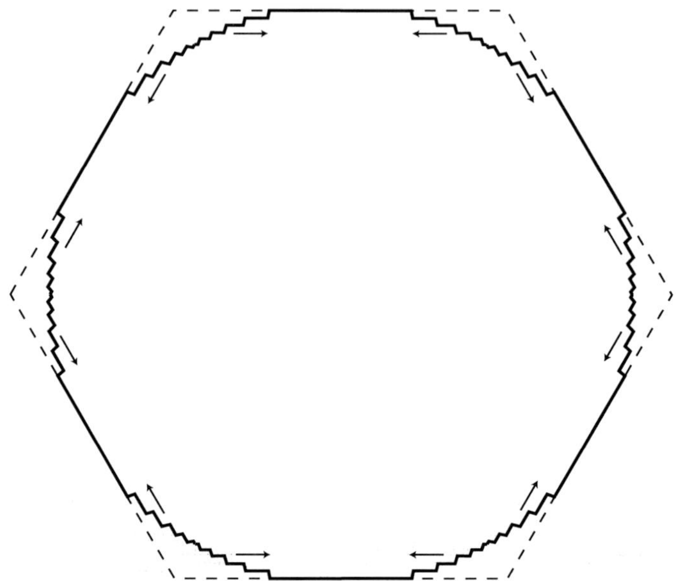

Figure 8.15. Schematic representation of the shrinking of an initially polygonal loop.

shapes, whereas growing ones (i.e. the largest ones) are very often polygonal. The only exceptions to this rule are observed in low stacking fault energy metals, such as Cu, in which shrinking loops remain polygonal. This difference can be explained in the following way: athermal and easy jog nucleation can take place at the corners of shrinking loops (Figure 8.15), whereas the reverse growing process requires the thermally activated and more difficult nucleation of jog-pairs along the rectilinear parts. This shows that the activation parameters are probably different for growing and shrinking loops. The shrinking process is easier for a given total driving force (including the line tension of the loop and eventually the surface tension of the fault). It must also be noted that the line tension approximation is well adapted to the rounded shrinking loops, not to the polygonal growing ones (see Section 8.3.3). In spite of this restriction, Bonafos et al. (1998) showed that the measured average radius $\bar{R}$ increases with time according to Eq. (8.49) (Figure 8.16). The authors have also verified that the total number of interstitials involved in the process remains constant, as expected. The activation energy of the average growth process is 4.4 eV, close to the self-diffusion energy ($U_{\text{sd}} = 4.8$ eV).

8.4.2.3 Growth of loops under high defect supersaturations. This situation corresponds to the very beginning of the recovery process of quenched metals and ion-implanted Si. According to the remark at the end of Section 8.3.1, the chemical driving

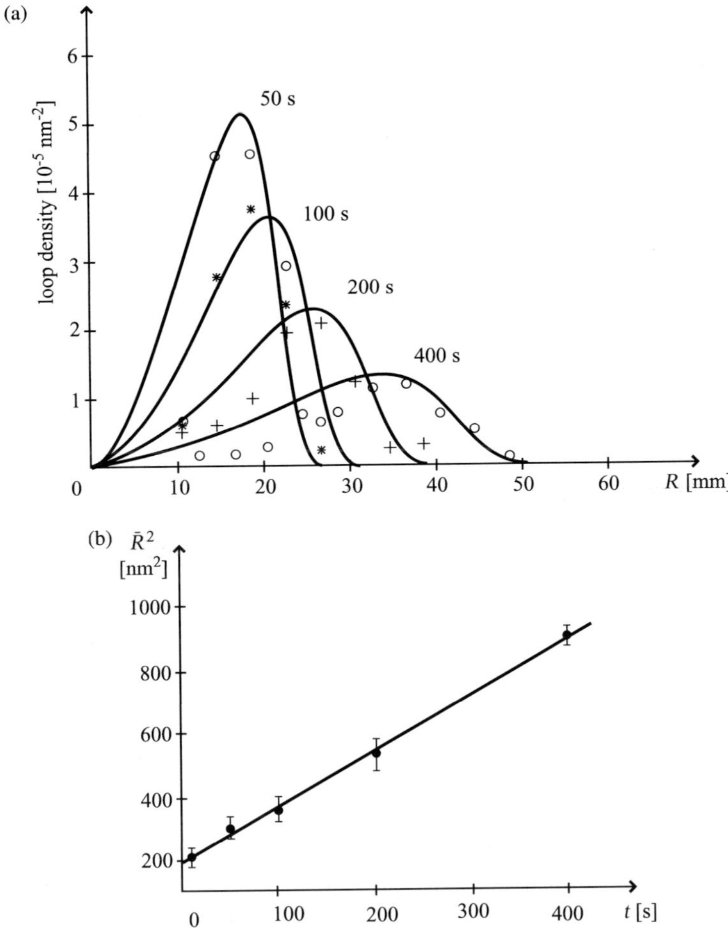

Figure 8.16. Experimental measurements of interstitial loop radii in implanted Si (Bonafos et al., 1998). (a) Radii distribution as a function of time. (b) $\bar{R}^2$ as a function of time.

force is much higher than the line tension and surface tension back forces, so that the activation energy of growth can be identified as U_d (or either $U_d + (1/2)U_{jp}^{(c)}$ or $U_d + U_{jp}^{(c)}$ in the low jog density approximation). Experiments in Si showed that the activation energy of interstitial loops growth is, in this regime, actually substantially smaller than in the competitive growth regime (respectively, 1–2 and 4.4 eV, according to Bonafos et al. (1998)).

8.4.2.4 Conclusions on the loop-annealing experiments. There is a fairly good quantitative agreement with the theory of climb:

(i) The shrinkage velocity is proportional to the total driving force when the jog density is high (rounded loops).
(ii) The activation energies have the right order of magnitude.

However, the case of growing polygonal loops needs to be modelled using the low jog density approximation.

8.4.3 Irradiation-induced creep

Irradiation provides another way of inducing dislocation climb in crystals. Irradiation, however, generates interstitial and vacancy point defects that both interact with dislocations. Interstitials are more mobile than vacancies, in such a way that two different regimes of interstitial loop growth can be observed (Kiritani, 1977). At low temperatures, vacancies remain immobile and accumulate in the crystal. They tend to slow down the climb movement of interstitial dislocation loops, which results in a non-linear growth process. At higher temperatures, vacancies coalesce into voids, and growth of interstitial loops takes place at a constant velocity depending on the irradiation flux. The transition between the two regimes allows the activation energy of vacancy diffusion to be determined.

When combined with an external applied stress, irradiation enhances the velocity of creep by several orders of magnitude. This phenomenon, called irradiation creep, plays an important role in the life-time of the core of nuclear reactors. Several types of experiments have been carried out under neutron irradiation and under electron irradiation in high voltage electron microscopes. The substructures that develop in such conditions have been reviewed by Wolfer (1980) and Caillard and Martin (1981):

- At low doses, interstitial loops form preferentially in dense planes of highest normal tensile stress in neutron-irradiated steel (Okamoto and Harkness, 1973, Brager et al., 1977, see Figure 8.17) and in electron-irradiated aluminium and steel (Tabata et al., 1977; Caillard et al., 1980). This anisotropic nucleation and growth contributes to strain in the direction of the applied stress.
- At higher doses, interstitial loops interact and form an isotropic three-dimensional network.

Two types of models have been proposed, based on these observations:

- The preferential nucleation of interstitial loops in planes with a high normal tensile stress (Brailsford and Bullough, 1973). This mechanism is assumed to take place at the very early stages of irradiation creep, when interstitial clusters collapse to form loops. It is followed by the more or less isotropic growth of this anisotropic distribution. It can account for the observed microstructure and very high creep rate at the beginning of irradiation creep.

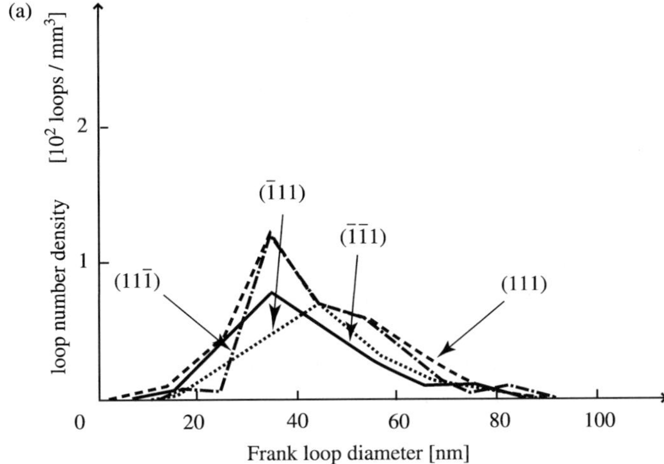

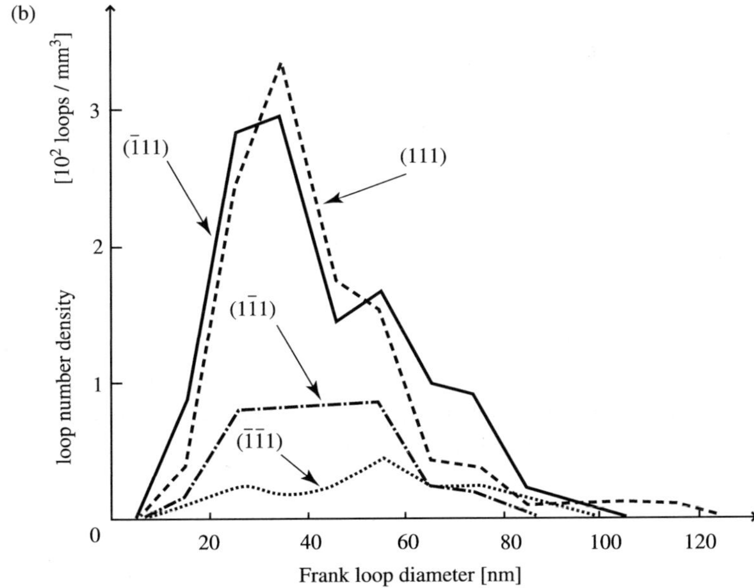

Figure 8.17. Densities of Frank loops in neutron-irradiated steel (a) without external applied stress and (b) with an external applied stress. From Brager et al. (1977).

– The stress induced preferential absorption (SIPA) mechanism (Heald and Speight, 1974). This model is based on the anisotropic migration of point defects towards dislocations climbing in the direction favoured by the external applied stress. This interaction is different from that considered in the conventional climb mechanism controlled by point defect absorption. This long-range elastic interaction results from the second term in Eq. (8.6a) that is neglected in the classical theory of climb. The gradient in the hydrostatic pressure around edge dislocations is modified by the applied stress, which results in an anisotropic climb of the dislocation network. This process should account for the later stages of irradiation creep, when the crystal contains a very dense and isotropic dislocation network (Bullough and Willis, 1975).

These two models appear to be complementary. According to Wolfer (1980), the whole process, including the anisotropic growth of loops, could, however, be accounted for by the SIPA mechanism alone, although the high creep rates observed at the beginning of creep would remain difficult to explain.

8.5. CONCLUSION

The above theoretical considerations show in particular that climb is a poorly documented process, especially in the high stress range. Indeed, under such conditions, the usual approximations are no longer valid. In particular, as summarized in Table 8.1, the stress exponent of the dislocation velocity is larger than 1 and the activation energy can be larger than the self-diffusion energy. Former climb theories were developed primarily for pure metals. These usually deform under low stresses at high temperatures. However, new materials, for high temperature applications, resist much higher stresses, which has necessitated the extension of the field of the theory. Indeed, in such materials designed to impede dislocation glide, climb becomes a competitive mechanism at high stresses.

Experiments in which dislocations move exclusively by climb are difficult to design. However, the fragmentary results available assess the theoretical predictions. It is clear that additional data would be welcome.

REFERENCES

Balluffi, R.W. (1969) *Phys. Stat. Sol.*, **31**, 443.

Balluffi, R.W. & Granato, A.V. (1979), in *Dislocations in Solids*, vol. **4**, Ed. Nabarro, F.R.N., North Holland Publishing Company.

Bonafos, C., Mathiot, D. & Claverie, A.G.L. (1998) *J. Appl. Phys.*, **83**, 3008.

Brager, H.R., Garner, F.A. & Guthrie, (1977) *J. Nucl. Mater.*, **66**, 301.

Brailsford, A.D. & Bullough, R. (1973) *Phil. Mag.*, **27**, 49.

Brunner, D., Plachke, D. & Carstangen, H.D. (1997) *Mater. Sci. Eng. A*, **234–236**, 310.

Bullough, R. & Willis, J.R. (1975) *Phil. Mag.*, **31**, 855.

Burton, B. & Speight, M.V. (1986) *Phil. Mag.*, **53**, 385.

Caillard, D. (2000), in *Quasicrystals: Current Topics*, Eds. Bellin-Ferré, E., Berger, C., Quiquandon, M. & Sadoc A., World Scientific, p. 387.

Caillard, D. & Martin, J.L. (1981) *J. Microsc. Spectrosc. Electron.*, **6**, 361.

Caillard, D., Martin, J.L. & Jouffrey, B. (1980) *Acta Met.*, **28**, 1059.

Caillard, D., Vanderschaeve, G., Bresson, L. & Gratias, D. (2000) *Phil. Mag.*, **80**, 237.

Dobson, P.S., Goodhew, P.J. & Smallman, R.E. (1967) *Phil. Mag.*, **16**, 9.

Edelin, G. (1971) *Phil. Mag.*, 1547.

Edelin, G. & Poirier, J.P. (1973a) *Phil. Mag.*, **28**, 1203; (1973b) *Phil. Mag.* **28**, 1973.

Eggeler, G. & Dlouhy, A. (1997) *Acta Mater.*, **45**, 4251.

Feuerbacher, M., Baufeld, B., Rosenfeld, R., Bartsch, M., Hauke, G., Beyss, M., Wollgarten, M., Messerschmidt, U. & Urban, K. (1995) *Phil. Mag. Lett.*, **71**, 91.

Feuerbacher, M., Metzmacher, C., Wollgarten, M., Urban, K., Baufeld, B., Bartsch, M. & Messerschmidt, U. (1997) *Mater. Sci. Eng. A*, **233**, 103.

Fraser, H.L., Loretto, M.H. & Smallman, R.E. (1973b) *Phil. Mag.*, **28**, 667.

Fraser, H.L., Smallman, R.E. & Loretto, M.H. (1973a) *Phil. Mag.*, **28**, 651.

Heald, P.T. & Speight, M.V. (1974) *Phil. Mag.*, **29**, 1075.

Hirth, J.P. & Lothe, J. (1982) *Theory of Dislocations*, 2nd Edition, Krieger Publishing Company, Malabars, Florida.

Kiritani, M. (1977), in *High Voltage Electron Microscopy*, Eds. Imura, T. & Hashimoto H., Japanese Society of Electron Microscopy, supplement of Journal of Electron Microscopy **26**, p. 505.

Kirschner, H.O.K. (1973) *Acta Metall.*, **21**, 85.

Le Hazif, R., Antolin, J. & Dupouy, J.M. (1968) *Trans JIM*, supplement, **9**, 247.

Louchet, F. & Ignat, M. (1986) *Acta Metall.*, **34**, 1681.

Messerschmidt, U., Geyer, B., Bartsch, M., Feuerbacher, M. & Urban, K. (1999) *Mat. Res. Soc. Symp. Proc.*, **553**, 319.

Mikulla, R., Gumbsch, P. & Trebin, H.R. (1998) *Phil. Mag. Lett.*, **78**, 369.

Mompiou, F., Caillard, D. & Feuerbacher, M. (2003) *Phil. Mag.*, submitted for publication.

Okamoto, P.R. & Harkness, S.D. (1973) *J. Nucl. Mater.*, **48**, 204.

Peach, M. & Koelher, J.S. (1950) *Phys. Rev.*, **80**, 436.

Philibert, J. (1985) *Diffusion et Transport de Matière dans les Solides*, Les Editions de Physique.

Schoeck, G. (1980) in *Dislocations in Solids*, vol. **3**, Ed. Nabarro, F.R.N., North Holland Publishing Company, p. 63.

Seidman, D.N. & Balluffi, R.W. (1966) *Phil. Mag.*, **13**, 649.

Silcox, J. & Whelan, M.J. (1960) *Phil. Mag.*, **5**, 1.

Smallman, R.E. & Westmacott, K.H. (1972) *Mat. Sci. Eng.*, **9**, 249.

Srinivasan, R., Savage, M.F., Daw, M.S., Noebe, R.D. & Mills, M.J. (1997) *MRS Symp. Proc.*, **460**, 505.

Tabata, T., Nakajima, Y., Kida, T. & Fujita, H. (1977) in *High Voltage Electron Microscopy*, vol. **26**, Eds. Imura, T. & Hashimoto H., Japanese Society of Electron Microscopy, supplement of Journal of Electron Microscopy, p. 519,.

Tartour, J.P. & Washburn, J. (1968) *Phil. Mag.*, **18**, 1257.

Washburn, J. (1972), in *Radiation Induced Voids in Metals*, Eds. Corbett, J.W. & Iannello L.C., Nat. Technical Information Service, US Dpt of Commerce, Springfield, Virginia, p. 647.

Wolfer, W.G. (1980) *J. Nucl. Mater.*, **90**, 175.

Chapter 9
Dislocation Multiplication, Exhaustion and Work-hardening

Chapter 9
Dislocation Multiplication, Exhaustion and Work-hardening

Plastic deformation can be described as the net result of the following sequential processes: (i) the creation of fresh dislocations, (ii) their motion through the crystal towards areas where they are arrested and (iii) their storage or annihilation in the latter areas. It can be anticipated that the slowest one of these processes is rate controlling.

As dislocations stop moving, they can either be stored in bundles, cell walls, grain or subgrain boundaries, thus contributing to work-hardening. They can also leave the crystal or annihilate with dislocations of opposite sign thus taking part in work-softening. Storage or annihilation causes a decrease in the mobile dislocation density.

In the following, the state of knowledge on multiplication (Section 9.1) and exhaustion processes (Section 9.2) will be presented in connection with the resultant work-hardening. The various dislocation mobility mechanisms are the subject of most other chapters. A few interrelations between mobile dislocation exhaustion or multiplication rates are presented, in particular in Section 9.3. Finally, the special dislocation multiplication mechanisms at surfaces will be exposed in Section 9.4 in the case of crack propagation in ductile fractures and at crystal interfaces.

9.1. DISLOCATION MULTIPLICATION

This section first deals with multiplication mechanisms, which were conceived in the early stages of dislocation physics. They are compared with observed mechanisms. We then review several aspects of multiplication in covalent crystals, where these processes are very active at the onset of deformation. This includes a description of deformation curves, for monotonic as well as transient tests, results of mesoscopic simulations of sources and a numerical analysis of deformation curves in terms of various multiplication laws.

9.1.1 Models of sources

A dislocation multiplication process by glide was suggested by Frank and Read (1950). A gliding dislocation segment is bulging between two pinning points. The critical configuration is a half circle centered midway between the two points. Beyond this stage, the dislocation bulges further in an unstable configuration which leads to a complete loop while restoring the original configuration. The same process can start anew, the source emitting a series of loops. The critical stress is proportional to the reciprocal of the segment length and is athermal. Variants of this mechanism are presented in Figure 9.1.

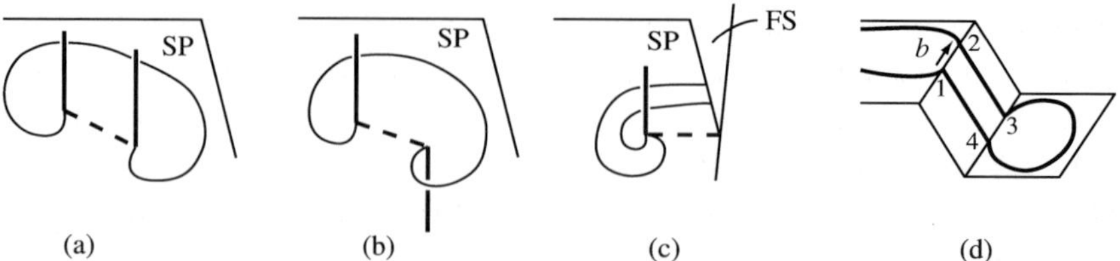

Figure 9.1. Schematics of possible configurations of dislocation mills. (a) The U mill. (b) The Z mill. (c) The L mill. SP: slip plane, FS: free surface. (After Nabarro, 1987). (d) The double cross-slip mechanism (Koehler, 1952). b is the Burgers vector.

In Figure 9.1(a) and (b), the revolving segment is pinned at both ends. Alternatively, it can be pinned at one end, the source operating in the crystal or in the vicinity of a free surface as illustrated in Figure 9.1(c). A practical way of creating the configuration given in Figure 9.1(a) is the double cross-slip mechanism of Figure 9.1(d).

An early experimental assessment of a Frank–Read source of Figure 9.1(a) or (b) type was provided by Dash (1956) in silicon decorated with copper. The double cross-slip mechanism has been claimed to account for successive etch pit observations on deformed lithium fluoride (Johnston and Gilman, 1960).

The Frank–Read process inspired Bardeen and Herring (1952) who proposed a configurationally similar type of source, in which the active dislocation segment bows out by climb.

Much less widely known is another multiplication process imagined by Bourdon et al. (1981) for a smectic A phase, submitted to a dilative or compressive strain, normal to the layers. The corresponding experiment is depicted in Figure 9.2. The sample is maintained

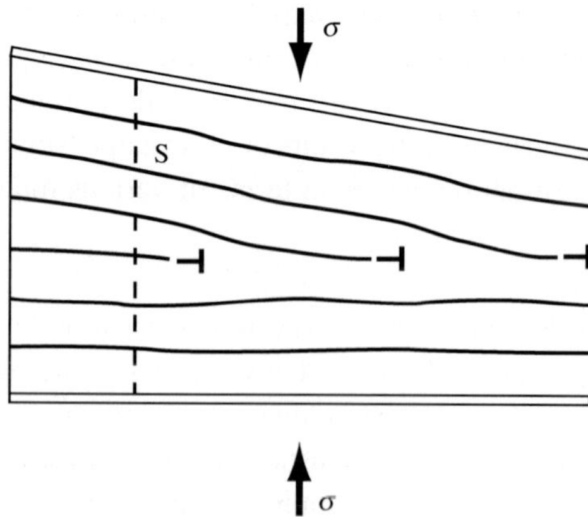

Figure 9.2. Schematics representing a deformation experiment of a smectic liquid crystal under a normal stress. S: screw dislocation. After Oswald and Kléman (1982).

between two glass plates forming a dihedron with a small angle. The number of layers of the liquid crystal increases towards large thicknesses. This configuration can be accounted for by a pure tilt dislocation wall lying in the bisector plane. A stress normal to the layers causes the edge dislocations to climb. It was proposed that screw dislocations, threading through the layers, with the same Burgers vector as the edges, would transform into a helix. At a critical stage, the helix would evolve towards a prismatic loop and a new straight screw segment. Such a mechanism is represented in Figure 9.3. The sign of the helix winding, together with that of the loop, depend on the normal stress: in compression a vacancy loop is emitted, while an interstitial one corresponds to tension. The critical stress for this helical instability has been estimated by Bourdon et al. (1981). It is worth noting that this process of helix formation is different from those suggested previously in crystals (see a review by e.g. Hirth and Lothe (1992)). Experimentally, this mechanism was claimed to account for one of two characteristic relaxation times measured by Oswald and Kléman (1984) when performing the experiment of Figure 9.2. The second one was assigned to edge dislocation climb.

Recent in situ experiments were performed using an optical microscope with polarized light and a device similar to that of Figure 9.2. This allows the dislocations in the liquid crystal to be observed at a temperature just above that of the SmA-SmC transition. For small sample thicknesses (close to 2 μm), dislocations become visible since the transition starts at their core, which affects light transmission and gives rise to a contrast along the line. The presence of a screw dislocation forest is confirmed by Lelidis et al. (2000). A similar experiment by Blanc et al. (2003) reveals fixed points in the liquid crystal around which dislocation loops are emitted (Figure 9.4). These correspond to dislocation sources.

It is proposed that such a mechanism of helical instability can also operate in crystals at screw dislocations submitted to the proper type of stress. An image of helix dislocation is presented by Appel et al. (2000) in deformed γ TiAl, although no indexation of the micrograph is proposed. Further investigations on this multiplication process are needed.

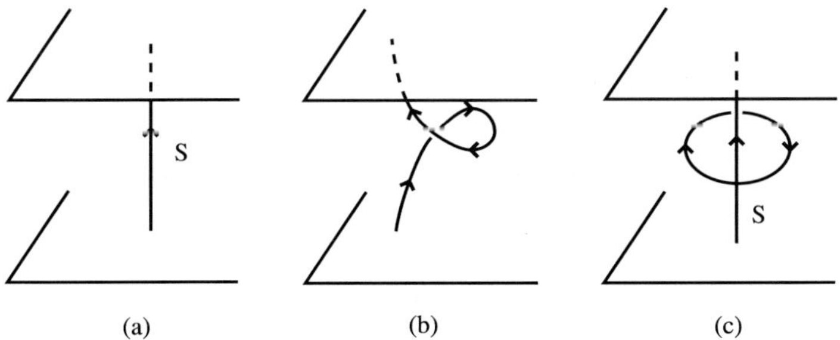

Figure 9.3. Schematics illustrating the evolution of a screw dislocation (S) under stress in the liquid crystal of Figure 9.2 (see text).

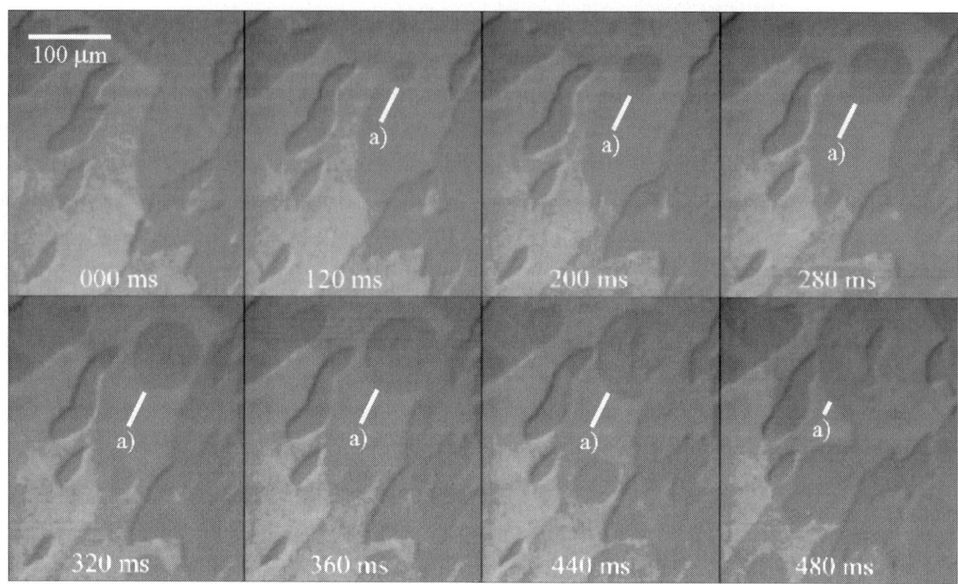

Figure 9.4. Dislocation loops emitted at fixed points in a smectic liquid crystal under a normal stress as a function of time. An expanding loop is indicated by a. Courtesy of C. Blanc and N. Zuodar.

9.1.2 *Observed dislocation sources*

The development of in situ experiments (see Chapter 2) has allowed the observation of several types of dislocation sources in a variety of crystals and deformation conditions. However, such observations are always unexpected, due to the random localization of sources and the small volume size investigated in TEM. Early information about static and dynamic observations of sources was provided by Whelan et al. (1957), Furubayashi (1969), Saka et al. (1970), Carter (1977), Kubin and Martin (1980) and Imura (1980). Observed multiplication processes can be classified as follows: stable sources emitting a large number of loops or temporary ones operating during dislocation motion (expansion of open and closed loops).

9.1.2.1 Glide sources with one pinning point. These are Frank Read sources of the L type, generating a spiral segment. This geometry is more favourable than that corresponding to a U or Z type mill, since only one pinning point is needed for multiplication to operate. Such configurations have been observed in a number of crystals and deformation conditions: see Kubin et al. (1980) for slip in niobium at 160 K or creep of aluminium at 200°C, Legros et al. (1996) for prism slip in Ti_3Al, Lagow et al. (2001) in high purity Mo at room temperature and Fnaiech et al. (1987) for glide in InSb at 250°C, a III–V compound, and in GaAs (see e.g. Figure 7.18).

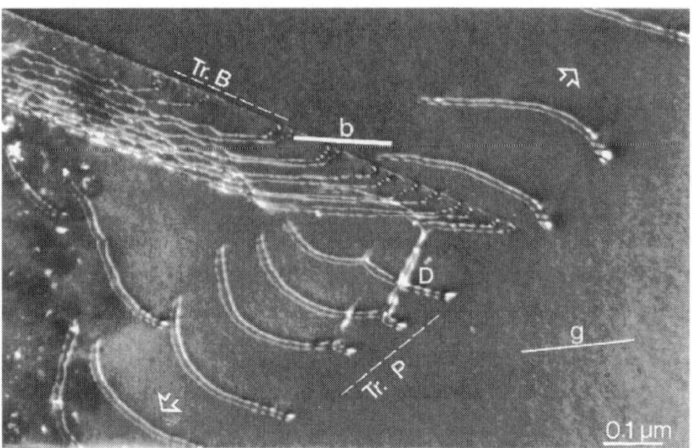

Figure 9.5. TEM in situ observation of a dislocation source in Ti_3Al with one pinning point operating on the prismatic plane, at the head of a slip band in the basal plane. The slip traces of the prismatic and basal planes are labelled Tr.P and Tr.B, respectively. *b* is the projection of the Burgers vector. $T = 300$ K. From Legros et al. (1996).

An example of such a source is shown in Figure 9.5, which operates in the prismatic plane of Ti_3Al at 300 K. The anchoring point for such sources is either a macrokink or a dislocation node. The emission of dislocations on both sides of the basal plane accounts for a rapid increase in the mobile dislocation density on the prism plane.

The study of dislocation glide in InSb at 250°C (Fnaiech et al., 1987) shows sources of this type which emit screw, α and β dislocations. The three corresponding velocities have been measured in different foil orientations. For a local stress of 50 ± 15 MPa, screw and 60° β rectilinear dislocation segments are slow, exhibiting a strong friction (see Section 7.3.1.2). Conversely, 60° α segments move rapidly, their velocity being almost 100 times higher than that of the two other species. A length effect is observed for the velocities of the slower dislocations.

9.1.2.2 Closed loop multiplication. This process is frequently observed in crystals which exert friction forces on screw dislocations under conditions which favour cross-slip. In such materials, multiplication takes place as a non-flexible screw encounters an obstacle strong enough to force the dislocation to go around it. This local pinning may induce cross-slip of both arms of the latter dislocation of unequal amplitude, thus inducing a closed dipole or prismatic dislocation loop. This is depicted in Figure 9.6. At a pinning point, the screw dislocation (1) is divided into two segments (2), the left one moving to a parallel slip plane by double cross-slip while a jog is formed (Figure 9.6(a)). The screw dislocation is thus trailing an edge dipole. It is released (3), provided the segment on the right cross-slips on its turn, while a closed prismatic loop is formed. Eventually, for large enough stresses, this loop can expand closed (Figure 9.6(b)) by simultaneous glide on the primary and

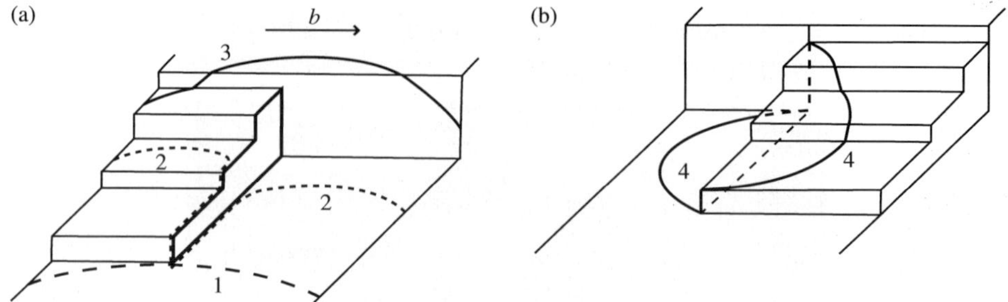

Figure 9.6. Schematics illustrating the closed loop multiplication process (see text). From Caillard and Couret (2002).

cross-slip planes. The mobile dislocation density thus increases, but less slowly than in the preceding multiplication process. Conversely, narrow dipoles do not expand and are observed to transform into debris.

This loop nucleation process has been observed in a number of situations: e.g. slip of ordinary dislocations in γ TiAl at intermediate temperatures (Couret, 1999), slip on $\{110\}$ in FeAl ordered alloys (Molénat et al., 1998) and for slip of $\langle 100 \rangle$ dislocations in NiAl crystals (Caillard et al., 1999). An example is shown in NiAl in the "soft orientation" at 143 K in Figure 9.7. Dislocation d, with a [100] Burgers vector, gets pinned at a small defect C (Figure 9.7(a) and (b)). It escapes on Figure 9.7(c) leaving a prismatic loop behind which expands along the (010) plane (Figure 9.7(d)–(f)). It exhibits rectilinear edge segments along [001]. It intersects the foil surface in Figure 9.7(f), while two new dislocations d$'$ and d$''$ have been formed.

9.1.2.3 Open loop multiplication. This process, which takes place under conditions similar to the preceding ones, is depicted in Figure 9.8. The first stages are similar to those of Figure 9.6, except that the cross-slip distance is larger in the open loop case. This allows the two edge segments of (2) to rapidly glide apart so that no dipole is formed, but instead an open loop. This mechanism has been observed under a variety of conditions such as prismatic slip both in magnesium by Couret and Caillard (1985) and titanium by Farenc et al. (1993), in high purity molybdenum by Lagow et al. (2000) at 300 K and in an aluminium lamella in a Al–$CuAl_2$ eutectic alloy during creep at 300°C by Kubin et al. (1980). It is illustrated in Figure 9.9, which corresponds to the former case. A rectilinear screw dislocation moves towards the right, across the Peierls valleys of the prism plane (see Section 6.1.3). A small dislocation loop appears on the screw at a pinning point on Figure 9.9(a) thus creating two segments labelled 1 and 2. The open loop 3 expands according to the mechanism of Figure 9.8, from (b) to (d), the screw portions being slower than the edge ones.

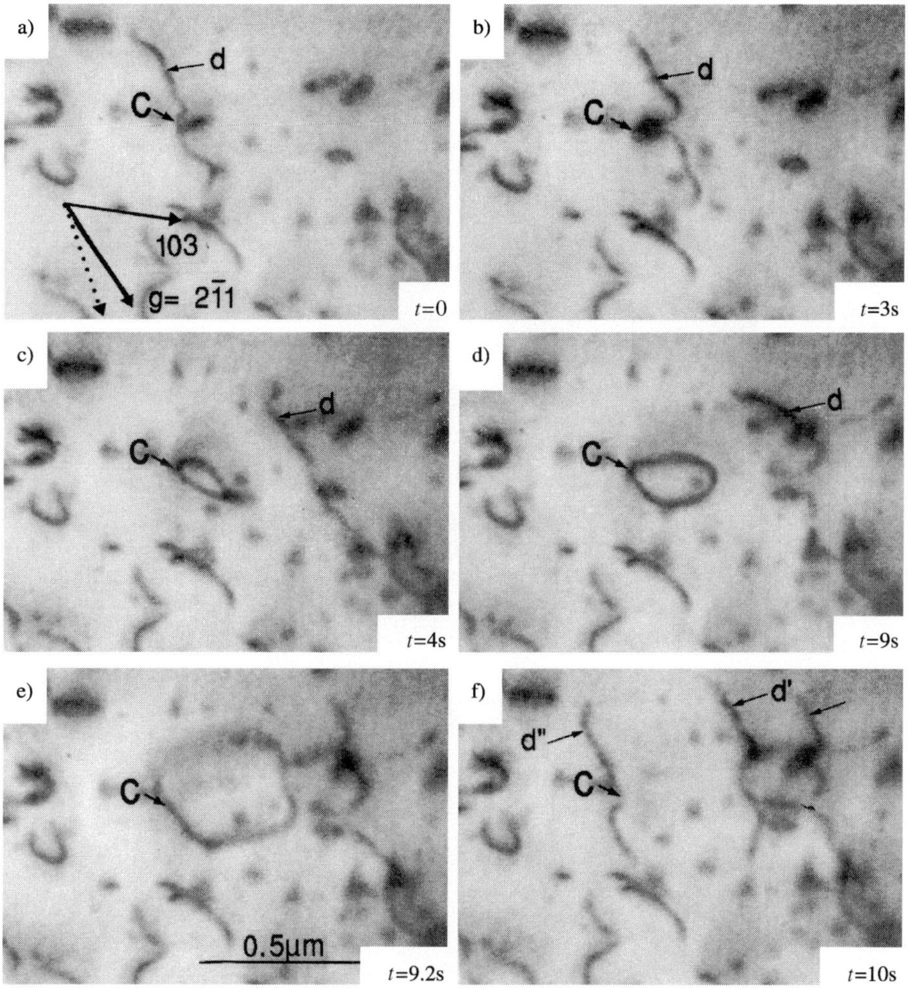

Figure 9.7. TEM in situ sequence of the closed loop multiplication process in stoichiometric NiAl single crystals at 143 K. See text for details. From Caillard et al. (1999).

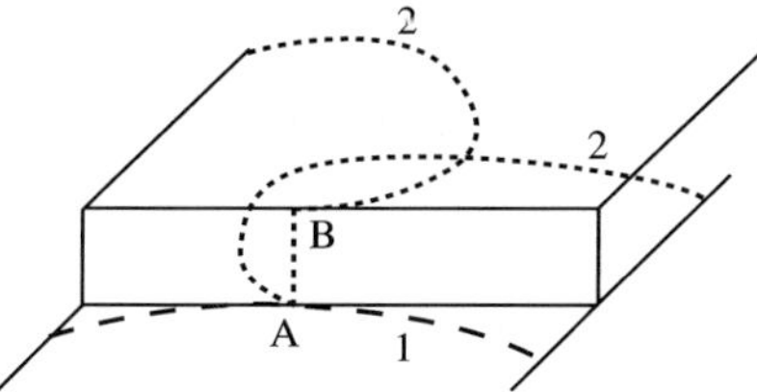

Figure 9.8. Schematics illustrating the open loop multiplication process. See text.

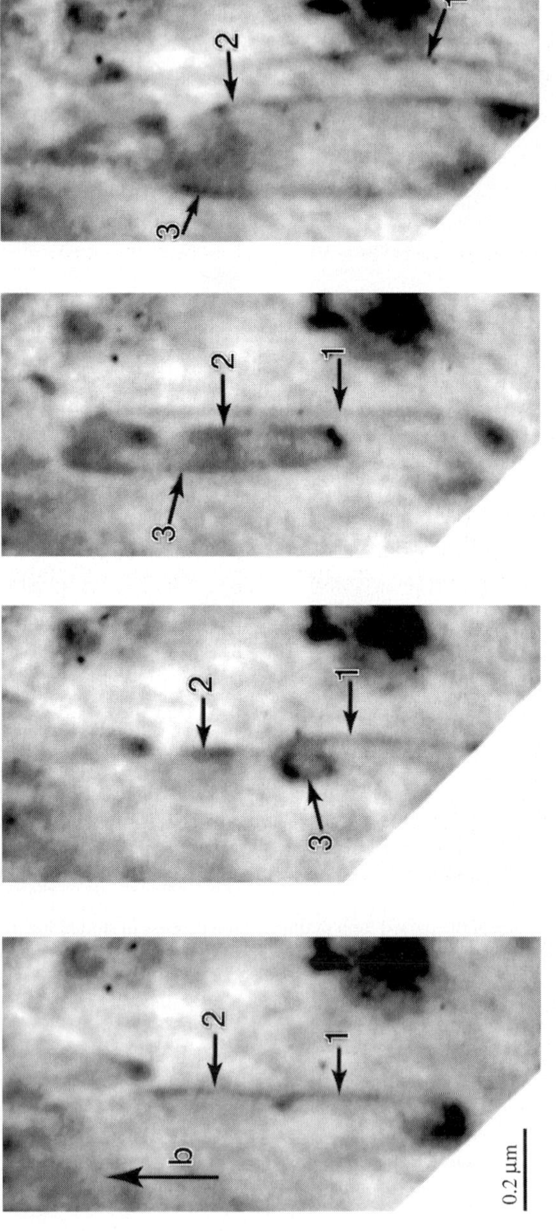

Figure 9.9. TEM in situ sequence showing dislocation multiplication by the open loop process on the prismatic plane in Mg. $T = 300$ K. See text. From Couret and Caillard (1985).

Open loop multiplication has also been observed under other circumstances. In BCC metals at low temperatures it is quoted by, for example, Louchet et al. (1979) in Nb and Fe below 220 K and in Mo below 350 K. In this case, the superjog AB of Figure 9.8 can move between two operations of the source, so that the authors use the word "*wandering sources*". Open loop multiplication is also reported. in a single grain quasicrystal of $Al_{70.6}Pd_{21.1}Mn_{8.3}$. The source operates by pure climb in a 2-fold (mirror) plane (see Figure 9.10). This figure provides a rare example of a Bardeen–Herring type source (see Section 9.1.1).

9.1.3 Multiplication processes in covalent materials

Covalent crystals can be considered as ideal materials to study dislocation multiplication processes, with regard to their very low initial dislocation densities. The main results about

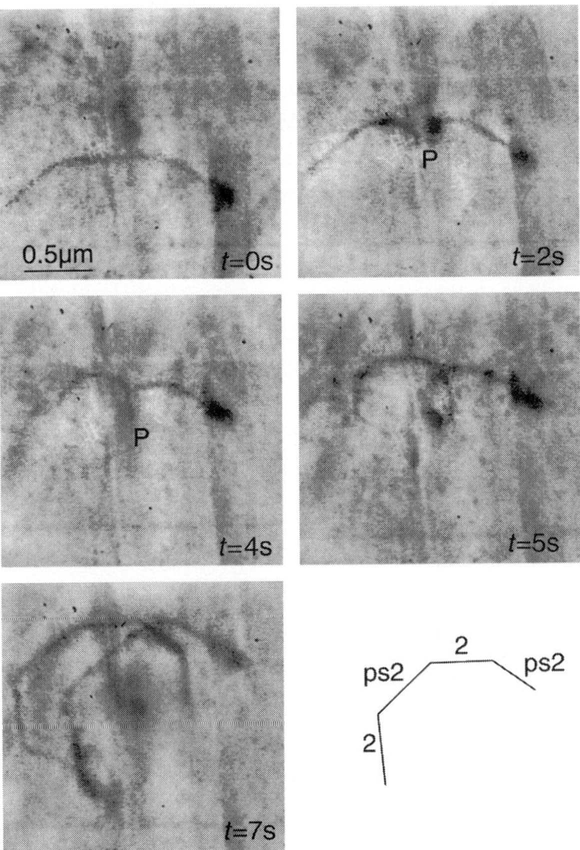

Figure 9.10. TEM in situ sequence of dislocation multiplication by the open loop process operating by climb in a single grain quasicrystal of $Al_{70.6}Pd_{21.1}Mn_{8.3}$. P refers to the pinning point, 2 and ps2 to two-fold and pseudo two-fold directions, respectively. Courtesy of Caillard and Mompiou.

their plastic behaviour can be found in excellent review papers by, e.g., Alexander and Haasen (1968) or George and Rabier (1987). There is a renewal of interest in these materials, centred on multiplication mechanisms (Siethoff et al., 1999). Various aspects of multiplication processes, both macroscopic and microscopic, are reviewed below for diamond cubic (dc) covalent crystals. These include the features of deformation curves which are related to multiplication and the search for proper dislocation multiplication laws. Since abundant data are available for silicium, the results in the following predominantly refer to germanium, which is far less documented, unless otherwise specified.

9.1.3.1 General features. The stress–strain curves exhibit a marked yield point, called the upper yield point (UYP) as illustrated in Figure 9.11. It is present in the temperature domain where the Peierls–Nabarro friction controls dislocation mobility. It corresponds to the stress increase necessary for dislocation source operation, which compensates for the low initial dislocation density. The underlying mechanisms are the subject of Sections 9.1.3.2 and 9.1.3.3. During this intense multiplication stage, heterogeneous glide is observed, as illustrated in Figure 9.12(a). Once a sufficient number of dislocations have been generated, their mutual interactions harden the crystal just after the lower yield point (LYP). As a consequence of the thermally activated nature of the lattice resistance to glide, the UYP and the LYP amplitudes decrease at rising temperatures. At a temperature between 850 and 920 K, for the investigated strain rate, the yield point vanishes, and glide becomes homogeneous as shown in Figure 9.12(b). The kink-pair mechanism is no longer controlling dislocation mobility.

For given deformation conditions, the upper yield stress is found to be constant within ±9%. This is due to differences in sample surface preparation which induce various densities of surface sources.

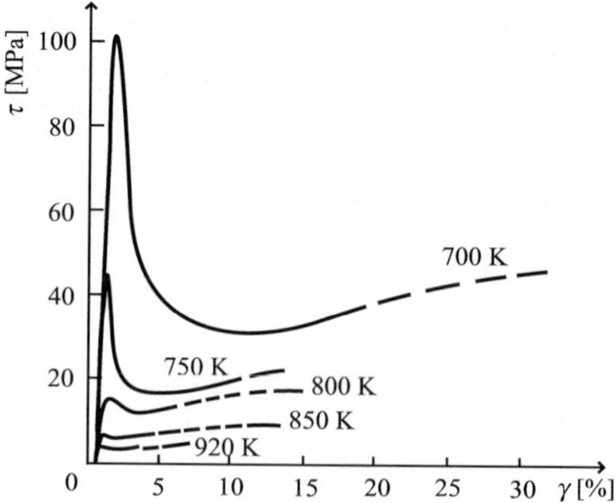

Figure 9.11. Stress–strain curves of ⟨123⟩ Ge single crystals as a function of temperature. Shear strain rate: $8.6 \times 10^{-5}\ s^{-1}$. Transient tests are performed where the curves are interrupted (Dupas et al., 2002).

(a)

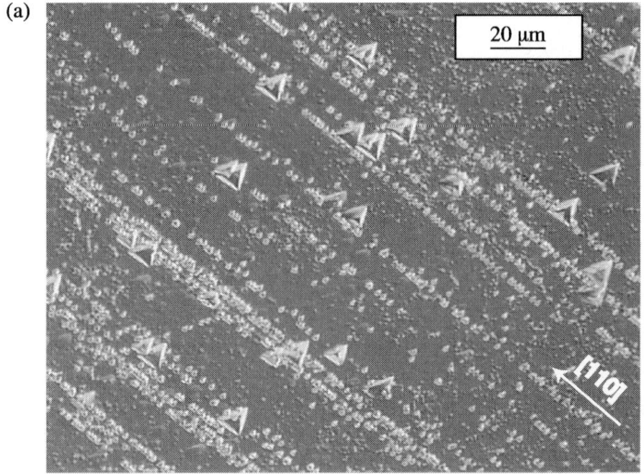

(b)

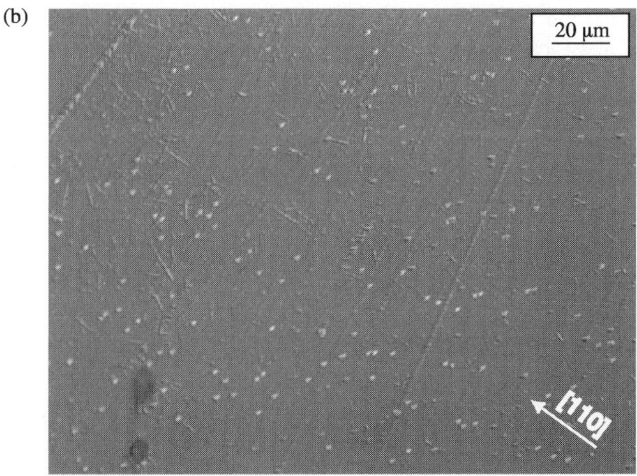

Figure 9.12. Etch-pit experiment showing the glide features in a ⟨123⟩ Ge single crystal. {145} sample face. (a) Heterogeneous glide during fast loading to 30 MPa, at 750 K (Dupas et al., 2002). (b) Homogeneous glide at 1000 K, under 1.6 MPa, at $\gamma = 0.3\%$ and $\dot{\gamma} = 4.3 \times 10^{-5}\ \mathrm{s}^{-1}$. (Courtesy of C. Dupas).

Dislocation multiplication at the onset of the monotonic curve can also be shown by performing transient tests before the UYP. This is illustrated in Figure 9.13 for a stress relaxation test starting at a stress of 66 MPa. This figure clearly shows that a reloading yield point is present following the relaxation test, the height of which is smaller than that of the UYP of the monotonic curve. This is an indication of intense dislocation multiplication during the transient. In addition, the relaxation curve of Figure 9.13(b)

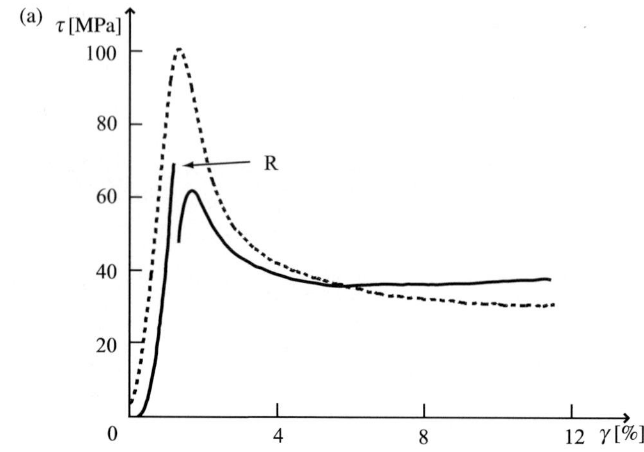

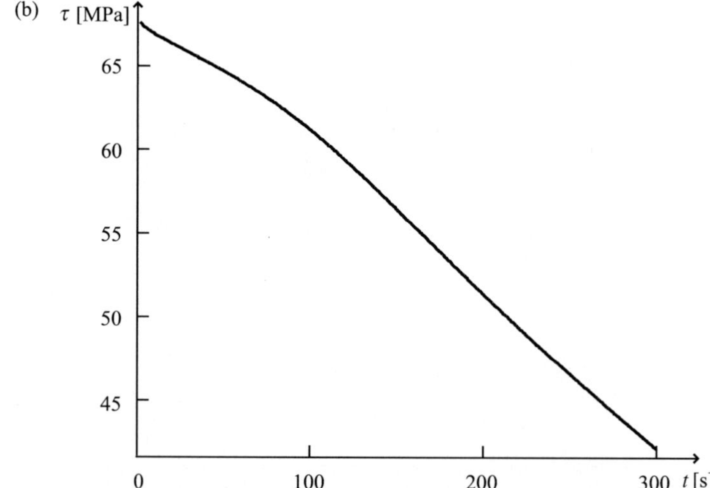

Figure 9.13. Deformation curve (full line) with a stress relaxation test R performed before the UYP. ⟨123⟩ Ge single crystal at 700 K. A monotonic curve (dotted line) is shown for comparison in (a). Details of the relaxation curve are shown in (b) (Dupas et al., 2002).

exhibits an almost constant stress rate, unlike in similar tests performed after the LYP. This peculiar feature is also thought to result from the multiplication process.

Performing a transient creep test before the UYP yields similar effects as shown in Figure 9.14. The creep test, which starts at 45 MPa, is followed by a lower reloading yield point than for relaxations. Inverse creep is observed, unlike for transient creep tests recorded after the LYP. These features are thought to be the signature of multiplication processes. The corresponding increase in dislocation density can be estimated as follows.

With regard to the low dislocation density before the UYP, the stress acting on dislocations equals the applied stress (45 MPa). The creep rate equals the slope of the curve in Figure 9.14(b) and is plotted as a function of time in Figure 9.14(c). Using the Orowan relation and an average value of the dislocation velocity at 45 MPa and 700 K from double etch pit experiment results by Schafer (1967), the mobile dislocation density can be estimated.

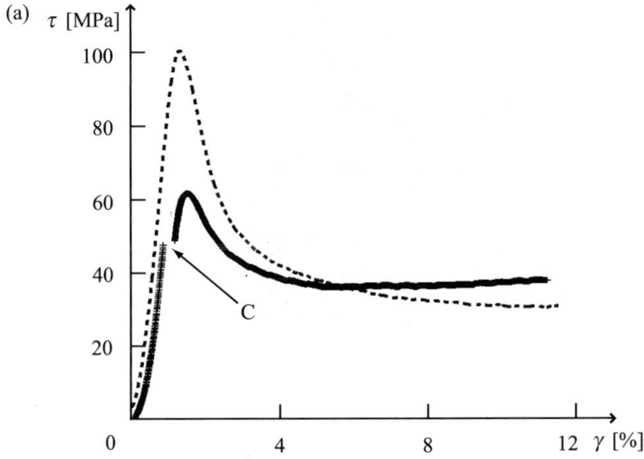

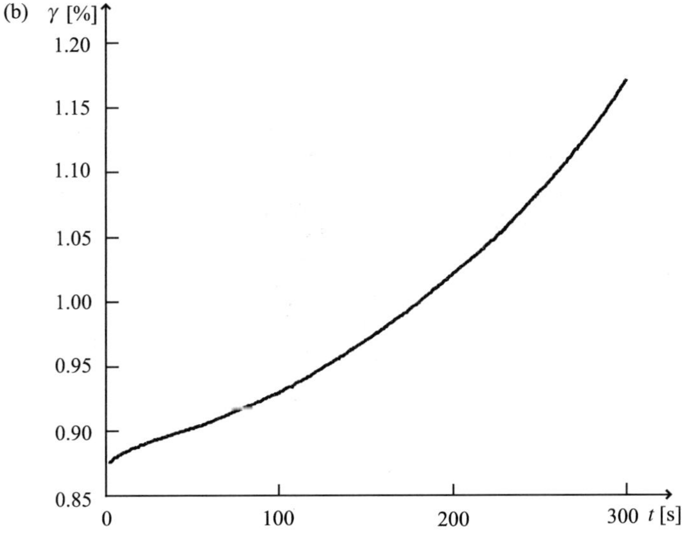

Figure 9.14. Same conditions as Figure 9.13 with a creep test C performed before the UYP in (a). Details of the creep curve in (b). Creep rate and estimated values of the mobile dislocation density in (c). The vertical bar indicates the experimental scatter around the average curve (Courtesy of C. Dupas).

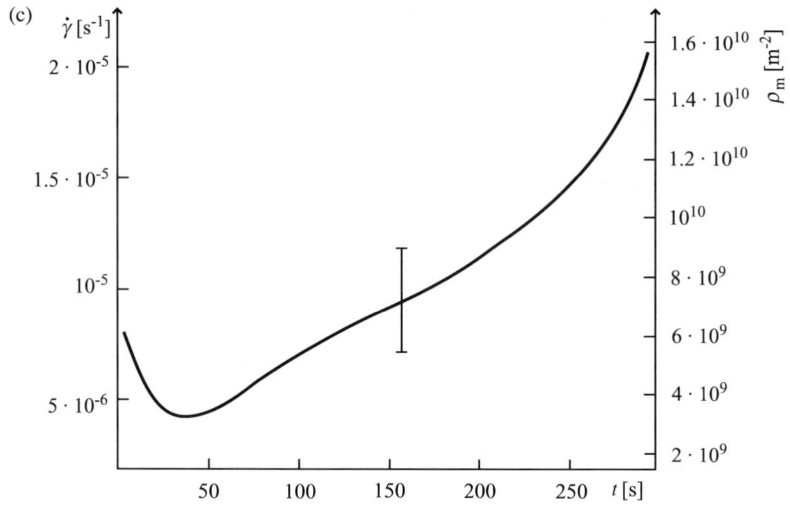

Figure 9.14. (*continued*)

As shown in Figure 9.14(c), it increases from about 5×10^9 to 16×10^9 m^{-2} over 300 s. In addition, the density at the onset of the transient is found to be about 20 times lower before the UYP as compared to that after the LYP for the same stress and temperature.

9.1.3.2 Three dimensional mesoscopic simulations of dislocation multiplication. From the early studies of covalent crystal plasticity, various efforts have been made to determine the dislocation multiplication laws that underlie the yield point. Alexander and Haasen (1968) were already emphasizing that the evolution of the mobile dislocation density with stress or strain was a key issue to the understanding of this plasticity. They proposed a model based on a simple set of coupled equations which describes the evolution of the dislocation density (assumed to be uniform). The equations are the following:

$$\dot{\gamma} = \dot{\tau}/M + \rho_m b v \tag{9.1}$$

which results from the machine equation (2.6) combined with the Orowan relation (1.1),

$$v = v_0(\tau^*/\tau_0)^m \exp(-Q/kT) \tag{9.2}$$

where v_0, τ_0, m and Q are constants. Eq. (9.2) refers to the length independent regime of the dislocation velocity (Section 7.2). The effective stress is:

$$\tau^* = \tau - \alpha\sqrt{\rho} \tag{9.3}$$

where Eq. (2.6) has been used together with the Taylor relation.

For modelling dislocation multiplication, they started from the general empirical law:

$$d\rho_{\mathrm{m}} = \rho_{\mathrm{m}} \mathrm{v}\, \mathrm{d}t\delta \tag{9.4}$$

δ being a multiplication coefficient. It expresses the increment $\mathrm{d}\rho$ of dislocation density during dt as proportional to the area swept by the dislocations ($\rho \mathrm{v}\ \mathrm{d}t$) or the plastic strain. It is used for metallic crystals (see e.g. Estrin, 1996). With the Orowan relation, Eq. (9.4) becomes:

$$\dot{\rho}_{\mathrm{multi}} = K_{\mathrm{multi1}} \dot{\gamma}_{\mathrm{p}} \tag{9.5a}$$

Here, and in the following, $K_{\mathrm{multi}\ i}$ (with i between 1 and 3) is a constant.

Alexander and Haasen transformed Eq. (9.4) by expressing $\delta = K\tau^*$, K being a constant. Under such conditions:

$$\dot{\rho}_{\mathrm{multi}} = K_{\mathrm{multi2}} \dot{\gamma}_{\mathrm{p}} \tau^* \tag{9.5b}$$

This law has been widely accepted and used, in spite of its phenomenological aspect, since it properly describes the upper yield point characteristics as a function of temperature and strain rate. However, its validity has been questioned (George and Rabier, 1987).

In addition, little was known about the specific properties of Frank–Read sources in dc covalent crystals, such as their dynamics in the presence of a strong lattice resistance to glide. To study these issues, three dimensional mesoscopic computer simulations of such sources in Si were performed by Moulin et al. (1997). They considered a single source, operating with a screw segment, at the centre of the model crystal. Because of the Peierls forces, the loop segments were along the $\langle 110 \rangle$ directions of the $\{111\}$ slip plane. The effective stress acting on each dislocation segment includes three contributions: (i) the resolved applied stress, (ii) the sum of the elastic interaction stresses and (iii) the back stress originating from the line tension force which opposes any increase in length. The velocity laws considered were equations 7.22(a) and (b) of the kink-pair mechanism for, respectively, the length dependent and independent regimes. With this set of rules, a source segment, under a constant applied stress, expands and emits hexagonal loops as illustrated in Figure 9.15. Such simulations reproduce the dynamical and geometrical aspects of dislocation sources in Si, as observed in TEM (see e.g. Figure 7.11).

Furthermore, Moulin et al. (1999) succeeded in reproducing the UYP properties in Si crystals by expanding the previous mesoscopic simulations as follows. The model specimen contains sources of the above type, randomly distributed and exhibiting a Gaussian length distribution. The dislocation velocity corresponds to equation 7.22(b) (low stresses and high temperatures). The effective stress is used and the activation energy is taken from dynamic etch pit data (see George and Rabier, 1987). Multiplication takes place exclusively at the initial sources (no cross-slip possibilities). Under these assumptions, the authors succeeded in reproducing dislocation structures reasonably similar to those observed by TEM. They also generated stress–strain curves that exhibit a UYP and a

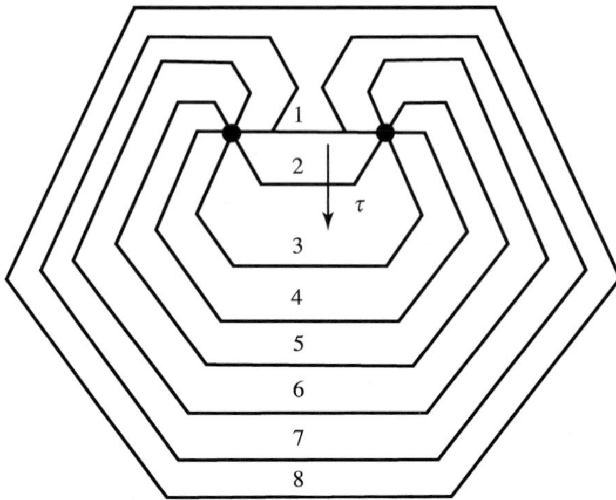

Figure 9.15. Simulated loop emission by a Frank–Read source in a {111} slip plane of Si. The screw source segment is 0.81 μm long. $\tau = 35$ MPa. $T = 1000$ K. From Moulin et al. (1997).

LYP, the features of which agree with the observed ones. The dislocation density increases rapidly from the onset of deformation until after the UYP. It still exhibits a moderate increase at the LYP. The average dislocation velocity follows the evolution of stress.

Moreover, Moulin et al. (1999) were able to provide a dislocation multiplication law based on the following arguments. The simulated sources emit approximately equidistant loops, the distance Δe between two successive loops scaling with $1/\tau^*$. Under such conditions, all the loops move at the same velocity v. During dt, the change in mobile density is $\mathrm{d}\rho_\mathrm{m} = \Phi N \mathrm{v}\mathrm{d}t/V$, for N loops emitted by a source, where Φ is a geometrical factor, V the crystal volume. ρ_m is related to the N by:

$$\rho_\mathrm{m} = (\Phi/V)\left(\sum_{i=1}^{N} i\,\Delta e\right) \approx (\Phi\Delta e/2V)N^2$$

where the ith loop is at distance $i\,\Delta e$ from the source.

Then, by eliminating N, dρ_m can be written $\mathrm{d}\rho_\mathrm{m} \propto \mathrm{v}\sqrt{\rho_\mathrm{m}}\mathrm{d}t/\sqrt{\Delta e}$. Expressing $\Delta e(\tau^*)$ and using the Orowan relation, a dislocation multiplication law is obtained:

$$\dot{\rho}_\mathrm{multi} = K_\mathrm{multi3}\,\dot{\gamma}_\mathrm{p}\sqrt{\tau^*/\rho_\mathrm{m}} \tag{9.5c}$$

The corresponding multiplication rate is illustrated in Figure 9.16, where it is compared with those related to the previously proposed laws. This figure shows that the peak in multiplication rate reproduced by Eqs. (9.5a) and (9.5b), does not properly coincide with the simulated UYP as compared to the more physically based predictions of the present simulations (law Eq. (9.5c)).

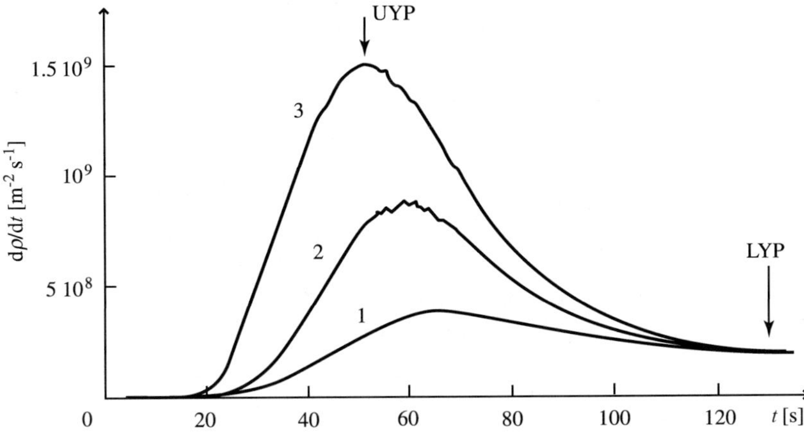

Figure 9.16. Simulated multiplication rate dρ/dt versus time for various multiplication laws in Si. Curves 1 and 2 correspond to Eqs. (9.5a) and (9.5b), respectively, and curve 3 to Eq. 9.5(c). $T = 1000$ K. Shear strain rate: $6.4 \times 10^{-5}\ \mathrm{s}^{-1}$. Initial dislocation density: $7.5 \times 10^7/\mathrm{m}^2$. The YPs are simulated ones. From Moulin et al. (1999).

9.1.3.3 Testing the proper multiplication laws. Consequences of Eqs. (9.5a)–(9.5c) can also be tested as follows, to determine which of them better fits the mechanical test results. A recent attempt with Germanium data (Figure 9.11) by Fikar et al. (2002) is presented, focussed on dislocation densities. The mobile dislocation density is the net result of multiplication and exhaustion (namely storage and escape at the free surfaces). This latter effect can be estimated, considering that during dt, the escaping defects are in a volume v dtS, (S is the escape section), their number being ρ_mv dtS. Therefore, using the Orowan relation yields:

$$\dot{\rho}_\mathrm{surf} = -\frac{2}{b\ell}\dot{\gamma}_\mathrm{p} \tag{9.6}$$

where ℓ is the crystal dimension along the direction of motion ($\ell = 2\mathrm{V/S}$, where V is the sample volume). For the other exhaustion processes, the three following laws have been considered:

$$\dot{\rho}_\mathrm{ex} = K_\mathrm{ex1}\dot{\gamma}_\mathrm{p} \tag{9.7a}$$

$$\dot{\rho}_\mathrm{ex} = K_\mathrm{ex2}\dot{\gamma}_\mathrm{p}\rho \tag{9.7b}$$

$$\dot{\rho}_\mathrm{ex} = K_\mathrm{ex3}\dot{\gamma}_\mathrm{p}\sqrt{\rho} \tag{9.7c}$$

In the three equations above, $K_{\mathrm{ex}j}$ (with j between 1 and 3) is a negative constant.

Under such conditions, the evolution of the mobile dislocation density is characterized by:

$$\dot{\rho}_\mathrm{m} = \dot{\rho}_\mathrm{multi} - \dot{\rho}_\mathrm{ex} - \dot{\rho}_\mathrm{surf} \tag{9.8}$$

and that of the total density by:

$$\dot{\rho} = \dot{\rho}_{\text{multi}} - \dot{\rho}_{\text{surf}} \tag{9.9}$$

For the sake of clarity, let us consider the combination of Eqs. (9.5a) and (9.7a) to be tested first. In the first step, these two equations are combined with Eqs. (9.1)–(9.3), (9.8) and (9.9). A system of coupled first order differential equations is obtained which express, successively, $\dot{\gamma}$, $\dot{\tau}$, $\dot{\rho}_m$ and $\dot{\rho}$ as a function of the quantities τ, ρ_m and ρ. It can be numerically integrated which yields stress, strain, mobile and total densities as a function of time. In the second step, the free parameters are fitted to mechanical test data points, each one consisting of a set of three quantities (time, stress, strain).

To reduce the number of adjustable parameters, v_o, m and Q in Eq. (9.2) have been determined from dislocation velocity data from etch pit experiments (Georges and Rabier, 1987). In Eq. (9.3), the coefficient α, which characterizes the type of dislocation interaction, has been taken as equal to $1/\pi$, from Berner and Alexander (1967). Therefore, two parameters only have to be adjusted, namely $K_{\text{multi}\ i}$ and $K_{\text{ex}\ j}$.

Since three multiplication and annihilation laws have been considered above, nine combinations of relations Eqs. (9.5) and (9.7) have to be tested.

The case of transient tests, being more simple, is treated first. Indeed, equation (2.6) is changed into (2.7) in the case of stress relaxations, while under creep conditions, the plastic strain rate equals the creep rate. Curves corresponding to the best fits only, are presented below. Figure 9.17 shows an excellent agreement for a stress relaxation test and a creep transient experiment, combining Eqs. (9.5c) and (9.7c). The corresponding mobile dislocation densities are represented in Figure 9.18.

To calculate a monotonic curve, the strain rate is fixed to the experimental value. Results are illustrated in Figure 9.19, using the same multiplication and annihilation laws as for the transient tests. The initial conditions are set at the UYP. The onset of the curve is

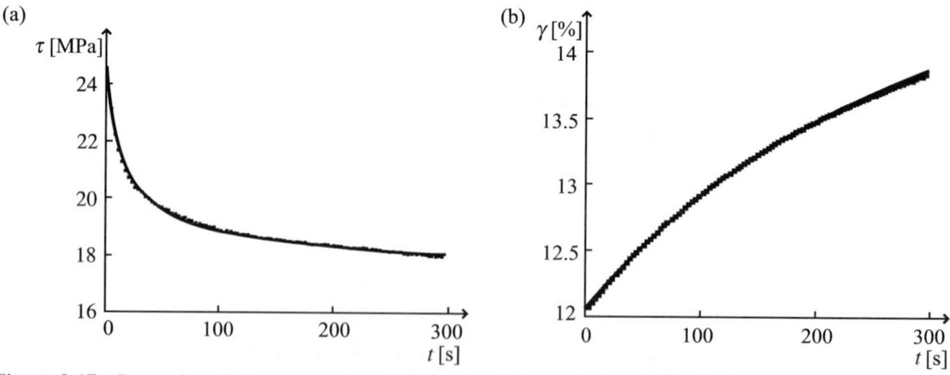

Figure 9.17. Comparison between experimental (dots) and best calculated (thin line) curves for transient tests performed after the LYP. ⟨123⟩ Ge single crystal at 750 K. (a) Stress relaxation experiment. (b) Creep transient at a 12% strain. $\tau = 25.4$ MPa. Monotonic shear strain rate $8.6 \times 10^{-5}\ \text{s}^{-1}$. Laws (9.5c) and (9.7c) have been combined in (a) and (b). From Fikar et al. (2002).

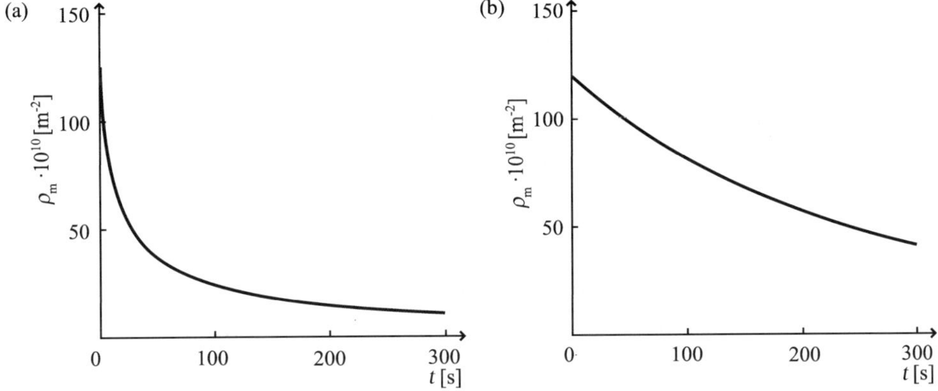

Figure 9.18. Mobile dislocation densities as a function of time during the tests of Figure 9.17. (a) Stress relaxation experiment. (b) Creep transient. Same laws as in Figure 9.17. From Fikar et al. (2002).

quite well fitted, at least to a shear strain of 24%. In particular, the stresses and strains corresponding to the UYP and the LYP are satisfactorily reproduced. The internal stress and the mobile dislocation density can be predicted and their variations with strain are represented in Figures 9.19 and 9.20, respectively.

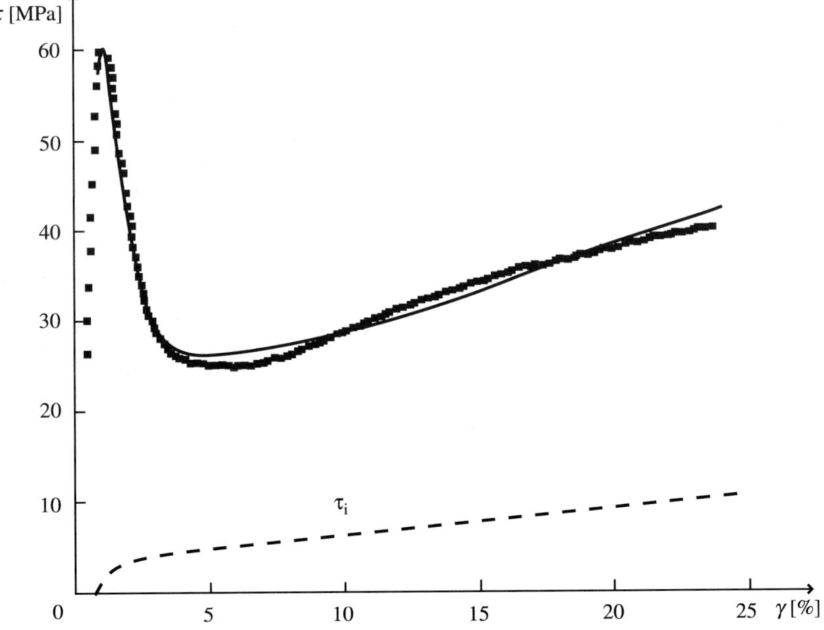

Figure 9.19. Experimental stress–strain curve (dots) and calculated curve (thin continuous line) for a ⟨123⟩ Ge single crystal at 750 K. The dotted line curve corresponds to the calculated internal stress. Same laws as in Figure 9.17. From Fikar et al. (2002).

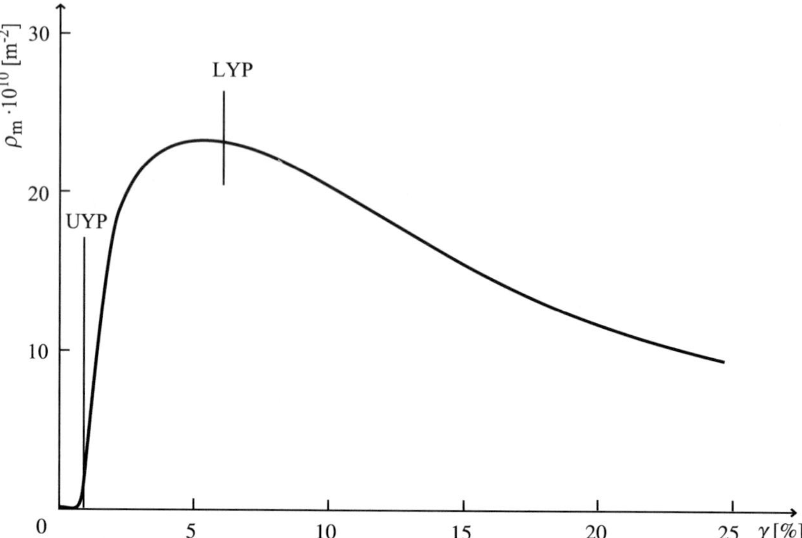

Figure 9.20. Mobile dislocation density as a function of strain corresponding to the experiment in Figure 9.19. The positions of the UYP and the LYP are indicated. From Fikar et al. (2002).

To conclude, with regard to the preceding simulations, the method used here has several advantages. Only two adjustable parameters have to be fitted. This was not the case in a comparable attempt of constitutive modelling of metallic crystal plasticity, based on dislocation densities (Kocks, 1976; Mecking and Kocks, 1981).

With regard to the choice of exhaustion laws, the various types used (Eqs. (9.7)) do not reveal drastic differences in the fit quality. For multiplication, law (9.5a) used for metals is the worst one, while the law proposed by Moulin et al. (1999) seems to be more appropriate (Eq. (9.5c)).

9.1.3.4 Conclusions about dislocation multiplication in covalent crystals. The analysis in Sections 9.1.3.2 and 9.1.3.3 confirms the microstructural phenomena that underlie the mechanical response under the various conditions of Section 9.1.3.1. At the onset of monotonic loading, a significant stress increase is necessary to enhance dislocation mobility and the source operation rate. At the UYP, the high multiplication rate acheived results in a zero stress rate, i.e. the plastic strain rate equals the applied one, according to equation (2.6). However, the dislocation density still increases (Figure 9.20), so that overshooting takes place, leading to a decrease in the applied stress. Therefore, the presence of the yield point is directly related to the low initial dislocation densities, together with the low stress sensitivity in the dislocation velocities as expressed by Eq. (9.2) and described in Section 7.3.2. The mobile dislocation density is still weakly increasing at the LYP. At that stage, dislocation interactions lead to crystal hardening as normal plastic flow sets in.

In addition, the mobile dislocation density in the range of $10^9\ m^{-2}$ at the onset of deformation, has increased by 2 orders of magnitude after the LYP (see Figure 9.20 for Ge at 750 K). Before the LYP, ρ_m and ρ are quite similar, while the internal stress is negligible. It then starts increasing with strain and is a small fraction of the total stress at 750 K (Figure 9.19). As far as transient tests are concerned, before the LYP, dislocation multiplication dominates, while exhaustion is negligible. However, after the LYP, exhaustion is more significant than multiplication, so that ρ_m is observed to decrease along the stress relaxation or the creep transient (Figure 9.18). The corresponding exhaustion mechanism is described in Section 9.2.4.

Finally, whilst the recent developments given in Sections 9.1.3.2 and 9.1.3.3 are a step forward in the clarification of the behaviour of dc covalent crystals, some improvements are still needed. For the mesoscopic simulations, the source density evolution with strain should be accounted for, as well as glide heterogeneity at low strains. For the numerical simulations, deformation curves corresponding to other temperatures and strain rates should be investigated.

9.2. MOBILE DISLOCATION EXHAUSTION

Among all possible exhaustion mechanisms, some which have been identified are now presented under various conditions. The following situations are quoted as examples: cell formation in copper deformed in stage II, Kear Wilsdorf locking in Ni_3Al, impurity or solute diffusion towards dislocation cores in a CuAl solid solution and dipole formation and annihilation in germanium.

9.2.1 Cell formation

The example of $\langle 110 \rangle$ copper single crystals deformed in stage II is described first. In this double slip orientation, the well known stages II and III are observed as stress increases, as illustrated in Figure 9.21. The hardening coefficient is constant along stage II with a value close to 185 MPa, which corresponds to θ/μ close to 1/230 (μ is the shear modulus). Such a value is commonly accepted (Kocks, 1966). Hardening decreases in stage III due to the activation of dislocation cross-slip. The microstructural parameters during stage II can be determined using successive relaxations (Sections 2.1.4 and 2.1.5). An example is illustrated in Figure 9.22. It shows that after 200s, the mobile dislocation density keeps 55% of its onset value ρ_{mo}, while average dislocation velocities have decreased substancially.

In the following, in order to compare different materials under various deformation conditions, a mobile dislocation exhaustion rate, labelled $\Delta\rho_m/\rho_{mo}$, has been defined as

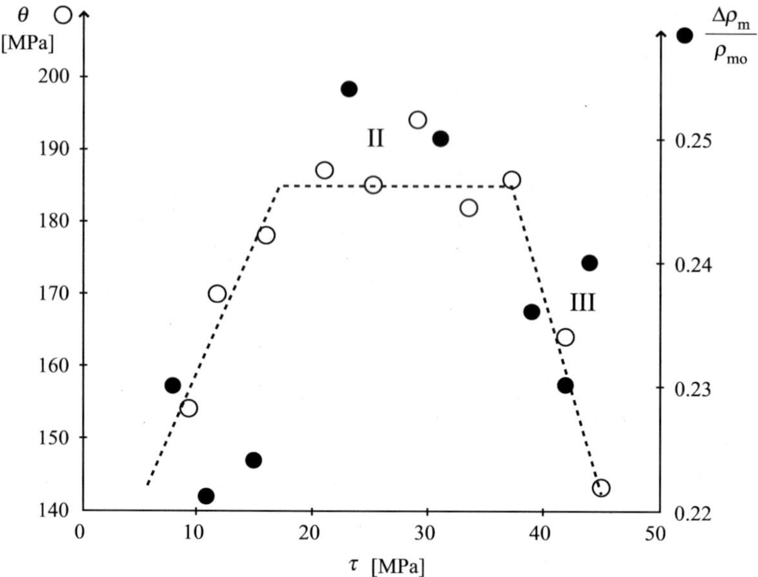

Figure 9.21. Work-hardening coefficient θ and mobile dislocation exhaustion rate $\Delta\rho_m/\rho_{mo}$ as a function of stress along the stress–strain curve. ⟨110⟩ Cu single crystal at 300 K. From Martin et al. (1999).

follows: $\Delta\rho_m$ is the mobile density decrease, which corresponds to a decay of one order of magnitude of the onset deformation rate (or to a time $t = 9c$ in relations (2.27) and (2.36)).

If a test similar to that of Figure 9.22 is performed along the stress–strain curve, the mobile dislocation exhaustion rate is observed to change as illustrated in Figure 9.21. It is remarkable that the parameter $\Delta\rho_m/\rho_{mo}$ follows the same trend as the work-hardening coefficient, within the experimental scatter.

Though no predictive and quantitative description of stages II and III is available in FCC metals (see e.g. Nabarro, 1986; Argon, 1996), it is accepted that the interaction of two dislocation families moving, respectively, along two different slip planes is responsible for stage II work-hardening (Seeger, 1957). These interactions lead to the formation of cells as illustrated in Figure 9.23.

Therefore, under the present stage II conditions, dislocation exhaustion through cell formation leads to θ/μ close to 1/230 and a mobile dislocation exhaustion rate of the order of 25% (Figure 9.21). This means that 25% of the mobile dislocations are coming to rest when the deformation rate is 10% of its onset value.

9.2.2 *Exhaustion through lock formation in Ni_3Al*

Screw dislocation locking in Ni_3Al is described in Section 10.1.4.1. This process leads to high mobile dislocation exhaustion rates $\Delta\rho_m/\rho_{mo}$. The values of this parameter are plotted

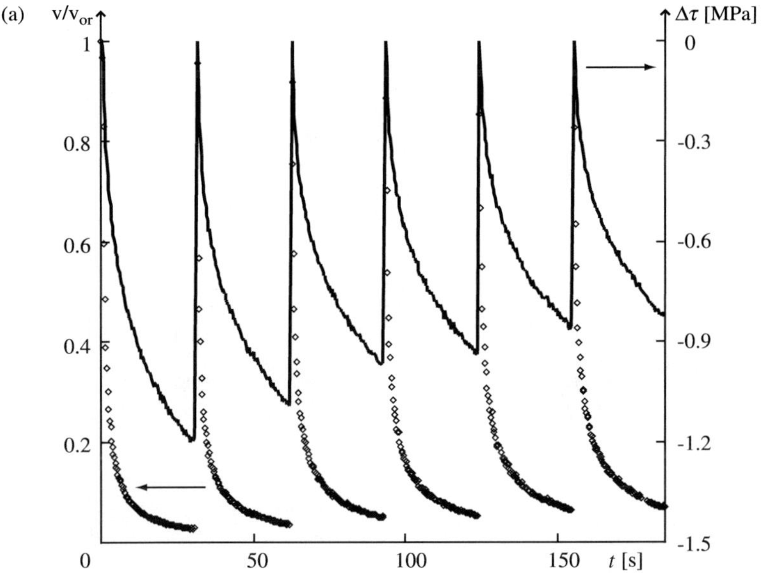

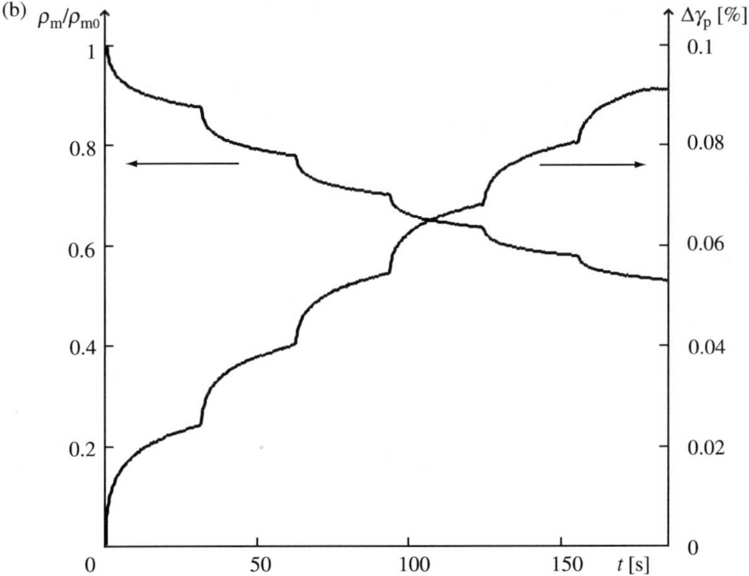

Figure 9.22. Repeated stress relaxation experiment (six tests) performed in stage II of a ⟨110⟩ Cu single crystal at 300 K. $\tau = 39.2$ MPa. (a) Amount of relaxed stress and dislocation velocity. v_{or} is the velocity at the onset of each relaxation. (b) Corresponding plastic strain and mobile dislocation density. ρ_{mo} is the mobile density at the onset of the transient (Courtesy of T. Kruml and O. Coddet).

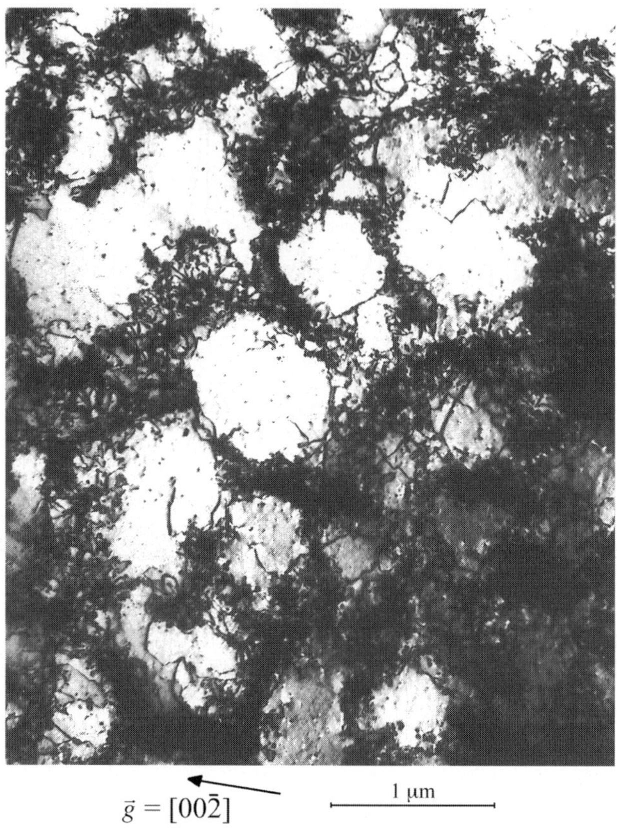

Figure 9.23. Typical cell structure corresponding to the end of stage II in a ⟨110⟩ Cu single crystal deformed at 300 K after Bonneville et al. (1988).

in Figure 9.24 along the stress–strain curve together with those of the work-hardening coefficient. In $Ni_3(Al,Hf)$, room temperature corresponds to the onset of the temperature range where the flow stress increases anomalously with temperature (see Figure 10.24). In Figure 9.24, a preplastic domain corresponds to a decrease of θ as stress increases. In the plastic domain, θ becomes constant but retains a high value. In the vicinity of the yield stress (170 MPa), θ is 5200 MPa (θ/μ close to 1/22) while $\Delta\rho_m/\rho_{mo}$ is 77% (Figure 9.24). It is worth noting that the $\theta(\tau)$ and $\Delta\rho_m/\rho_{mo}(\tau)$ curves look different in Figures 9.21 and 9.24 since the work-hardening mechanisms that operate in both types of crystals are not the same. The comparison between hardening in metals and in ordered intermetallics is considered and interpreted in Section 10.1.7.2.

The new fact here is that, as observed in Cu above, the variation of θ with stress parallels that of the parameter $\Delta\rho_m/\rho_{mo}$. Moreover, Matterstock et al. (1999) measured

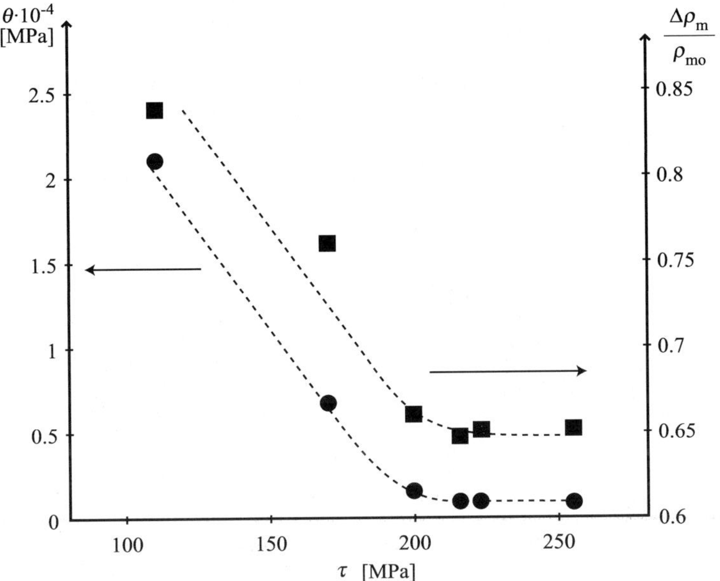

Figure 9.24. Work-hardening coefficient θ and mobile dislocation exhaustion rate $\Delta\rho_m/\rho_{mo}$ as a function of stress along the stress–strain curve. ⟨123⟩ $Ni_3(Al,Hf)$ single crystal at 300 K. From Martin et al. (1999).

these two parameters at a 0.2% plastic strain in these single crystals. They showed that they exhibit parallel trends as a function of temperature: they first increase and then decrease, the peak temperature being close to 550 K. Another example is shown on Figure 10.28, from Kruml et al. (2002), for $Ni_{74}Al_{26}$ polycrystals. The parameters are measured at, respectively, 3 and 5% plastic strains.

Therefore, there seems to be a fair correlation between hardening and dislocation exhaustion in the cases of Cu and Ni_3Al, despite different hardening mechanisms.

9.2.3 *Impurity or solute pinning (Cottrell effect)*

It has been shown in Section 3.2.2 that dislocations moving slowly or arrested at obstacles may attract impurities or solute atoms that diffuse towards the core, provided the temperature is adequate. Such an effect can also exhaust mobile dislocations. It is illustrated here in different types of crystals.

First, a Cu 7.5 at.%Al solid solution is examined. When deformed at room temperature, the single crystal stress–strain curves exhibit a marked stage of easy glide with almost no hardening, followed by a parabolic hardening stage (Neuhäuser and Schwink, 1993). Several aspects of Cottrell locking can be observed in this alloy. Figure 9.25 shows a repeated stress relaxation test performed after the stage of easy glide. The relaxations are logarithmic, so that $\Delta\rho_m/\rho_{mo}$ can be estimated through curves similar to those in

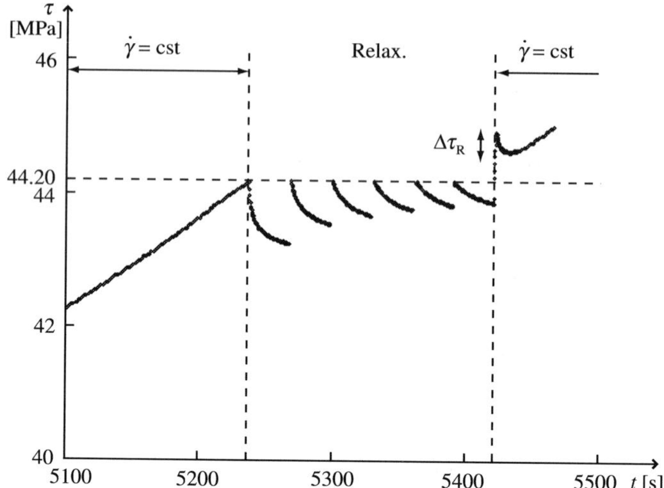

Figure 9.25. Repeated stress relaxation experiment on a Cu 7.5 at.%Al single crystal at 295 K. Constant shear strain rate $16.3 \times 10^{-5}\ s^{-1}$, $\gamma = 39\%$ (Courtesy of T. Kruml and O. Coddet).

Figure 9.22. At $\tau = 44.2$ MPa, $\Delta\rho_m/\rho_{mo}$ is close to 42%, about twice the value found above for Cu. A remarkable feature of Figure 9.25 is that a yield point at reloading after the transient test (a few tenth of MPa) can be observed, unlike in Cu under comparable conditions (Section 9.2.1). This is thought to be the signature of ageing aluminium atoms migrating towards dislocations as the crystal is held up at stresses close to 44 MPa during the transient test. Neuhäuser and Schwink (1993) consider that solute diffusion takes place in FCC solid solutions at temperatures as low as $T_M/5$, T_M being the melting point. At such low temperatures, they quote that this is not a long distance diffusive motion but, more likely, switching of solutes to favourable positions in the core region.

Another evidence of this phenomenon in the same alloy is obtained using dip test experiments (Section 2.1.7) along the monotonic curve. An example of results is presented in Figure 9.26. It represents the reduced strain rate as a function of the stress reduction $\Delta\tau$. The reduced strain rate is first observed to decrease as $\Delta\tau$ increases. For $\Delta\tau$ between 2 and 5.5 MPa, a zero strain rate is observed, while it becomes negative for larger $\Delta\tau$s. This trend is not observed in similar experiments performed by Milicka (1999) during creep of CuZn alloys. In this latter case, the curve corresponding to Figure 9.26 clearly intersects the horizontal axis at a precise $\Delta\tau$ value. The present data on CuAl can be understood as follows: for a stress reduction of 2 MPa, the applied stress is 30.6 MPa, close to the internal stress, since a zero strain rate is observed. Dislocations are slow enough to get pinned by the diffusing solute atoms. For larger stress reductions, the applied stress is to small for pinned dislocations to move back. This reverse motion becomes possible as $\Delta\tau$ approaches 6 MPa, which corresponds to the onset of a negative strain rate.

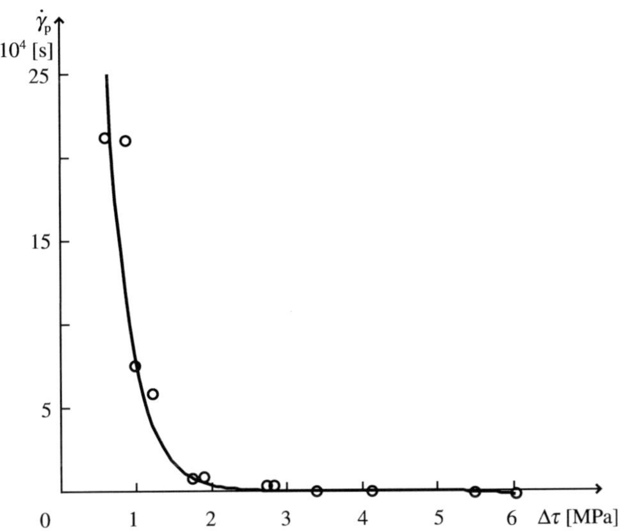

Figure 9.26. Data from a strain dip test experiment on a Cu 7.5 at.%Al single crystal at 295 K. $\tau = 32.6$ MPa $\gamma = 24\%$. (Courtesy of T. Kruml and O. Coddet).

Dislocation pinning by diffusive processes is also evidenced in a $Ni_3(Al,Hf)$ single crystal deformed around the peak temperature for the yield stress (about 750 K). A section of the motonic curve corresponding to constant strain rate compression is shown in Figure 9.27(a). Serrated yielding can be observed around 513 MPa, which corresponds to dynamic strain ageing (see Section 3.2.2). Another spectacular effect of a diffusive process is illustrated in Figure 9.27(b). A series of four successive relaxations is performed at 395 MPa. An unusual behaviour is observed when comparing the final and initial strain rates of neighbouring relaxation curves, which are indicated on the figure: in the four cases, the strain rate at the end of a relaxation is larger than the initial strain rate of the following relaxation (negative strain rate sensitivity). The reverse situation is observed in all relaxation series reported so far (see e.g. the schematics of Figure 2.5). Indeed, they provide evidence that reloading between two relaxations yields a larger strain rate under a higher stress. In the present case, it is believed that during the quasi-elastic reloading, immobile dislocations are pinned by diffusive processes, so that they move more slowly at the onset of the following relaxation. In addition, the subsequent constant strain rate curve exhibits a reloading yield point caused by dislocation unpinning. Note that the diffusive processes able to lock dislocations in such compounds, as well as the corresponding temperature range, are discussed in Section 10.1.5.5.

9.2.4 Exhaustion with annihilation

A different process of dislocation exhaustion is observed in the Ge single crystals of Section 9.1.3. Advantage is taken of the thermally activated friction forces in this compound.

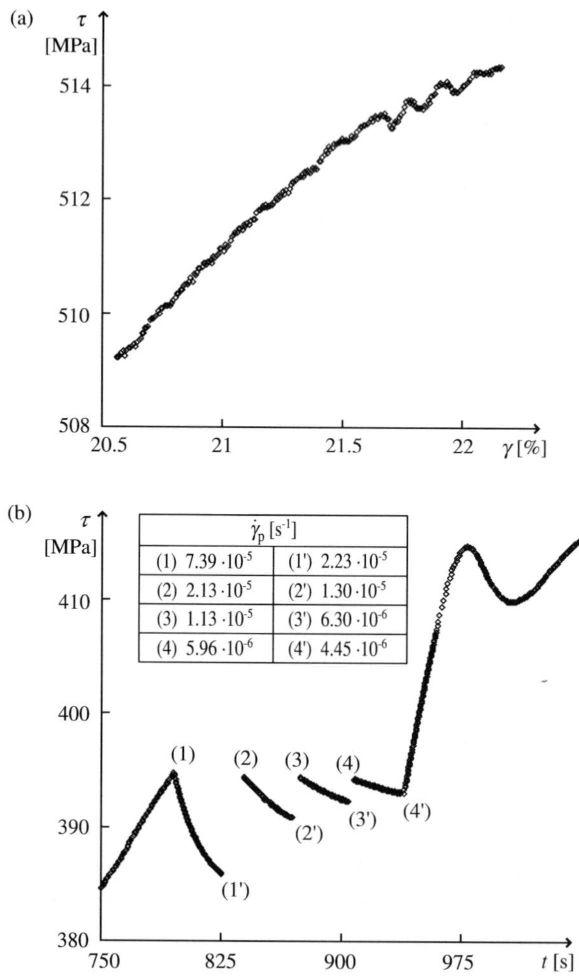

Figure 9.27. Data from deformation experiments in a ⟨123⟩ $Ni_3(Al,Hf)$ single crystal. (a) Stress–strain curve at 773 K. (b) Succession of four relaxations during monotonic loading at 873 K. Strain rates are indicated. (1)–(4) refer to initial ones and (1′)–(4′) to final ones (Courtesy of B. Viguier and T. Kruml).

These allow the stressed dislocations to be pinned by cooling the sample under load after deformation. Above the LYP, TEM observations reveal some amount of strain localization, as illustrated in Figure 9.28(a). Looking at the primary glide plane shows dislocation dipoles (Figure 9.28(b)). It is thought that dislocation groups of opposite signs, moving along neighbouring planes, are responsible for dipole formation. Figure 9.28(b) also shows evidence of small prismatic loops. A systematic investigation of their average sizes reveals that it decreases for rising deformation temperatures. This suggests that these loops are produced by dipole coalescence through cross-slip or climb, depending on dislocation character.

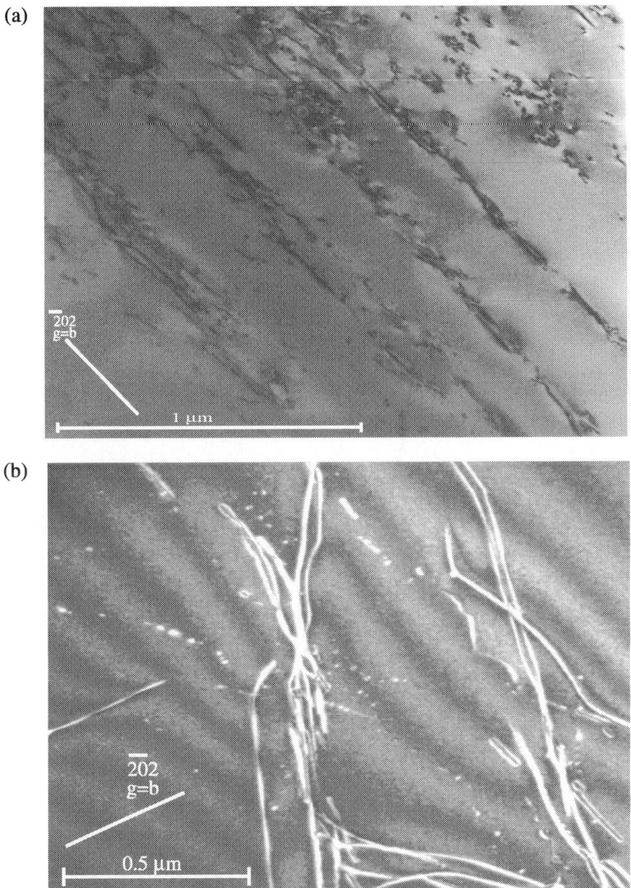

Figure 9.28. TEM observation of dislocation arrays in a ⟨123⟩ Ge single crystal after the LYP at 750 K. The primary glide plane is (a) end on ($\tau = 34$ MPa) and (b) parallel to the picture plane ($\tau = 25$ MPa). From Dupas et al. (2002).

These TEM observations indicate that dislocation annihilation takes place during plastic deformation. This phenomenon accounts for the multiplication yield point observed when reloading the sample after a relaxation series shown in Figure 9.29. It may look similar to the ones shown in Figures 9.25 and 9.27 for the CuAl alloy and Ni_3Al, respectively. However, the underlying mechanisms are different.

To conclude this section, dislocation exhaustion can occur through different mechanisms, depending on the crystal type and the deformation conditions. In most examples presented here, a fair correlation is found between the work-hardening coefficient and the mobile dislocation exhaustion rate. These preliminary considerations do not include yet a check of the proposed exhaustion laws (Eqs. 9.7).

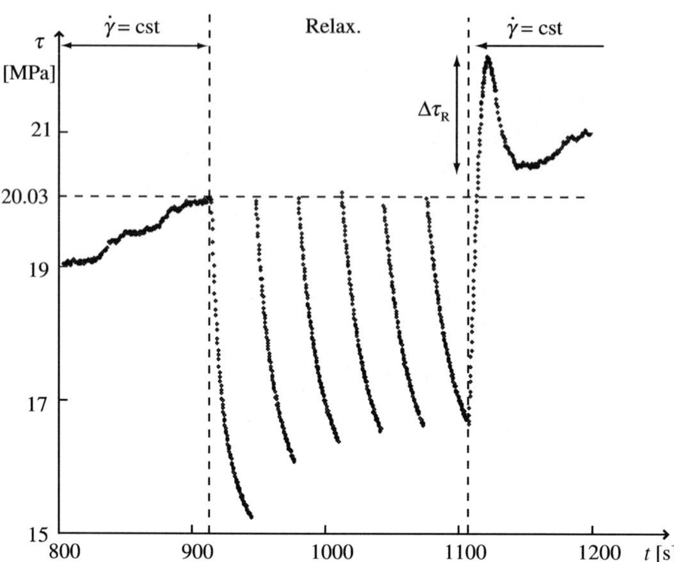

Figure 9.29. Stress relaxation series in a ⟨123⟩ Ge single crystal at 750 K. A reloading yield point, of amplitude $\Delta\tau_R$ is clearly visible. (Courtesy of C. Dupas and T. Kruml).

9.3. WORK-HARDENING VERSUS WORK-SOFTENING

Recent reviews concerning the state of development of a work-hardening theory (see e.g. Kocks and Mecking, 2003 or Nabarro et al., 2002) confirm an early statement by Cottrell (1953): "It was the first to be attempted by dislocation theory and may be the last to be solved". One difficulty for a general theory of work-hardening is that this phenomenon is the net result of dislocation multiplication and exhaustion processes. As illustrated in Sections 9.1 and 9.2, their description is far from being complete.

It is interesting to note that repeated stress relaxation experiments provide some insight into these phenomena. All experiments of this type presented so far (see Chapter 2 and Section 9.2) are related to positive work-hardening coefficients (see e.g. Figure 2.5). The corresponding relaxation curves exhibit a stress drop that decreases as the relaxation number increases along the series. It has also been shown that this provides evidence of a decrease of the mobile dislocation density with increasing strain.

Some examples are presented in which the work-hardening coefficient can be negative or close to zero. The Taylor relation expresses the stress as a linear relation of the square root of dislocation density. It says that positive θ values correspond to an increase of the total dislocation density with strain. In other words, dislocation multiplication takes

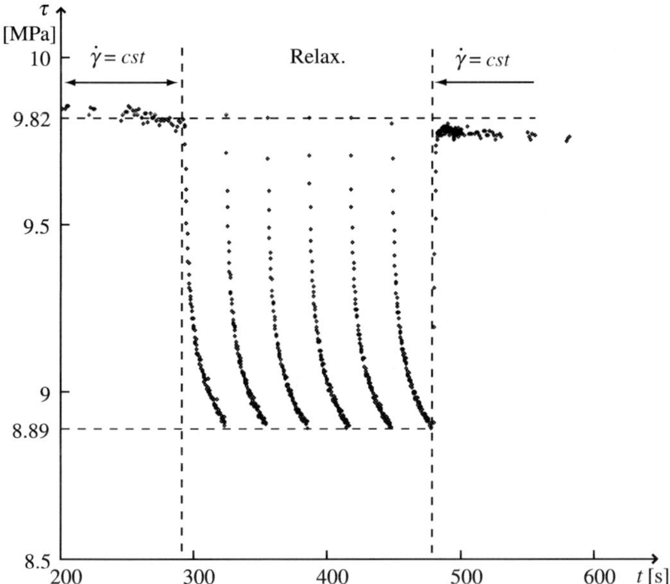

Figure 9.30. Monotonic stress–time curve in stage I of a Cu 7.5 at.%Al single crystal with a repeated stress relaxation experiment. $T = 300$ K. $\gamma = 4.5\%$, $\dot{\gamma} = 16.3\ 10^{-5}\ s^{-1}$ (Courtesy of T. Kruml and O. Coddet).

over annihilation. For work-softening, the total dislocation density decreases with strain, i.e. exhaustion processes dominate. For θ close to zero, multiplication counterbalances annihilation with an approximately constant dislocation density.

Figure 9.30 illustrates this latter case. The stress–strain curve clearly exhibits a marked stage of zero hardening that corresponds to easy glide. The stress relaxation curves are identical along the series. This can be interpreted considering a constant mobile dislocation density throughout the transient. The decrease of deformation rate along one curve reflects that of the dislocation velocity with stress. It is remarkable that the stress level of 9.8 MPa on the monotonic curves is the same just before and after the transient. This confirms that no change in dislocation density occurred.

Figure 9.31 illustrates a situation corresponding to work-softening. For some reason, the stress–strain curve of a predeformed copper crystal exhibits a negative slope at the onset of plasticity. It is clear that in the corresponding transient test, the amount of stress drop for each relaxation curve increases along the series (Figure 9.31). This reflects an increase in mobile dislocation density with time. Conversely, for positive work-hardening (point (B) of the monotonic curve), the stress relaxation has the same features as in Figure 9.25 (decreasing mobile density).

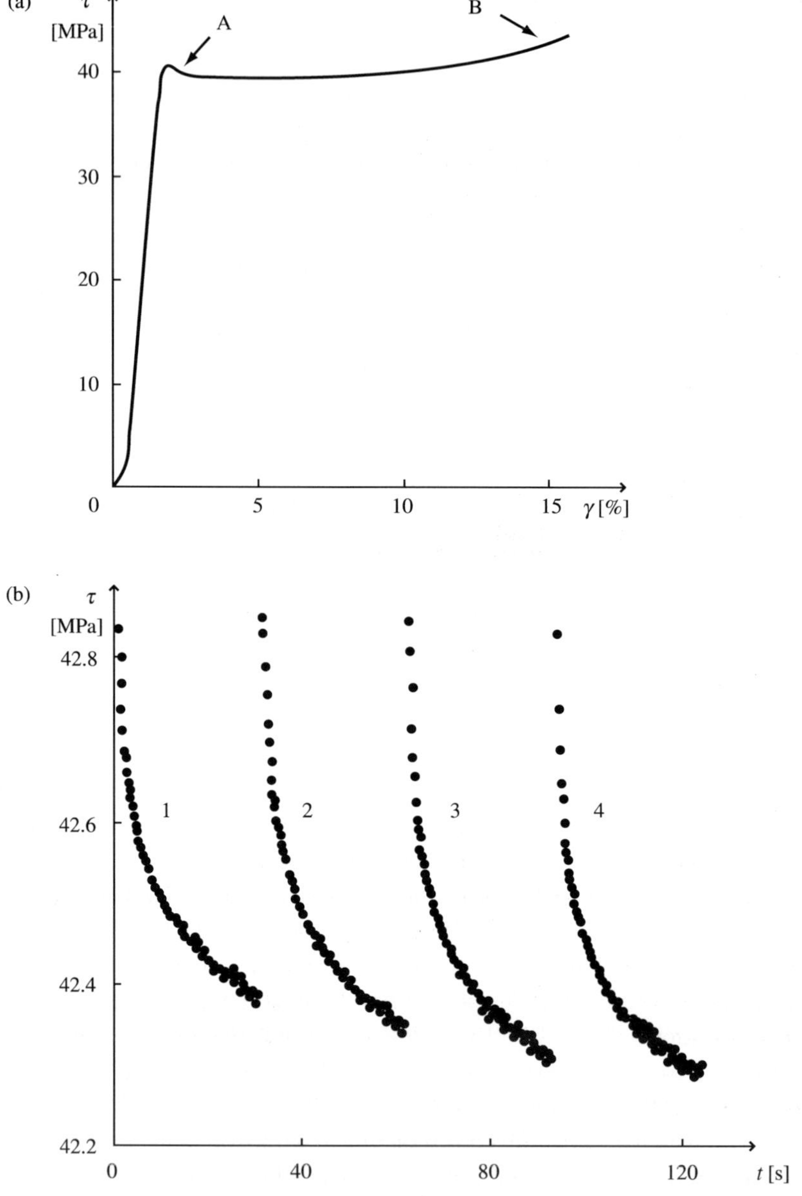

Figure 9.31. Mechanical behaviour of a predeformed Cu single crystal at 120 K. Shear strain rate: $3.8 \times 10^{-5}\ s^{-1}$ (a) Stress–strain curve exhibiting work-softening at point A and hardening in B. (b) Repeated stress relaxation curve in A (Couteau et al., 2001).

9.4. CONCLUSIONS ABOUT DISLOCATION MULTIPLICATION, EXHAUSTION AND SUBSEQUENT WORK-HARDENING

Many early models of sources have been observed experimentally, after they had been imagined. Selected dislocation exhaustion processes have been presented, respectively, with and without annihilation. After a stress relaxation test, two situations are observed in the case of an exhaustion transient: a yield point at reloading is observed or not, depending on the conditions. The latter case corresponds to easy operation of dislocation sources (e.g. Cu deformed in stage II, Ni_3Al below and above a given temperature interval). The former case concerns conditions in which sources operate less easily because a friction acts on dislocations (lattice resistance in Ge, diffusional effects in, respectively, CuAl and, inside a given temperature interval, Ni_3Al–see Martin et al., 2000).

Quantitative laws about dislocation multiplication rates are not straightforward. Computer simulations of sources in dc covalent crystals seem to yield appropriate laws for these compounds. However, for metallic compounds, such laws are quite phenomenological at present. Similarly, quantitative exhaustion laws are quite approximate.

Nevertheless, the above results illustrate the fact that the net rates at which mobile dislocations multiply or exhaust (depending on the conditions) can be approached experimentally and quantified. It appears that work-hardening coefficients are connected with these rates, when either exhaustion (work-hardening) or multiplication (work-softening) dominates.

9.5. DISLOCATION MULTIPLICATION AT SURFACES

Multiplication processes at surfaces have stimulated significant efforts in two domains, respectively, ductile fracture and interfaces between two different crystals. These are two fields with practical as well as theoretical issues which represent large communities.

9.5.1 Dislocation generation at crack tips

The fracture behaviour of many materials changes abruptly from brittle to ductile as the temperature rises, over a rather narrow range. This is a key issue in many structural applications.

Ductile fracture is expected to operate as soon as dislocation loops can be emitted at the crack front. The estimation of the proper conditions has stimulated many efforts both experimentally (see e.g. Georges and Michot, 1993) and theoretically (see e.g. Weertman 1996). The review below is not exhaustive but aims at presenting the main issues and the present state of understanding.

Rice and Thomson (1974) performed the first attempt at modelling dislocation activity at the crack tip. They have calculated the stability of a sharp crack against emission of a blunting dislocation for a number of crystals, as well as the energy to form a stable dislocation loop from the crack tip. They decompose the force on a dislocation near the crack tip into three components: the first one is related to the crack stress field, the second one is a surface tension force caused by creating more surface near the blunted crack, while the third one is an image force. The two last components tend to pull the dislocation back into the crack while the first one repels it. The estimation of these forces yields a significant parameter $\mu b/\gamma$, where γ is the surface energy. The authors conclude that crystals which have a small value of this parameter and dislocations with a wide core are ductile, while those with narrow cores and a large value of the parameter are brittle.

Schoeck (1996) extended these calculations by considering loop emission on glide planes tilted against the crack plane (oblique planes that do not contain the crack edge). He described the emanating loop in the framework of the Peierls model by a distribution of infinitesimal dislocations in which the interplanar atomic energy is included. He showed that such a loop emission requires energies which can be thermally activated. Such a geometry avoids the formation of ledges, unlike for a plane containing the crack plane and is thus more favourable.

Xu and Argon (1997) extended the previous calculations for inclined planes, oblique planes and on cleavage ledges using the Peierls approach. Figure 9.32 illustrates the geometry of the crack and that of the glide plane. The authors use a variational boundary integral method to solve for the saddle point configuration of nucleated loops and associated energies. The results are used to estimate the brittle to ductile transition temperature. They conclude that only loop nucleation on cleavage ledges leads to realistic values of this transition temperature. This conclusion is supported by the experimental observation that dislocation nucleation at a crack tip occurs preferentially at heterogeneities (see e.g. Hirsch et al., 1989; Georges and Michot 1993)

These considerations allow materials to be classified. Ductile ones, such as FCC metals, some HCP metals and BCC tantalum, cannot cleave. The remaining crystals are intrinsically brittle and therefore susceptible to a brittle ductile transition.

9.5.2 Dislocation nucleation at a solid free surface

This problem has received much attention in connection with misfit dislocation generation during epitaxial growth. These are formed to accommodate part of the misfit between the stress free lattice parameters of film and substrate. It is well known that the first atomic layers that are deposited are strained to match the substrate surface, thus forming a coherent interface (for not too large misfits). However, as the film thickens, the strain energy increases so much that it becomes energetically favourable for misfit dislocations to be introduced. The critical thickness for this to occur was first discussed by Van der Merwe (1962). This parameter has been estimated in several studies (see e.g. Matthews et al., 1976; People and Bean, 1985; Beanland et al., 1996, Mooney, 1996).

One of the processes that have been imagined consists of the nucleation of half shear loops at the free surface of the film and their expansion towards the interface. It is

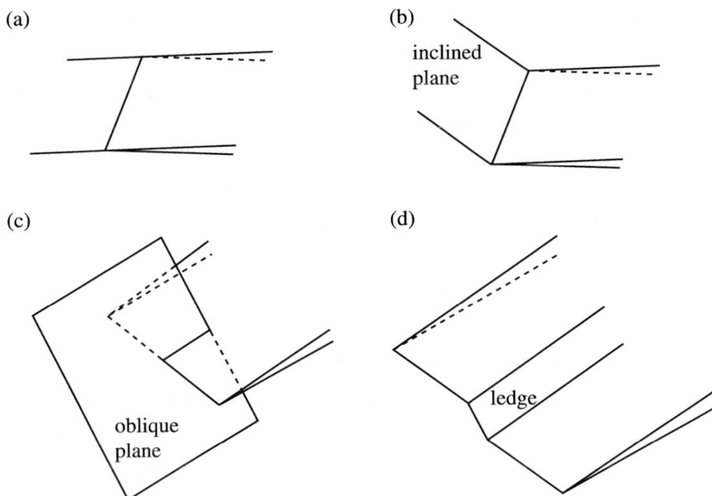

Figure 9.32. The different glide planes investigated by Xu and Argon (1997) for loop emission ahead of a crack.

illustrated in Figure 9.33. As the half loop meets the interface, a length of misfit dislocation is formed as well as two threading dislocations. The determination of the conditions for such a successful event to occur is a complex problem. Several solutions have been proposed with an increasing degree of sophistication.

To estimate the loop energy in the film, Matthews et al. (1976) considered that for half loop formation, as proposed by Frank (1950) and Hirth (1963), the energy released by the loop depends on the misfit and the free surface energy change. Brown et al. (1968) have emphasized the role of core energy which should also be included. The loop energy increases with radius R from zero to a maximum value that corresponds to the activation energy at a critical radius. At a given temperature the process is likely to operate provided (i) the misfit is large enough and (ii) the film thickness exceeds the critical radius.

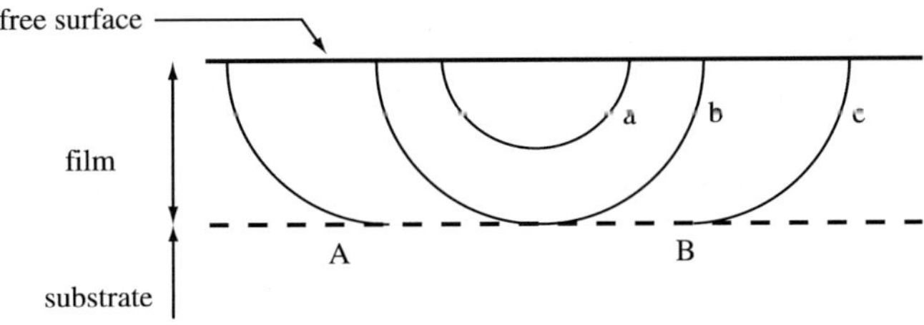

Figure 9.33. Schematics showing the formation of a half loop at a free surface to form a length AB of misfit dislocation. (a) Represents a subcritical half loop which becomes critical in (b) and stable in (c) under the misfit stress.

Dislocation generation at a free surface has been treated by a different method by People and Bean (1985) aiming at the determination of the critical thickness. The method is based on energy balance considerations. The critical thickness is estimated by a coincidence of the strain energy density and the energy density associated with the dislocation generating mechanism which is of minimum energy. For this, the energy density estimations of Nabarro (1967) are considered by the authors. In both treatments, the involved stresses are very high.

More recently, these estimations have been questioned by Junqua and Grilhé (1997). They revisit image forces, which are estimated traditionally, considering the elastic energy and not the surface energy. In such an approximation these forces tend to infinity as the dislocation reaches the surface. They consider that surface steps that form as a dislocation leaves a crystal relax to minimize surface energy. This results in an additional force which opposes the images forces. The authors predict a x^{-2} variation of the force for surface dislocation distances $x \leq 2b$. They suggest that activation energies needed to introduce a dislocation into the crystal are lower than the above estimates for dislocations nucleated at surface steps.

The relaxation of stresses in epitaxial films is now the subject of various computer simulations (see e.g. Lemarchand et al., 2000; Rao and Hazzledine, 2000).

It is also worth noting that in the case of large misfits, other mechanisms of strain relaxation have been revealed by Thibault et al. (2000) using high resolution TEM. Depending on the crystals, they include interfacial mixing, phase transformation and ordering, which operate as alternative processes to lower the energy.

9.5.3 Conclusion on dislocation multiplication at free surfaces

The two cases studied above show that the multiplication processes involved are very different from those operating in the crystal volume. The stresses necessary to generate dislocation embryos at surfaces are consequent ones and nucleation does not involve pre-existing dislocations. However, the presence of defects, such as surface steps at a free surface and cleavage ledges at crack tips, make nucleation easier.

REFERENCES

Alexander, H. & Haasen, P. (1968) *Solid State Phys.*, **22**, 28.

Appel, F., Oehring, M. & Wagner, R. (2000) *Intermetallics*, **8**, 1283.

Argon, A.S. (1996), in *Physical Metallurgy*, Eds. Cahn, R.W. & Haasen P., 4th Edition, Elsevier, Amsterdam, p. 1877.

Bardeen, J. & Herring, C. (1952) *Imperfections in Nearly Perfect Crystals*, Wiley, New York, p. 261.

Beanland, R., Dunstand, D. & Goodhew, P. (1996) *Adv. Phys.*, **45**, 87.

Berner, K. & Alexander, H. (1967) *Acta Metall.*, **15**, 933.

Blanc, C., Zuodar, N., Kléman, M., Martin, J.L. & Lelidis, I. (2003) *Phys. Rev. E*, submitted for publication.

Bonneville, J., Escaig, B. & Martin, J.L. (1988) *Acta Metall.*, **36**, 1989.

Bourdon, L., Kléman, M., Lejcek, L. & Taupin, D. (1981) *J. Physique*, **42**, 261.

Brown, M.L., Woolhouse, G.R. & Valdre, U. (1968) *Phil. Mag.*, **17**, 781.

Caillard, D. & Couret, A. (2002) *Mater. Sci. Eng. A*, **322**, 108.

Caillard, D., Vailhé, C. & Farkas, D. (1999) *Philos. Mag. A*, **79**, 723.

Carter, B. (1977) *Philos. Mag.*, **35**, 75.

Cottrell, A.H. (1953) *Dislocations and Plastic Flow in Crystals*, Oxford University Press, London.

Couret, A. (1999) *Philos. Mag., A*, **79**, 1977.

Couret, A. & Caillard, D. (1985) *Acta Mater.*, **38**, 1445.

Couteau, O., Charbonnier, C., Kruml, T. & Martin, J.L. (2001) in *Multiscale Modeling of Materials-2000*, vol. **653**, Eds. Kubin, L.P., Selinger, R.L., Bassani, J.L. & Cho K., MRS, Warrendale, p. ZI.7.I.

Dash, W.C., (1956) *J. Appl. Phys.*, **27**, 1193.

Dupas, C., Zuodar, N., Coddet, O., Kruml, T. & Martin, J.L. (2002) *J. Phys.: Cond. Matter.*, **14**, 12989.

Estrin, Y. (1996) *Unified Constitutive Laws of Plastic Deformation*, Academic Press, New York, p. 69.

Farenc, S., Caillard, D. & Couret, A. (1993) *Acta Met. Mater.*, **41**, 270.

Fikar, J., Kruml, T., Viguier, B. & Dupas, C. (2002) *J. Phys.: Cond. Matter.*, **14**, 12887.

Fnaiech, M., Reymond, F., Couret, A. & Caillard, D. (1987) *Phil. Mag. A.*, **55**, 405.

Frank, F.C. (1950) *Symposium on Plastic Deformation of Crystalline Solids*, Carnegie Institute of Technology, Pittsburgh, p. 89.

Frank, F.C. & Read, W.T. (1950) *Symposium on Plastic Deformation of Crystalline Solids*, Carnegie Institute of Technology, Pittsburgh, p. 44.

Furubayashi, E. (1969) *J. Phys. Soc. Japan*, **27**, 130.

George, A. & Michot, G. (1993) *Mater. Sci. Eng.*, **A164**, 118.

George, A. & Rabier, J. (1987) *Rev. Phys. Appl.*, **22**, 941.

Hirsch, P.B., Roberts, S.G., Samuels, J. & Warner, P.D. (1989) in *Advances in Fracture Research*, vol. **1**, Eds. Salama, K., Ravi-Chandar, K., Taplin, D.M.R. & Rama Rao P., Pergamon, Oxford, p. 139.

Hirth, J.P. (1963) *Relation Between Structure and Strength in Metals and Alloys*, H.M.S.O., London, p. 218.

Hirth, J.P. & Lothe, J. (1992) *Theory of Dislocations*, 2nd reprint edition, Krieger Publishing Company, Malabar, p. 623.

Imura, T. (1980) *Electron Microsc.*, **4**, 280.

Johnston, W.G. & Gilman, J.J. (1960) *J. Appl. Phys.*, **31**, 632.

Jungua, N. & Grilhé, J. (1997) *Phil. Mag., letters,* **75**, 125.

Kocks, U.F. (1966) *Phil. Mag.*, **13**, 541.

Kocks, U.F. (1976) *J. Eng. Mater. Technol.*, **98**, 76.

Kocks, U.F. & Mecking, H. (2003) *Prog. Mater. Sci.*, 1 **48**, 171.

Koehler, S. (1952) *Phys. Rev.*, **86**, 52.

Kruml, T., Conforto, E., Lo Piccolo, B., Caillard, D. & Martin, J.L. (2002) *Acta Mater.*, **50**, 5091.

Kubin, L.P. & Martin, J.L. (1980) in *Strength of Metals and Alloys*, vol. **3**, Eds. Haasen, P., Gerold, V. & Kostorz G., Pergamon, Toronto, p. 1639.

Kubin, L.P., Louchet, F., Caillard, D. & Martin, J.L. (1980) in *Electron Microscopy 1980*, vol. **4**, Eds. Brederoo, P. & Van Landuyt J., 7th European Congress on Electron Microscopy Foundation, Leiden, p. 288.

Lagow, B.W., Robertson, I.M., Jouiad, M., Lassila, D.H., Lee, T.C. & Birnbaum, H.K. (2001) *Mater. Sci. Eng. A*, **309–310**, 445.

Legros, M., Couret, A. & Caillard, D. (1996) *Phil. Mag. A*, **73**, 81.

Lelidis, I., Kléman, M. & Martin, J.L. (2000) *Mol. Cryst. Liq. Cryst.*, **351**, 187.

Lemarchand, D., Devincre, B., Kubin, L.P. & Chaboche, J.L. (2000) in *Mat. Res. Soc. Symp. Proc.*, vol. **578**, Eds. Robertson, I.M., Lassila, D.H., Devincre, B. & Philips R., MRS, Warrendale, p. 87.

Louchet, F., Kubin, L.P. & Vesely, D. (1979) *Phil. Mag.*, **39**, 433.

Martin, J.L., Lo Piccolo, B. & Bonneville, J. (2000) *Intermetallics*, **8**, 1013.

Martin, J.L., Matterstock, B., Spätig, P. & Bonneville, J. (1999) in *Proc. 20th Risø Int. Symp. Mater. Sci.*, Eds. Bilde-Sørensen, J.B., Cartensen, J.L., Hansen, N., Juul Jensen, D., Leffers, T., Pantleon, W., Pedersen, O.B. & Winther G., Risø Nat. Lab, Roskilde, p. 103.

Matterstock, B., Martin, J.L., Bonneville, J. & Kruml, T. (1999) in *Mat. Res. Soc. Symp. Proc.*, vol. **552**, Eds. George, E.P., Mills, M.J. & Yamaguchi M., Mater. Res. Soc, Warrendale, USA, p. KK5.17.1.

Matthews, J.W., Blakeslee, A.E. & Mader, S. (1976) *Thin Solid Films*, **33**, 253.

Mecking, H. & Kocks, U.F. (1981) *Acta Metall.*, **29**, 1865.

Milicka, K. (1999) *Acta Metall.*, **47**, 1831.

Molénat, G., Rosner, H. & Caillard, D. (1998) *Mater. Sci. Eng. A*, **258**, 196.

Mooney, P.M. (1996) *Mat. Sci. and Eng. R*, **17**, 105.

Moulin, A., Condat, M. & Kubin, L. (1997) *Acta Mater.*, **45**, 2339.

Moulin, A., Condat, M. & Kubin, L. (1999) *Acta Mater.*, **47**, 2879.

Nabarro, F.R.N. (1967) *Theory of Crystal Dislocations*, Clarendon, Oxford, p. 75.

Nabarro, F.R.N. (1986) in *Strength of Metals and Alloys*, vol. **3**, Eds. McQueen, H.J., Bailon, J.P., Dickson, J.I., Jonas, J.J. & Akben M.G., Pergamon, Oxford, p. 1667.

Nabarro, F.R.N. (1987) *Theory of Crystal Dislocations*, Dover Publ. Inc, New York, p. 24.

Nabarro, F.R.N. & Duesberry, M.S. (2002) in *Dislocations in Solids*, vol. **11**, Elsevier, Amsterdam.

Neuhäuser, H. & Schwink, C. (1993) in *Plastic Deformation and Fracture of Materials*, vol. **6**, Eds. Cahn, R.W.C., Haasen, P. & Kramer E.J., VCH, New York, p. 191.

Oswald, P. & Kléman, M. (1982) *J. Physique Lett.*, **43**, L411.

Oswald, P. & Kléman, M. (1984) *J. Physique Lett.*, **45**, L328.

People, R. & Bean, J.C. (1985) *Appl. Phys. Lett.*, **47**, 322.

Rao, S. & Hazzledine, M. (2000) in *Mat. Res. Soc. Symp. Proc.*, vol. **578**, Eds. Robertson, I.M., Lassila, D.H., Devincre, B. & Philips R., MRS, Warrendale, p. 389.

Rice, J.R. & Thomson, R. (1974) *Philos. Mag.*, **29**, 73.

Saka, H., Doi, M. & Imura, T. (1970) *J. Phys. Soc. Japan*, **29**, 803.

Schafer, S. (1967) *Phys. Stat. Sol.*, **19**, 297.

Schoeck, G. (1996) *Philos. Mag.*, **74**, 419.

Seeger, A. (1957), in *Dislocations and Mechanical Properties of Crystals*, Ed. Fischer, J.C., Johnston, W.G., Thomson, R., Vreeland & T., Jr.,, Wiley, New York, p. 243.

Siethoff, H., Ahlborn, K. & Schröter, W. (1999) *Phys. Stat. Sol (a)*, **174**, 205.

Thibault, J., Dressler, C. & Bayle-Guillemaud, P. (2000) in *Mat. Res. Soc. Symp. Proc.*, vol. **578**, Eds. Robertson, I.M., Lassila, D.H., Devincre, B. & Philips R., MRS, Warrendale, p. 235.

Van der Merwe, J. (1962) *J. Appl. Phys.*, **34**, 123.

Weertman, J. (1996) *Dislocation Based Fracture Mechanics*, World Scientific Publishing, New York.

Whelan, M.J., Hirsch, P.B., Horne, R.W. & Bollmann, W. (1957) *Proc. R. Soc. London*, **A240**, 524.

Xu, G. & Argon, A.S. (1997) *Philos. Mag.*, **75**, 341.

Chapter 10

Mechanical Behaviour of Some Ordered Intermetallic Compounds

Chapter 10
Mechanical Behaviour of Some Ordered Intermetallic Compounds

The volume of research that is being conducted on these compounds is directly related to the interest in their development for use as high temperature structural materials. The 1998 Materials Research Society Symposium entitled "High Temperature Intermetallic Alloys VIII" (Edited by George, et al, 1999, 120 papers), the International Symposium on "Intermetallics for the 3rd Millennium, a Symposium dedicated to Professor Robert Cahn (1999)" (40 papers) and the 2001 ISSI Conference (Edited by Hemker, et al, 2001, 93 papers) are only a few of the recent international conferences that have been dedicated to the study of intermetallic alloys. They illustrate the importance that these alloys have attained. The present section is not a complete review of their basic deformation mechanisms (for this, see e.g. *Dislocations in Solids*, Vol. 10, ed. by Nabarro and Duesberry, 1996). The present chapter deals mostly with the deformation mechanisms in Ni_3Al, but also in a few other compounds such as Co_3Ti, Cu_3Au, FeAl and β CuZn.

10.1. Ni_3Al AND $L1_2$ COMPOUNDS

Extensive investigations are available about the very pronounced yield strength anomaly in this crystal versus temperature. This phenomenon was first reported by Westbrook (1975). The anomalous behaviour of the work-hardening coefficient with the latter parameter has also been investigated but to a lesser extent. A number of mechanical test data have been produced. They concern constant strain-rate tests, as well as stress relaxation and creep transients, Cottrell–Stokes type experiments and a few creep tests. An impressive amount of data about the corresponding microstructural features is also available. They include both "post mortem" observations of slip traces, dislocation structures for various deformation conditions, and also in situ TEM experiments. Modelling of the flow stress behaviour as a function of temperature has also been undertaken. It has been the subject of numerous controversies. The main difficulty is that it has to be refined, because of the overabundance of data. The compounds of interest here consist of binary ones and ternary ones, single and polycrystals.

10.1.1 General considerations

Ni_3Al is an ordered crystal of the $L1_2$ structure which remains ordered up to the melting point. In such a lattice, when oriented for single slip, glide occurs either on the octahedral or the cube plane depending on the temperature (Figure 10.1). In the temperature domain which corresponds to the strength anomaly, dislocations are observed to glide along the primary octahedral plane (between about 300 and 600 K in Figure 10.1). A typical aspect of the microstructure is shown in Figure 10.2(a). Long straight screw dislocations are locked in Kear–Wilsdorf lock (KWL) configuration, a characteristic of this temperature domain (Kear and Wilsdorf, 1962). Such dislocation geometry indicates that the edge dislocations have a high mobility along the octahedral plane. The latter dislocations are not frequently observed and the majority of them (80% at 423 K) lie in the octahedral primary plane. The density of secondary dislocations is low.

Near the stress peak temperature (600–780 K on Figure 10.1), most dislocations are primary ones. As shown in Figure 10.2(b), they have a tendency to form either edge or screw segments so that almost regular square nets of dislocations can be observed. All dislocations lie, and are dissociated, in the cube cross-slip plane. The geometry of the nets suggests that the respective mobility of screw and edge segments is comparable.

Above the stress peak temperature, several dislocation families are observed. Among them, a substantial density belongs to the primary cube slip system. They appear as parallel short straight segments in Figure 10.2(c).

These microstructural features are common to all investigated Ni_3Al compounds. (For more details see Veyssière and Saada, 1996.)

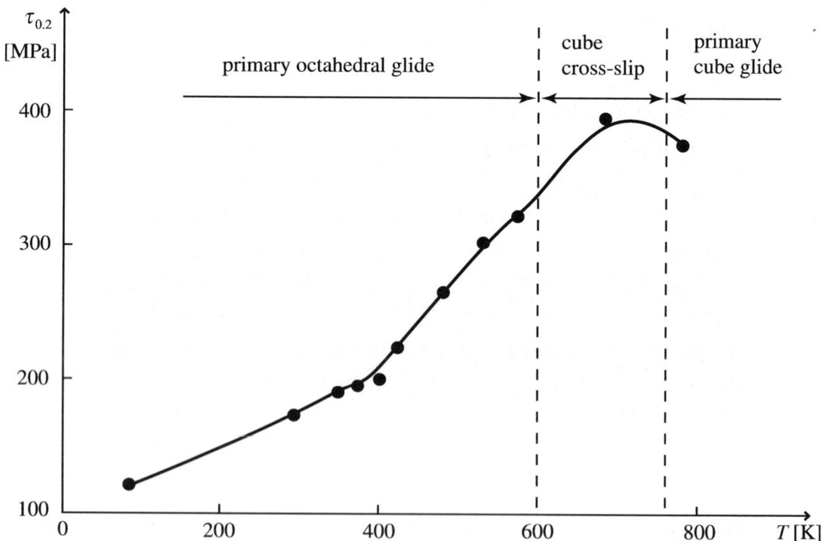

Figure 10.1. Yield stress (considered as the 0.2% offset stress $\tau_{0.2}$) as a function of temperature. ⟨123⟩ Ni_3(Al, 3 at.%Hf) single crystal. From Kruml et al. (1997). The active glide planes are indicated.

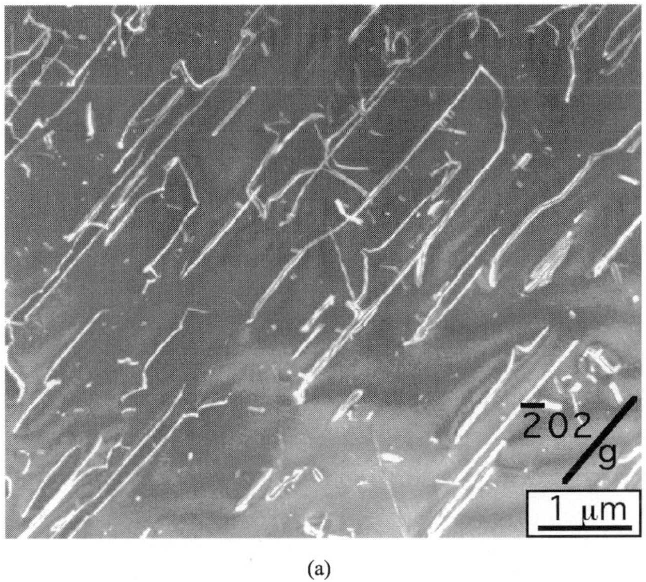

(a)

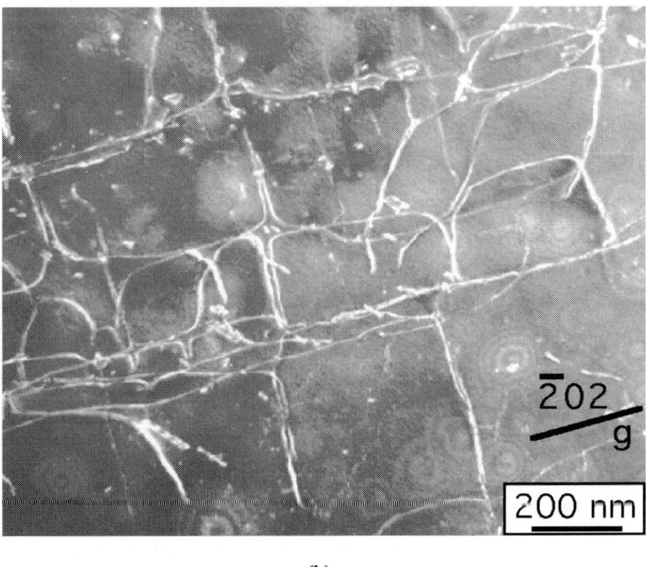

(b)

Figure 10.2. Dislocation features after compression at (a) room temperature: foil plane parallel to the primary octahedral plane; (b) 683 K: foil plane parallel to the cube cross-slip plane and (c) 780 K: same foil orientation as in (b). Same single crystal as in Figure 10.1. After Kruml et al. (1997).

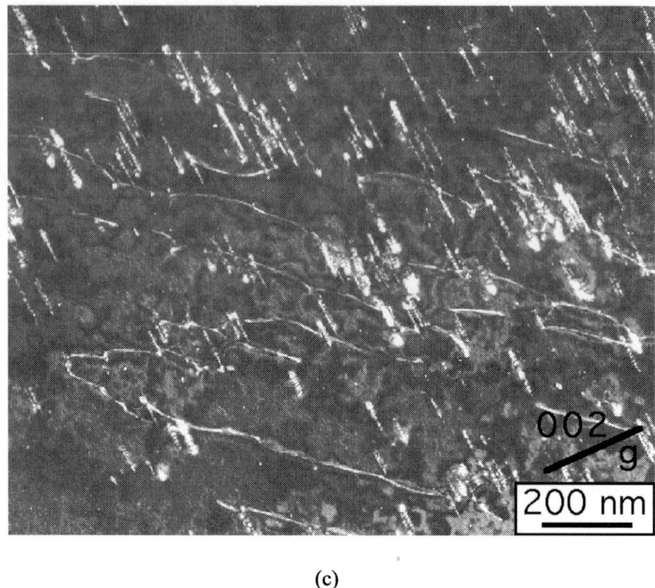

(c)

Figure 10.2. (*continued*)

10.1.2 Dislocation cores

There is a general agreement in the literature that the mechanical properties of Ni_3Al can be understood in terms of dislocation mobility and thus dislocation core geometry.

The Burgers vectors of the dislocations that control deformation are much larger than they are in ordinary metals. The associated strain energy leads to a dissociation into smaller partial dislocations that are connected by crystallographic faults. In the temperature range of the strength anomaly, dislocations glide on $\{111\}$ of Burgers vector $\langle 110\rangle$. These superdislocations are split into two superpartials (Burgers vector 1/2 $\langle 110\rangle$). They bind an antiphase boundary on the $\{111\}$ plane (APB_{111}). Each superpartial decomposes into two Shockley dislocations which bind a complex stacking fault (CSF). The corresponding dislocation geometry is shown in Figure 10.3(a). As glide proceeds along $\{111\}$, screw dislocations cross slip into the cube cross-slip plane thus forming a KWL (Kear and Wilsdorf, 1962). Its structure is represented in Figure 10.3(b, c).

The driving force for lock formation is a torque effect between the two screw superpartials, due to anisotropic elasticity effects. These lie in the fact that $\{111\}$ is not a mirror plane of the cubic structure (Yoo, 1986; Paidar et al., 1992). Some anisotropy in the APB energies also contributes to this force. The kinetics of the process are controlled by the thermally activated cross-slip of the leading superpartial from the octahedral to the cube plane.

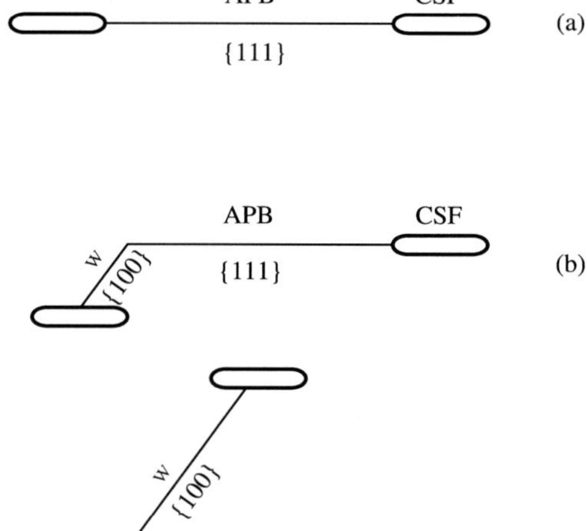

Figure 10.3. Schematics of the dislocation core geometry in Ni_3Al. (a) Glissile configuration on {111}. (b) Incomplete KWL configuration for the screw dislocation. (c) Complete KWL. w measures the APB width along {100}.

Such a mechanism leads to a decrease in mobility and/or density of dislocations which control plastic flow. The thermal activation of cross-slip leads to a flow stress anomaly since dislocation exhaustion becomes more pronounced at increasing temperatures. Consequently, the determination of the dislocation core geometry has been the subject of intense research. The aim was the determination of the complex stacking fault energy γ_{CSF}, a key parameter of the above process. The APB_{111} and APB_{100} energies, γ_o and γ_c, respectively, have also been determined for a range of compounds. Needless to say that experimentally measured values of the fault energies provide a much needed way to check and verify the results of first principles and atomistic calculations.

10.1.2.1 Technical difficulties bound to dislocation core characterization in Ni_3Al. Since the four Shockley dislocations are confined in a narrow core, the use of the weak-beam technique (Cockayne et al., 1969) is required. The fault energies are computed from the distances between dislocations. Consequently, the exact positions of the latter have to be known. This necessitates the comparison of experimental micrographs with computer simulated images. The actual dislocation distances are different from the distances between dislocation images on the micrographs. The CUFOUR software is frequently used for this purpose (Schäublin and Stadelmann, 1993). Additional details on the procedure are given by Kruml et al. (2000a). The separation of the four Shockley partials for determining γ_{CSF} is a difficult task. It can be done more easily on edge segments lying along {111} where the partial separation is larger. When the dislocation distances are determined, the fault energies are computed using a program developed by Baluc et al. (1991a,b) for APBs (rather large splitting). For narrow CSFs, the program of Schoeck (1997) based on the Peierls model is preferable.

Depending on the amount of core extension and diffraction conditions, two to four images or more, that correspond to the four Shockley partials, are visible on the micrograph. This has been shown by image simulations as well as TEM observations (see e.g. Baluc et al., 1991a,b; Hemker 1997; Kruml et al., 2000a,b). An example of dislocation images is presented in Figure 10.4.

The determination of γ_c also requires special care. Indeed, screw dislocations are frequently observed to be split on both the octahedral and the cube plane in the same sample (see Figure 10.5). An example is shown in Figure 10.5(a) of a screw dislocation in $Ni_{75}Al_{25}$ deformed at room temperature. The superpartials exhibit two different separation distances in the upper and lower parts of the figure, respectively, and they emerge at the two foil surfaces along two different traces. This indicates that the superdislocation is partly dissociated along the (010) cube plane (traces tr_c, apparent dissociation width d_c) and partly along the $(1\bar{1}1)$ octahedral plane (trace tr_o, apparent dissociation width d_o). The former configuration is a complete KWL (Figure 10.3(c)) and the latter is probably an incomplete one (Figure 10.3(b) with $w \sim b$). Figure 10.5(b) illustrates a similar dislocation behaviour

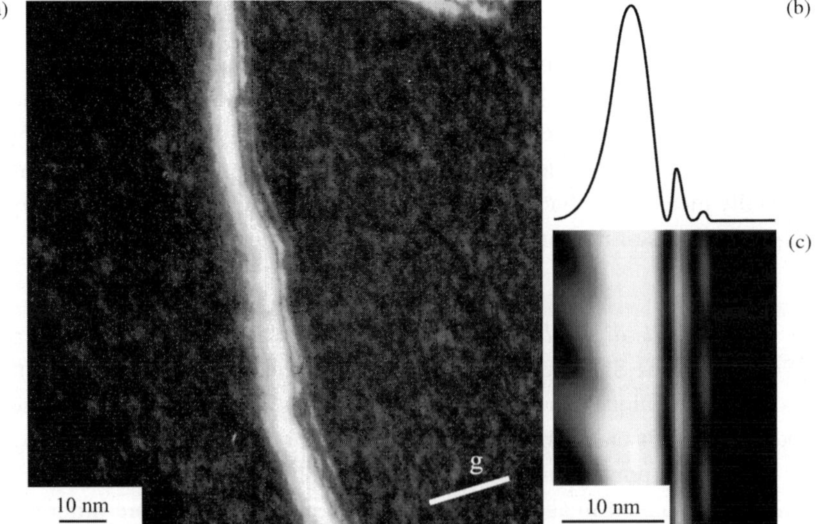

Figure 10.4. Weak-beam images of an edge superdislocation in (Ni_3Al,Hf). Due to the high value of γ_{CSF}, only three dislocation images are visible. They correspond to the four Shockley partials on {111}. (a) Micrograph. (b) and (c) Computer generated images. From Kruml et al. (1999).

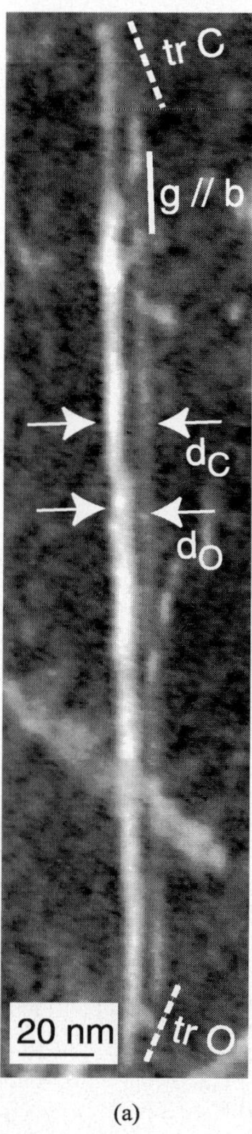

(a)

Figure 10.5. Weak-beam images of screw dislocations in Ni_3Al polycrystals after room temperature deformation. (a) $Ni_{75}Al_{25}$. The dislocation ($b = [\bar{1}01]$) is dissociated in the cube plane (010) in its upper part and in the octahedral plane $(1\bar{1}1)$ in its lower part. d_c and d_o are the corresponding apparent dissociation widths. tr_c and tr_o are the traces of the APB planes. (b) $Ni_{76}Al_{24}$. Three dislocations are dissociated along different planes. The one at the bottom is dissociated in (100), the one in the middle on $(\bar{1}11)$ and the one at the top in (111). The corresponding apparent APB widths are labelled: d_c, d_{o1}, d_{o2} and APB traces: tr_c, tr_{o1}, tr_{o2}. $\vec{b} = [01\bar{1}]$. From Kruml et al. (2002).

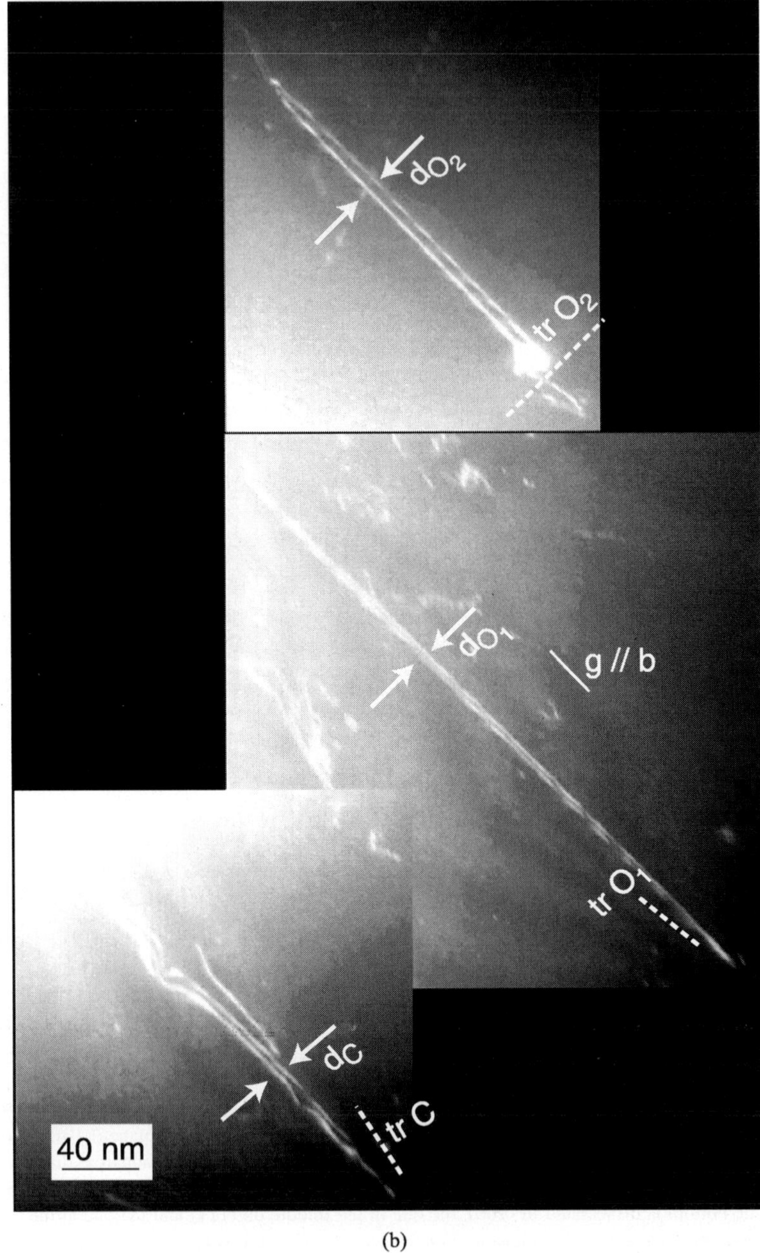

(b)

Figure 10.5. (*continued*)

in $Ni_{76}Al_{24}$ deformed at room temperature. Three screw superdislocations are successively dissociated in (100) (tr_c, d_c), in $(\bar{1}11)$ (tr_{o1}, d_{o1}) and in (111) (tr_{o2}, d_{o2}). Only screw superdislocations with clear indications of a cube dissociation plane have to be considered for a safe determination of γ_c.

10.1.2.2 Data about fault energies. Several attempts at determining fault energies can be found in the literature. Table 10.1 lists data that can be considered as reliable, i.e. have been determined using the procedures of the preceding section unless otherwise specified. Indeed, ignoring the correction for image shifts inherent to dislocation contrast leads to an underestimation of APB energies by about 20%, the correction increasing with γ values. It will be shown below that the parameters of the table can be related to macroscopic parameters such as the yield stress and the work-hardening coefficient.

Table 10.1 shows that alloying elements such as Hf, Ta or B increase fault energies substantially. Moreover, γ_c, γ_o and γ_{CSF} are observed to increase simultaneously. This is illustrated in Figure 10.6 where γ_{CSF} and γ_c are represented as a function of γ_o. Linear dependences are clearly observed with a steeper increase of γ_{CSF} with γ_o as compared to γ_c. In particular a value of $\gamma_c/\gamma_o \sim 0.83$ is found. Only data point number 9 differs substantially from the linear dependence. But these are the only γ_{CSF} data available for any alloy with a solute interstitial. The correlation between fault energies shown in Figure 10.6 is not surprising. Indeed a change in chemical nature is expected to alter both γ_{APB} values and even γ_{CSF}, since a CSF is the combination of a stacking-fault with an antiphase boundary.

Table 10.1. Survey of fault energies of $L1_2$ compounds.

Alloy (in at.%)	γ_c (mJ/m^{-2})	γ_o (mJ/m^{-2})	γ_{CSF} (mJ/m^{-2})	Data points number[a]	References
Ni_3(Al, 3%Hf)	250 ± 25	300 ± 25	460	1	Kruml et al. (2000a)
Ni_3(Al, 1%Ta)	200 ± 25	237 ± 30	352 ± 50	2	Baluc and Schäublin (1996)
$Ni_{74}Al_{26}$	184 ± 16	219 ± 17	277 ± 49	3	Kruml et al. (2002)
$Ni_{75}Al_{25}$	160 ± 20	195 ± 13	236 ± 29	4	Kruml et al. (2002)
$Ni_{76}Al_{24}$		180 ± 20	206 ± 30	5	Hemker and Mills (1993)
	135 ± 18	176 ± 11	206 ± 27	6	Kruml et al. (2002)
$Ni_{78}Al_{22}$	104 ± 15	175 ± 15	235 ± 40	7	Karnthaler et al. (1996)
Ni_3(Al, 1%Pd)	102 ± 11[b]	144 ± 20[b]		8	Sun et al. (1999)
Ni_3(Al, 1%B)		173 ± 15	335 ± 60	9	Hemker and Mills (1993)
Ni_3(Al, 0.25%Hf)	120 ± 20[b]	150 ± 20[b]		10	Bontemps-Neveu (1991)
Ni_3(Al, 2%Hf)	150[b]	185[b]		13	Bontemps-Neveu (1991)
Ni_3(Al, 1.5%Hf, 0.2%B)	150[b]	185[b]		15	Ezz and Hirsch (1994)
Ni_3Ga		110[b]		16	Suzuki et al. (1979)

[a] Refer to Figures 10.6 and 10.27.

[b] No image simulation.

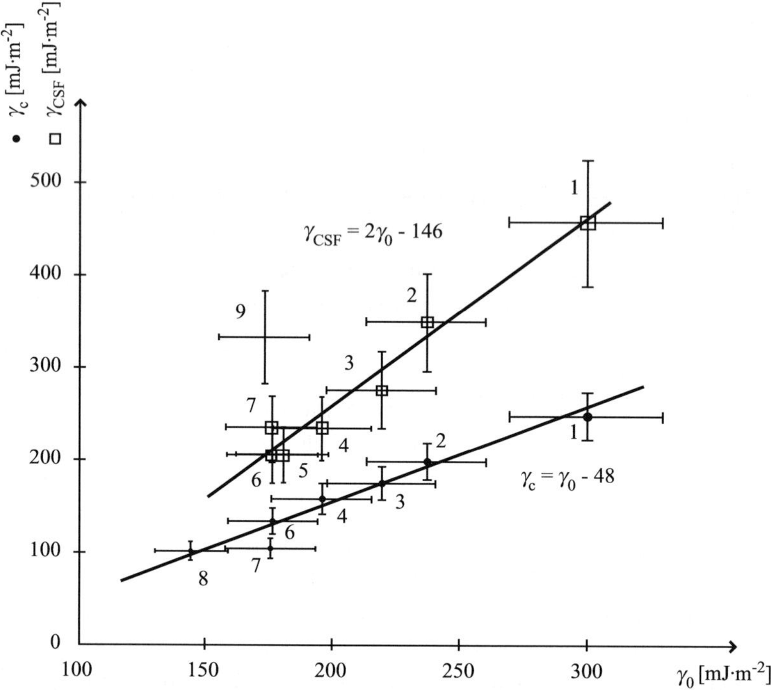

Figure 10.6. APB energy on the cube plane and CSF energy represented as a function of the APB energy on the octahedral plane for various Ni_3Al compounds. Error bars correspond to typical values of 10 and 15% for APB and CSF energies, respectively. Numbering of data is done according to Table 10.1 From Kruml et al. (2000a, 2002).

10.1.3 Cube glide

Cube glide can be observed in bulk samples with specific orientations or in TEM in situ experiments. An anomalous behaviour of the CRSS for cube slip with temperature is observed, as illustrated in Figure 10.7, in a variety of Ni_3Al compounds. This will be discussed in Section 10.1.5.5. It can be seen in particular that the CRSS for cube glide is quite comparable in bulk and in situ experiments.

The temperature which corresponds to the peak stress changes with alloy composition. To characterize this process safely, {100} dislocation cores have first to be observed post mortem. Then TEM in situ experiments are necessary to observe the motion of dislocations on the cube plane.

10.1.3.1 Dislocation cores. Most of the dislocations which participate in cube glide have ⟨110⟩ Burgers vectors. In addition, some dislocations with ⟨100⟩ Burgers vectors and a 45° character are sometimes observed (Veyssière and Douin 1985; Korner 1989; Sun and Hazzledine 1989; Baluc, 1990). TEM observations made by Veyssière (1984),

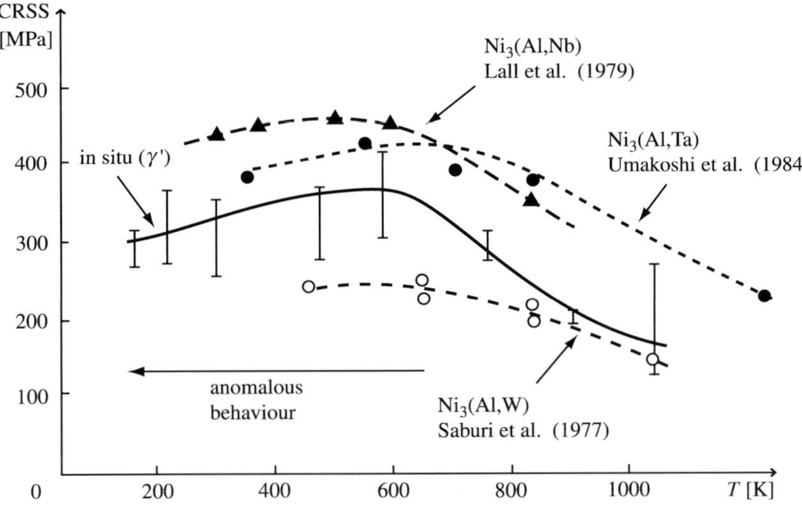

Figure 10.7. Anomalous temperature variation of the cube RSS for which substantial cube slip is observed in some Ni_3Al alloys. In situ data are from Clément et al. (1991) on the γ/ phase of a superalloy. After Caillard (2001).

Sun and Hazzledine (1989), Korner (1989), Sun et al. (1991a,b) and Baluc and Schäublin (1996) have revealed that dislocation loops on the cube plane exhibit straight segments in both the screw and edge directions. This suggests a lower dislocation mobility along these two directions.

The lower mobility of screws is clearly related to their core geometry (Figure 10.3(c)) which is sessile with respect to cube glide. Similarly, the lower mobility of edge dislocations can be related to several non-planar configurations which have been determined. Three-fold dissociations have been observed for the latter, for which several geometries have been proposed. A detailed study by Sun et al. (1991a,b) using weak-beam and lattice resolution imaging in Ni_3Ga and $Ni_3(Al,Ti)$ revealed two types of dissociation schemes:

(i) After low temperature deformation (673–873 K), rectilinear edge dislocations are dissociated by cube glide into two superpartial dislocations separated by an APB ribbon. Each superpartial dissociates into a Lomer–Cottrell lock. The corresponding core is shown in Figure 10.8(a) and is called a double Lomer–Cottrell lock.

(ii) After higher temperature deformation (973 K), edge superdislocations are dissociated by climb into two Frank partials and a stair-rod dislocation (Figure 10.8(b)). Baluc and Schäublin (1996) also reported such a dissociation in $Ni_3(Al,Ta)$. An atomic resolution image of the obtuse fault dihedron of Figure 10.8b can be found in Baluc (1990a and b).

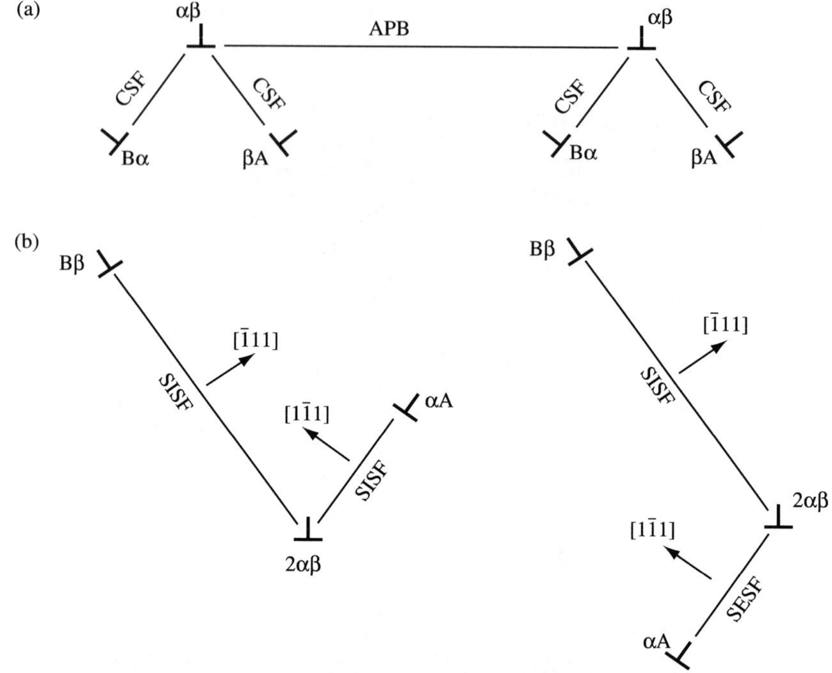

Figure 10.8. Schematics of core configurations for an edge dislocation on the cube plane in Ni_3Ga (after Sun et al. (1991)). Dislocations are seen end on along a ⟨110⟩ direction. SISF and SESF are super intrinsic and extrinsic stacking-faults, respectively. Bα, αβ, etc. designate partial dislocations using Thompson notations. (a) Low and (b) high temperatures.

10.1.3.2 Dislocation mobility. TEM in situ experiments were performed between 573 and 1013 K on Ni_3(Al, 0.25at.%Hf) single crystals (Molénat and Caillard, 1992).

The first set of experiments was performed at 573 and 673 K, just below the flow stress peak temperature. Cube slip results from the cross-slip of dislocations gliding in the octahedral plane. The glide distance along the cube plane is small and dislocation loops tend to be rectilinear along screw directions. They are always dissociated into two superpartial dislocations.

Another set of experiments was performed at 873 K above the flow stress peak temperature. The substructure that develops in thin foils consists of rectangular loops gliding on {100} planes with segments oriented along ⟨110⟩ directions. Under weak beam conditions (see Figure 10.9), the screw segments appear to be dissociated into two superpartial dislocations (dissociation width 6 nm). Edge segments exhibit two distinct configurations (fig. 10.8b): (1) two superpartials with a separation in the range of 9 nm (fig. 10.8a); (2) a more complex configuration. The dynamic sequences reveal that the velocity of dislocation loops is controlled by friction forces acting on both screw and

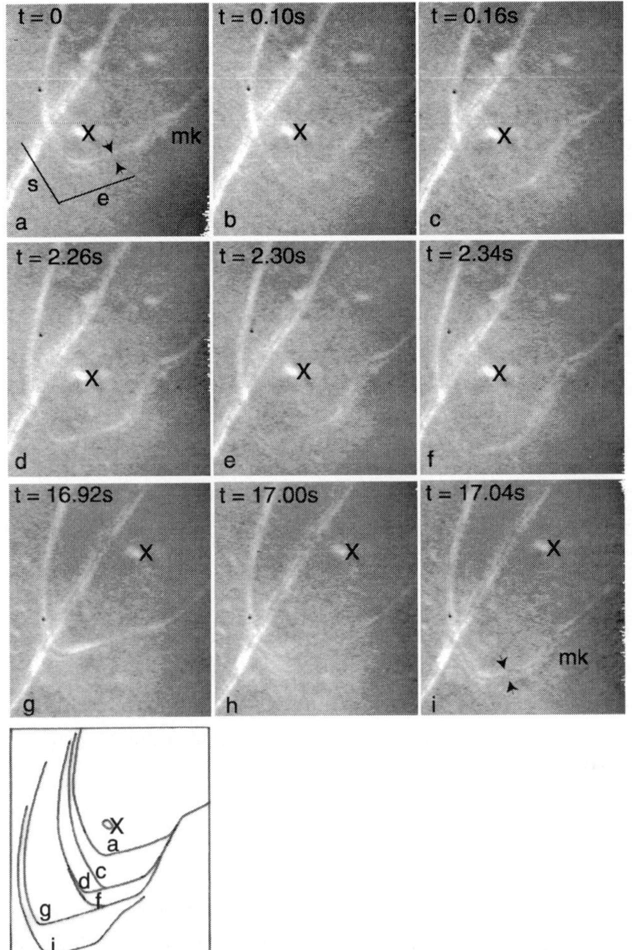

Figure 10.9. Sequence of cube glide at $T = 873$ K in Ni_3(Al, 0.25 at.% Hf). Weak beam conditions. *s* and *e* are screw and edge orientations, respectively. *X* is a fixed point. Note that the edge segment remains locked in frames f, g, h during 15 s. It then jumps further within one video frame, creating the macrokink mk. From Molénat and Caillard, 1992.

edge segments. However, the screw part exhibits a slow and viscous motion, unlike the edge part which moves jerkily: very rapid jumps are observed, faster than 1/50 s (1 video frame) creating superkinks, separated by long waiting times in locked positions. During a stress relaxation of a few minutes, the aspect of edge segments is progressively modified. The dissociation into two superpartials is replaced by a more complex one. The latter is sessile and resists stress increase experiments. Similar glide features are observed at 1013 K.

These experiments reveal that in this weakly alloyed compound cube glide proceeds as follows: (i) screw dislocation motion is controlled by a kink pair mechanism acting on the Kear–Wilsdorf configurations; (ii) the jerky glide of edge dislocations corresponds to a locking–unlocking mechanism. The core exhibits a double Lomer–Cottrell configuration in the Peierls valleys (Figure 10.8a) and a metastable glissile configuration while jumping over the Peierls hills. The Lomer–Cottrell type of lock subsequently transforms by climbing into another non-planar configuration which appears to be completely sessile (Figure 10.8b). This transformation is more rapid as the temperature increases. This locking–unlocking mechanism is described in detail in Chapter 4. The origin of the anomalous increase of the CRSS for cube glide (Figure 10.7) as a function of temperature is discussed in Section 10.1.5.5.

10.1.4 Octahedral glide

10.1.4.1 General considerations. This process has been extensively studied over the last 20 years in properly oriented single crystals. Some significant features of the mechanical behaviour are recalled:

- the strain-rate sensitivity, S, is rather low and slowly decreases from 80 K to the peak temperature in polycrystals of $Ni_3(Al,Cr)$ (Thornton et al., 1970). The corresponding activation areas deduced from stress-relaxation experiments (Chapter 2) are rather high and are stress and temperature dependent (Baluc et al., 1988). In $Ni_3(Al,Hf)$ single crystals, they are in the range of several hundred b^2 (b being the Burgers vector of superpartial dislocations). This indicates that thermal activation plays a role in the temperature domain of the strength anomaly (Spätig et al., 1993). It is worth noting that two different types of S are measured, according to the authors, who consider: (i) the transient response just after the strain-rate jump (see e.g. Thornton et al., 1970) or (ii) the long term one (see e.g. Demura and Hirano, 1997). The former S values are usually found to be positive and small. They correspond to those measured in stress-relaxation tests. The latter S values are close to zero or negative.
- A violation of the Schmid law and a traction compression asymmetry can sometimes be observed (see Lall et al. (1979) for $Ni_3(Al,Nb)$, Umakoshi et al. (1984) for $Ni_3(Al,Ta)$, Ezz et al. (1987) for Ni_3Ga). This proves that obstacles to dislocations for one stress direction are either altered or inefficient at the beginning of reverse straining.

Numerous microstructural observations have been performed "post mortem" or "in situ". They report a high density of screw dislocation segments in the strength anomaly domain. They are sometimes dissociated on the {100} cross-slip plane. These are called complete KWL. They are connected by macrokinks dissociated along {111}. However, a significant number of locks appear to be dissociated either on a plane close to {111} or a plane lying between the octahedral and the cube plane (Korner, 1989; Bontemps and Veyssière, 1990; Lin and Wen, 1993). An example has been shown in Figure 10.5. These are incomplete locks described in Figure 10.3b, (see Sections 10.1.4.2 and 10.1.4.3).

10.1.4.2 Microscopic aspect of {111} glide. To understand these various features and to characterize dislocation mobility on {111}, TEM in situ experiments have been performed. Pioneer work was undertaken by Nemoto et al. (1977), Suzuki et al. 1977 and Lours et al. (1991). The experiments by Molénat and Caillard (1991, 1994) on Ni_3(Al, 0.25 at.% Hf) between 120 and 1100 K, choosing various orientations have shed new light on the mechanism of octahedral glide. Figure 10.10 compares the CRSS for octahedral glide for mechanical tests on various crystal orientations and for these in situ experiments. The latter values appear to be rather scattered. Indeed the measurement of local stress in the foil (Chapter 2) suffers here from the lack of suitably curved dislocations between the straight screw segments. Nevertheless, the average stress is seen to increase with temperature below 800–1000 K and to decrease above 1000 K for bulk samples and thin foil experiments.

At the onset of the stress anomaly domain, straight screw superdislocations are observed to glide jerkily along {111}. Two distinct types of jerks can be observed.

(i) *Jumps over variable distances:* Starting from an incomplete lock configuration, the screw superdislocation glides along {111} over a rather large distance before being locked again. The process of locking and unlocking is represented in Figure 10.11 in two dimensions. The glissile configuration of Figure 10.11(a) is only observed dynamically. Locking occurs as follows: the leading superpartial cross-slips at point *P* on the cube plane

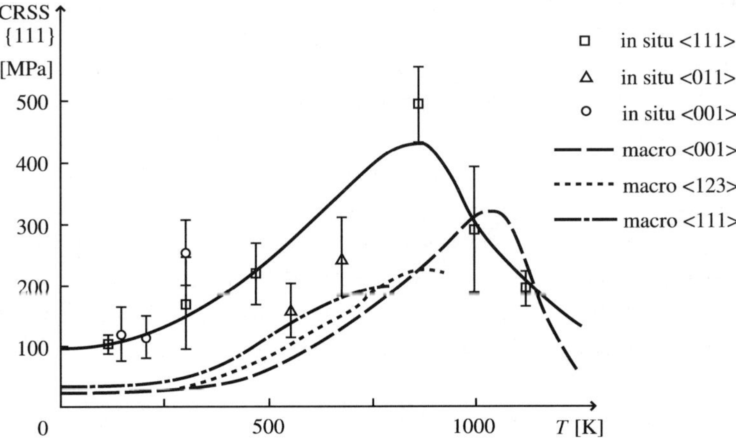

Figure 10.10. RSS projected on {111} in Ni_3(Al, 0.25 at.% Hf) as a function of temperature. Various crystal orientations have been tested. In situ experiments by Molénat and Caillard (1994). "Macro" refers to mechanical tests by Bontemps–Neveu (1991).

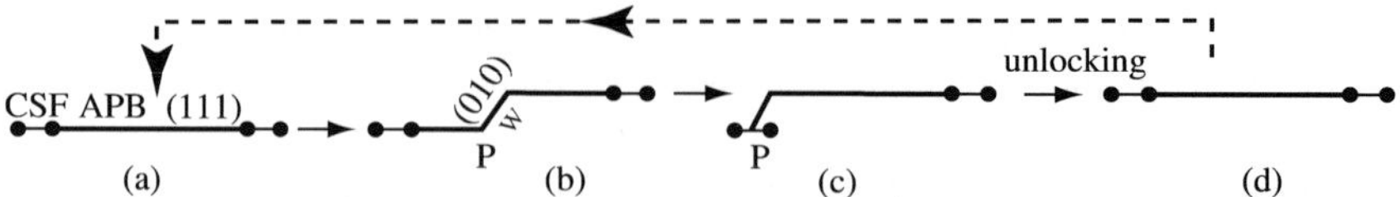

Figure 10.11. The locking–unlocking of screw superdislocations gliding along {111} jumps over variable distances. The glissile configuration in (a) is restored in (d). See text (Molénat and Caillard, 1994). Dislocation seen end on, along a ⟨110⟩ direction.

over a short distance w and back to the primary {111} plane (Figure 10.11(b)). The trailing superpartial gets locked at P (Figure 10.11(c)). Unlocking occurs after constriction of the latter dislocation and its subsequent cross-slip to the {111} plane. The superdislocation will be locked again after gliding any distance along {111}. In contrast to the model of Paidar et al. (1984), the cross-slip process does not occur locally along the dislocation, but leads to a sessile configuration extended along the screw. Under such conditions, the homogeneous cross-slip of the trailing superdislocation controls the unlocking process observed. Figure 10.12 represents the same process in three dimensions. The resulting macrokinks exhibit variable lengths along {111} and are called "simple" macrokinks (Veyssière and Saada, 1996). The split configuration in Figures 10.11(c) is the "incomplete" KWL shown in Figure 10.3(b). Its width of splitting, w, along the cube plane is of a few interatomic distances, i.e. smaller than the natural width of splitting.

(ii) *APB jumps:* the rectilinear screw dislocation dissociated along {111} jumps over a distance that scales with the dissociation width. This experimental observation is illustrated in Figure 10.13. Video weak beam images of moving screws are analysed as follows.

Intensity profiles are obtained by scanning the frame in a direction perpendicular to the dislocation. The two large peaks in Figure 10.13 correspond to the superpartials lying on {111} (Section 10.1.2.1). A comparison of the intensity profiles in Figure 10.13(a) and (b) shows that the image of the leading superpartial before the jump is superimposed on

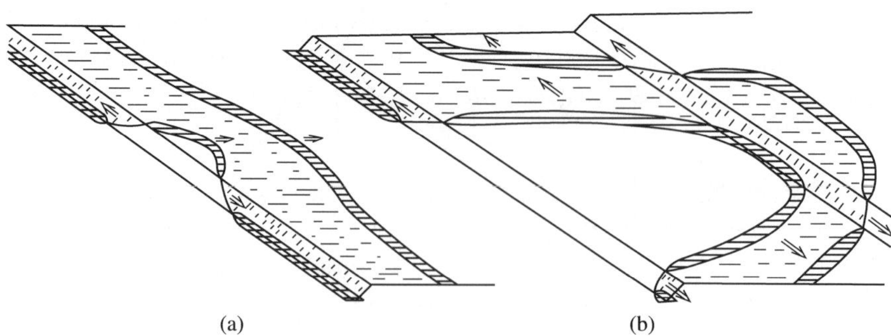

Figure 10.12. 3D schematics of the process represented in Figure 10.11. Formation of simple macrokinks.

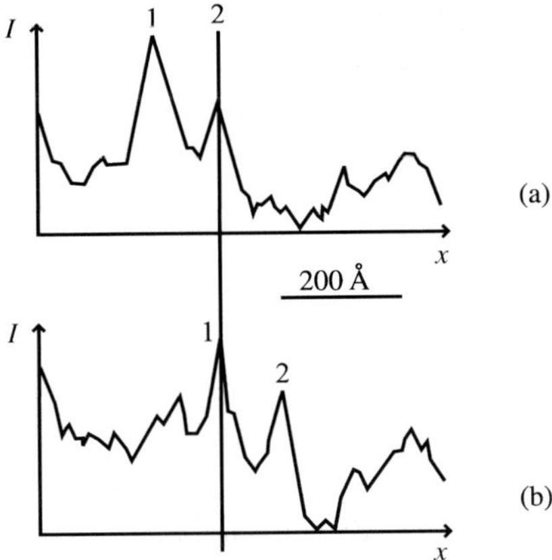

Figure 10.13. Intensity profile of a screw dislocation moving through an APB jump. TEM in situ experiment at 300 K. Weak beam observations. 1 and 2 are the intensity peaks of each superpartial. x is the distance along the slip plane (a) before and (b) after the jump. From Molénat and Caillard (1994).

the image of the trailing superpartial after the jump, within less than 1 nm. The schematics that corresponds to this type of jump is depicted in Figure 10.14. The sessile screw of Figure 10.14(a) unlocks by cross-slip of the trailing superpartial in Figure 10.14(b). It gets locked again in Figure 10.14(c), the trailing superpartial being stopped in QR where the leading superpartial was initially. Eventually, the leading superpartial can cross-slip again during the waiting time for unlocking, as shown in Figure 10.14(d). The lock formed is identical to the initial one. The three-dimensional aspect of this process is illustrated in Figure 10.15. It shows in particular that the "elementary" macrokinks that form have a height equal to the width of splitting of the APB on {111}. Elementary and simple macrokinks have also been observed post mortem (see below).

10.1.4.3 Complete versus incomplete KWL. The existence of incomplete KWL is assessed by the in situ experiments of the preceding section. Moreover, it is supported by a number of other experimental observations. Several studies report large numbers of elementary macrokinks: in Ni_3Ga (Couret et al., 1993), Ni_3(Al, 0.25%Hf) and

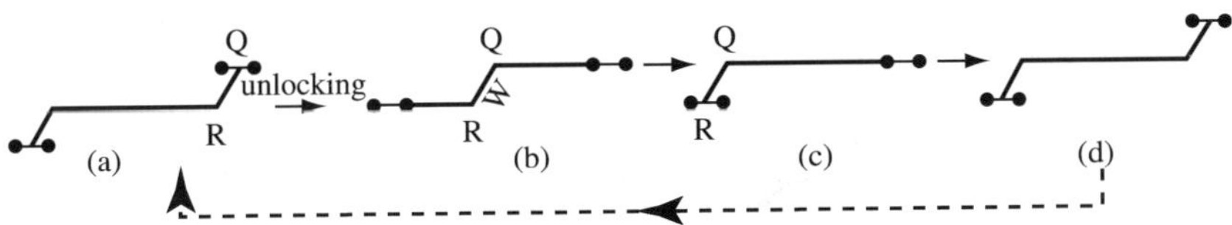

Figure 10.14. Schematics describing an "APB" jump. The glissile configuration in (a) is restored in (d). See text. Compare with Figure 10.11. From Molénat and Caillard (1994).

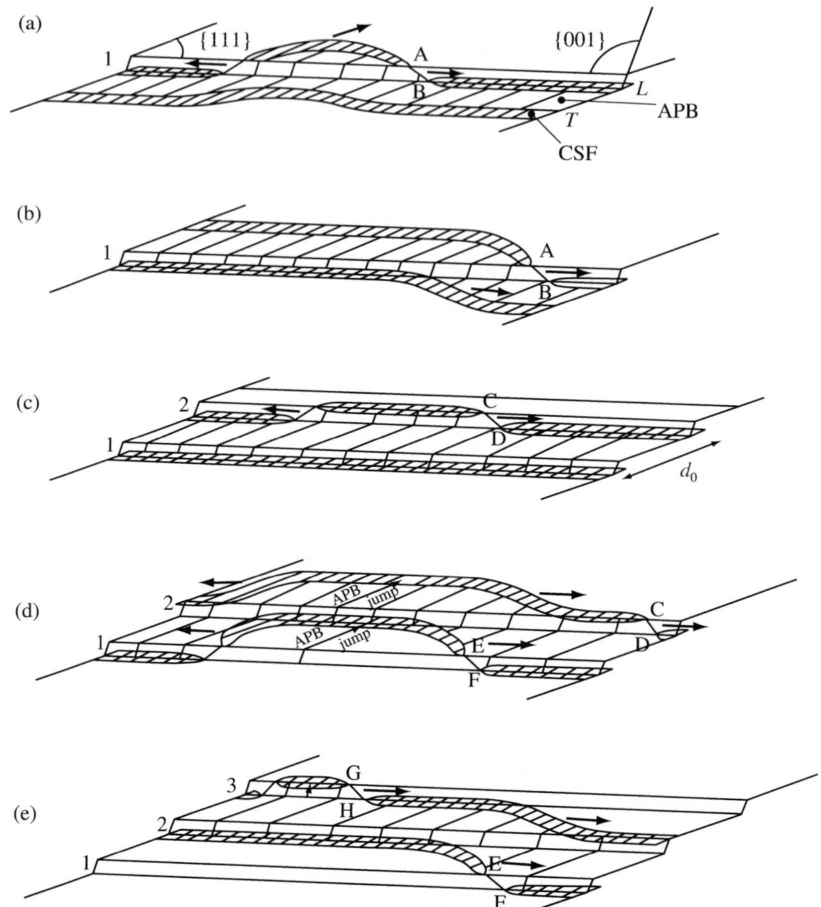

Figure 10.15. Three-dimensional representation of the process described in Figure 10.14. Formation of "elementary" macrokinks. L and T are the leading and trailing superpartials, respectively. Both are dissociated along {111}. d_o is the APB width of splitting along {111}. From Caillard (2001).

Ni_3(Al, 1.5%Hf) (Veyssière and Saada, 1996). These observations confirm that the cross-slip distance w along the cube plane is often very small, in the range of one atomic distance (Molénat et al., 1993; Coupeau et al., 1999). Such locks are also responsible for the microstructural features reported in Section 10.1.4.1. Moreover, the observation of APB jumps demonstrates that unlocking is an intrinsic process (Figure 10.15). In particular, it shows that unlocking through the bulging of a macrokink does not take place. This contrasts with a previous popular assumption which was not based on any observation (Mills et al., 1989; Sun et al., 1991a,b; Hirsch, 1992; Mills and Chzran, 1992). Therefore, the incomplete locks play an important role in the plasticity of Ni-based $L1_2$ alloys.

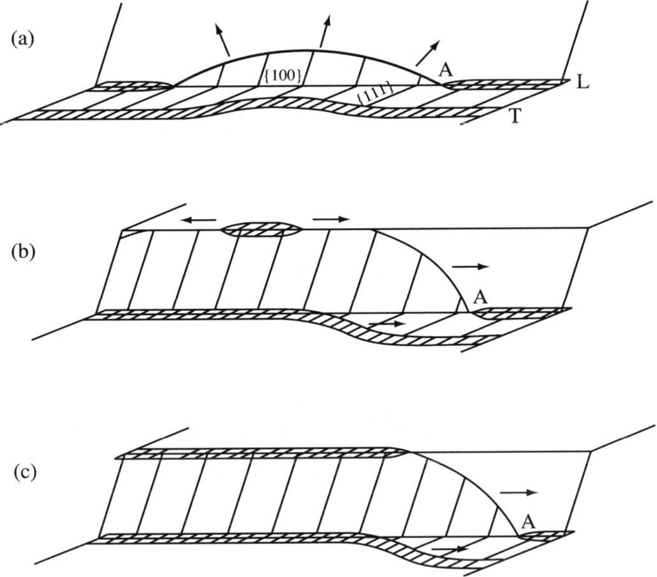

Figure 10.16. Formation of a complete KWL. Schematic representation of the corresponding cross-slip process. (a) Recombination and cube glide of the leading superpartial, (b) its redissociation on the primary {111}, (c) extension of the lock. From Caillard (2001).

In addition, complete locks are always present. They are formed in a single step cross-slip process illustrated in Figure 10.16: the leading superpartial glides over the maximum distance allowed by γ_c (100 APB width) before dissociating again in an intersecting {111} plane. This mechanism is described by Hirsch (1992) and Caillard (1996). The activation energy of the process corresponds to the recombination of the leading superpartial over a significant length and bulging along the cube plane (Section 4.3.1). This is larger than the activation energy for the formation of an incomplete lock. Figure 10.15 shows that the latter energy equals the energy of a kink pair of short height.

The conditions leading to the simultaneous formations of a bulge or a kink pair on the cube plane have been discussed at the end of Section 4.3.4.

10.1.5 Understanding the mechanical properties of Ni_3Al compounds

The important characteristics are as follows.

10.1.5.1 Definition of the yield stress. It has been emphasized that a smooth transition is observed between the preplastic and plastic stages. This corresponds to a rounded shape of the stress/strain curve at low strains (Hemker et al., 1992; Shi et al., 1996). It is commonly accepted that at plastic strains lower than about 2×10^{-3} deformation is produced via the rapid motion of edge dislocations. For larger strains, flow is associated with the more

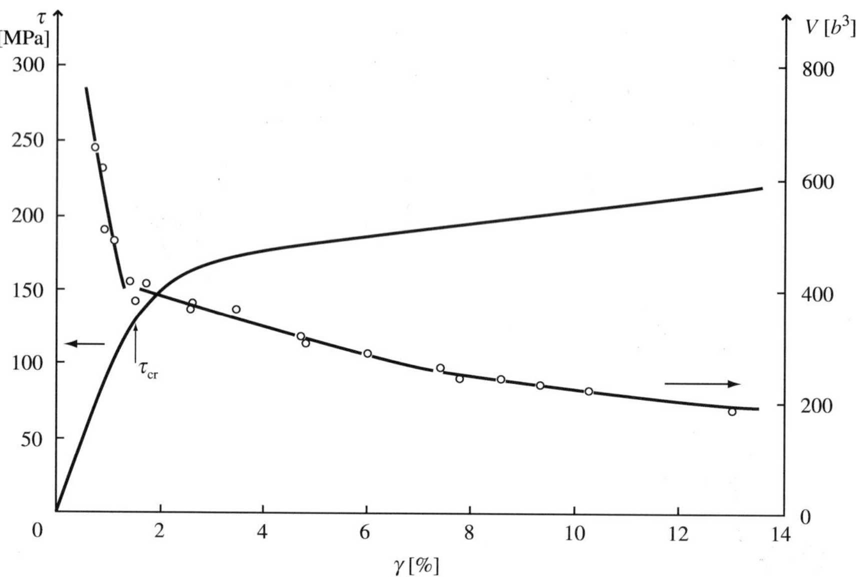

Figure 10.17. Stress–strain curve of a ⟨123⟩ Ni_3(Al, 3.3 at.%Hf). $T = 293$ K. $\dot{\gamma} = 10^{-4}\ s^{-1}$. The microscopic activation volume V is also represented as a function of strain. The position of τ_{cr} is indicated. From Spätig et al. (1993a).

difficult propagation of screws that yield the anomalous stress/temperature dependence (Mulford and Pope, 1973). However, no clearly defined yield point is observed on the stress–strain curve. To provide evidence of a possible critical stress between the micro and macroplastic domains, microscopic activation volumes V have been measured as a function of strain using the technique of stress relaxation series (see Chapter 2). ⟨123⟩ Ni_3(Al, 3.3 at.%Hf) single crystals have been used at three temperatures in the anomaly domain of the flow stress: 293, 423 and 573 K. A typical stress–strain curve is shown in Figure 10.17. V values are also reported as a function of strain.

The $V(\gamma)$ curve shows two distinct domains separated by a critical stress, $\tau_{cr} = 130 \pm 15$ MPa. τ_{cr} can be compared to the mean value of $\tau_{0.2} = 140 \pm 6$ MPa (0.2% offset stress). The same type of observation holds for the other temperatures investigated. The results are summarized in Figure 10.18. This figure shows a fair agreement between the conventional $\tau_{0.2\%}$ and the critical stress deduced from the strain dependence of the activation volume. Therefore, $\tau_{0.2\%}$ is considered as the yield stress.

10.1.5.2 Temperature variations of the yield stress and work-hardening rate. From the above, it is natural to consider the work-hardening coefficient that corresponds to a 0.2% plastic strain, $\theta_{0.2}$. As shown by Spätig et al. (1993b), the experimental determination of

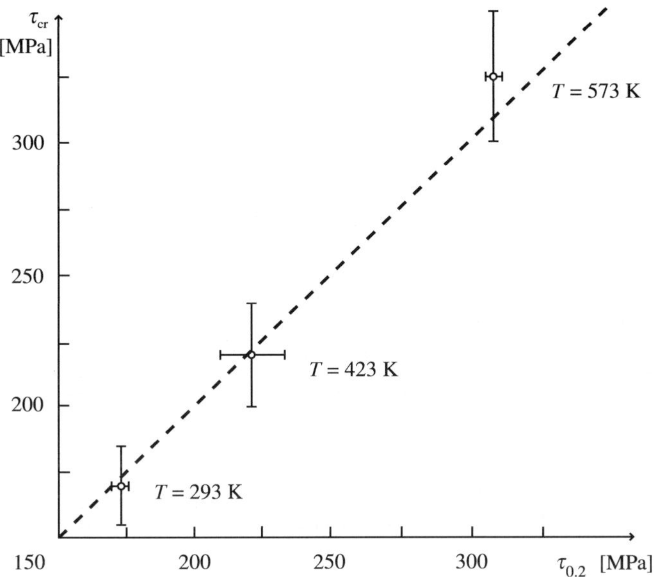

Figure 10.18. Transition stress τ_{cr} as a function of $\tau_{0.2}$ at three temperatures. Same crystal as in Figure 10.17. After Spätig et al (1993b).

$\tau_{0.2}$ and $\theta_{0.2}$ as a function of temperature requires special care. A quick procedure consists of deforming the same specimen several times, up to about a 0.3% plastic strain, at increasing temperatures. This procedure has been used for $L1_2$ crystals (see e.g. Copley and Kear, 1967; Mulford and Pope, 1973). The reliable procedure consists of deforming a virgin specimen at each temperature. The $\tau_{0.2}$ (T) curves are different depending on the procedure as illustrated in Figure 10.19(a). At each temperature, the reliable value of the yield stress is lower.

The $\theta_{0.2}$ (T) curves, look quite different, the reliable curve exhibiting one maximum, as compared to two maxima for the other one. The following data refer to the reliable procedure.

A typical variation of the yield stress and the corresponding work-hardening coefficient is represented in Figure 10.20 for a ternary compound. In this case, the temperature that corresponds to the yield stress peak is close to 1000 K, while that for the work-hardening coefficient is close to 450 K. These features can be observed in a variety of such compounds (see Section 10.1.5.6).

10.1.5.3 Yield stress peak temperature (single crystals). As recalled in Section 10.1.1, early studies have shown that this peak corresponds to the transition between the anomalous regime of {111} slip and thermally activated {100} slip at high temperatures, except for ⟨100⟩ orientations. The corresponding temperature is labelled $T_{p,\tau}$ and $T_{p,\sigma}$ for

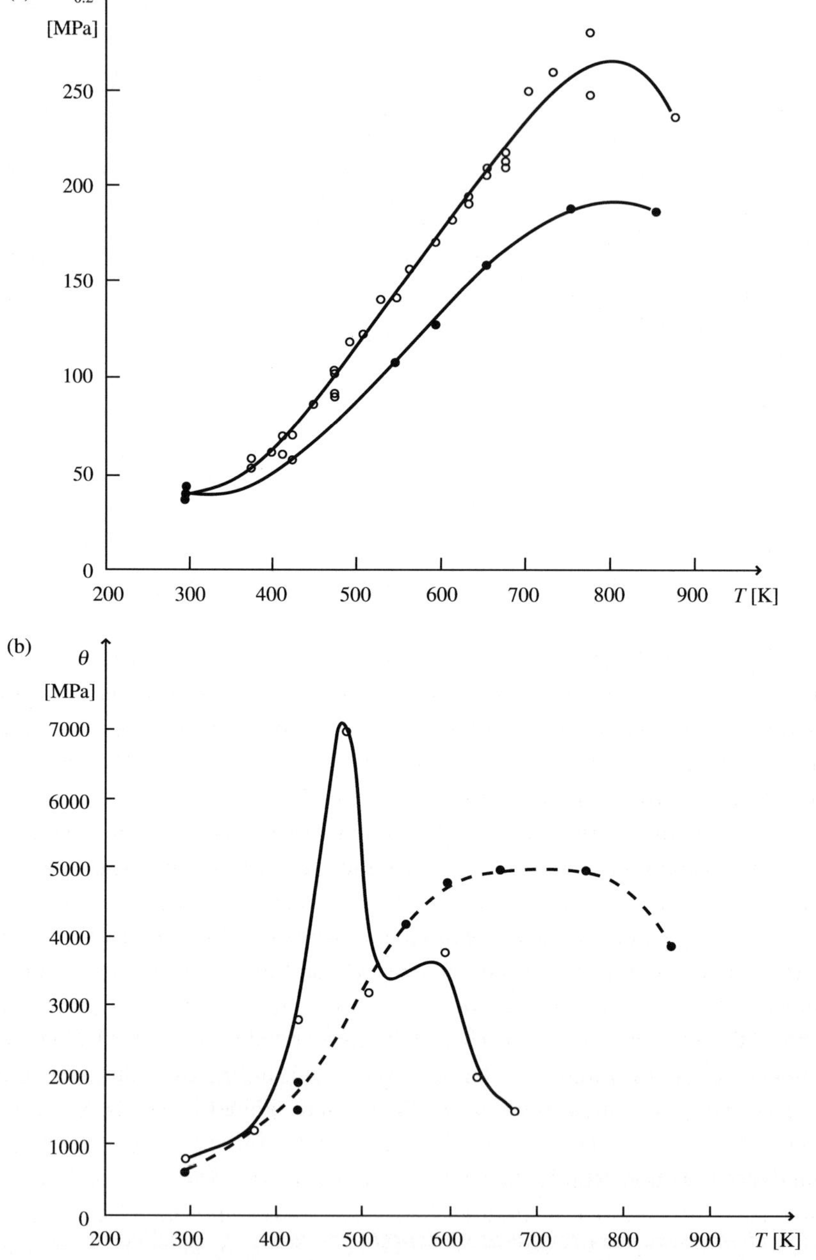

Figure 10.19. Mechanical parameters as a function of temperature in ⟨123⟩ Ni_3(Al, 1 at.%Ta) measured with the quick procedure (○) and the reliable procedure (●), respectively (see text). (a) Yield stress $\tau_{0.2}$. (b) Work-hardening coefficient $\theta_{0.2}$ (After Spätig, et al., 1995).

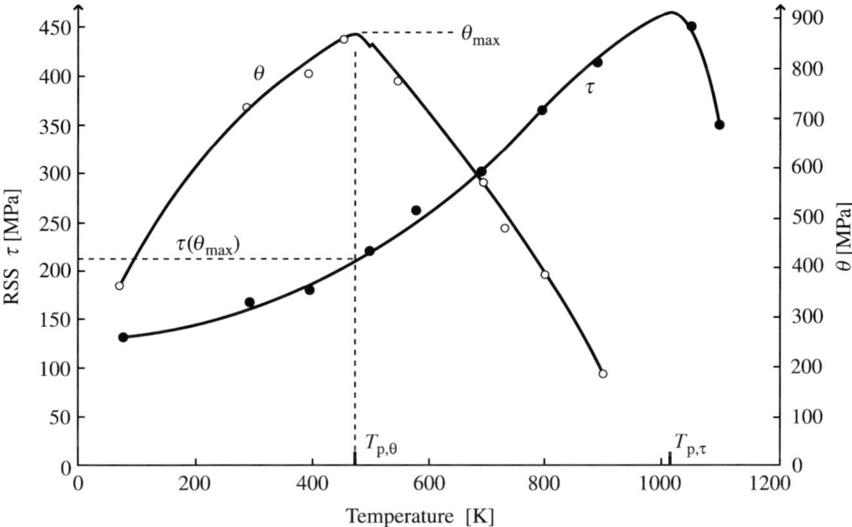

Figure 10.20. Variation with temperature of the yield stress, τ, and the associated work-hardening coefficient, θ, in single crystalline Ni_3(Al, 1.5 at.% Hf). From Ezz and Hirsch (1994).

single and polycrystals, respectively. Since the microscopic mechanisms of {001} slip (Section 10.1.3) and {111} slip (Section 10.1.4) have been described, they are considered here to model the yield stress peak temperature. It is natural to pay attention to the complex stacking fault energy which is a key parameter for both {111} and {001} slip. Mechanical data are considered for binary and ternary compounds successively.

The behaviours of the first set of single crystals are compared. They correspond to the following compositions in the ⟨123⟩ orientation: $Ni_{76.6}Al_{23.4}$, Ni_3(Al, 1 at.% Ta) and Ni_3(Al, 3.3 at.% Hf). The yield stress $\tau_{0.2}$ as a function of temperature is represented in Figure 10.21. γ_{CSF} values of these compounds can be found in Table 10.1. Figure 10.21 shows that the three crystals exhibit a strength anomaly over the temperature range 300–700 K. Moreover, at a given temperature, the binary compound appears to be the weakest and the Hf compound the strongest. Table 10.1 indicates that the Hf compound has the highest CSF energy value while the binary compound has the lowest one. These data can be correlated, considering the cross slip mechanism that leads to the formation of KWL, complete and incomplete ones. At a given temperature the cross-slip frequency is controlled by the constriction of the leading superpartial (see Chapter 5 and step (a) to (b) in Figure 10.11 or (a) to (b) in Figure 10.12 or (c) to (d) in Figure 10.14). The highest frequency, i.e. the highest dislocation exhaustion rate, corresponds to the compound with a large CSF energy. Therefore, to keep a constant strain-rate, a higher stress is required for deformation to proceed via the motion of shorter dislocation segments. Hemker and Mills (1993) were the first to point out such a correlation between γ_{CSF} and strength. In their

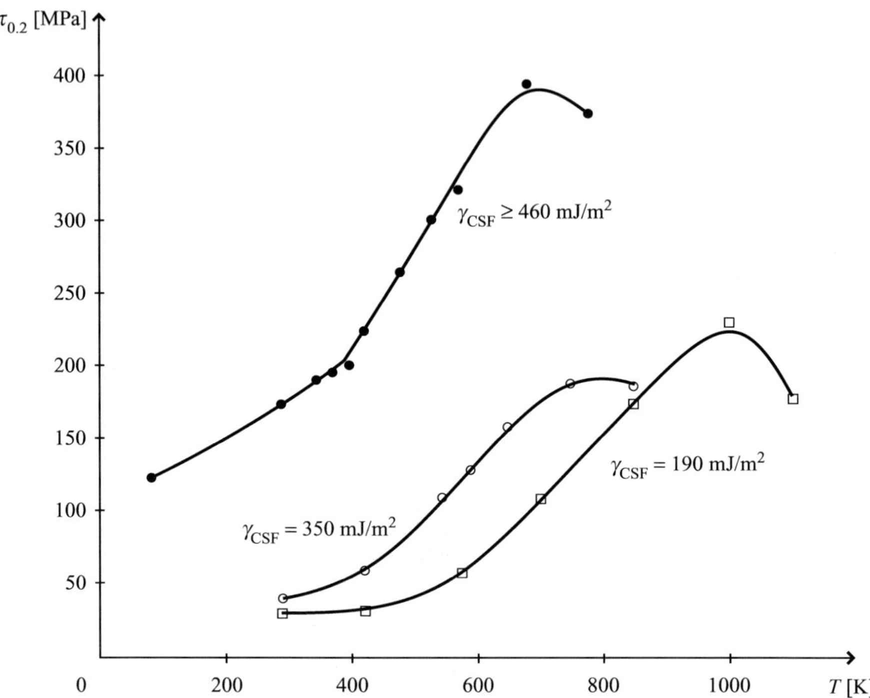

Figure 10.21. The 0.2% offset stress $\tau_{0.2}$ as a function of temperature for: □ $Ni_{76.6}Al_{23.4}$, ○ Ni_3(Al, 1 at.% Ta), ● Ni_3(Al, 3.3 at.% Hf). ⟨123⟩ single crystals. The CSF energies are also indicated. From Spätig (1995) and Kruml et al. (2000).

experiments, they were comparing a binary and a boron doped Ni_3Al, for which $\tau_{0.2}$ was measured at one temperature only. The correlation is established here over several hundred degrees of temperature and for large differences in stresses. A confirmation of this interpretation comes from the direct measurement of mobile dislocation exhaustion rates (or cross-slip frequency). They are obtained from repeated stress relaxation data using relation (2.27), in which ρ_m is measured after a time, t, for which the deformation rate becomes half the size. They are measured at a plastic strain of 0.2%. At 573 K, measured values of ρ_m/ρ_{mo} are 69% for Ni_3Al, 65% for Ni_3(Al,Ta) and 60% for Ni_3(Al,Hf), i.e. the mobile dislocation exhaustion rates are successively 31, 35 and 40%. These values show that the cross-slip frequency increases with the yield stress at a given temperature for the three compounds of interest.

It is worth noting another effect due to this cross-slip mechanism which concerns some dislocation features. It is illustrated in Figure 10.22. It summarizes TEM data from various authors observing superdislocations split along {111}. It is remarkable that no data are reported for dislocations near screw orientations. In the following, α_{min} is the minimum angle reported by the authors between the superdislocations and their Burgers vector.

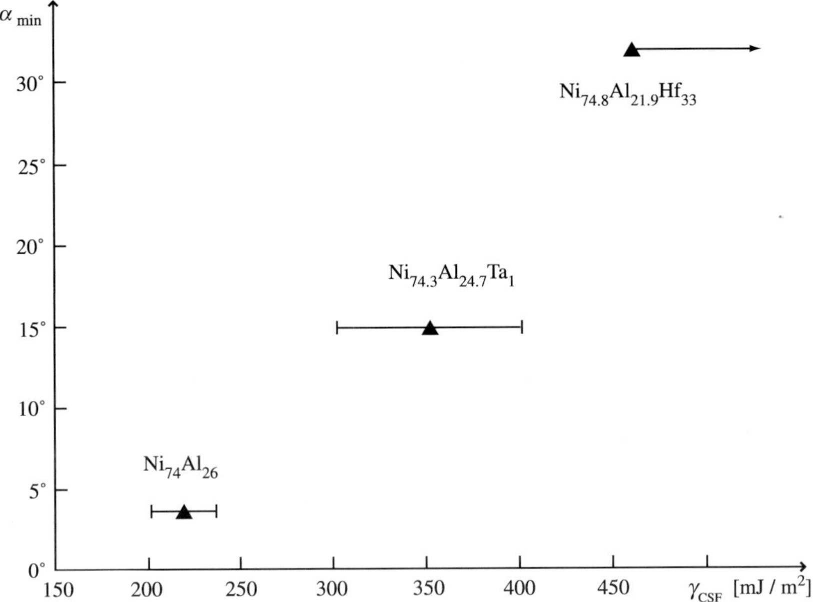

Figure 10.22. Values of parameter α_{min} as a function of γ_{CSF} for superdislocations on {111}. Various Ni_3Al compounds. See text for definition of α_{min} and references.

The fact that α_{min} never approaches zero is thought to be the consequence of superdislocation cross-slip from the octahedral to the cube plane as it approaches screw orientation. Following this line, α_{min} is plotted as a function of the γ_{CSF} values given in Table 10.1. For Ni_3(Al, 3.3 at.% Hf) α_{min} is close to 32° (Kruml et al., 2000a). For Ni_3(Al, 1 at.% Ta) α_{min} is 15° (Baluc et al., 1991a,b; Baluc and Schäublin, 1996) and 3° for $Ni_{74.1}Al_{25.9}$ (Dimiduk et al., 1993). Figure 10.22 clearly shows that the complex stacking fault energy and the parameter α_{min} increase simultaneously.

After this interpretation of the strength as a function of γ_{CSF} below the stress peak temperature, the stress necessary for cube slip is now examined. From the microscopic study of Section 10.1.3, cube glide above $T_{p,\tau}$ proceeds via the viscous motion of screws controlled by a kink-pair mechanism and a jerky motion of edges according to the locking–unlocking mechanism. This agrees in particular with the low values measured for the activation areas: In $Ni_{76.6}Al_{23.4}$ the yield stress peak temperature is close to 1000 K (Figure 10.21), while activation area values are 100 b^2 at 1000 K and 96 b^2 at 1100 K (b is the superpartial Burgers vector). In Ni_3(Al, 3.3 at.% Hf) $T_{p,\tau}$ is close to 630 K (Figure 10.21), while activation areas are 400 b^2 at 680 K and 30 b^2 at 780 K (Bonneville et al., 1997a,b). For the mobility mechanisms of screws and edges the friction is again directly related to γ_{CSF}. The friction force is expected to be small for large CSF energies (the constriction energy is smaller for cube glide to proceed). Consequently, for a given temperature and strain rate, a lower stress is required. Therefore, the stress–temperature curve is shifted towards lower temperatures. For the same compound, the stress–temperature curve in the stress anomaly domain is also shifted in the same direction and so, consequently, is $T_{p,\tau}$. Figure 10.21 agrees fairly well with this qualitative description: the Hf compound with the highest γ_{CSF} value has the lowest $T_{p,\tau}$. The opposite is observed for the binary compound.

10.1.5.4 Yield stress peak temperature (polycrystals). A similar type of behaviour and interpretation also applies to polycrystals. This has been checked for binary compounds of various compositions: $Ni_{74}Al_{26}$, $Ni_{75}Al_{25}$ and $Ni_{76}Al_{24}$. The fault energy values are listed in Table 10.1, while Figure 10.23 illustrates the variation with temperature of $\sigma_{0.2}$. Before comparing the curves of Figure 10.23, a grain size correction has to be done. Indeed, this parameter is smaller in $Ni_{74}Al_{26}$ (500 μm as compared to 800 μm for the two other compounds). The "corrected" yield stress should be slightly lower in Figure 10.23 for this Al rich compound (indeed, a 25% yield stress decrease is observed by Lopez and Hancock (1970), as the 500 μm grain size is doubled). In spite of the correction, $Ni_{74}Al_{26}$ still exhibits the highest strength over the temperature range. From Figure 10.23 and Table 10.1, it appears that, as the aluminium content increases, γ_{CSF} rises and (i) at a given temperature below $T_{p,\sigma}$, the yield stress increases and (ii) the yield stress peak temperature

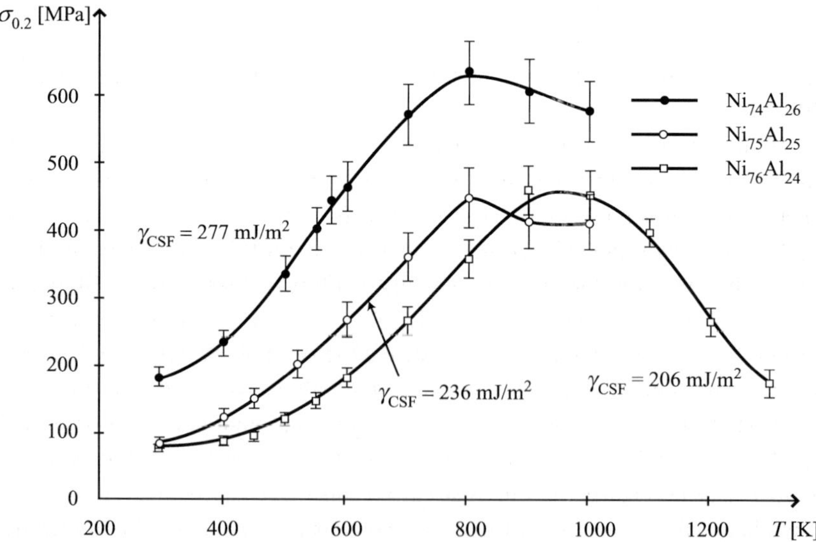

Figure 10.23. 0.2% offset stress as a function of temperature for binary Ni_3Al polycrystals. The compositions are indicated. From Kruml et al. (2002).

is shifted to the left ($T_{p,\sigma}$ is close to 800 K for $Ni_{74}Al_{26}$ and to 950 K for $Ni_{746}Al_{24}$). Similar measurements in polycrystals of compositions ranging between $Ni_{76.5}Al_{23.5}$ and $Ni_{73.5}Al_{26.5}$ confirm the above trend of variation of the yield stress in the strength anomaly temperature range (Noguchi et al., 1981). Therefore, thc correlation between γ_{CSF} and the strength and the stress-peak temperature, respectively, appears to be the same as for single crystals, with a better accuracy in the latter case.

10.1.5.5 Conclusion about the peak temperature for the yield stress. The constriction frequency of the core split outside the cube plane accounts fairly well for the variation with γ_{CSF} of the anomalous yield stress at given temperature, as well as the magnitude of the stress-peak temperature. Nevertheless, this description is qualitative. It accounts for the main trends, observed in Figures 10.21 and 10.23 but it is likely that other effects, not considered so far, somehow influence the mechanical behaviour. Indeed, the above constriction process of the CSF also depends on (i) the elastic constants of each compound and their variation with temperature, (ii) the temperature evolution of γ_{CSF}, and (iii) the various stress levels at which the compounds are compared. These parameters or effects are poorly documented.

Apart from the cross-slip locking mechanism in the strength anomaly domain, other diffusion controlled locking processes do take place. These have different signatures such as: (i) serrations on the stress–strain curves over limited temperature ranges, (ii) a measurable yield point at reloading after successive relaxations (see Chapter 9) over the same temperature range and (iii) a zero or negative strain-rate sensitivity of the stress which was already made evident in the early work of Thornton et al. (1970). All these well identified effects cannot be accounted for by the cross-slip mechanism described above.

Various diffusion processes have been proposed that lead to dislocation locking: (i) solute atom diffusion to the superdislocation core (Cottrell mechanism, see Section 3.2.2), (ii) relaxation of the atomic structure along the APB (Brown mechanism, see Figure 10.35) and (iii) change of APB plane by climb (climb dissociation). The net dislocation behaviour is, however, the same. For the three processes, the friction stress on the mobile dislocations increases with temperature which contributes to the strength anomaly (see Section 10.2.1). In addition the induced friction is higher for slow dislocations and lower for fast dislocations. The latter can be more mobile, which accounts for a negative strain-rate sensitivity on stress and serrated yielding as explained by Kubin and Estrin (1991) and developed in Section 3.2.2. Let us note that thc strcss anomaly observed for cube glide (Figure 10.7) can be interpreted only in terms of diffusion processes.

To conclude this section, the cross-slip mechanism, plus some diffusional processes, account, at least qualitatively, for all the features of the strength anomaly and of the temperature of the yield stress peak in Ni_3Al.

For a deeper understanding of deformation processes a study of work-hardening is necessary.

10.1.5.6 The temperature of the work-hardening peak in single crystals. It is worth noting that the anomalous behaviour of the work-hardening rate with temperature has only recently stimulated research. The same remark holds for the discrepancy between the peak temperature for stress, $T_{p,\tau}$, and that for work-hardening, $T_{p,\theta}$, at given strain. Consequently, in spite of a large effort devoted to the mechanical properties of $L1_2$ compounds, only a few data on work-hardening have been published. Some of them are illustrated in Figure 10.24. It shows that for the three compounds of Figure 10.21, $T_{p,\theta}$ is systematically lower than $T_{p,\tau}$. This trend is illustrated for other $L1_2$ crystals by Bontemps-Neveu (1991), Couret et al. (1993), Ezz and Hirsch (1994, see Figure 10.20), Ezz (1996) and Masahiko and Hirano (1997). Although the lack of data prevents the relative values of $T_{p,\tau}$ and $T_{p,\theta}$ from being checked, it seems that the latter is systematically lower than the former, at least when τ and θ are measured at or above a 0.2% plastic strain. It has been suggested by Saada and Veyssière (1993) that the peak in work-hardening corresponds to the temperature at which the $\tau(T)$ curve is inflected. It

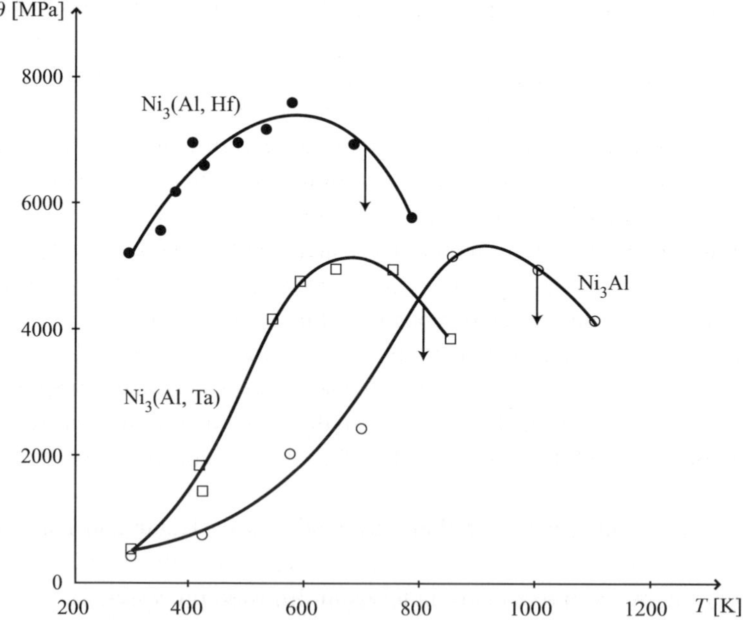

Figure 10.24. Work-hardening coefficient measured at a 0.2% plastic strain, $\theta_{0.2\%}$ versus temperature for the three ⟨123⟩ Ni_3Al compounds of Figure 10.21. Vertical arrows indicate the stress-peak temperature, $\tau_{p,\tau}$, in each case. The exact compositions are those in Figure 10.21. From Spätig (1995).

is worth noting that such properties of the work-hardening rate are not observed anymore at too low strains. This parameter increases as strain decreases. In the transient domain where microplasticity dominates, work-hardening is connected with phenomena different from those discussed below (Veyssière and Saada, 1996).

The interpretation that we propose for the presence of a peak temperature for work-hardening is illustrated by the schematics of Figure 10.25. For the sake of simplification, the stress–strain curves are assumed to consist of two distinct hardening stages, with a weaker slope at high stresses. The corresponding crystal exhibits a strength anomaly ($T_1 < T_2 < T_3$). The positive variation of stress with temperature (at a given plastic strain γ_{po}) is represented in Figure 10.25(b). The curve $\theta(T)$ at γ_{po} of Figure 10.25(c) exhibits a maximum at temperature $T_{p,\theta}$. In the case of Ni_3Al, the two hardening stages along the stress–strain curve can be interpreted in terms of the stability of KWL with respect to stress.

The motivation of early studies of this question was to predict whether these locks were stable at the stress-peak (Saada and Veyssière, 1992; Chou and Hirsch, 1993). Given the existence of two types of locks established in Section 10.1.4.3, the stability study had to be refined (Caillard and Paidar, 1996). Let us note that a set of locks is likely to exist, which differ by their core extension w along the cube plane, w ranging between b and the natural width of splitting along {100}. Their stability is estimated, referring to the schematics of Figure 10.26, considering infinitely long dislocations interacting in the frame of anisotropic elasticity.

Under the applied stress, the leading superpartial is pushed forward along {111} while the trailing one is locked with respect to cube glide. For unlocking to take place, the trailing superpartial has to yield, i.e. $\tau_{cs}^{(t)}$ must be larger than the friction stress $\tau_{cs}^{(min)}$ on the cube plane. However, the stress component $\tau_{cs}^{(l)}$ can push the leading superpartial along {100}, which further locks the superdislocation. Therefore, for successful unlocking, {001} cross-slip of the trailing superpartial must be faster than that of the leading one. This condition is expressed via a second threshold stress, τ_{ul} (Caillard and Paidar, 1996), which has also been computed. The critical stresses for the lock to yield are found to be of the order of:

$$\tau_i = \frac{\gamma_o}{b}\left[1 - \frac{\gamma_c}{\gamma_o}\frac{1 + 2/A}{\sqrt{3}}\right] \tag{10.1}$$

for an incomplete lock with $w \sim b$, and:

$$\tau_c = \frac{\gamma_o}{b}\left[1 - \frac{\gamma_c}{\gamma_o}\frac{1}{\sqrt{3}}\right] \tag{10.2}$$

for a complete lock. A is the elastic anisotropy ratio and b the Burgers vector of the superpartial dislocation.

The rare values of A in the literature are found to be close to 3.3. With the γ_o and γ_c data of Table 10.1, it is obvious that incomplete locks are less stable than complete ones

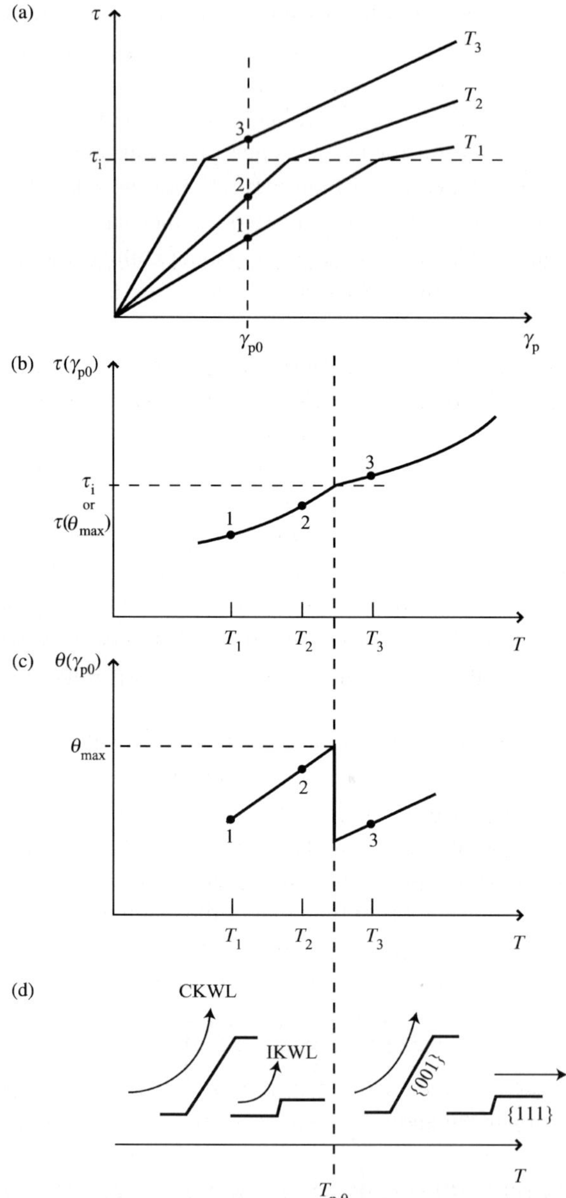

Figure 10.25. Schematics illustrating the existence of a peak in work-hardening at a given temperature for a material that exhibits a strength anomaly. (a) Stress–(plastic)strain curves at three temperatures, with two hardening stages. (b) Stress at given plastic strain (γ_{po}) as a function of temperature. (c) Work-hardening coefficient (at γ_{po}) as a function of temperature. τ_i is the transition stress between the two hardening mechanisms. $T_{p,\theta}$ is the temperature of the maximum of work-hardening. From Kruml et al. (2002).

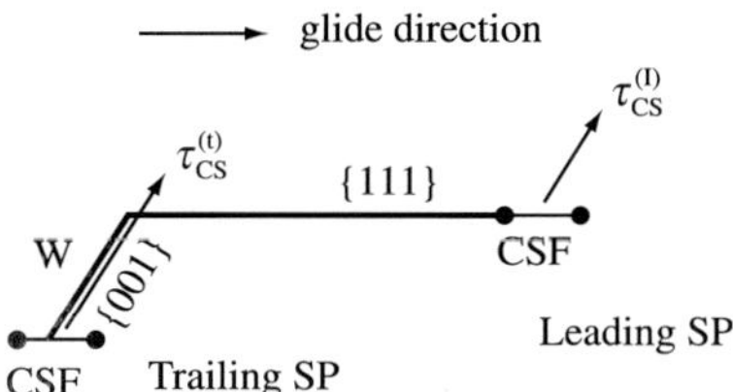

Figure 10.26. Schematics of an incomplete KWL seen end on. SP: superpartial. τ_{cs} is the shear stress acting on superpartials along {001}. The superscripts t and l refer to trailing and leading SPs, respectively. From Caillard (2001).

($\tau_i < \tau_c$). It has been proposed that the stress corresponding to the peak in work-hardening $\tau(\theta_{max})$ (see Figure 10.20) corresponds to the yielding of incomplete locks with $w \sim b$ (Caillard, 2001). Under such conditions, the two hardening stages in Figure 10.25 can be interpreted as follows: (i) at low stresses $\tau < \tau_i$ in Figure 10.25(a) or low temperatures $T < T_{p,\theta}$ in Figure 10.25(c), dislocation exhaustion by complete and incomplete lock formation dominates hardening and results in high θ values; (ii) at high stresses and high temperatures, the same mechanism operates but in competition with the yielding of incomplete locks ($\tau > \tau_i$), which results in lower θ values. In this type of interpretation, the maximum work-hardening value θ_{max} corresponds to a stress $\tau(\theta_{max})$ which equals τ_i (relation (10.1)). An example of determination of $\tau(\theta_{max})$ is shown in Figure 10.20.

To assess this description, data have been reviewed for Ni_3Al compounds in which information is available about (i) APB energy values (Table 10.1) and (ii) mechanical parameters such as $T_{p,\theta}$ and $\tau(\theta_{max})$ (see Table 10.2). The comparison between τ_i and $\tau(\theta_{max})$ is made as follows. Data about γ_c and γ_o indicate that the ratio γ_c/γ_o ranges between 0.77 and 0.84. A reasonable average value of 0.8 can be adopted (see Veyssière and Saada, 1996). This allows us to simplify relations (10.1) and (10.2):

$$\tau_i \approx 0.26\gamma_o/b \tag{10.3}$$

$$\tau_c \approx 0.54\gamma_o/b \tag{10.4}$$

These relations predict a linear dependence of τ_i and τ_c on γ_o. Therefore, $\tau(\theta_{max})$ values are plotted as a function of γ_o in Figure 10.27. For comparison, two straight lines going through the origin of respective slopes $0.24/b$ and $0.54/b$ represent relations (10.3) and (10.4), respectively. In spite of the experimental scatter of data in Tables 10.1 and 10.2, a fair agreement is found with the line of slope $0.24/b$ and a complete disagreement with the other line. This shows that $\tau(\theta_{max})$ and τ_i exhibit similar values and the same variation as a function of γ_o. This strongly supports the above interpretation that the maximum in work-hardening corresponds to the onset of yielding of incomplete KWL.

Table 10.2. Survey of mechanical test data for single crystals of $L1_2$ compounds.

Alloy (in at.%)	Straining conditions	$T_{p,\theta}$ (K)	$\tau(\theta_{max})$ (MPa)	Data points number[c]	References
Ni_3(Al, 3%Hf)	⟨123⟩ axis $\gamma_p = 0.2\%$	573 ± 50	310 ± 30	1	Spätig (1995)
Ni_3(Al, 1%Ta)	⟨123⟩ axis $\gamma_p = 0.2\%$	700 ± 50	180 ± 20	2	Spätig (1995)
$Ni_{76}Al_{24}$	⟨123⟩ axis $\gamma_p = 0.2\%$	900 ± 50	190 ± 20	5	Spätig (1995)
Ni_3(Al, 0.25%Hf)	τ at 0.2%				Bontemps-Neveu (1991)
	θ at 1%				
	⟨111⟩ axis	600 ± 100	160 ± 40	10	
	⟨123⟩ axis	700 ± 100	150 ± 40	11	
	10° from ⟨001⟩	750 ± 100	150 ± 40	12	
Ni_3(Al, 2%Hf)	τ at 0.2%				Bontemps-Neveu (1991)
	θ at 1%				
	⟨123⟩ axis	500 ± 100	210 ± 40	13	
	10° from ⟨001⟩	550 ± 100	175 ± 40	14	
Ni_3(Al, 1.5%Hf 0.2%B)	τ at 0.2%				Ezz and Hirsch (1994)
	θ at 1%				
	⟨123⟩ axis	500 ± 100	210 ± 20	15	
Ni_3Ga	τ at 0.05%				Ezz (1996)
	θ at 1%				
	O_a axis	500 ± 100	120 ± 10	16	
	O_b axis	470 ± 100	125 ± 10	17	

[c] Refer to Figures 10.6 and 10.27.

10.1.5.7 The temperature of the work-hardening peak in polycrystals. For comparison, the binary polycrystals of Section 10.1.5.4 have been submitted to similar investigations to further test the validity of the preceding interpretation. Some peculiar properties in this case limit the field of investigation. For example, intergranular failure severely limits the ductility above 1000 K so that maximum plastic strains of 8% were achieved (Matterstock et al., 1999). The transient at the onset of the stress–strain curve is also more extended than for single crystals. Therefore, the σ and θ values measured at a 0.2% plastic-strain are not meaningful. These quantities have been measured at plastic strains of 3 and 5%, respectively. In what follows they are successively referred to as $\sigma_{3\%}$, $\sigma_{5\%}$, $\theta_{3\%}$ and $\theta_{5\%}$. The $\theta(T)$ curves for polycrystals exhibit a shallow maximum as illustrated in Figure 10.28.

Since the evolution of work-hardening with temperature parallels that of the mobile dislocation exhaustion parameter $\Delta\rho_m/\rho_{mo}$ (see definition in Section 9.2.1), the latter parameter has been measured as well, for a better definition of $T_{p,\theta}$. An example of variation with temperature of parameters σ, θ and $\Delta\rho_m/\rho_{mo}$ is presented in Figure 10.28 at 3 and 5% plastic strains, respectively. It clearly shows that θ and $\Delta\rho_m/\rho_{mo}$ exhibit

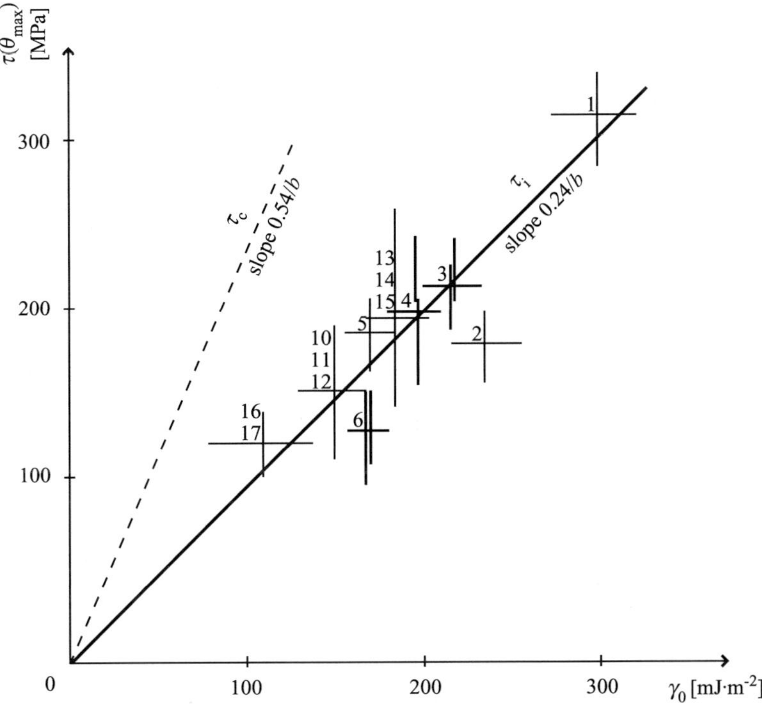

Figure 10.27. Stress corresponding to the maximum in work-hardening, $\tau(\theta_{max})$, as a function of the APB energy on {111}, γ_o. The data point numbers refer to Tables 10.1–10.3. Two vertical bars at the same composition (points number 3, 4 and 6) correspond to plastic strains of 3 and 5% (see Table 10.3). The straight lines correspond to relations 10.3 and 10.4. After Caillard (2001) and Kruml et al. (2002).

a peak, while the stress does not, over the temperature range investigated. Figure 10.28 shows, in particular, that for polycrystals $T_{p,\theta}$ is smaller than $T_{p,\sigma}$, as in single crystals (Section 10.1.5.3).

To test the above interpretation of the peak in hardening, the parameters of interest were measured for the three compositions investigated, on curves similar to those of Figure 10.28. The results are presented in Table 10.3. To allow for a comparison with relations (10.3) and (10.4) established for single crystals, the stresses σ of Table 10.3 have been converted into an "estimated" resolved shear stress assuming a Schmid factor of 0.45. The corresponding data are also plotted in Figure 10.27. It shows that all points fall along the straight line with the weakest slope, as for single crystals. It is worth noting that the stress corresponding to the maximum of θ is the same, within the experimental scatter, at 3 and 5% plastic strains. All these characteristics are in favour of the above model for the maximum in work-hardening.

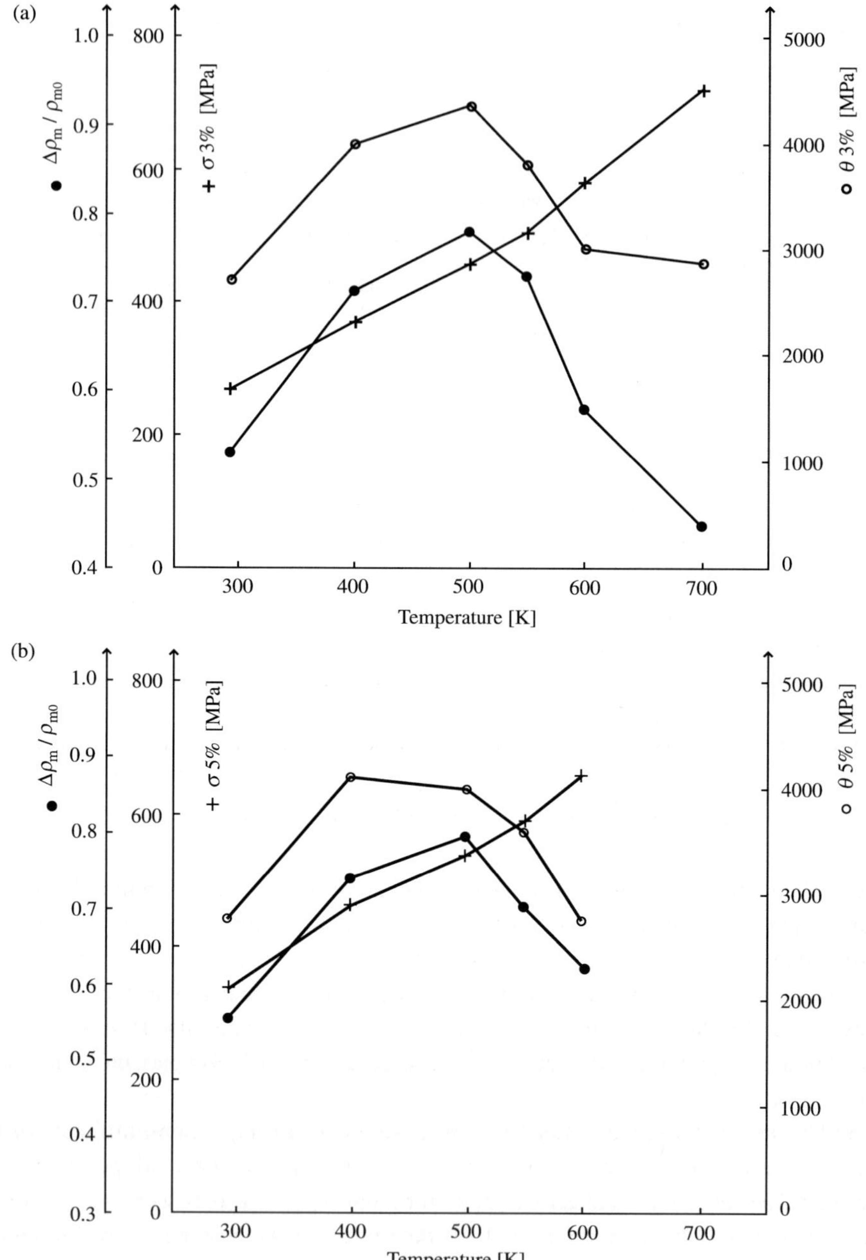

Figure 10.28. Variation as a function of temperature of stress σ, work-hardening coefficient θ and mobile dislocation exhaustion rate $\Delta\rho_m/\rho_{mo}$. $Ni_{74}Al_{26}$ polycrystals. Corresponding plastic strains: 3% in (a); 5% in (b). For sake of clarity error bars are not indicated. Uncertainties are $\pm$10% for θ and $\pm$12% for $\Delta\rho_m/\rho_{mo}$. From Kruml et al. (2002).

Table 10.3. Survey of mechanical test data of polycrystals of $L1_2$ binary compounds.

Compound	Plastic strain (%)	Peak temperatures (K)		Corresponding stress (MPa)	Estimated RSS (MPa)[d]	Data points number[e]
		θ	$\Delta\rho_m/\rho_{mo}$			
$Ni_{76}Al_{24}$	3	500	500–650	230–340	100–150	6
	5	450–600	450–600	250–330	110–150	6
$Ni_{75}Al_{25}$	3	450–600	420–490	300–470	135–210	4
	5	480–550	450–520	450–550	200–247	4
$Ni_{74}Al_{26}$	3	450–520	450–550	410–500	185–255	3
	5	400–480	420–500	450–530	200–240	3

[d] RSS computed with an average Schmid factor of 0.45.
[e] Refer to Figures 10.6 and 10.27.

10.1.5.8 Conclusions about the peak in work-hardening. The description of the $\theta(T)$ curves, given in the preceding section, emphasizes the role of incomplete KWL in the deformation processes of $L1_2$ compounds. In particular, these explain in situ observations of the jerky glide of superdislocations along {111} (Section 10.1.4.2). In agreement with the estimation of τ_i (relation (10.1)), a large number of elementary macrokinks, signature of APB jumps, were observed in samples predeformed at stresses higher than τ_i, i.e. at a temperature higher than $T_{p,\theta}$: 423 K for Ni_3(Al, 1.5 at.%Hf) and 673 K for Ni_3(Al, 0.25%Hf) (Veyssière and Saada, 1996). Conversely, post mortem observations of Ni_3(Al, 0.25 at.%Hf) deformed at 300 K ($T < T_{p,\theta} \sim 700$ K) do not exhibit a large number of elementary macrokinks. This is due to a corresponding stress much smaller than τ_i.

In a few rare cases, e.g. Ni_3Ga, the $\theta(T)$ curve increases again after an initial maximum and eventually exhibits a second peak at a higher temperature (Figure 10.29). The corresponding stress is close to τ_c (relation (10.2)), the yielding stress for complete KWL (Caillard and Molénat, 1999). In this particular case, complete locks are expected to yield at stresses lower than the peak stress. Additional research is needed in this area to confirm the proposed interpretation of yielding by octahedral glide before primary cube glide operates.

Complete and incomplete locks have different critical formation configurations that correspond to a kink-pair and a recombined superpartial bulging in the cube plane, respectively (see Figures 10.15 and 10.16). They also have different energies, the higher one for the complete lock. It is worth noting that the above two domains of the anomalous temperature range have been quoted earlier in the enlightening work of Thornton et al. (1970) on polycrystals. The existence of $T_{p,\theta}$ was not explicitly formulated. However, their Figure 3(b), which is a schematics that represents the anomalous $\sigma(T)$ curve, exhibits an inflexion point. The two temperature domains below and above this point are labelled

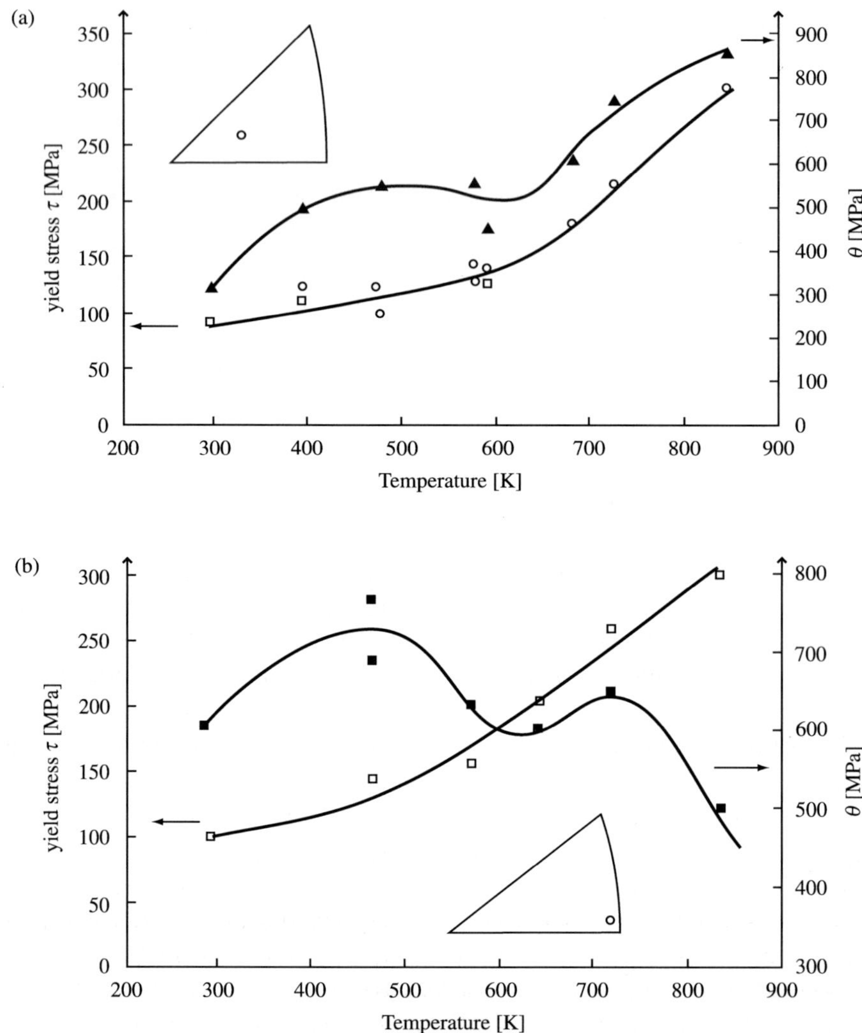

Figure 10.29. Variation with temperature of stress τ and work-hardening coefficient θ in Ni_3Ga single crystals. (a) and (b) correspond to two different orientations. From Ezz (1996).

I and II, respectively, by the authors. Strain-rate jump experiments in both domains exhibit a different response according to the domain and two types of creep transients are also observed: in domain I a rapid decrease of the creep-rate is observed, contrary to domain II. The authors propose two different glide mechanisms, based on microstructural observations: (i) mobile dislocation exhaustion in domain I (called exhaustion hardening) in agreement with the above interpretation for $T < T_{p,\theta}$ and (ii) a "debris mechanism" in domain II attributed to the temperature enhanced activity of the cube cross-slip system. Much later, the operation of this mechanism just below or at the stress-peak temperature has been extensively evidenced and studied by TEM on single crystals (Douin et al., 1986; Kruml et al., 1997). Domain II can also be interpreted by the yielding of incomplete locks described above.

Another reason has been claimed to explain the maximum of work-hardening, considering the presence of APB tubes near the lower end of the anomalous temperature regime. Such tubes have been observed to form during octahedral glide of the superdislocation, thus slowing down its motion (Bonneville et al., 1991; Shi et al., 1996). TEM observations in a variety of compounds also show that these tubes annihilate with increasing temperature, close to $T_{p,\theta}$. A subsequent increase of the dislocation velocities is thus anticipated which should influence work-hardening. However, no quantitative evaluation of this effect is available at the moment, unfortunately.

Finally, recent three-dimensional computer experiments were used to simulate the mechanical behaviour of Ni_3Al, in which several types of KWL were taken into account (Devincre et al., 1999; Devincre, 2000). In these studies, incomplete locks were also considered with a temperature dependent extension along the cube plane. The simulations yielded a monotonic temperature increase of the work-hardening rate. In the model in Section 10.1.5.4, incomplete locks have a small extension along $\{001\}$ which accounts for a peak in work-hardening at the experimentally observed temperature and stress.

10.1.6 The role of different fault energies

Although it was accepted long ago that in Ni_3Al compounds with a $L1_2$ structure, mechanical properties are directly related to the core geometry, the exact connection was still a matter of mystery. It was known that the addition of ternary elements such as Hf, Ta, B had a marked influence on the strength. One of the reasons claimed was an alteration of the core (see e.g. Heredia and Pope (1991) for Hf additions) without any further details. A systematic attempt at correlating APB energies with the yield stress was performed experimentally by Dimiduk et al. (1993) for a range of $L1_2$ compounds. The conclusion was that no obvious correlation was found.

After the results and interpretations presented in Section 10.1, the role of the different fault energies is clearly established. The complex stacking fault energy is directly involved in the activation energy of cross-slip from $\{111\}$ onto $\{100\}$. Therefore, this is a key parameter for the strength, including orientation effects as represented schematically in Figure 10.30.

The APB energies and their ratio γ_c/γ_o (which is close to 0.8) determine the stress at which strain-hardening peaks. It also corresponds to the inflexion point on the stress–temperature curve. These results are also valid for single and polycrystals.

It is then possible to tailor the mechanical properties of Ni_3Al compounds, bearing in mind these simple rules, as a function of the fault energies available in Table 10.1.

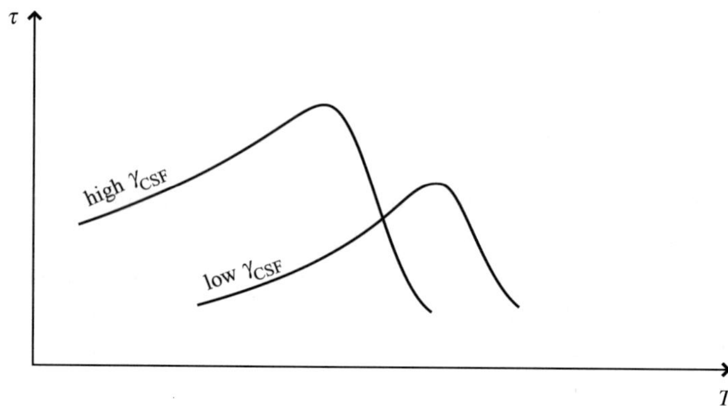

Figure 10.30. Schematic representation of the influence of the complex stacking fault energy on the $\tau(T)$ curve for Ni_3Al.

10.1.7 Strength and dislocation density

In view of the very high values of strength and work-hardening in ordered intermetallics, it is natural to check whether these parameters can be accounted for by dislocation interactions. Answers to this question appear to be controversial: Ezz et al. (1995) support forest hardening; Greenberg and Ivanov (1997) invoke dislocation interactions for the interpretation of temperature change experiments; Louchet (1995) assumes a high rate of dislocation storage to explain high work-hardening coefficients. Conversely, the observed dislocation densities are sometimes claimed to be too low to explain these unusual mechanical properties and instead exhaustion of dislocation sources is proposed as an alternate mechanism (Mills and Chrzan, 1992; Veyssière and Saada, 1996; Devincre et al., 1999; Demura and Hirano, 2000; Kruml et al., 2002).

This section reviews a few available data about dislocation densities to try and correlate them with the mechanical behaviour. Ni_3Al is considered because of more abundant information.

10.1.7.1 Values of dislocation densities in Ni_3Al. Surprisingly, this quantity has not been considered thoroughly as a rule. Some studies provide one measurement only at a given strain and temperature. Data have been provided by Kruml et al. (2000b, 2001) in $\langle 123\rangle$ and $\langle 145\rangle$ Ni_3(Al, 3.3 at.% Hf) single crystals as a function of stress, strain and temperature and to a lesser extent by Baluc (1990) in Ni_3(Al, 1 at.% Ta). The TEM methods used and related uncertainties are exposed by Kruml et al. (2000b, 2001).

10.1.7.2 Dislocation densities and mechanical parameters. The well known relation due to Taylor (1934) is considered here, which relates the stress τ and the dislocation density:

$$\tau = \tau_o + \alpha\mu b\sqrt{\rho} \tag{10.5}$$

where τ_o is a stress component different from that which arises from dislocation interactions, α is a dimensionless constant and b the Burgers vector of a superdislocation in the present case. α is related to the strength of dislocation interactions. Such a type of relation is valid for several kinds of situations such as: long range elastic interactions (with the Frank network, with dislocations in parallel slip planes), overcoming of junctions with forest dislocations, stress necessary to operate a source. The values of α have been the subject of a long and controversial debate (see e.g. Basinski and Basinski, 1979). A review of such values based on theoretical estimations and experimental measurements has been published by Lavrentev (1980). They range between 0.05 and 1.3 for FCC, BCC and HCP structures.

The data of Kruml et al. (2000b, 2001) are summarized in Figure 10.31. The main comments on these data are the following: As strain (or stress) increases at given temperature, or as temperature increases at approximately constant strain, the dislocation density increases. The uncertainties on ρ values measured by TEM are rather large. However, the agreement shown in Figure 10.31 between the data and relation (10.5) with $\alpha = 1$ is satisfactory. Indeed, such comparisons between τ and $\sqrt{\rho}$ are usually tested on log/log plots (see Viguier (2003) for intermetallics and Orlova (1991) for metals) unlike in the present case. The data points away from the straight line are those measured at low plastic strains (1 and 3%) and the one at 84 K. This means that the Taylor relation does not describe the onset plastic stages and, according to Figure 10.1, 84 K is at the very beginning of the strength anomaly.

The value found, $\alpha = 1$, is comparable with the high values of Lavrentev (1980). The same remark holds for the density values, which range here between 10×10^{12} and 160×10^{12} m^{-2} for plastic shear strains lower than or equal to 22% (see Orlova (1991) for data on metals). According to relation (10.5), the high strength of Ni_3Al is related to higher elastic constants, a larger Burgers vector and possibly a higher τ_0 value.

Although, work-hardening in Ni_3Al is connected with the increase of dislocation density, the rates are very high ($\leq 10^{-1}$ μ) as compared to those in metals ($\sim 10^{-3}$ to 5×10^{-3} μ in stage II–see e.g. Martin et al, 1999). This difference has to be explained in spite of the above similarities. The answer can be found in considering dislocation mean free paths Λ. These can be estimated by using the integral form of the Orowan equation:

$$\gamma_p = \rho b \Lambda \tag{10.6}$$

Using the values of γ_p and ρ from Figure 10.31, Λ values are found to lie between 1 and 16 μm (Viguier et al., 2002, Viguier, 2003). These are much smaller than the corresponding ones in metals which are estimated to be in the millimeter range (Nabarro et al., 1964). Consequently, comparing Ni_3Al with metals, relation (10.6) predicts a lower γ_p

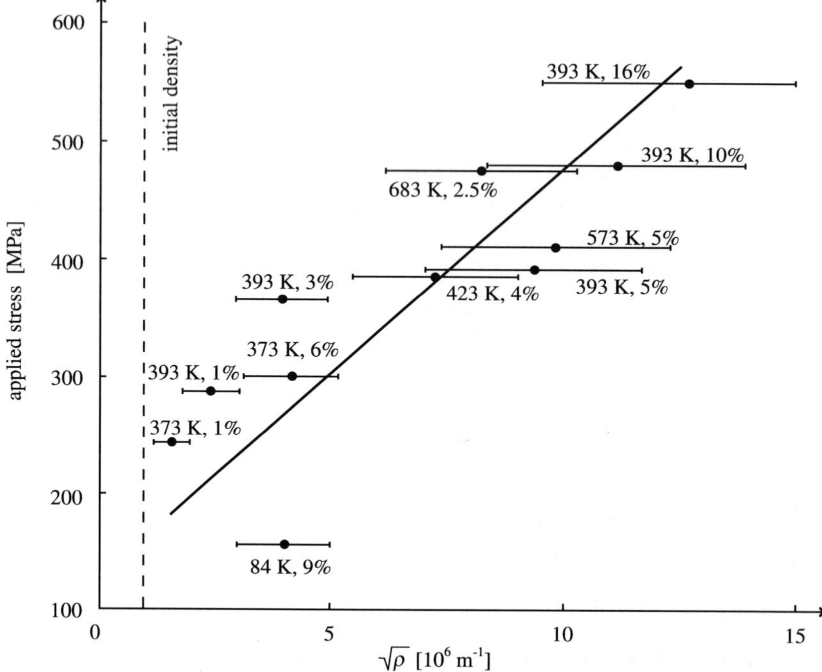

Figure 10.31. Applied stress versus square root of dislocation density. ⟨123⟩ and ⟨145⟩ Ni_3(Al, 3 at.% Hf) single crystals. The deformation temperature and the plastic strain are indicated. The straight line corresponds to relation (10.5) with $\alpha = 1$. From Kruml et al. (2001).

corresponding to similar ρ values (Figure 10.32) and a lower mean free path Λ. This results in a strong work-hardening coefficient for Ni_3Al.

To conclude, in the stress anomaly domain, the strength and the strain-hardening rate appear to be connected with the dislocation density. The high values of θ in Ni_3Al can be accounted for by a much smaller dislocation mean free path, due to the cross-slip locking mechanism.

10.2 STRESS ANOMALIES IN OTHER INTERMETALLICS

Several other intermetallic compounds exhibit an anomalous behaviour of strength versus temperature (see a review by Wee et al., 1980). As a rule the corresponding mechanisms are far less documented than for Ni_3Al. The state of knowledge is exposed for $L1_2$ alloys, such as Cu_3Au and Co_3Ti, and then for B_2 alloys, such as FeAl and β CuZn. For these compounds sufficient information is available on mechanical properties, glide systems and dislocation structures. The properties of γ TiAl are not covered in the present material

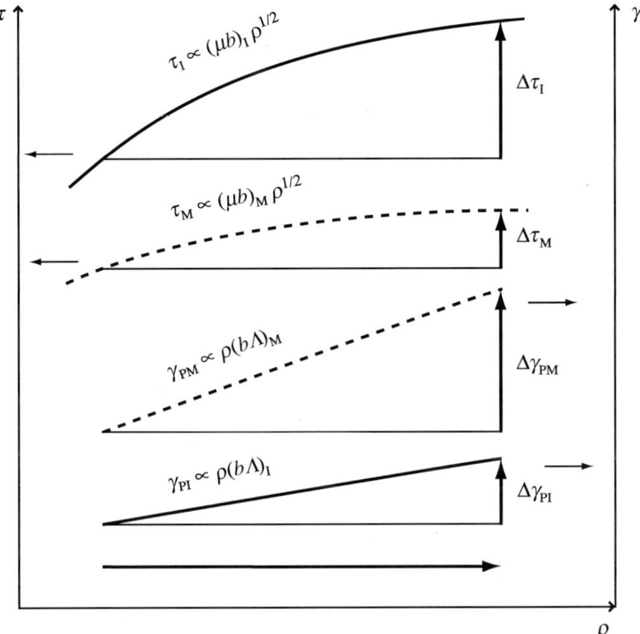

Fig. 10.32. Schematics illustrating the difference in work-hardening between Ni_3Al and metals. Stresses and plastic strains are represented as a function of the dislocation density. Subscripts M and I refer to a metal and intermetallics. Λ is the dislocation mean free path (see text).

because of controversial interpretations proposed to the abundant experimental data available.

10.2.1 Other $L1_2$ crystals

The yield stress temperature curves for Cu_3Au and Co_3Ti are presented in Figures 10.33 and 10.34, respectively. In Cu_3Au, the yield stress increases steeply with temperature up to the order–disorder transition temperature (670 K), above which it decreases abruptly. In addition, the orientation effects are far less important than in Ni_3Al over the whole temperature range. For Co_3Ti, the active glide system is, as a rule, the octahedral one, with pronounced orientation effects. There is a stress anomaly on this glide system as for Ni_3Al (compare Figures 10.10 and 10.34). A small stress–strain rate sensitivity is observed. Cube slip is more difficult to activate than in Ni_3Al: it is only reported above the peak temperature for the $\langle 111 \rangle$ orientation.

The following interpretations have been proposed to account for these data. In Cu_3Au, Pope (1972) attributed the strength anomaly to an interaction between dislocations and local regions of disorder. Yamaguchi and Umakoshi (1990) claim a climb dissociation in which the APB plane changes from {111} to {100} as the temperature rises. Brown's

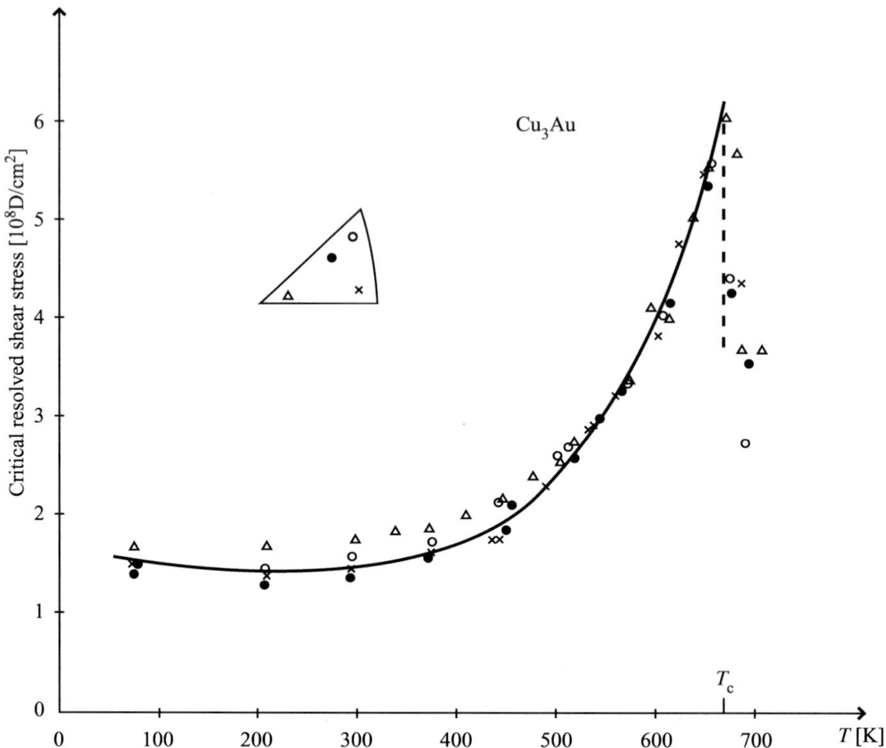

Figure 10.33. Temperature dependence of the CRSS on {111} in Cu_3Au for different orientations. From Kuramoto and Pope (1976). T_c is the order–disorder transition temperature.

model proposes another process in which glide is hindered by a diffusion mechanism (Brown, 1959). It is depicted in Figure 10.35. The APB lies in the glide plane. As it moves, the leading dislocation creates a fresh APB (energy γ_1) which relaxes by diffusion, thus lowering its energy to γ_1'. Then the trailing dislocation flows, without restoring the perfect crystal and thus creating a fault of energy γ_1''. The latter subsequently vanishes by a diffusional process again. Under such conditions, the stress required to move the dislocations is:

$$\sigma = (\gamma_1 + \gamma_1'' - \gamma_1')/2b.$$

This model may explain the stress anomaly close to the order–disorder transition temperature T_c. Diffusional effects may indeed strongly modify the structure and energy of APBs above 0.75 T_c in $L1_2$ alloys as claimed by Sanchez et al. (1987). However, TEM observations (Kear and Wilsdorf, 1962; Sastry and Raswamany, 1976) reveal a high density of screw dislocations in the strength anomaly domain. The above interpretations of

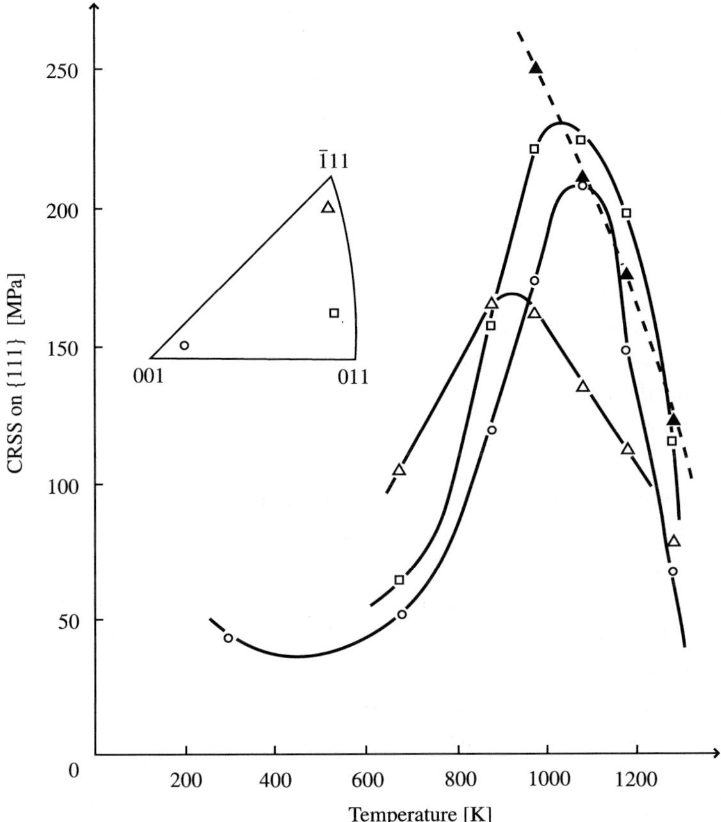

Figure 10.34. Temperature and orientation dependence of the CRSS on {111} for Co_3Ti. From Takasugi et al. (1987). Closed symbols correspond to cube slip.

Pope (1972), as well as diffusion models, would lead to curved dislocations without any particular role of the screws.

For Co_3Ti, TEM observations also reveal screw superdislocations dissociated into two superpartials and an APB ribbon presumably on the cube plane, in the anomalous domain (873 K) (Liu et al., 1989). In the low temperature regime (below 500 K on Figure 10.34), the normal stress–temperature dependence has been explained by a planar dissociation of superdislocations. Two superpartials are observed with a ribbon of SISF (Liu et al., 1989).

Figure 10.35. Schematics illustrating the Brown mechanism of APB glide (see text).

For Cu_3Au and Co_3Ti, it is thus tempting to consider that the observed straight screw dislocations are locks of the Kear–Wilsdorf type. Therefore, the yield stress anomaly can be interpreted as in Ni_3Al. It is clear that more refined TEM observations on dislocation structures and cores are needed to support this interpretation.

10.2.2 B2 alloys

The B2 structure is derived from the BCC lattice. Accordingly, superdislocations have been found in some crystals of this group. Their ⟨111⟩ Burgers vector decomposes into 1/2 ⟨111⟩ Burgers vectors (identical to those in BCC metals—see Section 6.3.2), the corresponding superpartials binding an APB ribbon. In addition, ordinary dislocations with ⟨100⟩ Burgers vectors are also found in some of those crystals.

Among these alloys, β CuZn has been studied most extensively. Its properties are reviewed, together with some other crystals of the same ordered structure.

10.2.2.1 Deformation mechanisms in β CuZn. The yield stress as a function of temperature is represented in Figure 10.36. It increases up to a maximum at a temperature around 200°C which is well below the order–disorder transition temperature of 460°C. Figure 10.36 shows a strong orientation effect, while a tension–compression asymmetry has also been evidenced (Nohara et al., 1984). Slip lines correspond to {110} planes in the strength anomaly domain while {112} slip is observed above the stress peak temperature (Umakoshi et al., 1976; Nohara et al., 1984).

These orientation effects cannot be explained as in $L1_2$ structures since ⟨111⟩ dislocations are not dissociated into partials and the Escaig effect (see Chapter 5) does not apply. Additional pure shear experiments have been performed by Matsumoto and Saka (1993) along the ⟨111⟩ direction in the {112} and {110} planes, respectively. No stress asymmetry is observed when the direction of shear is reversed. The difference between compression and shear tests may be explained by the effect of stresses normal to the slip plane, which do not operate in shear. Duesberry (1983) discussed the effect on the core of 1/2 ⟨111⟩ dislocations in BCC metals of the various components of the stress tensor. These may play a role in orientation effects evidenced in β CuZn compression tests (Matsumoto and Saka, 1993).

Surprisingly, a CRSS anomaly is observed in the shear test results for ⟨111⟩ dislocations gliding along {110} and {112} with a stress peak temperature around 200°C. This value is close to that evidenced in the compression tests of Figure 10.36. A weak CRSS anomaly is also found for the ⟨100⟩ {010} system in pure shear experiments again with a stress maximum close to 200°C.

TEM observations show that ⟨111⟩ superdislocations are active in the stress anomaly domain, while other Burgers vectors are active above the stress peak temperature (Zhu and Saka, 1989; Dirras et al., 1992). ⟨111⟩ loops appear elongated along the screw orientation but are not as straight as in $L1_2$ crystals. These superdislocations are separated into

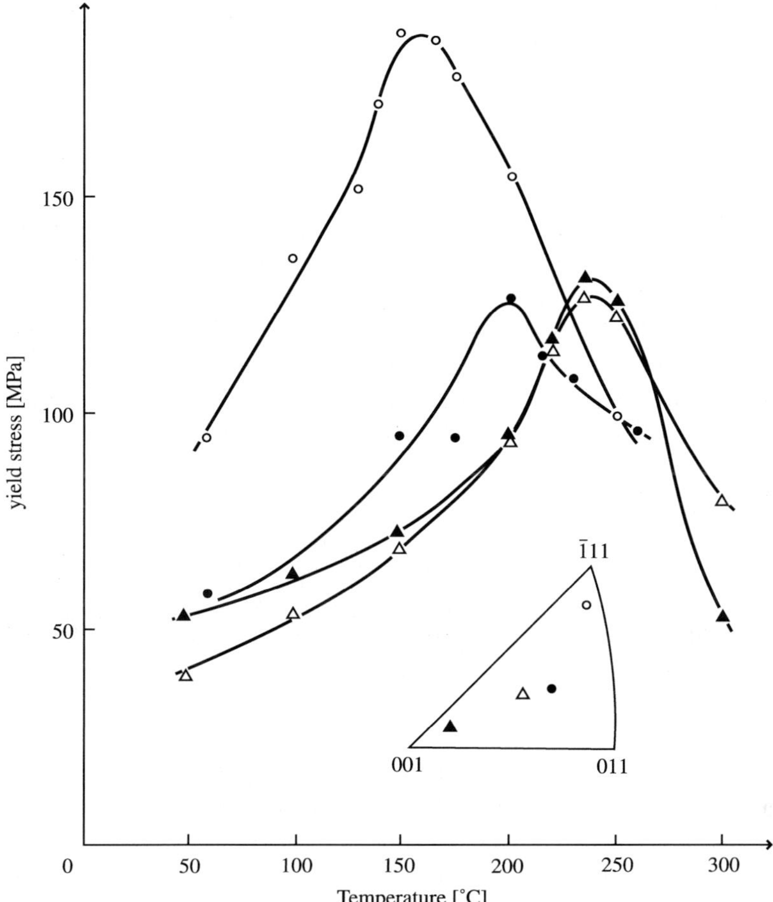

Figure 10.36. Yield stress of β CuZn as a function of temperature for various orientations. From Umakoshi et al. (1976).

two 1/2 ⟨111⟩ partials binding an APB ribbon. As the temperature rises, the APB extends out of the glide plane, whatever the dislocation character. This is the signature of a climb dissociation process, except for the screws which cross-slip onto the {112} plane. The in situ experiments by Nohara (1991) reveal a jerky motion of ⟨111⟩ dislocations on {110} below $T_{p,\tau}$ and a viscous motion of other types of dislocations above $T_{p,\tau}$.

The interpretations proposed for the above properties are somewhat contradictory and confusing. The orientation effects in compression are likely to be related to dislocation core effects, while the strength anomaly seems to be bound to diffusive effects (see Section 3). Indeed, it seems to be the only available way to account for a stress anomaly for ⟨111⟩

and ⟨100⟩ dislocations gliding on {110} and {112}. $T_{p,\tau}$ appears to be linked to the appearance of thermally activated {112} slip at high temperatures.

10.2.2.2 FeAl compounds. For compositions close to that of Fe_3Al, these crystals exhibit BCC related ordered structures: DO_3 below the transition temperature T_c and B2 above. The mechanical properties of polycrystals were studied by Stoloff and Davies (1964) and of single crystals by Hanada et al. (1981) and Schröer et al. (1993). The stress–temperature curves exhibit anomalous behaviour above 600 K. The stress peak temperature is situated either below or above T_c, depending on the orientation ($T_c = 825$ K for Fe 24.8 at.% Al). Strong orientation effects are observed as well as a zero stress–strain rate sensitivity just below $T_{p,\tau}$ (Schröer et al., 1993). A diffusion controlled mechanism is expected to account for this behaviour (as in β CuZn), supported by the following arguments:

(i) Several evidences of static and dynamic ageing are observed. Cottrell atmospheres around dislocations (see Section 3.2 and 10.1.5.5) have been reported in three-dimensional atomic scale images by Blavette et al. (1999)
(ii) In situ straining experiments in TEM (Molénat et al., 1997) reveal bursts of gliding dislocations in Fe 30 at.%Al at 643 K (i.e. below the stress peak temperature)
(iii) Cross-slip locking from {112} onto {110} planes is less efficient than in $L1_2$ alloys, because the driving torque is much weaker in BCC related structures.

10.2.3 Conclusion on strength anomalies in ordered intermetallics

The preceding review shows that strength anomalies are present in a number of intermetallic compounds. They differ by their crystal structures, dislocation core geometries and the presence or not of an order/disorder or phase change temperature.

Depending on the compound, the strength anomaly can be very pronounced (Ni_3Al) or less pronounced (FeAl). It can be orientation dependent (e.g. β CuZn) or not (e.g. Cu_3Au). It is, in general, observed in a temperature range where static and dynamic ageing phenomena are also present. It, therefore, seems that the diffusive processes listed in Section 10.1.5.5 are present as a rule. They account for zero or negative strain-rate sensitivities in particular and also TEM observations of bursts of dislocation glide. In some cases, a dislocation cross-slip mechanism that induces dislocation locking can be superimposed, in crystal structures which allow dissociated cores ($L1_2$). This accounts for orientation effects.

The two types of mechanisms exhibit additional differences. Although they are enhanced by an increase in temperature, they react in opposite directions to a stress increase in the anomaly domain. Indeed, if cross-slip is more frequent at higher stress, diffusion locking is less efficient under the same conditions because of faster dislocations.

Therefore, the strength anomaly due to pure diffusion processes is less pronounced. The strain-rate sensitivity S can be negative in this last case because of the two dislocation populations with low and high velocities, respectively. For cross-slip locking, S can approach zero but remains positive.

10.3 CREEP BEHAVIOUR OF Ni_3Al COMPOUNDS

In contrast to constant strain-rate experiments on ordered intermetallics that exhibit a strength anomaly, their creep behaviour has stimulated much less interest. However, creep experiments may be easier to interpret since the applied stress exerted on dislocations does not change appreciably during the course of experiment. Ni_3Al creep resistance is exposed here, since it is better documented as compared to other intermetallics.

Apart from the creep experiments of Thornton et al. (1970), it is only in the past 15 years that creep testing has been used to explore the time dependence of plastic flow of Ni_3Al in the anomalous flow regime (see e.g. Hemker et al., 1991; Rong et al., 1995; Zhu et al., 1998; Uchic and Nix, 2001).

Intermediate temperature requirements combined with detailed TEM observations were conducted in tension on single crystals of various orientations (Hemker et al., 1992). The compositions were Ni_3(Al, 1 at.% Hf, 0.24 at.%B) and Ni_3(Al, 1 at.%Ta) oriented near $\langle 001 \rangle$ and near $\langle 123 \rangle$. Stresses were between 325 and 745 MPa and temperatures between 530 and 973 K. This is the temperature regime where the yield strength increases anomalously with increasing temperature. Examples of creep curves are shown in Figure 10.37. They all exhibit the same basic shape: in the initial portion of the curve, the creep rate decreases, indicating a "normal" primary creep response. The creep rate then reaches a minimum and gives way to an extended region where it continually increases with strain. Orders of magnitude are $5 \times 10^{-8}\ s^{-1}$ in the primary stage and $3 \times 10^{-6}\ s^{-1}$ in the accelerating stage. Primary creep does not lead to a steady state but is instead exhausted. It is followed by an inverse type of transient.

A careful examination of the primary part of the curves shows an anomalous behaviour of the creep strain with temperature. For example, under an applied stress of 323 MPa, the creep strain after 25 h is found to be 9×10^{-3} at 630 K and 4×10^{-3} at 823 K (Hemker et al., 1991). However, this effect is so subtle that it is not visible at a scale similar to that of Figure 10.37 and most of the strain is achieved following a normal temperature behaviour as illustrated in the same figure.

Slip trace analysis reveals that octahedral glide is activated during primary and inverse creep together with cube glide in the accelerating part of the curve. TEM observations show numerous KWL at the end of primary creep. These contribute to the exhaustion of octahedral glide and consequently primary creep and explain the anomaly in the creep strain as a function of temperature. As time proceeds, these locks act

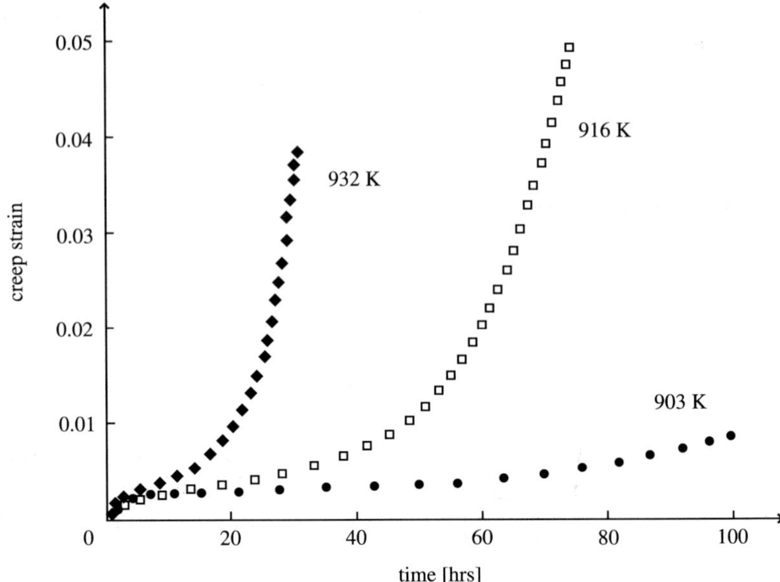

Figure 10.37. Creep curves for Ni_3(Al,Hf, B) single crystals at intermediate temperature. $\sigma = 745$ MPa. Orientation: 10° away from ⟨001⟩. From Hemker and Nix (1989).

as sources for the cube cross-slip plane. Given sufficient time and temperature, these cross-slipped dislocations bow out and glide on the cube plane. This leads to an ever increasing creep rate and inverse creep. As seen in Section 10.1.3, cube glide is controlled by a Peierls type process and as such has a strong normal temperature dependence. Consequently, the intermediate temperature creep strength of Ni_3Al decreases with increasing temperatures in a normal manner. This description of the process agrees with TEM observations of dislocations curved along the {001} cross-slip plane, having the same Burgers vector as the KWLs of primary creep. Let us note that a comparison of the two curves at 916 and 932 K, respectively, of Figure 10.37, yields a creep activation energy of 255 kJ/mol (2.67 eV/atom). This should be related to the cube slip mechanism, according to the microstructural observations.

In contrast to the sigmoidal creep observed at intermediate temperatures, steady-state creep of Ni_3Al single crystals has been observed at 1273 K, i.e. above the peak in the yield stress–temperature curve (Hemker and Nix, 1993). The steady state is reached in less than 1% creep strain and observed to last for the duration of the test (about 25% creep strain). The activation energy determined in these tests is very high: 378 $\pm$ 40 kJ/mol (or 4.96 eV/atom) (Hemker and Nix, 1997). According to the authors, it is larger than the activation energy for Ni diffusion in Ni_3Al. It is closer to the activation energy for Al diffusion which is about 408 kJ/mol. Since both atomic species have to diffuse for dislocation climb to take place, this value suggests that climb-controlled processes are important in Ni_3Al at high temperatures.

10.4 CONCLUSIONS

This survey shows that a number of intermetallic compounds of different crystallographic structures exhibit strength anomalies under constant strain-rate conditions. For Ni_3Al, which is well documented, the flow-stress peak as well as the work-hardening peak temperatures are rather well understood in terms of dislocation core properties. Additional information is needed for the other compounds. There is evidence of diffusive processes taking place at temperatures close to the stress-peak temperature which certainly play a role. The nature of these processes is not determined unambiguously.

Some information is available about the creep properties of Ni_3Al but is very scarce for other compounds. At intermediate temperatures that correspond to the yield stress anomaly, sigmoidal creep is observed. At the onset of the creep curve, octahedral glide exhaustion accounts for the anomalous creep strain as a function of temperature. But this corresponds to a narrow portion of the curve. After this short primary stage, creep accelerates, with a normal behaviour with respect to temperature. This corresponds to dislocation activity on the cube cross-slip plane. High temperature creep seems to be connected with dislocation climb.

Data about activation energies are needed for the characterization of the above deformation processes, the kinetics of which are not properly quantified at the moment. An additional effort in modelling creep behaviour is also necessary. In particular, the spectacular result by Hemker and Nix (1989), in which cooling down the creeping sample from 913 to 300 K drastically increases the strain rate, deserves more attention.

REFERENCES

Baluc, N. (1990a) PhD thesis no 886, Ecole Polytechnique, Lausanne, Suisse.

Baluc, N. (1990b) in "*Electron Microscopy 1990*", vol. **4**, Eds. Peachey, L.D., & Williams, D.B., San Francisco Press, San Francisco, p.474.

Baluc, N. & Schäublin, R. (1996) *Phil. Mag. A*, **74**, 113.

Baluc, N., Stoiber, J., Bonneville, J. & Martin, J.L. (1988) *Isr. J. Technol.*, **24**, 269.

Baluc, N., Karnthaler, H.R. & Mills, M.J. (1991a) *Phil. Mag. A*, **64**, 137.

Baluc, N., Schäublin, R. & Hemker, K.J. (1991b) *Phil. Mag. Lett.*, **64**, 327.

Basinski, S.Z. & Basinski, S.J. (1979) in *Dislocations in Solids*, vol. **4**, Ed. Nabarro, F.R.N., North Holland, Amsterdam, p. 261.

Blavette, D., Cadel, E., Fraczkiewicz, A. & Menaud, A. (1999) *Science*, **286**, 2317.

Bonneville, J., Baluc, N. & Martin, J.L. (1991) in *6th Int. Symp. on Intermetallic Compounds (JIMI S-6)*, Ed. Izumi, O., The Japan Inst. of Metals, Tokyo, p. 323.

Bonneville, J., Martin, J.L., Spätig, P., Viguier, B. & Matterstock, B. (1997a) in *High Temperature Ordered Intermetallic Alloys VII*, vol. **460**, Eds. Koch, C.C., Liu, C.T., Stoloff, N.S. & Wanner A., Materials Res. Soc, Warrendale, p. 419.

Bonneville, J., Kruml, T., Martin, J.L., Matterstock, B., Viguier, B. & Spätig, P. (1997b) *Mater. Sci. Eng. A*, **234-236**, 770.

Bontemps, C. & Veyssière, P. (1990) *Phil. Mag. Lett.*, **61**, 259.

Bontemps-Neveu, C (1991) PhD thesis no 1623, Université Paris Sud (Orsay), France.

Brown, N. (1959) *Phil. Mag.*, **4**, 693.

Caillard, D. (1996) *Acta Mater.*, **44**, 2773.

Caillard, D. (2001) *Mater. Sci. Eng. A*, **319–321**, 74.

Caillard, D. & Paidar, V. (1996) *Acta. Mater.* **44**, 2759.

Caillard, D. & Molénat, G. (1999) in *Proc. of the 20th Risø Int. Symp. on Mater. Sci.*, Eds. Bilde Sorensen, J.B., Cartensen, J.V., Hansen, N., Juul Jensen, D., Leffers, T., Pontleon, W., Pedersen, O.B. & Winther G., Risø Nat. Lab, Roskilde, p. 1.

Chou, C.T. & Hirsch, P.B. (1993) *Phil. Mag. A*, **68**, 1097.

Clément, N., Molénat, G. & Caillard, D. (1991) *Phil. Mag.*, **64**, 697.

Cockayne, D.J.H., Ray, I.L.F. & Whelan, M.J. (1969) *Phil. Mag.*, **20**, 1265.

Copley, S.M. & Kear, B.H. (1967) *Trans. Metall. AIME*, **239**, 977.

Coupeau, C., Bonneville, J., Matterstock, B., Grilhé, J. & Martin, J.L. (1999) *Scripta Mater.*, **41**, 945.

Couret, A., Sun, Y.Q. & Hirsch, P.B. (1993) *Phil. Mag. A*, **67**, 29.

Demura, M. & Hirano, T. (2000) *Intermetallics*, **8**, 1005.

Devincre, B. (2000) in *Multiscale Phenomena in Plasticity*, NATO Science Series E, vol. **367**, Eds. Lépinoux, J., Mazières, D., Pontikis, V. & Saada G., Kluwer Acad. Pub., Dordrecht, p. 319.

Devincre, B., Veyssière, P. & Saada, G. (1999) *Phil. Mag. A*, **79**, 1609.

Dimiduk, D.M., Thompson, A.W. & Williams, J.C. (1993) *Phil. Mag. A*, **67**, 675.

Dirras, G., Beauchamp, P. & Veyssière, P. (1992) *Phil. Mag. A*, **65**, 815.

Douin, J., Veyssière, P. & Beauchamp, P. (1986) *Phil. Mag. A*, **54**, 375.

Duesberry, M.S. (1983) *Acta Metall.*, **31**, 429.

Ezz, S.S. (1996) *Acta Mater.*, **44**, 4395.

Ezz, S.S. & Hirsch, P.B. (1994) *Phil. Mag. A*, **69**, 105.

Ezz, S.S., Pope, D.P. & Vitek, V. (1987) *Acta Metall.*, **35**, 1879.

Ezz, S.S., Sun, Y.Q. & Hirsch, P.B. (1995) *Mater. Sci. Eng. A*, **192–193**, 45.

George, E.P., Mills, M.J., & Yamaguchi M., Editors (1999) in *High Temperature Ordered Intermetallic Alloys VIII*, vol. **552**, Mater. Res. Soc., Warrendale.

Greenberg, B.A. & Ivanov, M.A. (1997) *Mater. Sci. Eng. A*, **239–240**, 813.

Hanada, S., Watanabe, W., Sato, T. & Izumi, O. (1981) *Scripta Met.*, **15**, 1345.

Hemker, K.J. (1997) *Phil. Mag. A*, **76**, 241.

Hemker, K.J. & Mills, M.J. (1993) *Phil. Mag. A*, **68**, 305.

Hemker, K.J. & Nix, W.D. (1989) in *High Temperature Ordered Intermetallic Alloys III. Mater. Res. Soc. Symp. Proc.*, vol. **133**, Eds. Liu, C.T., Taub, A.I., Stoloff, N.S. & Koch C.C., MRS, Warrendale, p. 481.

Hemker, K.J. & Nix, W.D. (1993) *Metall. Trans. A*, **24A**, 335.

Hemker, K.J. & Nix, W.D. (1997) in *Structural Intermetallics 1997*, Ed. Nathal, M.V., et al., TMS, Warrendale, p. 21.

Hemker, K.J., Mills, M.J. & Nix, W.D. (1991) *Acta Metall.*, **39**, 1909.

Hemker, K.J., Mills, M.J. & Nix, W.D. (1992) *J. Mater. Res.*, **7**, 2059.

Hemker, K.J., Dimiduk, D.M., Clemens, H., Darolia, R., Invi, H., Larsen, J.M., Sikka, V.K., Thomas, M. & Whittenberger J.D., Editors (2001) in *Structural Intermetallics, Proceedings of the 3rd Int. Symp. on Structural Intermetallics (ISSI)*, TMS, Warrendale.

Heredia, F.E. & Pope, D.P. (1991) *J. Phys.*, **1**, 1055.

Hirsch, P.B. (1992) *Phil. Mag. A*, **65**, 569.

Intermetallics for the 3rd Millennium, a Symposium dedicated to Professor R. W. Cahn (1999) Proceedings in *Intermetallics*, (2000) **8**.

Karnthaler, H.P., Mühlbacher, E.T. & Rentenberger, C. (1996) *Acta Mater.*, **44**, 547.

Kear, B.H. & Wilsdorf, H.G.F. (1962) *Trans. Met. Soc. AIME*, **224**, 362.

Korner, A. (1989) *Phil. Mag. Lett.*, **60**, 103.

Kruml, T., Martin, J.L., Viguier, B., Bonneville, J. & Spätig, P. (1997) *Mater. Sci. Eng. A*, **239–240**, 174.

Kruml, T., Martin, J.L. & Bonneville, J. (2000a) *Phil. Mag. A*, **80**, 1545.

Kruml, T., Paidar, V. & Martin, J.L. (2000b) *Intermetallics*, **8**, 729.

Kruml, T., Martin, J.L. & Paidar, V. (2001) in *Structural Intermetallics 2001*, Ed. Hemker, K.J., TMS, Warrendale, p. 457.

Kruml, T., Conforto, E., Lo Piccolo, B., Caillard, D., Martin, J.L. (2002) *Acta Mater.*, **50,** 5091.

Kubin, L.P. & Estrin, Y. (1991) *J. Phys. III*, **1**, 929.

Kuramoto, E. & Pope, D.P. (1976) *Phil. Mag.*, **33**, 675.

Lall, C., Chin, S. & Pope, D. (1979) *Metall. Trans. A*, **10**, 1323.

Lavrentev, F.F. (1980) *Mat. Sci. Eng.*, **46**, 191.

Lin, D. & Wen, M. (1993) *Mater. Sci. Eng. A*, **164**, 384.

Liu, Y., Takasugi, T., Izumi, O. & Takahashi, T. (1989) *Phil. Mag. A*, **59**, 423–427.

Lopez, J.A. & Hancock, G.F. (1970) *Phys. Stat. Sol (a)*, **2**, 469.

Louchet, F. (1995) *J. Phys. III*, **5**, 1803.

Lours, P., Coujou, A. & Coulomb, P. (1991) *Acta Metall.*, **39**, 1787.

Martin, J.L., Matterstock, B., Spätig, P. & Bonneville, J. (1999) in *Proceedings of the 20th Risø Int. Symp. On Materials Science*, Eds. Bilde-Sørensen, J.B., Cartensen, J.V., Hansen, N., Huul Jensen, D., Leffers, T., Pantleon, W., Pedersen, O.B. & Winther G., Risø Nat. lab, Roskilde, p. 103.

Masahiko, M. & Hirano, T. (1997) *Phil. Mag. Lett.*, **75**, 143.

Matsumoto, A. & Saka, H. (1993) *Phil. Mag. A*, **67**, 217.

Matterstock, B., Conforto, E., Kruml, T., Bonneville, J. & Martin, J.L. (1999) in *High Temperature Ordered Intermetallic Alloys VIII,* vol. **552**, Eds. George, E.P., Mills, J.M. & Yamaguchi M., Materials Res. Soc., Warrendale, p. KK10.3.1.

Mills, M.J. & Chrzan, D.C. (1992) *Acta Met. Mat.*, **40**, 3051.

Mills, M.J., Baluc, N. & Karnthaler, P. (1989) in *High Temperature Ordered Intermetallic Alloys III*, vol. **133**, Eds. Liu, C.T., Taub, A.I., Stoloff, N.S. & Koch C.C., Mater. Res. Soc., Warrendale, p. 203.

Molénat, G. & Caillard, D. (1991) *Phil. Mag.*, **64**, 1291.

Molénat, G. & Caillard, D. (1992) *Phil. Mag. A*, **65**, 1327.

Molénat, G. & Caillard, D. (1994) *Phil. Mag.*, **69**, 939.

Molénat, G., Caillard, D., Sun, Y. & Couret, A. (1993) *Mater. Sci. Eng. A*, **164**, 407.

Molénat, G., Couret, A. & Caillard, D. (1997) *Mater. Sci. Eng. A*, **234–236**, 660.

Mulford, R.A. & Pope, D.P. (1973) *Acta Metall.*, **21**, 1375.

Nabarro, F.R.N. & Duesberry M.S., Editors (1996) in *Dislocations in Solids: "$L1_2$ ordered alloys"*, vol. **10**, North Holland, Amsterdam.
Nabarro, F.R.N., Basinski, S.Z. & Holt, D.B. (1964) *Adv. Phys.*, **13**, 193.
Nemoto, M., Echigoya, J. & Suto, H. (1977) *J. Electron Microsc.*, **26**, 467.
Noguchi, O., Oya, Y. & Suzuki, T. (1981) *Met. Trans.*, **12A**, 1647.
Nohara, A. (1991) in *Proceedings Int. Symp. on Intermetallic Compounds*, Ed. Izumi, O., Jap. Inst. of Metals, Tokyo, p. 561.
Nohara, A., Izumi, M., Saka, H. & Imura, T. (1984) *Phys. Stat. Sol (a)*, **82**, 163.
Orlova, A. (1991) *Acta Met.*, **39**, 2805.
Paidar, V., Pope, D.P. & Vitek, V. (1984) *Acta Metall.*, **32**, 435.
Paidar, V., Molénat, G. & Caillard, D. (1992) *Phil. Mag. A*, **66**, 1183.
Pope, D.P. (1972) *Phil. Mag.*, **25**, 917.
Rong, T.S., Jones, I.P. & Smallmann, R.E. (1995) *Acta Metall. Mater.*, **43**, 1385.
Saada, G. & Veyssière, P. (1992) *Phil. Mag. A*, **66**, 1081.
Saada, G. & Veyssière, P. (1993) *Proc. Int. Symp. on Struct. Intermetallics*, Champion, USA, p. 379.
Saburi, T., Hamana, T., Nenno, S. & Pak, H. (1977) *Jpn. J. Appl. Phys.*, **16**, 267.
Sanchez, J.M., Eng, S., Wu, J.P. & Tien, J.K. (1987) *MRS Symp. Proc.*, **81**, 57.
Sastry, S.M.L. & Raswamany, B. (1976) *Phil. Mag.*, **33**, 375.
Schäublin, R. & Stadelmann, P. (1993) *Mater. Sci. Eng. A*, **164**, 373.
Schoeck, G. (1997) *Phil. Mag. Lett.*, **75**, 7.
Schröer, W., Hartig, C. & Mecking, H. (1993) *Z. Metallkd.*, **84**, 294.
Shi, X., Saada, G. & Veyssière, P. (1996) *Phil. Mag. A*, **73**, 1419.
Spätig, P. (1995) PhD thesis no 1407, EPF Lausanne, Switzerland.
Spätig, P., Bonneville, J. & Martin, J.L. (1993a) *Mater. Sci. Eng. A*, **167**, 73.
Spätig, P., Bonneville, J. & Martin, J.L. (1993b) in *High Temperature Ordered Intermetallic Alloys V*, vol. **288**, Eds. Baker, I., Dariola, R., Whittenberger, J.D. & Yoo M.H., Mater. Res. Soc., Pittsburgh, p. 713.
Spätig, P., Bonneville, J. & Martin, J.L. (1995) in *Euromat 95*, vol. **D-E**, Assoc. Ital. Metallurgica, Milano, p. 97.
Stoloff, N.S. & Davies, R.G. (1964) *Acta Metall.*, **12**, 473.
Sun, Y.Q. & Hazzledine, P.M. (1989) in *High Temperature Ordered Alloys,* vol. **133**, Eds. Liu, C.T., Taub, A.I., Stoloff, N.S. & Koch C.C., Mater. Res. Soc., Pittsburgh, p. 197.
Sun, Y.Q., Crimp, M.A., Hezzeldine, P.M. & Couret, A. (1991a) *Phil. Mag. A*, **64**, 311.
Sun, Y.Q., Hazzledine, P.M., Crimp, M.A. & Couret, A. (1991b) *Phil. Mag. A*, **64**, 311.
Sun, J., Lee, C.S., Lai, J.K.L. & Wu, J.S. (1999) *Intermetallics*, **7**, 1329.
Suzuki, K., Ichihara, M. & Takeuchi, S. (1977) *J. Electron Microsc.*, **26**, 463.
Suzuki, S., Ichiara, M. & Takeuchi, S. (1979) *Acta Metall.*, **27**, 193.
Takasugi, T., Hirakawa, S., Izumi, O., Ono, S. & Watanabe, S. (1987) *Acta Metall.*, **35**, 2015.
Taylor, G. (1934) *Proc. R. Soc. A*, **145**, 362.
Thornton, P.H., Davies, R.G. & Johnston, T.L. (1970) *Metall. Trans.*, **1**, 207.
Uchic, M.D. & Nix, W.D. (2001) *Intermetallics*, **9**, 1053.
Umakoshi, Y., Yamaguchi, M., Namba, Y. & Murakami, K. (1976) *Acta Metall.*, **24**, 89.
Umakoshi, Y., Pope, D.P. & Vitek, V. (1984) *Acta Metall.*, **32**, 449.
Veyssière, P. (1984) *Phil. Mag. A*, **50**, 189.
Veyssière, P. & Douin, J. (1985) *Phil. Mag. A*, **51**, L1.

Veyssière, P. & Saada, G. (1996) in *Dislocations in Solids*, vol. **10**, Eds. Nabarro, F.R.N. & Duesberry M.S., North Holland, Amsterdam, p. 253.
Viguier, B. (2003) *Mater. Sci. Eng. A.*, *Mat. Sci. Eng.* A349/1–2, 132.
Viguier, B. Martin, J.L. & Bonneville, J. (2002) in *Dislocation in Solids*, vol. **11**, Eds. Nabarro, F.R.N. & Duesbery, M.S., North Holland, Amsterdam, p. 459.
Wee, D.M., Noguchi, O., Oya, Y. & Suzuki, T. (1980) *Trans. Jpn. Inst. Met.*, **21**, 237.
Westbrook, J.H. (1975) *Trans. Metall. Soc. AIME*, **209**, 898.
Yamaguchi, M. & Umakoshi, Y. (1990) *Prog. Mater. Sci.*, **34**, 1.
Yoo, M.H. (1986) *Scripta Met.*, **20**, 915.
Zhu, Y.M. & Saka, H. (1989) *Phil. Mag. A*, **59**, 661.
Zhu, W., Fort, D., Jones, I.P. & Smallman, R.E. (1998) *Acta Mater.*, **46**, 3873.

Conclusion

Following this review on the description of thermally activated mechanisms of dislocation mobility, we would like to emphasize the following points.

The various examples presented above demonstrate how deformation mechanisms can be experimentally characterized. Mechanical test data (of both monotonic and transient types) have to be analyzed in the framework of the thermal activation theory. First, stress and temperature domains that correspond to a given mechanism are determined by inspection of the stress/temperature curves, the activation volume/temperature and activation volume/stress curves, at given strain. Any change of slope or peak or discontinuity on these curves is a hint for a change in mechanism. For a given process, we have shown the usefulness of repeated transient tests (relaxation and creep tests) to measure reliably the true activation volume, as opposed to the apparent one. This latter parameter is of key importance, since it is directly linked to the number of atoms involved in the activated process. Knowing V also allows one to determine the activation energy. We have also shown the necessity of directly observing dislocation motion through in situ techniques to unambiguously determine deformation processes. These bring very useful information complementary to the conventional "post mortem" substructure observations, provided the possible artefacts are born in mind.

The experimental observations, in a variety of crystals, under a range of applied external conditions, show that in almost all cases, dislocations have to overcome a lattice resistance to glide. The corresponding thermally activated frictional force can be of different nature. It obeys:

- the Peierls–Nabarro mechanism in covalent materials, as illustrated by data related to Si, Ge, InSb, ZnS and GaAs.
- the kink-pair mechanism in metals, e.g. in BCC metals at low temperature, in close-packed structures under conditions of glide along non-close-packed planes. This is the case of prismatic slip in Mg, Be, Ti, Zn and glide on $\{100\}$, $\{110\}$, etc. planes studied in Al and Cu.
- the locking–unlocking mechanism, recently suggested by in situ observations of dislocation glide along the prism plane of Mg and Be, over well-defined temperature ranges. This mechanism is also observed to operate in $L1_2$ compounds, controlling superdislocation mobility along the cube plane in Ni_3Al, as well as in FeAl and Fe_3Al. Depending on the applied conditions, the locking–unlocking or the kink-pair mechanisms in metals can control the velocity of the same dislocations. The Peierls mechanism appears now as a special case of locking–unlocking.

According to the observations which have been reported, the concept of lattice resistance to glide in metals, which has been formerly developed for BCC crystals at low temperature, can be successfully extended to more complex materials.

The materials in which dislocations do not encounter any friction force are restricted to FCC crystals in the low temperature glide regime where dislocations are held up at forest dislocations, impurities or solute atoms in dilute solid solutions. The same situation can also take place in other materials as the friction force becomes athermal. At rising temperatures, localized obstacles can take over, the viscous motion of dislocations becoming jerky.

A concept which has survived the years remarkably is the description of dislocation cores in terms of partial dislocations. It has proved to be extremely useful in predicting correct orders of magnitude of energy barriers in the case of the kink pair and the locking–unlocking mechanisms in metals and intermetallic compounds as well as in the case of cross-slip in FCC metals. However, surprisingly, the corresponding dissociation has not been proved experimentally in all cases such as for screw dislocations in Al, superpartial dislocations in Ni_3Al compounds exhibiting a high complex stacking fault energy, screw dislocations in BCC metals. Atomistic simulations of dislocation cores at given stress and temperature appear as a promising route to more realistic descriptions of core transformations and associated energy barriers.

If crystal hardening can be described (but not predicted) by the existing theories, some progress has been achieved in the field of real dislocation sources, in the measurement of mobile dislocation exhaustion rates and in understanding some of the connections between multiplication, exhaustion and work-hardening in various types of crystals.

As the temperature becomes high enough to enhance vacancy mobility, dislocation climb becomes significant. Equations describing the climbing velocity are proposed here, taking into account both pipe and volume diffusion. The number of data in this field is too scarce and additional experiments are highly desirable.

Finally, glide localization, a deformation feature which is well established for given deformation conditions, is not implemented yet by any theoretical description. This limits the estimation of the densities of mobile dislocations and of the effective stress acting on them.

However, for the dislocation mechanisms that we have reviewed, the agreement between the experimental data and the modelling attempts appears to be quite satisfactory.

Glossary of Symbols

a		Interatomic distance
b		Burgers vector
	b_p	case of a Shockley partial
c		Atomic concentration of
	c_k	kinks along a dislocation
	c_{kp}	kink pairs along a dislocation
	c_j	jogs
	c_v, c_i	vacancies, interstitials, respectively
	$c^{(0)}$	at equilibrium
	$c^{(j)}$	at a jog
	$c^{(p)}$	in the dislocation core (pipe diffusion)
	$c^{(s)}$	supersaturation value
	c_b	obstacles
c_r, c_c		Time constants for stress relaxation and creep tests, respectively
d		Dissociation width of a dislocation
	d_o	at zero stress
h		Distance between atomic rows (e.g. Peierls valleys, kink or jog height)
j		Test number in a creep or stress relaxation series
k		Boltzmann constant
m		Stress exponent of dislocation velocity (or strain rate)
p		Hydrostatic pressure
t		Time
	t_ℓ	waiting time in a locking position
	Δt	duration of a creep or stress relaxation test
v		Dislocation velocity
	v_{kp}	controlled by the kink-pair mechanism
	v_{or}	at the onset of a transient test
	v_g	free glide velocity of a dislocation
	v_k	velocity of a kink along a dislocation
	v_j	velocity of a jog along a dislocation
	$\mathrm{v}_p^{(\ell)}$, $\mathrm{v}_p^{(t)}$	velocity of the leading and trailing partials, respectively

v_v, v_i Drift velocity of vacancies and interstitials, respectively

x Coordinate along the dislocation line
- Δx bulge or kink-pair width, mean-free path of vacancies along the dislocation core (pipe diffusion)
- x_c critical value of Δx
- x distance between jogs (case of climb)
- x' defined in Figure 7.7

y Coordinate perpendicular to dislocation line
- y_p dislocation displacement corresponding to the maximum slope of the Peierls potential
- y_c critical bulge extension
- y_g free glide distance
- y_o dislocation displacement near the trough of a Peierls valley

w Obstacle width, {100} APB width in Ni_3Al

A Activation area
- A_b case of a bulge mechanism
- A_{ik} case of elastically interacting kinks

APB Antiphase boundary

BCC Body centered cubic structure

CRSS Critical resolved shear stress

D Diffusion coefficient of
- D_k kinks along a dislocation
  - $D_k^{(p)}$ case of partial dislocations
- D_v, D_i vacancies, interstitials
  - $D^{(sd)}$ self-diffusion
  - $D^{(p)}$ pipe diffusion

E Dislocation energy per unit length
- E_o at the bottom of a Peierls valley
- ΔE amplitude of the Peierls potential

F Force
- $F_c^{(j)}$ chemical force on a jog

FCC Face centered cubic structure

HVEM High-voltage electron microscope

HCP Hexagonal close-packed structure

K^r, K^c Work-hardening coefficients during stress relaxation and creep tests

KWL Kear–Wilsdorf lock

L Dislocation length
L_F Distance between obstacles along the dislocation line (Friedal length)

M Elastic modulus of specimen and machine
M Dislocation mobility
- M_l, M_t case of leading and trailing partials
  - M_{30}, M_{90} case of 30° and 90° partials

P Probability per unit dislocation length of
- P_{kp} kink-pair nucleation
- P_{ul} unlocking
- P_l locking
- P_{cs} cross-slip

Q Experimental activation energy

S Strain-rate sensitivity of stress

SEM Scanning electron microscope

R Radius of curvature

T Absolute temperature
T_a Athermal temperature
T_a^{th} Estimated
T_a^{exp} Experimental

$T_{p,\tau}$, $T_{p,\sigma}$ Stress peak temperatures (single and polycrystals, respectively)

$T_{p,\theta}$ Work-hardening peak temperature

T_M Melting temperature
$\hat{T}$ Line tension
TEM Transmission electron microscope

U Energy
- U_k of a kink
- U_{kp} of a kink-pair
  - $U_{kp}^{(c)}$ critical (maximum) value (activation energy)
- U_b of a bulge
  - $U_b^{(c)}$ critical value
- U_{ik} of elastically interacting kinks
  - $U_{ik}^{(c)}$ critical value
  - $U_{ik}^{(c,p)}$ case of Shockley partial
- U_j of a jog
- U_{jp} of a jog pair

$U_{jp}^{(c)}$	critical value
U_{const}	of a constriction
U_{cs}	cross-slip (activation energy)
U_{l}	locking (activation energy)
U_{m}	kink migration (activation energy)
$U_{f}^{(v)}$, $U_{f}^{(i)}$	of formation of vacancies and interstitials, respectively
$U_{f}^{(v,p)}$, $U_{f}^{(i,p)}$	in the core of dislocations (pipe diffusion)
$U_{d}^{(v)}$, $U_{d}^{(i)}$	of vacancy and interstitial diffusion, respectively
$U_{d}^{(v,p)}$, $U_{d}^{(i,p)}$	along the core of dislocations (pipe diffusion)
$U_{sd}^{(v)}$, $U_{sd}^{(i)}$	of vacancy and interstitial self-diffusion, respectively
$U_{sd}^{(v,p)}$, $U_{sd}^{(i,p)}$	in the core of dislocations
$\Delta U_{sd}^{(v,p)}$	$U_{sd}^{(v)} - U_{sd}^{(v,p)}$
U_{d}	Solute diffusion activation energy
U_{E}	Activation energy of the kink-pair mechanism (Escaig approximation)
U_{int}	Dislocation–solute interaction energy
V	Activation volume of dislocation velocity
V_{a}	Apparent activation volume
V_{c}, V_{r}	Phenomenological activation volumes in relations (2.8) and (2.14), respectively
W	Work done by the stress
X	Mean-free path of kink-pairs in covalent crystals
β, β^{c}, β^{r}	Exponents of relation 2.22
γ	Stacking fault energy
γ_{CSF}	Complex stacking fault energy
γ_{o}	APB energy on {111}
γ_{c}	APB energy on {001}
γ	Shear strain
γ_{p}	Plastic shear strain
$\Delta\gamma_{j}$	Creep strain after Δt for test number j
$\dot{\gamma}$	Shear strain-rate
$\dot{\gamma}_{p}$	Plastic strain-rate
$\dot{\gamma}_{po}$	at the onset of a transient
$\dot{\gamma}_{ij}$, $\dot{\gamma}_{fj}$	Plastic shear strain-rate at the onset and end of the transient number j, respectively
μ	Shear modulus

ν_D	Debye frequency
ν	Vibration frequency of a dislocation segment and Poisson's ratio
σ	Applied stress
$\dot{\sigma}$	Applied stress-rate
τ	Shear stress (case of glide) or normal stress (case of climb)
$\Delta\tau$	Stress increase between two successive creep tests
$\Delta\tau_j$	Amount of stress relaxation after Δt for test number j
$\Delta\tau_R$	Yield point amplitude at reloading after a transient test
τ_p	Peierls shear stress
$\tau_{min}^{(b)}$	minimum shear stress for the bulge critical configuration, in the kink-pair mechanism
$\tau_{max}^{(ik)}$	maximum shear stress for the elastically interacting kinks approximation
τ_i	elastic interaction between partials and yielding stress of incomplete KWLs
τ_c	yielding stress for complete KWLs
τ_l, τ_t	component of applied shear stress on leading and trailing partials (τ_{lt} when equal), respectively
$\tau_{d\infty}$	shear stress to separate two partials
$\tau_{eff}^{(l)}$, $\tau_{eff}^{(t)}$	effective stress on leading and trailing partials
$\dot{\tau}$	shear stress-rate
τ^*	effective shear-stress
τ_μ	athermal component of the stress
θ	Work-hardening coefficient
θ_p	plastic work-hardening
ρ	Dislocation density (total)
ρ_m	mobile dislocation density
ρ_{mo}	at the onset of a transient test
Φ	Flux
Λ	Dislocation mean free path
Ω	Atomic volume
Ω_r, Ω_c	Parameters in relations 2.25 and 2.35, respectively
ε	Total strain (polycrystals)
ε_p	Plastic strain
$\dot{\varepsilon}$	Strain-rate
$\dot{\varepsilon}_p$	Plastic strain-rate

$\dot{\varepsilon}_{po}$	Plastic strain-rate at the onset of a creep or stress relaxation test
ΔG	Gibbs free energy of a deformation mechanism
ΔG°	Gibbs free energy at zero stress
ΔH	Activation enthalpy for a deformation mechanism
ΔH_{a}	Apparent activation enthalpy
ΔZ	Variation of parameter Z
$\Delta \Omega$	Local change of atomic volume

Index